TOUGHENED COMPOSITES

Symposium on Toughened Composites
sponsored by
ASTM Committee D-30
on High Modulus Fibers
and Their Composites
Houston, Texas, 13-15 March 1985
in cooperation with
NASA-Langley Research Center

ASTM SPECIAL TECHNICAL PUBLICATION 937
Norman J. Johnston, NASA-Langley
Research Center, editor

ASTM Publication Code Number (PCN)
04-937000-33

1916 Race Street, Philadelphia, PA 19103

Library of Congress Cataloging-in-Publication Data

Symposium on Toughened Composites (1985: Houston, Tex.)
Toughened composites.

(ASTM special technical publication; 937)
"ASTM publication code number (PCN) 04-937000-33."
Includes bibliographies and index.
1. Composite materials—Congresses. I. Johnston, Norman J. (Norman Joseph), 1934– . II. ASTM Committee D-30 on High Modulus Fibers and Their Composites. III. Title. IV. Series.
TA418.9.C6S94 1985 620.1′18 87-1387
ISBN 0-8031-0934-2

Library of Congress Catalog Card Number: 87-1387/

NOTE

The Society is not responsible, as a body,
for the statements and opinions
advanced in this publication.

Printed in Baltimore, MD
May 1987

Foreword

The Symposium on Toughened Composites was held in Houston, Texas, on 13-15 March 1985. ASTM Committee D-30 on High Modulus Fibers and Their Composites was sponsor in cooperation with NASA-Langley Research Center. Norman J. Johnston, NASA-Langley Research Center, served as symposium chairman and has edited this publication.

Related ASTM Publications

Composite Materials: Fatigue and Fracture, STP 907 (1986), 04-907000-33

Elastic-Plastic Fracture Mechanics Technology, STP 896, (1985), 04-896000-30

High Modulus Fiber Composites in Ground Transportation and High Volume Applications, STP 873 (1985), 04-873000-33

A Note of Appreciation to Reviewers

The quality of the papers that appear in this publication reflects not only the obvious efforts of the authors but also the unheralded, though essential, work of the reviewers. On behalf of ASTM we acknowledge with appreciation their dedication to high professional standards and their sacrifice of time and effort.

ASTM Committee on Publications

ASTM Editorial Staff

Contents

INTERLAMINAR FRACTURE

THERMOPLASTICS

Overview

In May, 1983, NASA-Langley Research Center held a Workshop on Tough Composite Materials which resulted in NASA conference publication CP-2334. The conference reviewed most of the NASA-sponsored research on this subject, holding workshop sessions in three areas: fracture toughness/impact, constituent property-composite property relationships, and matrix synthesis and characterization. The workshop was of great benefit to NASA and its industrial and university peers. It was hoped a second meeting on this topic could be held again in two years.

The ideal forum for such an encore was an ASTM national meeting. Under the perceptive leadership of Chairman W. W. Stinchcomb, ASTM Committee D-30 on High Modulus Fibers and Their Composites invited NASA to cooperate in sponsoring a Symposium on Toughened Composites. It was held in Houston, Texas, 13–15 March 1985.

The purpose of the symposium was to provide a state-of-the-art perspective of the on-going research to develop tougher high performance continuous graphite fiber reinforced composite materials. A second objective was to make the symposium as multidisciplinary as possible from a materials standpoint. To do this, papers were invited and classified in five categories:

- toughened composites: Prospectives From Industry,
- micromechanics,
- interlaminar fracture,
- thermoplastics, and
- thermosets.

Twenty-six papers presented at the symposium are included in this STP which brings together a wide range of disciplines currently involved in the research and development of tough high performance composites.

Research that is focused on some ultimate end use is usually multidisciplinary in nature, for example, the development of composite materials for aerospace applications. What is often missing in the process is the interdisciplinary activity. Researchers from different disciplines need to make a conscious effort to communicate and exchange ideas with each other. Such intercourse is indispensable to the ultimate success of the activity. It also helps one shed the myopic view so easily gained and retained in the practice of highly technical skills. The papers in this volume will challenge the reader to make a conscious effort at this much needed interdisciplinary communication.

The papers in this volume should also provide the reader with some answers, or at least suggest approaches, to the following questions of general import to the development of toughened composite materials.

- What is toughness and how is it measured?
- How much toughness is needed for a particular application?
- What methods, structural, mechanical, and chemical, can be used to optimize toughness and damage tolerance with acceptable property trade-offs?
- What generic research should be pursued to understand the behavior of tough composites, for example, viscoelastic effects?
- What new concepts need to be developed and pursued?
- Do tougher composites possess long-term durability under a real-world environment?
- What correlations are needed between resin properties and composite properties to help guide synthetic efforts on new matrices?
- What part does fiber-resin interfacial adhesion play in controlling composite properties?

Often, an interdisciplinary approach is required to develop solutions to the challenges raised by these questions. For example, toughness is approached by the chemist at the molecular level, by the fracture mechanician at the ply and sublaminate level, and by the structural engineer at the subcomponent and component level. Ultimately, contributions from all three disciplines must be combined to bring about the successful development of a tough, damage tolerant composite structure.

The reader probably will not be completely satisfied with the answers provided herein. Nor should he be! A paper may fall short of his expectations or may not focus on certain aspects of the problem he perceives as important. Hopefully, the results (or lack of them) published in this STP will stimulate the search for more and better solutions and generate the desire and need for improved research cooperation across disciplines. The fruitage from such activities may well require a third multidisciplinary meeting on toughened composites within the next several years.

A summary of the contents of this volume follows.

Toughened Composites: Prospectives from Industry

Kam and Walker presented basic guidelines for the selection of a composite material system for application to aircraft structures. Two major selection criteria, structural characteristics and process characteristics, were discussed. *Griffin* concluded, on the basis of a variety of mechanical tests on advanced toughened composites, that increased values of design allowable tensile strength can be easily obtained. However, similar increases in compressive strength allowables cannot be achieved without additional improvements in matrix properties.

Micromechanics

Sohi, Hahn, and Williams investigated the effect of resin modulus and toughness on compressive behavior of unidirectional and quasi-isotropic graphite/epoxy composites. Mechanisms of failure and failure propagation were also studied using undamaged, impact damaged, and open hole specimens. *Hirschbuehler* attempted to correlate mechanical properties of neat resins and their respective composites in order to develop predictive relationships. Resin flexural modulus, resin strain-work-to-failure, as well as composite compressive strength after impact, interlaminar fracture toughness (G_{Ic}), short beam shear strength after impact, and open hole compressive strength were studied. *Hunston, Moulton, Johnston, and Bascom* established a relationship between values of neat resin G_{Ic} and composite interlaminar G_{Ic} as measured by the double cantilever beam (DCB) specimen. They found that for a variety of reasons tougher resins translate less than 50% of their toughness to the composite, and that many thermoplastic composites exhibit poor interfacial bonding and lower than expected G_{Ic} values.

Jordan and Bradley conducted a detailed in situ scanning electron micrograph (SEM) study of the delamination failure of brittle and rubber toughened epoxy laminates tested under a variety of Mode I/Mode II ratios. The nature and extent of the crack tip deformation/damage zone, and the development of microcrack zones ahead of the crack tip were observed. *Hibbs, Tse, and Bradley* determined the Mode I, Mode II, and mixed mode delamination fracture toughness and controlling micromechanisms of fracture for five graphite-epoxy composites containing a systemmatic variation in toughness and interfacial bonding. Toughness was highly dependent on interfacial adhesion and mode of loading. In the brittle systems, an increase in the percentage of Mode II loading led to a dramatic increase in delamination toughness through the development of hackles. *Bascom, Boll, Hunston, Fuller, and Phillips* also presented SEM fractography of delaminated surfaces observing features such as fiber pullout, hackle markings, resin fracture, and resin shear yielding. The effects of changing fiber mechanical properties and matrix resin fracture energy on the fractography of delamination were described.

Schwartz and Hartness attempted in a rather Edisonian manner to improve interlaminar fracture toughness, 90° tensile strength, and 0° compressive strength by applying model tough polymeric thin coatings on the fiber surface. *Weinberg* used liquid wetting studies on graphite and glass filaments to calculate solid surface energies which, when combined with thermoplastic polymer surface tensions, were used to predict resin wetting of the fiber in terms of work of adhesion. Notch tensile strength of carbon fiber-thermoplastic composites correlated with the predicted work of adhesion. The results appear to confirm the importance of employing wetting studies in interfacial investigations. *Whitney and Drzal* developed a model for predicting the axisymmetric stress distribution around an isolated fiber fragment. This analytical model should be an improvement over existing "shear lag" models for stress analysis of the single-fiber interfacial shear strength test specimen so commonly used today.

Interlaminar Fracture

O'Brien, Johnston, Raju, Morris, and Simonds further refined the edge delamination test for measuring interlaminar fracture toughness of composites by studying Mode I and mixed Mode I/II versions, varying coupon size and matrix resins, and determining the contribution of residual thermal and moisture stresses on strain energy release rates. *Poursartip* addressed the problem of edge delamination growth under fatigue loading for a brittle and a toughened epoxy graphite composite by developing power law correlations for the two materials. His study showed that energy release must be compared to the increasing resistance to further growth caused not only by new delamination but also by associated off-axis matrix cracking. *Adams, Zimmerman, and Odom* also fatigue tested edge delamination specimens for two toughened epoxies, one an interesting interleaf composition, in two different layup orientations at two test frequencies and two load ratios. Importantly, all laminates exhibited a significant decrease in strain energy release rate with increasing cycles to failure.

Daniel, Shareef, and Aliyu determined the effects of loading rates on Mode I DCB interlaminar fracture toughness of a graphite elastomer-modified epoxy composite. Crack extension rates up to 21 mm/s (49.6 in./min) were used; over three decades of crack velocity, a 20% decrease in G_c was observed. Using the end-notched flexure specimen, *Russell and Street* investigated the static and fatigue behavior of delaminations subjected to pure Mode II shear loading. Their results indicated that increasing matrix toughness improves Mode II shear fracture energy less than Mode I tensile fracture energy, and further, tougher systems are more sensitive to Mode II fatigue crack growth than are the brittle systems. *Johnson and Mangalgiri* presented Mode I, Mode II, and mixed mode interlaminar fracture toughness data for seven composites made with brittle and toughened matrix materials. The study showed that brittle fracture is controlled by the G_I component, while tough resin fracture is controlled by total strain-energy release rate.

Thermoplastics

Beever, O'Connor, Ryan, and Lou discussed the characterization of semicrystalline polyphenylene sulfide (Ryton®-PPS) neat resin moldings and composites. Molding conditions, especially cool down rates and annealing conditions, influence percent crystallinity and crystalline size which directly influence mechanical properties. *Nairn and Zoller* experimentally determined the residual thermal stresses in crystalline and amorphous thermoplastic composites and also predicted their magnitudes from the properties of the matrix. This work emphasizes the need to determine the effects of residual thermal stresses on composite properties. The next two papers dealt with polyetheretherketone (PEEK). *Cebe, Hong, Chung, and Gupta* conducted isothermal and rate-dependent crystallization of PEEK neat resin to achieve films with varying degrees of crystallinity and crystalline morphology. The room temperature mechanical properties of films varied with crystal size and size distribution but not significantly with degree of crystallinity. Using semicrystalline PEEK APC-2® composites, *Leach, Curtis, and Tamblin* investigated delamination behavior, damage as a result of low energy impact, and post-impact compressive strength.

Thermosets

Yee critically reviewed the reasons for lower-than-expected toughness in composites with toughened matrices. He discussed various mechanisms for toughening thermoset matrices, including use of second phases, strain softening combined with a high degree of orientation hardening, and low temperature relaxation processes. *Garcia, Evans, and Palmer* fabricated a hybrid composite from a 350°F cure epoxy by adding 3 to 15 parts of silicon carbide whiskers. The 90° tensile strengths and strains and tensile edge delamination strain levels were significantly increased while in-plane fiber dominated properties were all significantly reduced as a result of fiber damage incurred during fabrication. This approach appears promising if the fiber damage can be minimized. *Evans and Masters* disclosed the properties of new one-phase 350°F (176.6°C) cure epoxies with vastly improved mechanical, toughness, and damage tolerance properties over standard brittle systems. Novel improved interleafing materials were also described whose properties exceed the ultimate design strain target for post-impact compression. The interleaf concept exemplifies the ability to "engineer" solutions to critical problems if the mechanics of failure are understood. *Boschan, Tajima, Forsberg, Hull, and Harper-Tervet* describe the application of their recently developed computer assisted process models to facilitate fabrication of high quality graphite-epoxy composites by the standard autoclave/vacuum bag technique with minimum rejection rate. The thermal analyzer and chemiviscosity models were used to fabricate successfully 96- and 384-ply brittle 350°F (176.6°C) cure graphite-epoxy composites, respectively. The models have been extended to toughened matrix materials displaying non-Newtonian melt-flow behavior. *Hartness* explored the use of a novel semi-interpenetrating polymer network as a composite matrix by combining a dicyanate thermoset with a copolyester-carbonate thermoplastic. Composites exhibited substantially high interlaminar fracture toughness than brittle epoxies and excellent flexure properties.

Norman J. Johnston
NASA-Langley Research Center, Hampton, VA;
symposium chairman and editor

Toughened Composites: Prospectives From Industry

Clifford Y. Kam [1] *and Jeff V. Walker* [2]

Toughened Composites Selection Criteria

REFERENCE: Kam, C. Y. and Walker, J. V., **"Toughened Composites Selection Criteria,"** *Toughened Composites, ASTM STP 937,* Norman J. Johnston, Ed., American Society for Testing and Materials, Philadelphia, 1987, pp. 9–22.

ABSTRACT: The selection of composite material systems for use in large commercial transport aircraft depends on a number of material characteristics such as mechanical and physical properties, behavior in a hot-wet environment, resistance to delamination, ease of processing, and whether use of thick plies to reduce lay-up time is beneficial.

This paper reviews the design requirements and the key material characteristics needed by the structural designers. It has been well established that the current epoxy systems can be easily delaminated by the impact of external forces. This ease of delamination has resulted in a requirement to increase the toughness of the resin system so that its delamination resistance is two to four times higher than that of current epoxies. Other parameters include hot-wet requirements, low assembly cost potential, long shelf life, ease of processing, drapability requirements, and smoke generation.

KEY WORDS: composite materials, tough resins, design requirements processing, smoke generation

During the past decade, we have witnessed a large growth in the use of composite materials in aircraft structural applications. In particular, our typical advanced fighters are baselined with composite wing structure and much of the fuselage structure made from composites. Currently, this design trend toward composites is moving into the large cargo/transport and new military cargo airplanes to reduce structural weight. Studies show that by the mid 1990s, as much as 40% of the structural weight could be composites, and by the year 2000, composites would account for more than half the structural weight (Fig. 1).

Design Requirements

This surge in use of composite laminates has led to a desire to have a resin system that is far more resistant to impact damage than the current resins such as

[1]Branch manager, Airframe Research and Technology, Douglas Aircraft Co., Mail Station (36-52), 3855 Lakewood Blvd., Long Beach, CA 90846.

[2]Engineer, Composite Technology, Douglas Aircraft Co., 3855 Lakewood Blvd., Long Beach, CA 90846.

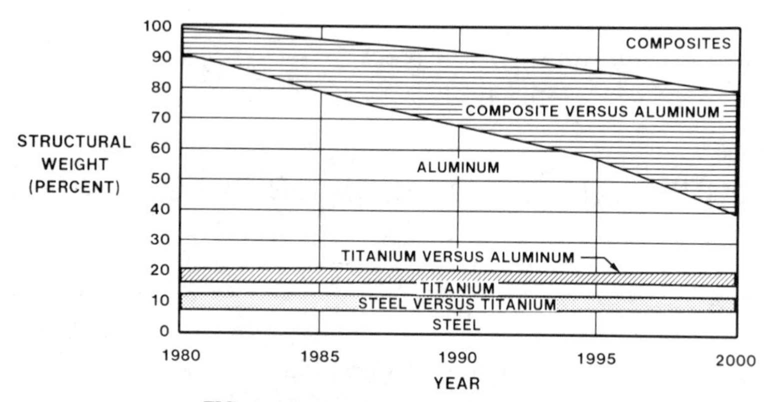

FIG. 1—*Structural material distribution.*

Narmco 5208 and Hercules 3501. Because coupon tests showed a degradation in structural strength after impact damage, the composite design community established a design strain limitation of 3000 μin./in. (3000 μcm/cm). Weight trade studies with this strain limitation show that when the structure is strength-critical, aluminum would be very competitive with composites, and when the added cost of manufacturing composite structure was taken into account, the composite structure was not competitive for many components. However, these data were enough to prod the resin producers to look for a resin system that would not show as much damage after impact. Thus, the class of "tough resins" was developed.

The tough resins are generally formulated by adding elastomeric or thermoplastic compounds to the more brittle MY-720/DDS resin base. With these additives, the new class of resins demonstrated that the design strain limits could be increased possibly to 6000 μin./in. (6000 μcm/cm).

The structural designer always checks a design for two major conditions: (1) Is the structure strong enough? (2) Will it meet the stiffness requirements? With metals, the designers need only be concerned with the amount of material needed (strength) for the loading condition or the bending (EI) or torsional (G_J) stiffness requirements. The designer of composite structures can select fibers for strength or modulus; he can orient the stacking sequence to optimize laminate strength or stiffness. He must limit the stress level because of the delamination potential of the laminate, and he needs to know how the operating and processing environment affects the strength and stiffness of the structure.

The application of composite materials to primary wing structure to reduce structural weight is forcing the structural designers and materials engineers to look for a new, toughened resin system that is far more tolerant of impact damage than the current resin systems. Preliminary design studies of new transports with high-aspect-ratio wings show that the bending stiffness requirements that are needed for flutter are dictating the material thickness. In Fig. 2, the wing box weight is shown relative to the wingspan for this type of advanced transport. Two

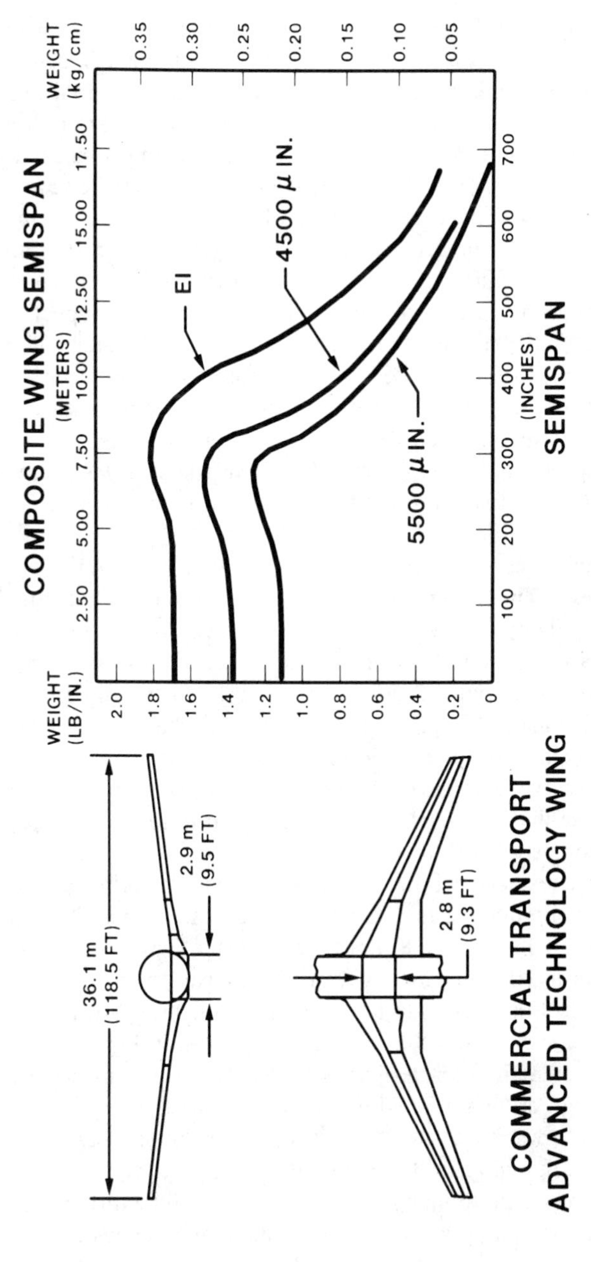

FIG. 2—*Weight comparison for bending stiffness and strength requirements.*

design constraints are noted: (1) bending stiffness thickness for flutter considerations and (2) strength thickness as constrained by strain limits of 4500 and 5500 μin./in. (4500 and 5500 μcm/cm). These strain limits were established as acceptable limits from the results of compressive test data of impact-damaged panels made from existing epoxy resin systems. The optimum structure obviously results when the strength and stiffness require the same material thickness.

Since a wing design can be constrained by the allowable strain, it is clear that we need a resin system that is more resistant to delamination after impact. It seems that the resin suppliers have gotten the message as they are putting improved resin systems on the market. This proliferation of resin systems makes the material selection even fuzzier. Adding to the improved resin systems, the fiber producers also are claiming that they can supply fibers that have better strength and modulus characteristics and that the fibers also fail at a higher strain level. A comparison of the unidirectional laminate tensile strength and stiffness is presented in Fig. 3. As indicated by the figure, the IM6 fiber composite is approximately 50% stronger than the T300, but its tensile modulus was increased by only 10%. However, when the improved fibers were used in a cross-plied laminate, it was noted that the improved strength and stiffness characteristics in the unidirectional laminate are not translated into the cross-plied laminates, as shown in Fig. 4.

With the new, improved fibers, there is an accompanying change in the filament diameter. The current filaments have a diameter of 7.5 μm, while the high-elongation fibers have a diameter of 5.0 μm. As a result of the smaller diameter, the cross-sectional tow area has been reduced from 0.48 mm^2 for T300 to 0.29 mm^2 for AS6 and 0.27 mm^2 for the IM6 fiber. Compression tests of unidirectional laminates (Fig. 5) show no noticeable change when IM6 fibers are used with the tough resin system. However, the thermoplastics tested by Douglas Aircraft have shown a large reduction in compressive strength, as noted in Fig. 5. Also note that thermoplastics tend to result in lower cross-ply compressive strength compared with thermoset systems.

The new tough resins may bring a reduction in manufacturing costs through use of thick plies. Preliminary tests of cross-plied laminates, where the ply thickness was varied from 0.127 to 0.508 mm (0.005 to 0.020 in.), have shown only a minor reduction in strength, as noted in Fig. 6. However, the thick plies can sharply reduce the time it takes to lay up a thick structural part. A comparison of the lay-up time for thick and thin plies is also shown in Fig. 6.

The hot-wet strength characteristics are also of concern to the structural designer. The design criteria applied to commercial aircraft require that the resin systems not lose any strength at 82°C (180°F) after a 30-day soak at 60°C (140°F) and 90% relative humidity. Military aircraft designers have a much higher temperature requirement. Preliminary test data show very little difference in strength between the tough resin systems under evaluation and the basic brittle systems. The compressive strength of the thermoplastics reviewed by Douglas Aircraft seem to be unaffected by the hot-wet environment, as noted in Fig. 7. However, the compressive strength is approximately 60% of current epoxy resin systems.

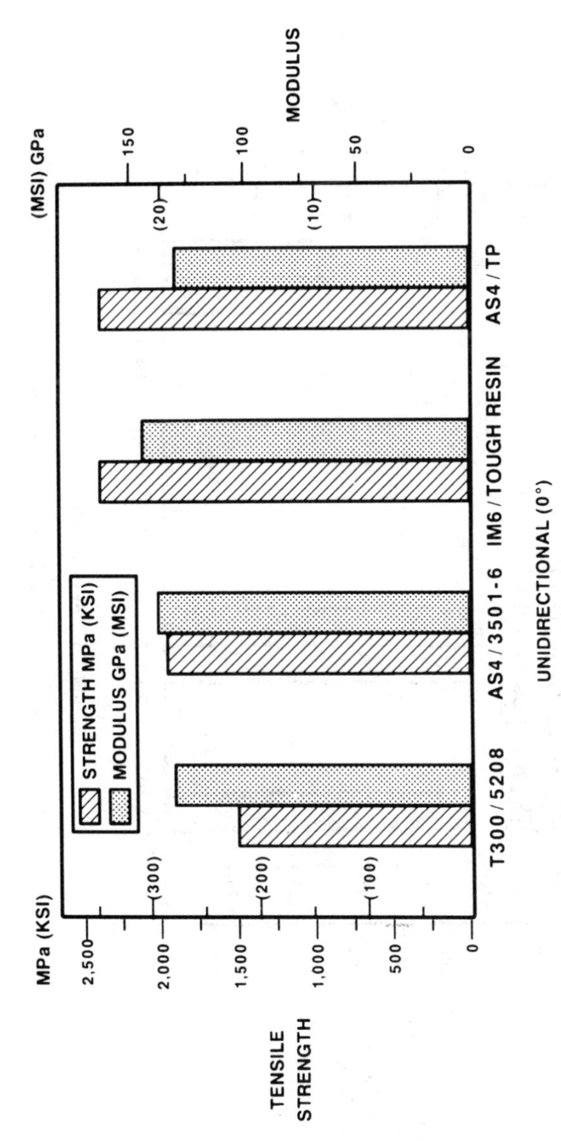

FIG. 3—*Strength-modulus comparison.*

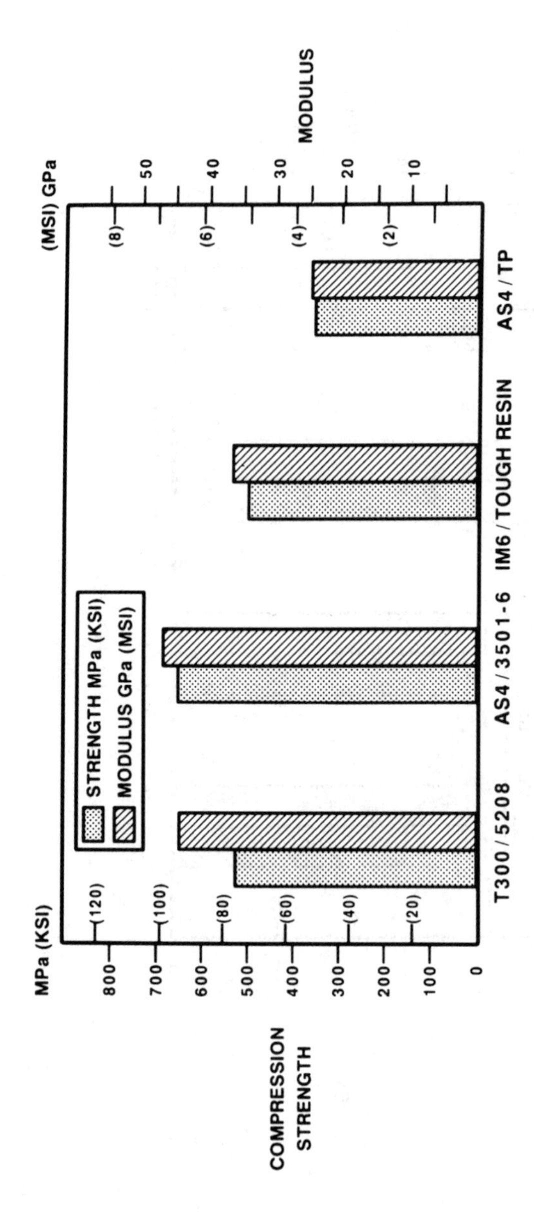

FIG. 4—*Compressive strength-modulus comparison.*

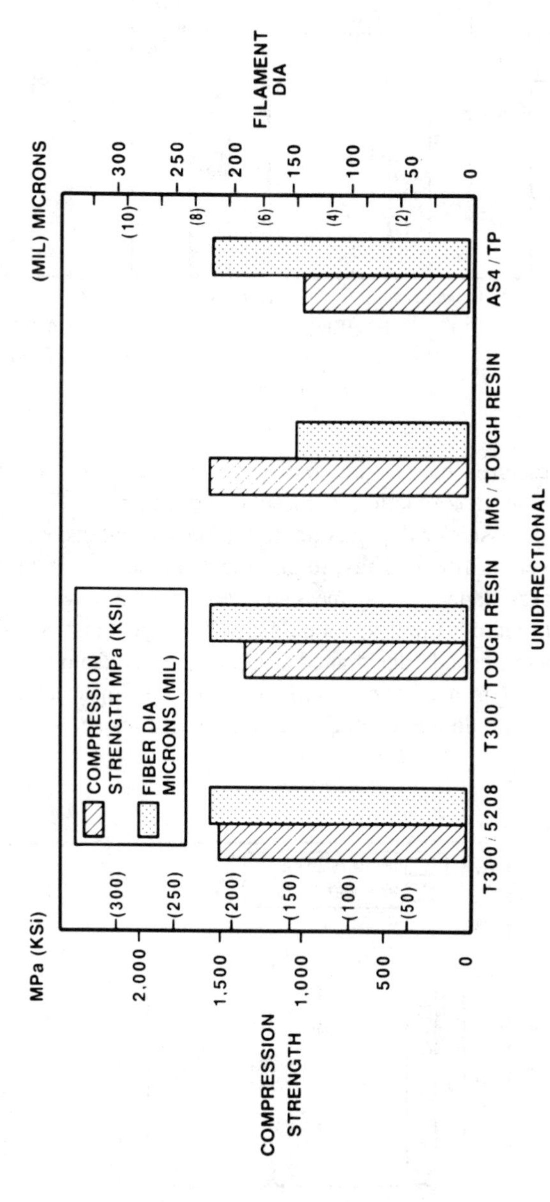

FIG. 5—*Effects of fiber diameter on compressive strength.*

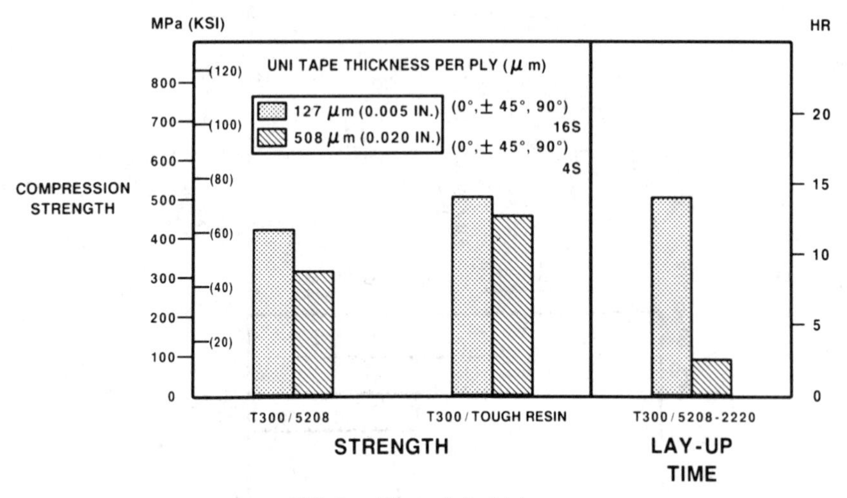

FIG. 6—*Effect of ply thickness.*

Composite structures have made the designers painfully aware of a new critical-failure mode: delamination of the plies after impact. This newly recognized failure mode basically depends upon the transverse tensile strength of the resin between layers of fibers. Thus, to provide a simple coupon test to evaluate the delamination resistance or toughness, the development engineers adopted an old adhesive test method called the double cantilever peel test (see Fig. 8). The toughness parameter, usually designated as G_{Ic}, is derived from the double cantilever beam test and is used for judging whether one material is better than another. However, the double cantilever beam test results show reasonable cor-

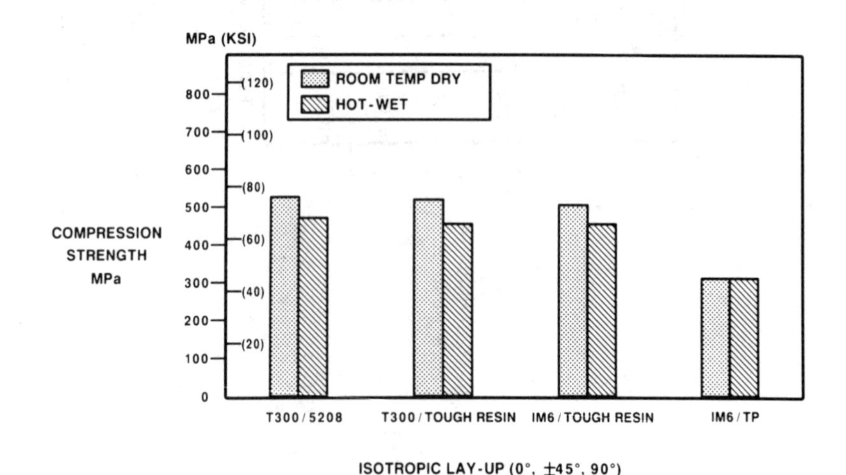

FIG. 7—*Environmental effects.*

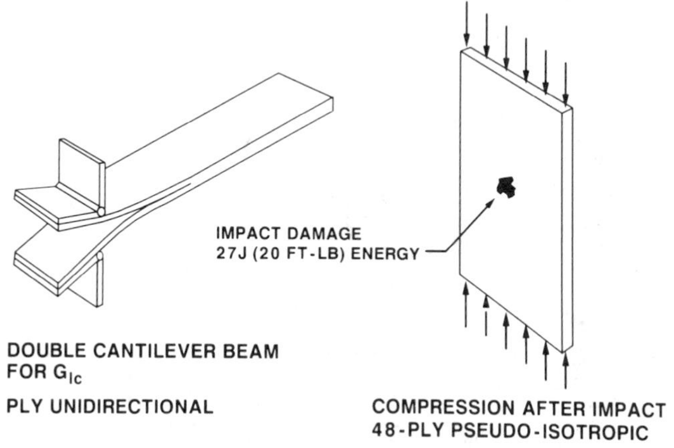

FIG. 8—*Composite toughness tests.*

relation of G_{Ic} with void and resin content. Tests have shown that high G_{Ic} values can be obtained for laminates with high void content. Accordingly, the compression strength test after impact is also used to evaluate toughness, as this test more closely represents the failure mode of an actual structure. The current method is to use the industry accepted impact standard of 27 J (20 ft · lbf) on a 48-ply pseudo-isotropic laminate.

When the G_{Ic} values are compared with compressive strength values after impact, the increases in G_{Ic} do not always translate into the same increases in compressive strength. As noted in Fig. 9, for the thermosets such as 5208, 3501-6, and 1808I, the G_{Ic} values seem to translate into the same relative increases in compressive strength after impact. However, with some of the thermoplastics, the large increase in G_{Ic} does not translate into the same increase in compressive strength. For example, the compression after impact tests of poly-(etheretherketone) (PEEK) specimens show a value of 0.38 MPa (55 ksi) for a G_{Ic} of 1620 J/m^2 (9 in. · lb/in.^{2}). Tests of 1808I show a compressive strength after impact of 0.38 MPa (55 ksi) for G_{Ic} of 450 J/m^2 (2.5 in. · lb/in.^{2}).

Process Requirements

Processing parameters such as cure cycle, exothermic activity, resin bleed requirements, and drapability are also of concern when selecting new tough resin systems. The cure cycle must be as simple as possible. Figure 10 shows a Narmco 5208-type cure which requires that pressure be applied at the correct time during the cure, or the resin and void content in the part will not be acceptable. For new resin systems, the material engineers have designated that pressure be applied at the beginning of the cure cycle and then held throughout the cure, and that the part temperature have a single step to the cure temperature, as also noted in Fig. 10. A single-step cure cycle is most acceptable since many subcontractors do not have sophisticated control equipment for governing multiple-step cure cycles.

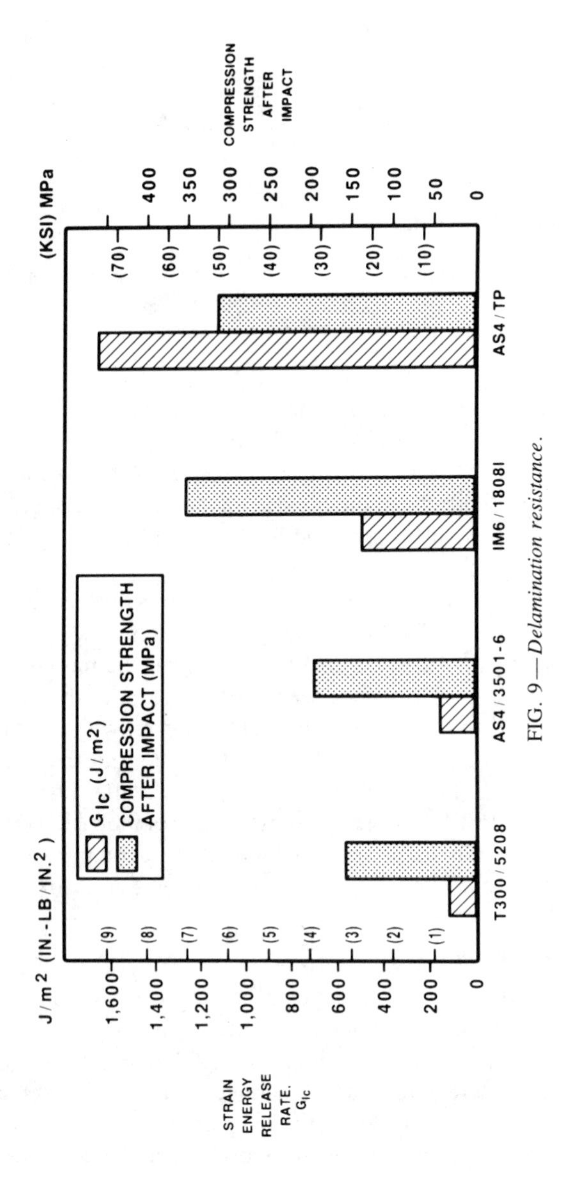

FIG. 9—*Delamination resistance.*

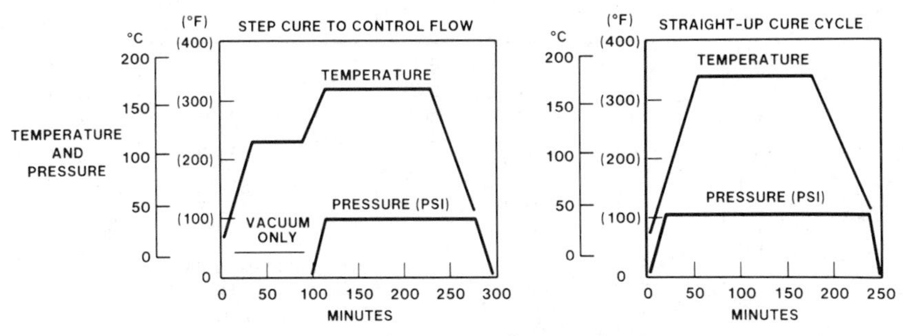

FIG. 10—*Comparison of cure cycles.*

The single-step pressure cycle will also show early bag leaks before the part gets too hot to abort the cure.

Another desirable resin characteristic is the no-bleed cure since it will not result in discarding costly resin. However, it is necessary to provide some means of allowing the volatiles to escape. Usually, some glass filament yarn strings are placed between the part and the bagging material to provide escape routes, especially with large parts.

To assess a material's forgiveness or versatility under different cure cycle conditions, the dynamic viscosity profile is measured. The acceptable viscosity range is shown in Fig. 11. It is most desirable that the resin viscosity be low enough to aid in volatile migration but high enough to prevent excessive resin flow from the laminate.

As composite materials encroach on primary structural elements, there is a need for thicker laminates. It would not be uncommon to see laminate thicknesses of 25.4 mm (1 in.) or more. As the parts get thicker and the size increases, the cure cycle of the structural component must be adjusted to account for exothermic activities. Calorimetric measurements of panels thicker than 12.7 mm (0.5 in.) show exothermic activity as temperatures approach 135 °C (275 °F) (Fig. 12).

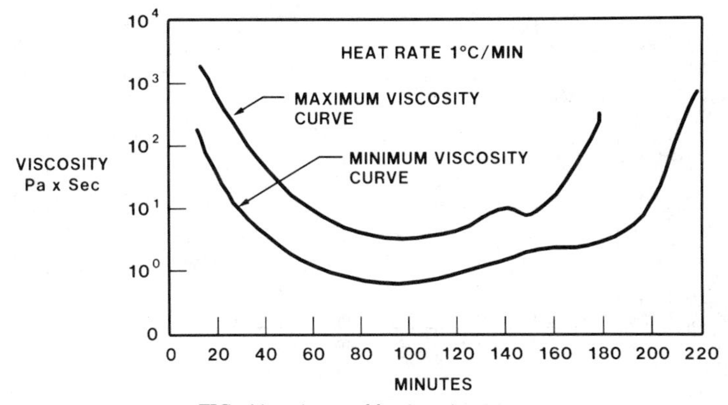

FIG. 11—*Acceptable viscosity range.*

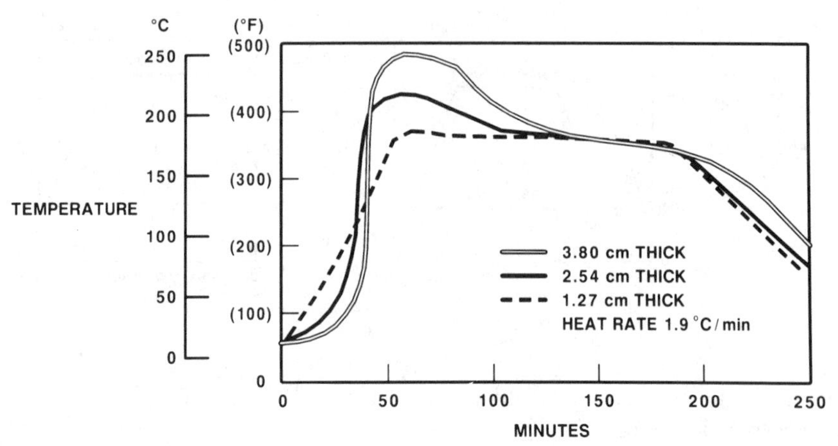

FIG. 12—*Exotherm versus laminate thickness.*

This means that parts thicker than 12.7 mm (0.5 in.) must have several isothermal dwell cycles during the heat-up phase, as shown in Fig. 13, to control exotherm and part temperature profile.

Another key material characteristic that is applicable for biwoven preimpregnated material is drapability. The drapability measurement also helps define a resin's practical shelf life and therefore is an important criterion for selecting resin systems. The measurement for drapability is a pull test of 45° biwoven prepreg material, as shown in Fig. 14. The material is pulled at a speed of 12.7 mm (0.5 in.) per minute until 38 mm (1.50 in.) of displacement is achieved. The load should not exceed 0.68 kg in the test region, as noted in the load-deflection plot shown in Fig. 14, for the material to have good drapability.

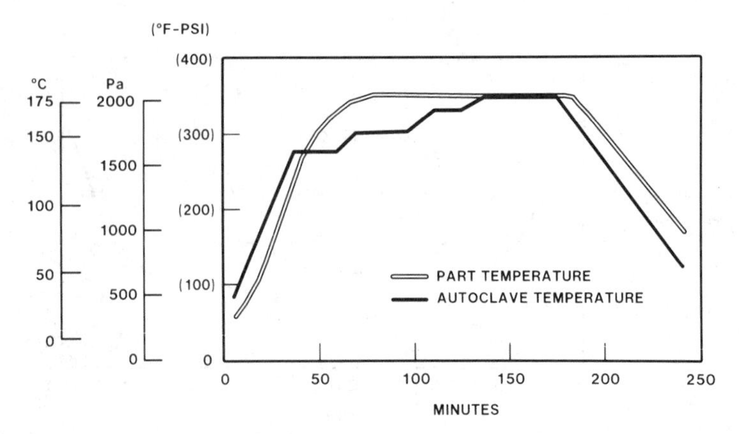

FIG. 13—*Controlling exotherm.*

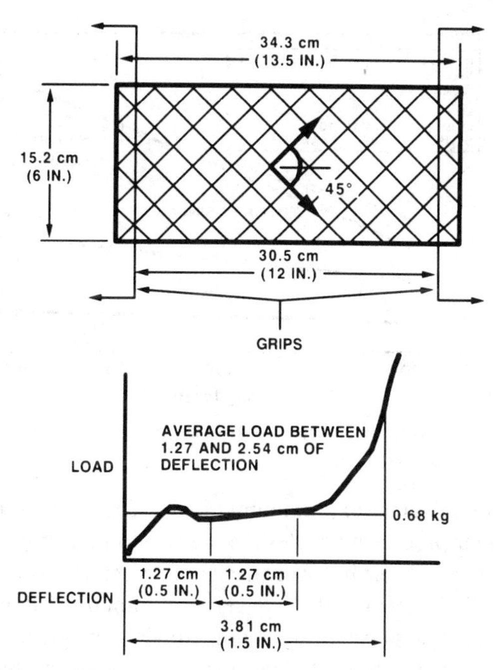

FIG. 14—*Drapability test for woven cloth.*

To lay up skins the size of transport aircraft wings, the preimpregnated material must be stable for as long as three weeks at near-room temperature. The new tough resins appear to be less stable as a result of their higher catalyst levels that work to decrease gel time and restrict resin flow. This reduced out-time requires a fabricator to schedule a two-shift workday which is not needed with the more stable materials.

Smoke generation is a coming issue for new resin systems, especially for fuselage structural components. Smoke generation is of major concern in selection of materials for fuselage structure, and especially for interior panels since the smoke will significantly reduce visibility during a fire. New materials are screened using a National Bureau of Standards smoke-density chamber. Figure 15 shows the results of smoke generation tests for several resin systems. The new tough resin systems that use butyl compounds will produce more smoke than the present 3501-6 or 5208 resins because of the modifiers used to make the systems tough. It appears that thermoplastics would be the most desirable class of resin system when smoke generation is a major concern. High resistance to impact damage also makes the thermoplastics a strong candidate for interior panels.

Conclusion

The material systems to be used in composite structures that will fly on our commercial transports in the mid 1990s must have mechanical and physical

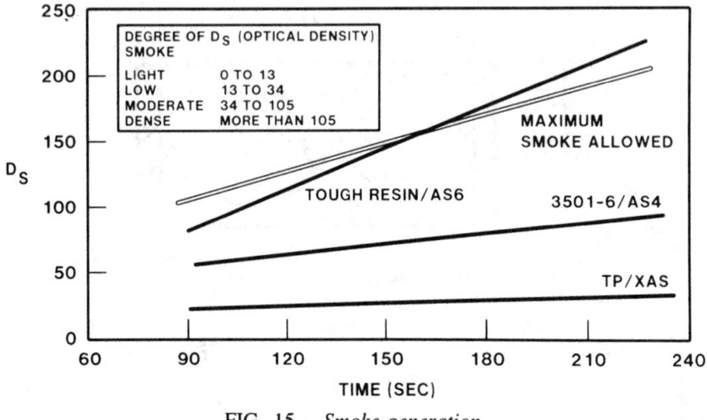

FIG. 15—*Smoke generation.*

properties at least equivalent to aluminum alloys. These future material systems need to have simpler processing cycles and be two to four times tougher than the current epoxy systems. Their hot-wet strength characteristics need to be acceptable to at least 82°C (180°F) at 90% relative humidity. It would be highly desirable to have a single pressure step coupled with a single temperature step so that multiple parts can be cured together in an autoclave. A no-bleed resin system would result in less waste and reduced disposal cost. To be useful for large, thick parts such as wing skins, the resins need to have several weeks of out-of-freezer life and yet be priced about the same as today's epoxies.

Charles F. Griffin [1]

Damage Tolerance of Toughened Resin Graphite Composites

REFERENCE: Griffin, C. F., **"Damage Tolerance of Toughened Resin Graphite Composites,"** *Toughened Composites, ASTM STP 937,* Norman J. Johnston, Ed., American Society for Testing and Materials, Philadelphia, 1987, pp. 23–33.

ABSTRACT: The weight savings potential of graphite composites for wing structures in transport aircraft is directly related to design strength values. These values must reflect the influence of defects, notches, impact damage, and environmental conditions on material strength. Within the last several years, improvements have been made in the constituents of graphite/epoxy composites. Fiber tensile strengths have increased and tougher resins have been formulated. Tension, impact, and compression tests were conducted on several of these advanced materials to evaluate their performance. Based on these tests, it was concluded that increased values for design allowable tensile strength can be obtained. However, similar increases in compressive strength allowables cannot be obtained until additional improvements are made to resin matrices.

KEY WORDS: composite materials, graphite/epoxy, toughened resins, impact damage, post-impact compression, notched tension, mechanical properties, tests

Numerous research programs have evaluated the deleterious effects of impact damage on the strength of graphite/epoxy structures [*1–3*]. Increasing the resin toughness has been shown to increase post-impact compressive strength of graphite/epoxy composites [*4,5*]. Within the last three years, many toughened resin graphite composites have been introduced by material suppliers. The objective of the investigation reported herein was to determine the static strength of several improved toughness composites and to evaluate the potential for increased design strength allowables. Data presented in this report were obtained as part of National Aeronautics and Space Administration (NASA) contract NAS1-16856 and a Lockheed-California Company independent research and development program.

[1] Research and development engineer, Lockheed-California Co., Department 76-20, Bldg. 199, P. O. Box 551, Burbank, CA 91520.

The Impetus for Improved Graphite/Composites

Advanced composite materials are being used for construction of aircraft structures because of the significant weight savings and potential acquisition cost reductions compared to aluminum structures. Application of advanced composites to transport aircraft secondary and medium primary components have demonstrated cost competitive weight savings of 25%. To achieve a significant reduction in transport aircraft structural weight requires the use of composites for construction of the wing and fuselage, which together comprise about 70% of the airframe weight. It should be emphasized that many current applications for composites are for stiffness critical components; large portions of a wing box would be strength critical.

Graphite/epoxy composites are brittle materials. Tensile strength reductions of 50% are typical for coupons containing a fastener hole. Test data also indicate that impact damage, even when not visible, seriously degrades laminate compressive strength. Structural criteria [*6,7*] require composite structures to be damage tolerant, or more specifically, that they have the ability to resist failure as a result of visible defects or damage for a specified period of time. For the case of nonvisible impact damage, structures must be able to carry design ultimate load without failure for the service life of the structure.

Figure 1 shows schematically the effects of various test conditions on the tensile and compressive strength of graphite/epoxy composites. Critical test conditions, those selected for determination of design allowable strains, were notched tension (open hole) and impacted compression. For a typical untoughened resin graphite/epoxy, T300/5208, design allowable strains for these conditions are 4500-μin./in. (4500-μcm/cm) tension and 4000-μin./in. (4000-μcm/cm) compression [*3*].

Preliminary design studies [*8*] predict that wing surfaces constructed with graphite/epoxy composites offer a large weight savings compared to an aluminum baseline, if design allowable strains can be increased from the current levels to

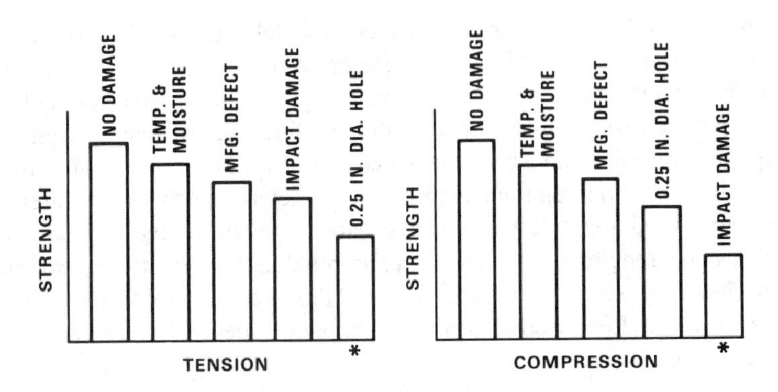

FIG. 1—*Effect of various test conditions on the strength of graphite/epoxy.*

6000 μin./in. (6000 μcm/cm). For example, for upper wing surfaces, which are typically designed by axial compression loading, predicted weight savings increases from 25.6 to 37.8% if the higher design allowable strain is used in the analysis. Similar increases in weight savings would be achieved for lower wing surfaces.

Materials Evaluated and Experimental Procedures

Three graphite fibers and seven resin systems were tested in this investigation (see Table 1). Materials were obtained as 12-in. (30-cm) wide unidirectional tape with an areal weight of 145 g/m^2. Crossplied laminates were laid-up by hand and cured in an autoclave using the supplier recommended cure cycle. Following cure, all laminates were ultrasonically inspected and the resin content measured. Resin contents of the test laminates ranged from 28 to 36% with the exception of the HST-7 material which had a resin content of 47% because of the duplex matrix construction [*9*].

Test data reported herein include laminate tensile and compressive strength, and impact behavior. Detail procedures for these tests are described in Ref *10*.

Static Tensile-Behavior

Tensile strength of the composites investigated was determined by conducting tests on coupons machined from crossplied laminates. Figure 2 shows the test fixture, a failed coupon, and the coupon geometry. Axial strain gages were used to determine far-field strains within the coupon. Notched specimens contained an open hole drilled on the coupon centerline.

Figure 3 presents a comparison of the tensile strain-to-failure for several composites investigated. Notched and unnotched data are presented for laminates having two fiber orientations. The data presented represents an average of at least three tests. Comparing the behavior of an untoughened resin composite, AS4/3502, to a toughened resin composite, AS4/2220-1, indicates that the toughened material displays a 27% increase in failure strain. These increases

TABLE 1—*Materials evaluated.*

	FIBERS
AS4	intermediate strain fiber
AS6	high strain fiber
Celion ST (HSC)	high strain fiber
	RESINS
3502	untoughened epoxy
2220−1	toughened epoxy
584	toughened epoxy
1806	toughened epoxy
5245C	modified bismaleimide
974	toughened epoxy
HST−7	duplex epoxy

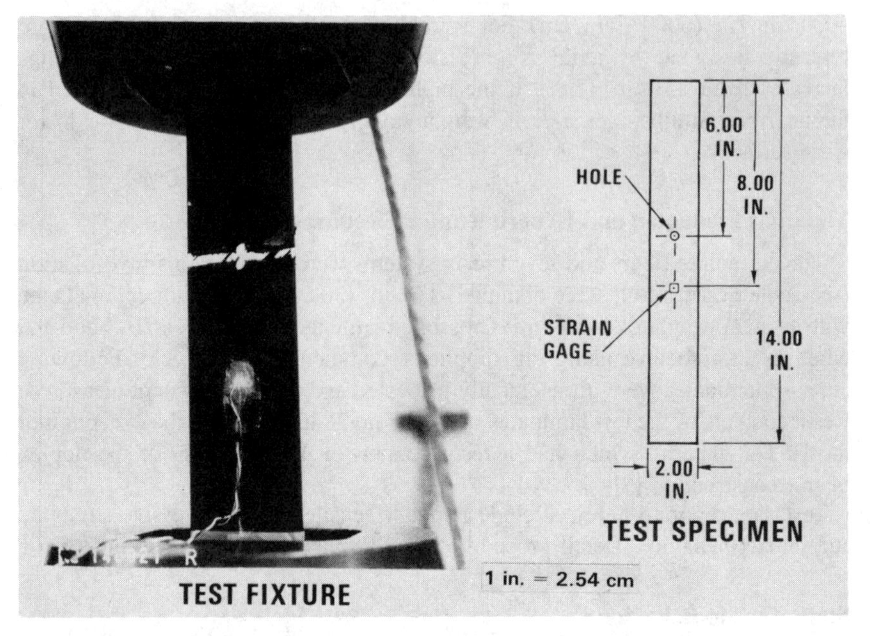

FIG. 2—*Tension test fixture and coupon geometry.*

in strength are due to superior fiber property translation within the more ductile resin matrix.

A comparison of the tensile strain-to-failure of quasi-isotropic laminates constructed with several toughened resin composites is also presented in Fig. 3. Note that laminates fabricated with the higher strain-to-failure fibers, such as AS6 and high strain Celion®, do perform proportionally better than the composite with the intermediate strain-to-failure fiber.

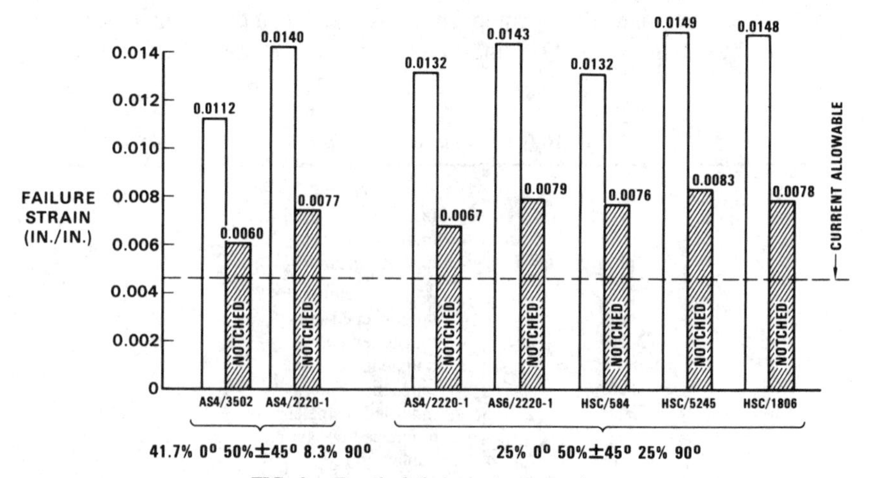

FIG. 3—*Tensile failure strain comparison.*

Based on the data presented, tensile design strain allowables of 6000 μin./in. (6000 μcm/cm) are attainable for composites constructed with high strain-to-failure graphite fibers combined with tougher resins.

Impact Response

Impact damage can occur during assembly as a result of dropped tools or other objects, during transportation or handling, and in service. Typical impact threats to wing and fuselage structures are generally high mass items traveling at low velocities. An impact test fixture, shown in Fig. 4, that duplicates typical structural support conditions was designed and fabricated. The size and thickness of the test laminate was selected based on the desire to have a stable post-impact compression coupon that would have visible impact damage which did not exceed 50% of the coupon width.

To conduct impact tests, a 25-in. (63.5-cm) long by 7-in. (18-cm) wide panel was clamped to a steel base plate with a 5- by 5-in. (13-by 13-cm) opening. The test panel was struck in the center of the opening with a 12-lb (5.4 kg) impactor which had a ½-in. (1-cm) hemispherical diameter hardened steel tip. Impact energy is computed as the product of impactor mass multiplied by drop height. After being impacted, the panels were inspected visually and ultrasonically to ascertain the amount of damage.

Data from impact tests conducted on quasi-isotropic laminates constructed with the same resin system but different fibers are presented in Fig. 5. Threshold damage (delamination of 0.25 in.^{2} [1.6 cm^2] or greater) energy for all laminates was identical, as was the amount of energy required to cause barely visible front

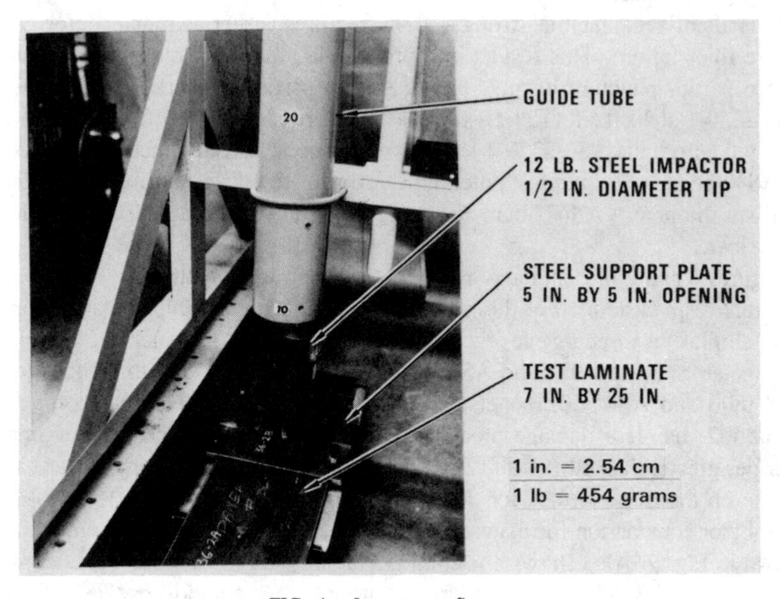

FIG. 4—*Impact test fixture.*

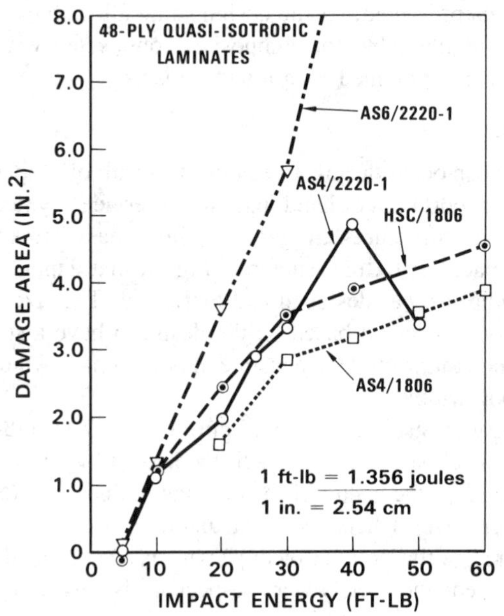

FIG. 5—*Impact response—fiber effects.*

surface damage, 20 ft · lbf (27 J). However, at impact energies greater than 10 ft · lbf (14 J), the composites fabricated with high strain-to-failure fibers displayed greater amounts of damage.

It is theorized that the stronger fiber requires larger laminate deformation before fiber failure. This results in more interlaminar and translaminar damage before partial puncture occurs. For example, partial puncture occurred in the AS4/2220-1 at 40 ft · lbf (54 J), whereas an energy level of 60 ft · lbf (81 J) was required before the AS6/2220-1 showed evidence of partial puncture. In the case of AS4 versus AS6, another potential reason for the different impact responses is fiber diameter. AS6 fibers are approximately 75% the diameter of the AS4 fiber.

Figures 6 and 7 present data on laminates constructed with the same fiber and different resin systems. For the materials shown in Fig. 6, impact damage areas were similar up to energy levels of 20 ft · lbf (27 J). At greater energy levels, damage areas in AS4/974 and AS4/2220-1 panels were greater than those in the AS4/1806 and AS4/3502 panels. Note that the untoughened resin composite, AS4/3502, has less damage area than toughened resin composites for impact energies greater than 20 ft · lbf (27 J). For all these materials, damage was barely visible on the front surface for 20-ft · lbf (27-J) impacts.

Another comparison for materials having the same fiber but different resins is shown in Fig. 7. Also shown for comparison are data for the untoughened resin composite AS4/3502. For this group of materials, two composites, HSC/974 and HSC/HST-7, had a damage threshold energy twice that of the other materials.

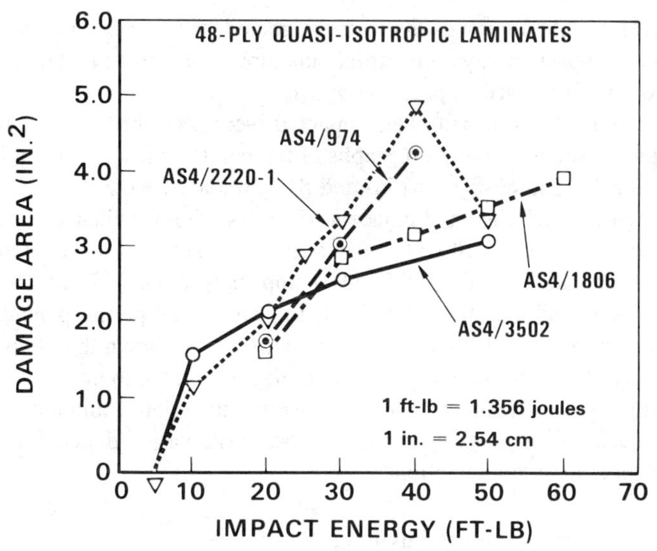

FIG. 6—*Impact response—resin effects.*

With exception of HSC/HST-7, the duplex resin material, all other materials tested had impact damage greater than the untoughened resin composite for impact energies greater than 30 ft · lbf (41 J). Partial puncture of the panel occurred at higher energies for toughened resin composites than the untoughened material. Again, it is theorized that the increased amount of local deformation which occurs before penetration is the cause of increased damage in toughened resin composites.

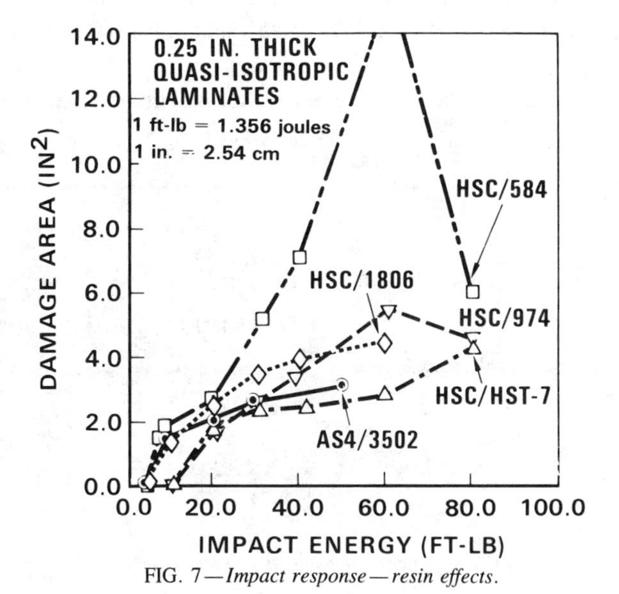

FIG. 7—*Impact response—resin effects.*

For several of the materials that were impacted, the damage area increased as a function of impact energy until partial puncture occurred. At that point, the area decreased with increased impact energy.

Several laminates that had been impacted were sectioned parallel to the 90° plies. Figure 8 shows photomicrographs of typical damage caused to a HSC/974 laminate for 40- and 60-ft · lbf (54- and 81-J) impacts. Note that delaminations run between the 90° plies and adjacent 45° plies. Translaminar cracks near the impact point and broken fibers on front and back surfaces are evident in both photomicrographs. The 40-ft · lbf (54-J) impact resulted in large visible front surface damage and the 60-ft · lbf (81-J) impact caused partial puncture.

Impact response study results discussed above have shown that graphite composites fabricated with tougher resins exhibit a greater amount of damage area than the untoughened resin composite. However, the relative amounts of delaminations, translaminar cracking, and fiber breakage varied depending upon the particular fiber/resin combination.

Post-Impact Compressive Behavior

Tests were conducted on each material to determine effects of impact damage on compressive strength. The compressive test fixture, Fig. 9, simply supported the coupon at the sides and clamped it at the loaded edges. This technique of stabilizing the coupon allows out-of-plane deflections associated with delamination growth. Each coupon was instrumented with back-to-back axial strain gages located away from the damage area.

One group of post-impact compression tests compared the performance of an untoughened resin, 3502, composite to a toughened resin composite, 2220-1. Both laminates had the same reinforcing fiber, AS4. The laminate orientation used for these tests was 41.7% 0°, 50% ± 45°, and 8.3% 90° plies. Post-impact

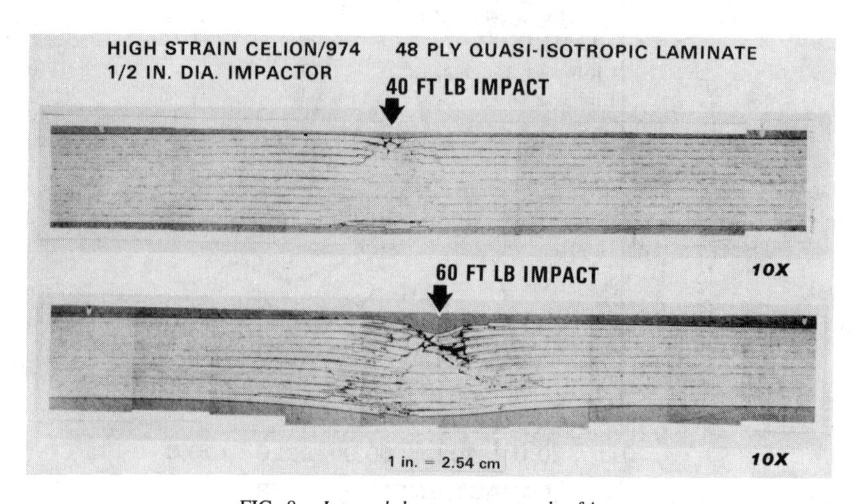

FIG. 8—*Internal damage as a result of impact.*

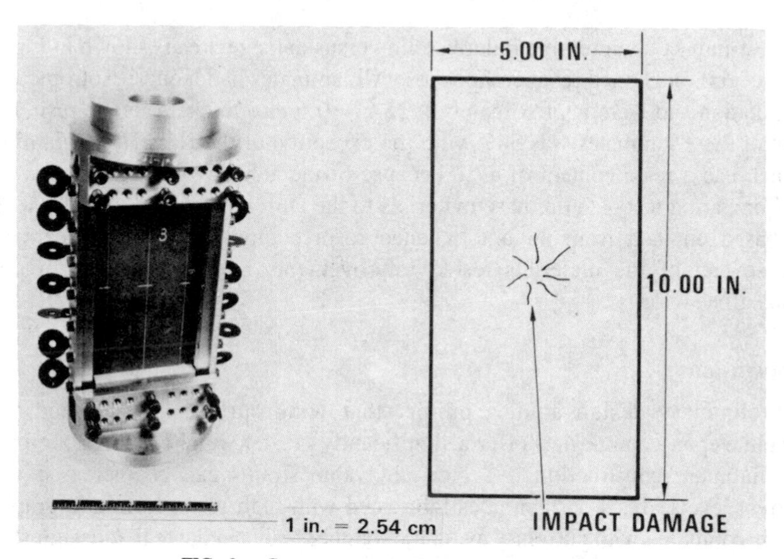

FIG. 9—*Compression test fixture and specimen.*

compressive failure strain versus ultrasonic inspection damage area are shown in Fig. 10. Each data point represents the average of at least three tests. For a (20-ft · lbf (27-J) impact energy, AS4/2220-1 had less impact damage than did AS4/3502 material and its post-impact failure strain was 19% greater. Damage on the impacted surface of both laminates was barely visible. At a (30-ft · lbf (41-J) energy level, where damage could be seen easily on the impacted surface, the damage area of AS4/2220-1 laminates was greater than that of AS4/3502 laminates. However, the AS4/2220-1 laminate still outperformed the AS4/3502 laminate, although only by 9%. Thus, only a small improvement was obtained with this particular toughened resin.

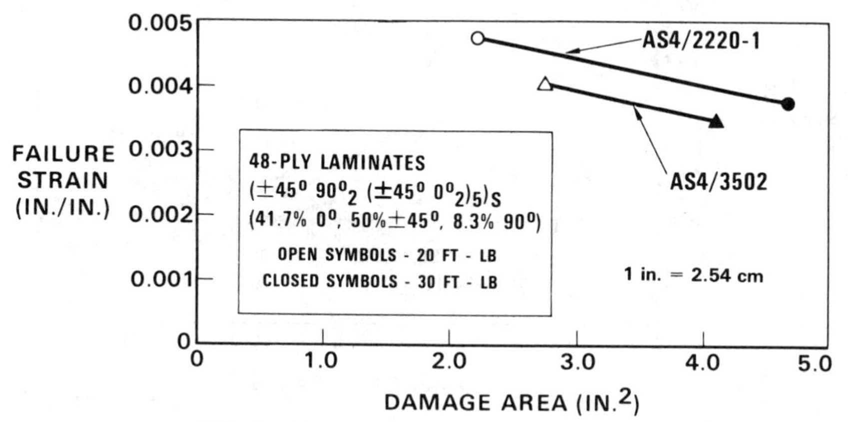

FIG. 10—*Post-impact compressive—resin effects.*

Post-impact compressive failure strain versus damage area is shown in Fig. 11 for several toughened resin composites. All laminates had a quasi-isotropic fiber orientation and were approximately 0.25 in. (0.6 cm) thick. Average resin content of these laminates was 34% with the exception of the HSC/HST-7 laminate which had a resin content of 47% because of the adhesive interleave.

Comparing test data on these materials to the current design allowable, which is based on data from an untoughened resin composite [3], it appears that for several of the materials tested improvements have been made in post-impact behavior.

Conclusions

Preliminary design studies predict that wing surfaces constructed with graphite/epoxy composites offer a significantly greater weight savings compared to aluminum construction if design allowable strains can be increased from current levels. Tests on laminates fabricated with high strain-to-failure graphite fibers combined with currently available tougher resins indicate that design strain allowables of 6000 μin./in. (6000 μcm/cm) for tensile loads can be obtained. Post-impact compression tests indicate that toughened epoxy composites offer improvements in strength compared to untoughened materials.

Although progress made in achieving improved performance by toughening the resin is encouraging, further toughness improvements in graphite fiber reinforced composite materials are required to enable a substantial increase in design allowable strains for compressive loading.

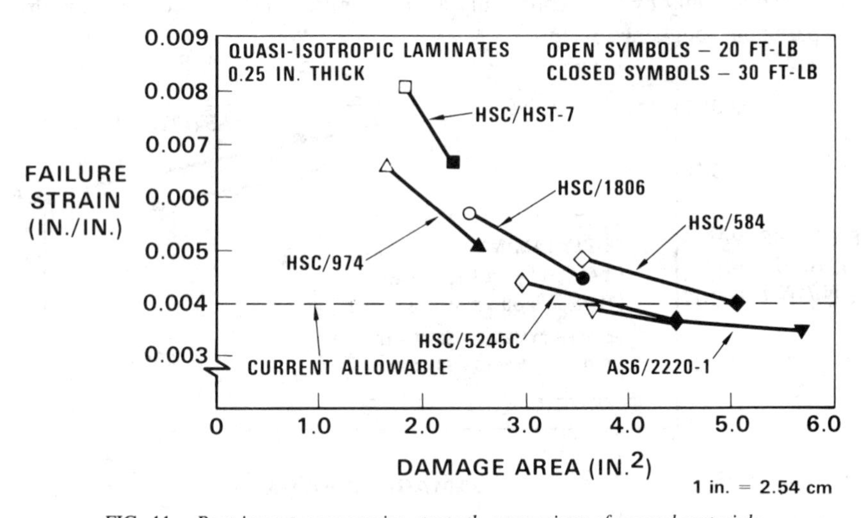

FIG. 11—*Post-impact compressive strength comparison of several materials.*

References

[1] Greszczuk, L. B. and Chao, H., "Investigation of Brittle Fractures in Graphite-Epoxy Composites Subjected to Impact," USAAMRDL-TR-75-15, McDonnell Douglas Astronautics Co., Huntington Beach, CA 92647, May 1975.

[2] Rhodes, M. D., Williams, J. G., and Starnes, J. H., "Low-Velocity Impact Damage in Graphite-Fiber Reinforced Epoxy Laminates," presented at the 34th Annual Conference of Reinforced Plastics/Composite Institute, sponsored by the Society of the Plastic Industry, Inc., Brookfield Center, CT 06805, Feb. 1979.

[3] Griffin, C. F. and Ekvall, J. C., "Design Allowables for T300/5208 Graphite/Epoxy Composite Materials," *Journal of Aircraft,* Vol. 19, No. 8, Aug. 1982, p. 661.

[4] Williams, J. G. and Rhodes, M. D., "The Effect of Resin on the Impact Damage Tolerance of Graphite-Epoxy Laminates," NASA TM 83213, Langley Research Center, Hampton, VA 23665, Oct. 1981.

[5] Byers, B. A. "Behavior of Damaged Graphite/Epoxy Laminates Under Compression Loading," NASA-CR-159293, National Aeronautics and Space Administration, Jan. 1980.

[6] McCarty, J. E., Ratwani, M. M., et al., "Damage Tolerance of Composites Interim Report No. 3," Air Force Contract F33615-82-C-3213, Boeing Military Airplane Co., Seattle, WA 98124, March 1984.

[7] Anonymous, FAA Advisory Circular 20-107A on Composite Aircraft Structure, Federal Aviation Administration, Washington, DC, 25 April 1984.

[8] Griffin, C. F., "Fuel Containment and Damage Tolerance in Large Composite Primary Aircraft Structures," NASA CR 166083, National Aeronautics and Space Administration, March 1983.

[9] Krieger, R. B., Jr., "The Relation Between Graphite Composite Toughness and Matrix Shear Stress-Strain Properties," *29th National SAMPE Symposium,* Society for the Advancement of Material and Process Engineering, Covina, CA 91722, 1984, pp. 1570–1581.

[10] "Standard Tests for Toughened Resin Composites—Revised Edition." NASA RP-1092, ACEE Composites Project Office, Langley Research Center, Hampton, VA 23665, July 1983.

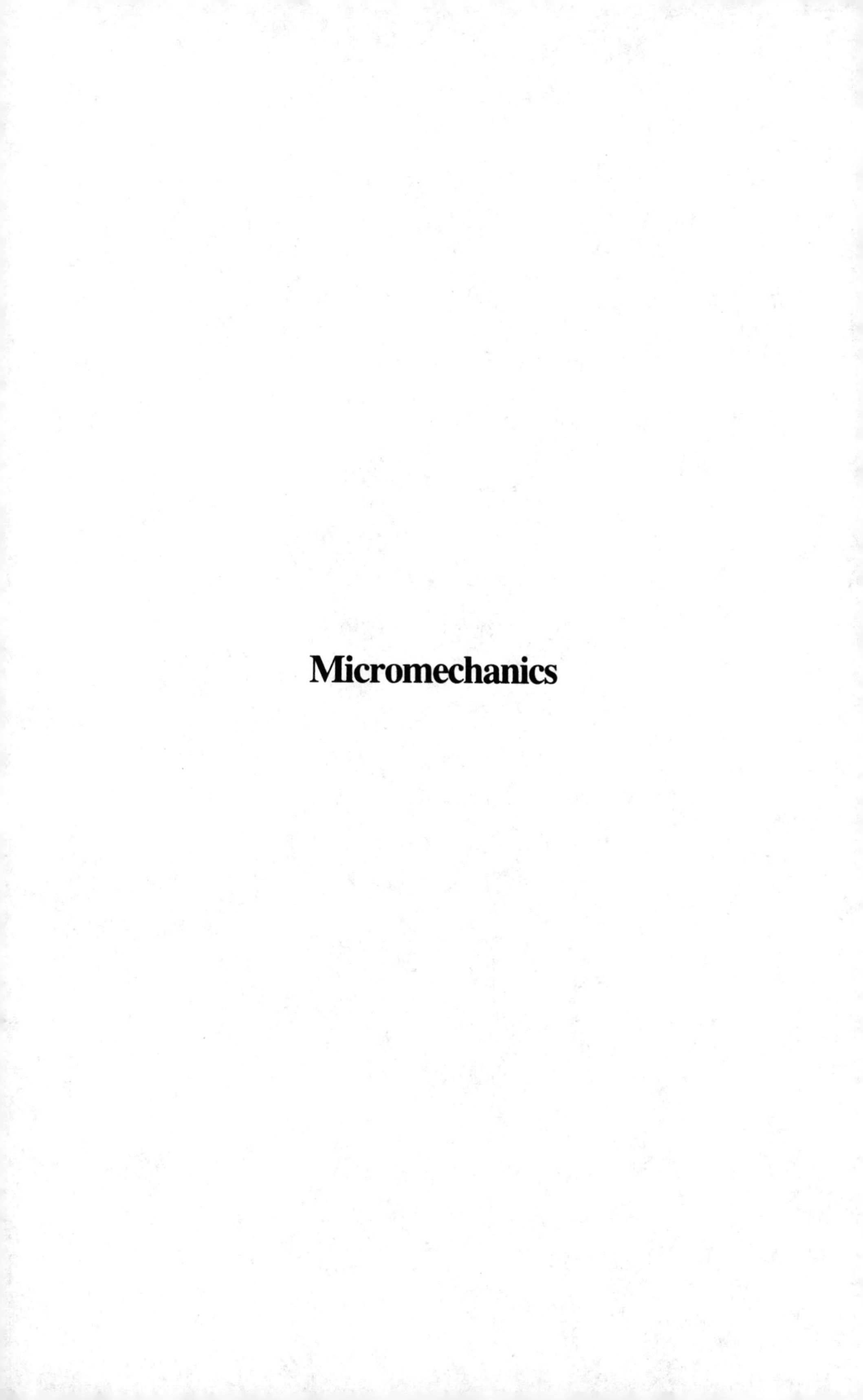

Micromechanics

Mohsen M. Sohi,[1] *H. Thomas Hahn,*[1] *and Jerry G. Williams*[2]

The Effect of Resin Toughness and Modulus on Compressive Failure Modes of Quasi-Isotropic Graphite/Epoxy Laminates

REFERENCE: Sohi, M. M., Hahn, H. T., and Williams, J. G. **"The Effect of Resin Toughness and Modulus on Compressive Failure Modes of Quasi-Isotropic Graphite/Epoxy Laminates,"** *Toughened Composites, ASTM STP 937,* Norman J. Johnston, Ed., American Society for Testing and Materials, Philadelphia, 1987, pp. 37–60.

ABSTRACT: Compressive failure mechanisms in quasi-isotropic graphite/epoxy laminates were characterized for both unnotched and notched specimens and also following damage by impact. Two types of fibers (Thornel 300 and 700) and four resin systems (Narmco 5208, American Cyanamid BP907, and Union Carbide 4901/MDA and 4901/mPDA) were studied. The widely used T300/5208 served as the baseline composite system. For all material combinations, failure of unnotched specimens was initiated by kinking of fibers in the 0° plies. A major difference was observed, however, in the mode of failure propagation after the 0° ply failure. In laminates made with Narmco 5208 resin, the 0° ply failure was immediately followed by delamination and catastrophic failure of the specimen. In BP907 resin, the fiber kinking was well contained without delamination, and still allowed further increase of load. The remaining two resins lay in their resistance to delamination between BP907 and Narmco 5208. The strength of quasi-isotropic laminates in general increased with increasing resin tensile modulus. The laminates made with Thornel 700 fibers exhibited slightly lower compressive strengths than did the laminates made with Thornel 300 fibers. The notch sensitivity as measured by the hole compressive strength was lowest for the BP907 resin and highest for the 5208 resin. For the materials studied, however, the type of fiber had no effect on the notch sensitivity. The area of impact damage was smallest for the BP907 resin. The 4901 resins were comparable to the 5208 resin in their impact resistance. Of the two fiber types, the T700 fiber consistently gave smaller damage area. The strength reduction after impact could be explained from the impact damage area and the unnotched strength.

KEY WORDS: graphite/epoxy composites, quasi-isotropic laminates, compressive failure, fiber kinking, delamination, impact damage, resin modulus, notch sensitivity

[1]Graduate student and professor, respectively, Center for Composites Research, Washington University, Campus Box 1087, St. Louis, MO 63130.

[2]Aerospace engineer, Structures and Dynamics Division, NASA Langley Research Center, MS 190, Hampton, VA 23665.

One of the drawbacks in the application of the current graphite/epoxy composites in structures is their high susceptibility to low velocity impact damage [*1–5*]. Because of the brittleness of the matrix resins used, low velocity impact often results in matrix cracking, delamination, and fiber fracture. Under compressive loading, the damage caused by impact leads to premature failure and results in an unacceptable loss of strength.

Recent studies have indicated that the extent of impact damage can be reduced by using tougher resins [*3–6*]. However, the use of tougher resins may result in the reduction of compressive strength because toughness is frequently obtained at the sacrifice of stiffness. Therefore, a proper compromise must be maintained between toughness and stiffness to improve impact resistance without risking a compressive failure.

The effect of resin properties on compressive behavior of unidirectional composites was studied in Refs *7, 8,* and *9*. It was found that the compressive strength increased and the mode of fiber failure changed from buckling to kinking, as the resin tensile modulus increased.

In the present paper, the same material systems that were used in Refs *7* and *8* were used to study the effect of resin modulus and toughness on compressive failure of quasi-isotropic laminates with and without a hole and after damage by impact. Notch sensitivity was measured using holes of different diameters. Impact resistance was characterized by the impact damage area and post-impact strength retention.

Materials

The fibers used were Thornel 300 (T300) and Thornel 700 (T700). These two fibers have approximately the same Young's modulus (about 230 GPa). The T700 fiber, however, has a higher tensile failure strain and a smaller diameter. These two fibers were combined with Narmco 5208, American Cyanamid BP907, and Union Carbide 4901/MDA and 4901/mPDA to provide seven different material systems. All laminates were fabricated according to the manufacturers' suggested cure cycles at the National Aeronautics and Space Administration (NASA) Langley Research Center. The Narmco 5208 resin, which was the baseline resin system, has the lowest tensile failure strain among all the resins used. The BP907 resin has the lowest tensile modulus and the Union Carbide 4901/mPDA resin the highest tensile modulus. Mechanical properties of the constituent materials are presented in Table 1. The BP907 resin is known to be tougher than the 5208 resin [*5*]; however, the toughness of 4901 resins is not known.

Specimen Preparation and Testing

Unnotched Specimens

Ultimate compressive strength and failure modes were studied using 24-ply laminates with a stacking sequence of $[45/0/-45/90]_{3S}$. The laminates were cut and tabbed into a geometry appropriate for an Illinois Institute of Technology

TABLE 1—*Properties of the constituent materials.*

Material	Diameter, 10^{-6} m	Modulus,[a] GPa	Failure Stress,[a] MPa	Failure Strain,[a] %
T300	7.0	230.00	3310.0	1.4
T700	5.1	233.00	4550.0	1.9
BP907	...	3.10	89.5	4.8
5208	...	4.00	57.2	1.8
4901/MDA	...	4.62	103.4	4.0
4901/mPDA	...	5.46	115.1	2.4

[a]In tension.

Research Institute (IITRI) compressive fixture. Testing was done on an Instron testing machine at a crosshead speed of 1 mm/min. During loading, specimens were monitored through a stereo microscope at magnifications up to ×100 for indications of failure. Some specimens were tested in a loading-unloading mode; that is, the specimen was loaded to a certain level and then unloaded. Upon unloading, the specimen was removed and examined for failure using an optical microscope at magnifications up to ×500. The stress-strain response for each material was determined using a strain gage bonded to one of the lateral faces of two specimens in each material system.

Specimens With Open Hole and Impact Damage

Two types of specimens were used to study the notch sensitivity. One group of specimens were $[45/0/-45/90]_{3S}$ laminate coupons 25 mm wide by 44.5 mm long with a center hole ranging from 3.18 to 9.54 mm in diameter. A special plate fixture was used to grip the specimen at the ends with slight side pressure. Axial load was introduced to the two edges bearing on the plates.

The other group of specimens were $[45/0/-45/90]_{6S}$ laminates. They were 127 mm wide by 254 mm long and had a 25- or 50-mm-diameter center hole. These specimens were tested according to the procedure described in Ref *5*.

The impact specimens were also $[45/0-45/90]_{6S}$ laminates 127 mm wide by 254 mm long. Impact damage was inflicted by striking the specimen with a 1.27-cm-diameter aluminum sphere propelled by compressed air. Two impact energies were employed: 17 and 34.4 J which corresponded to projectile speeds of approximately 110 and 154 m/s, respectively. Following the impact, the specimen was inspected visually and ultrasonically to determine the extent of damage. Details of the test procedure are reported in Ref *3*.

Results and Discussion

Unnotched Specimens

Failure Modes—Compressive failure of unnotched quasi-isotropic specimens was usually catastrophic. Occasional arrest of partially failed specimens was possible only by the loading-unloading procedure described previously. Failure was usually initiated in the vicinity of the tab ends, although subsequent failure was spread over the entire gage length.

Examinations of partially failed specimens suggest that compressive failure of quasi-isotropic laminates is triggered by the kinking of fibers in the 0° plies. This is followed by delamination and subsequent buckling of sublaminates.

Figure 1 shows a partially failed T300/BP907 specimen which was loaded to 81% of the average ultimate compressive strength (UCS). Figure 1*a* is the failure of a 0° ply as seen on an edge of the specimen. Initiation of failure at the edge may be due to specimen geometry, that is, edge effect, and also due to smaller lateral support provided to the fibers at the edge. Propagation of the failure in the plane of the 0° ply is seen in Fig. 1*b*, where the top nine plies were ground away to expose the failed 0° ply. Figure 1*a* and *b* clearly shows kinking of fibers in the 0° ply. No delamination is present and the adjacent off-axis plies are intact. It is therefore concluded that kinking of the 0° fibers precedes any other failure event. In this specimen, only the 0° ply shown in the figure had failed by kinking, and all the other 0° plies were intact.

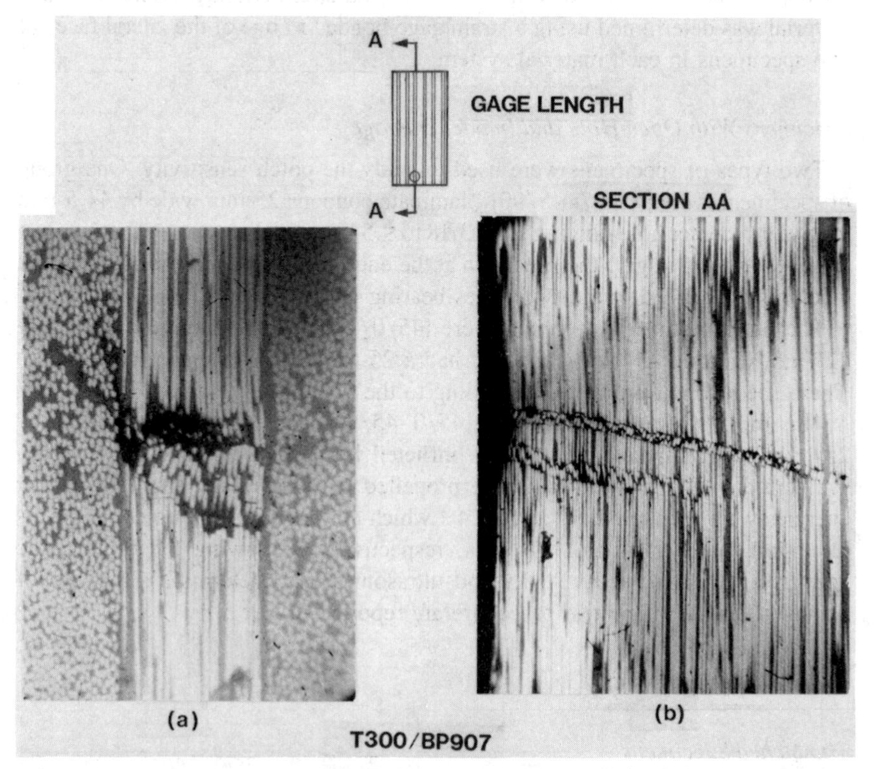

FIG. 1—*Partial failure in a T300/BP907 specimen loaded to 81% of the average ultimate compressive strength (UCS).*

Figure 2 shows a similar partial failure of a T700/4901/MDA specimen loaded to 93% UCS. The 0° ply shown in Fig. 2*b* has had the outer plies removed by grinding and indicates fiber kinking initiated at the free edge since failure exists only in this region.

Partial failures in the other laminates are shown in Fig. 3. All laminates except T700/4901/mPDA have only one 0° ply failure. As Figs. 1 to 3 show, failure in IITRI specimens usually initiates at one end of the gage length near the tabs.

The maximum prestress levels indicated in the figure, which were applied before the detection of kinking, are very close to the respective ultimate compressive strengths except for the laminates with the BP907 resin. The BP907 resin is the toughest of the resins used in the present study, and hence appears to be better able than the other resins to resist delamination following fiber kinking. Yet, this resin allows fiber kinking to occur at lower strains than the other resins do. The same trend was observed in unidirectional composites [*7*].

The T700/4901/mPDA laminate was loaded to the highest relative prestress level and shows failure in two 0° plies. Therefore, the sequential failure of 0° plies may occur before ultimate failure of the laminate. Fiber kinking in a 0° ply is progressive, as shown in Fig. 1, and therefore, one might expect multiple failure initiation sites.

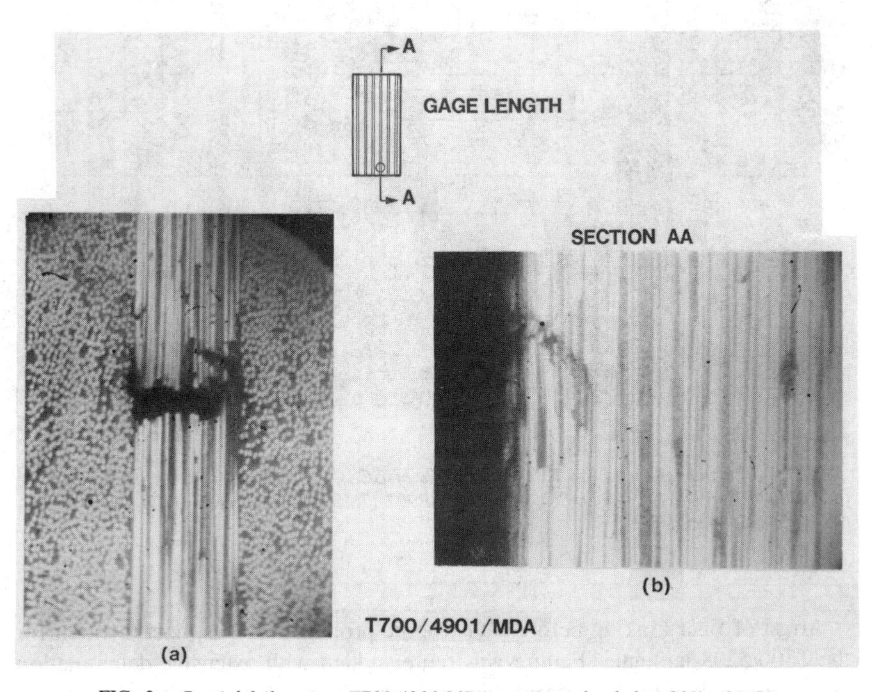

FIG. 2—*Partial failure in a T700/4901/MDA specimen loaded to 93% of UCS.*

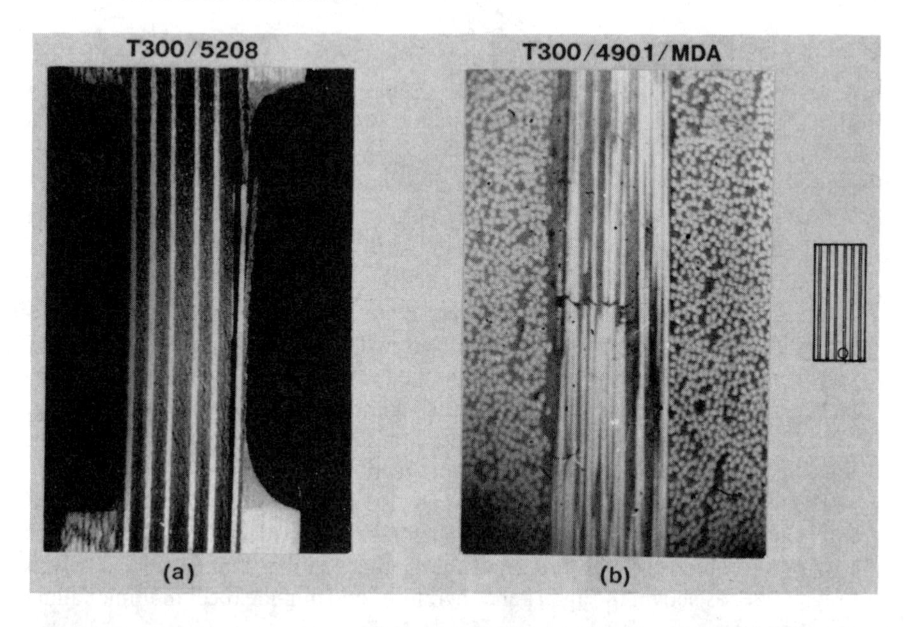

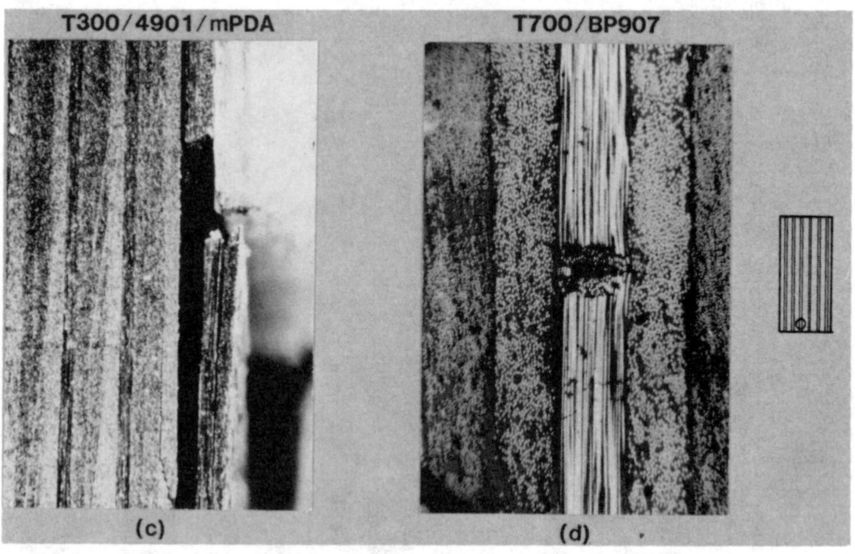

FIG. 3—*Partial failure in* (a) *T300/5208 (96% of UCS),* (b) *T300/4901/MDA (93% of UCS),* (c) *T300/4901/mPDA (92% of UCS),* (d) *T700/BP907 (78% of UCS), and* (e) *T700/4901/mPDA (98% of UCS) specimens.*

Arrest of fiber kinking before catastrophic propagation was most difficult for the T300/5208 laminate. Failure was quite sudden with extensive delamination after ultimate failure. Therefore, this laminate was judged to be the most brittle.

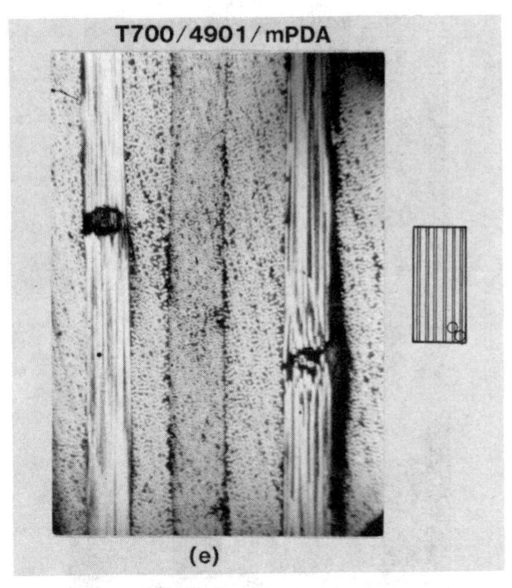

FIG. 3—*Continued.*

Although the laminates differ in the progression of damage, failure initiation in all seven composite systems is believed to be governed by the same mechanism, namely, the fiber kinking in 0° plies. The following is the concluded sequence of events that leads to the final failure of quasi-isotropic graphite/epoxy laminates:

1. As the compressive load increases, fiber kinking occurs in a 0° ply at an edge and propagates inward. The rotation of broken fiber segments in a kink band is both in and out of the plane of the 0° ply. The fiber kinking is of the same type as observed in unidirectional graphite/epoxy composites described in Ref *7*.
2. Failure of a 0° ply not only transfers the load to other 0° plies, but also results in load eccentricity since the specimen loses stiffness on the side with the failed ply. This enhances a sequential rather than random failure of the remaining 0° plies starting with the one closest to the failed ply and propagating outward (Fig. 3*e*).
3. A 0° ply with fiber kinking may move relative to the neighboring angle plies. This brings about delamination between the failed 0° ply and the angle plies.
4. After delamination, the sublaminates are more susceptible to buckling than the original laminate because of thinner thicknesses. The global buckling of sublaminates leads to final failure as shown in Fig. 4.

Compressive Behavior—Typical stress-strain curves for the different laminates are shown in Fig. 5. All the stress-strain curves exhibited strain softening which suggests the specimens did not experience global buckling. The reason is that the lateral surface on which the strain gage was to be bonded was chosen randomly and it is highly unlikely that all the gages would have been placed on the compressive side by mere coincidence. The absence of global buckling was

FIG. 4—*Typical global buckling of sublaminates following delamination.*

furthermore confirmed by using back-to-back strain gages on both sides of a T700/BP907 specimen. Figure 6 shows the strain outputs in which no strain reversal is observed.

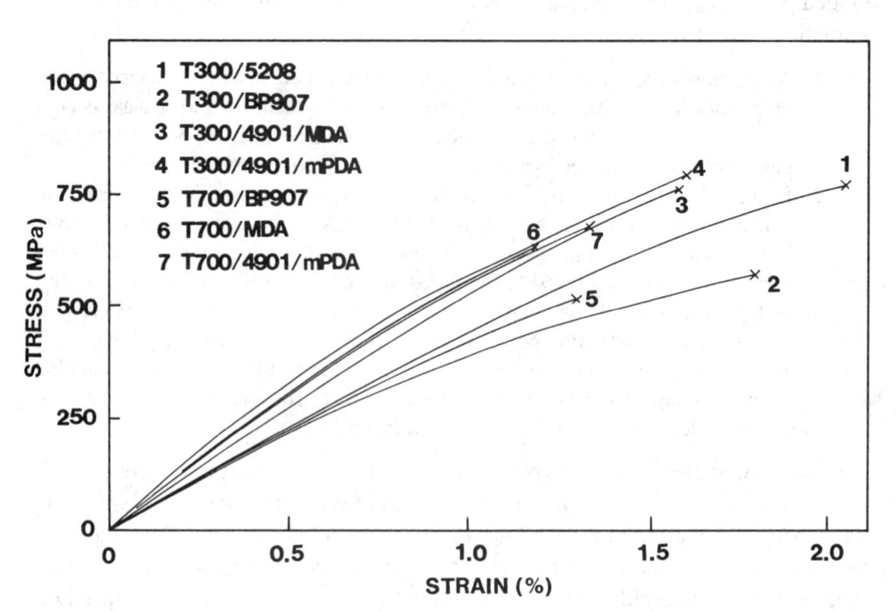

FIG. 5—*Typical stress-strain curves of different laminates.*

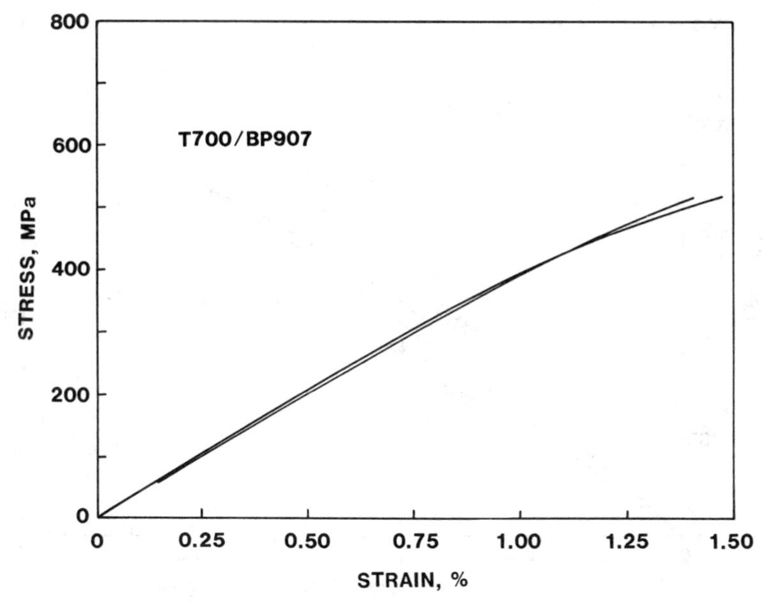

FIG. 6—*Stress-strain curves for a T700/BP907 specimen with back-to-back strain gages.*

The critical buckling stress, including the diminishing effect of composite transverse shear modulus, was calculated for a T300/5208 laminate (see Appendix A). The same stacking sequence as in unnotched specimens was assumed. Since no specimen-tab debonding was observed in the experiments, a column length equal to the specimen gage length was used in the calculations. To get a conservative estimate of the buckling stress, the column ends were assumed to be simply supported.

Including the effect of induced transverse shear forces was seen to reduce the critical buckling stresses by 52% (see Appendix). The resulting required stress of 1774 MPa to cause global buckling was, however, much higher than the 726 MPa obtained as the average failure stress. The other six laminates are expected to behave similarly since their mechanical properties are similar to those of the T300/5208 laminate. Therefore, it is concluded that the quasi-isotropic laminates tested failed in compression long before the critical buckling stress was reached.

Some of the specimens were loaded well into the nonlinear region and then unloaded. The stress-strain relation during unloading was almost the same as during loading (Fig. 7). Thus, the nonlinearity is not believed to have been caused by damage. In fact, the corresponding unidirectional composites also showed similar nonlinear behavior [*7*].

Although fiber kinking occurred in 0° plies before ultimate failure, as discussed earlier, it could not be detected on the stress-strain curves. No abrupt change in stress or strain was observed up to final failure.

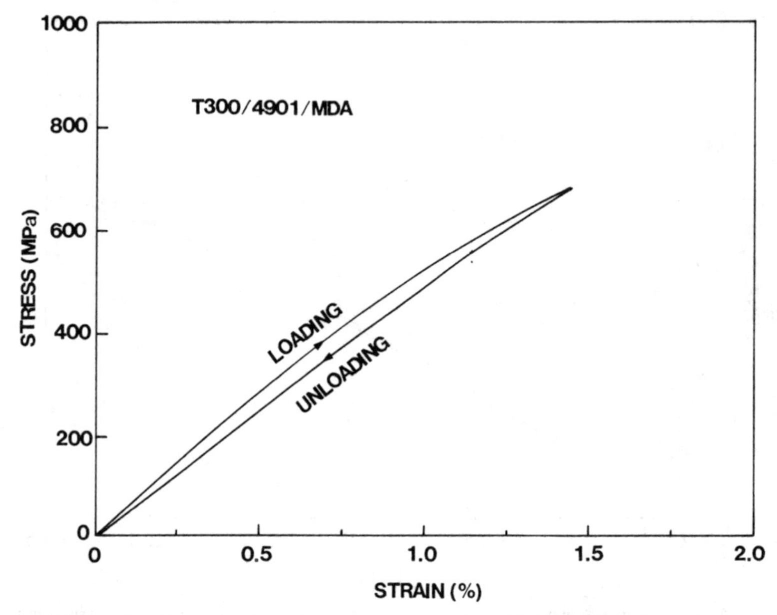

FIG. 7—*Loading-unloading stress-strain curves for a T300/4901/MDA specimen.*

Compressive properties of all seven quasi-isotropic laminates are listed in Table 2. In general, the quasi-isotropic strengths have a lower scatter band than the unidirectional strengths reported in Refs *7* and *8*. Even the T300/5208 composite, which showed considerable scatter in unidirectional strength, is fairly consistent in quasi-isotropic strength. A smaller scatter in strength indicates less sensitivity to defects. Although the initial fiber kinking acts as a defect after its inception, it is less critical to ultimate failure of quasi-isotropic laminates than that of unidirectional laminates. This conclusion is borne out by the ease with which fiber kinking can be monitored in quasi-isotropic laminates.

TABLE 2—*Properties of the composite laminates.*

Composite	Fiber weight Content, %	Failure stress[a] Average MPa	C.V., %	Failure Strain,[b] %	Modulus[b] GPa
T300/5208	57	726	3.4	2.05	46.5
T300/BP907	58	529	5.3	1.64	43.4
T300/490/MDA	67	734	6.6	1.51	58.3
T300/4901/mPDA	69	774	3.4	1.57	64.3
T700/BP907	59	526	2.8	1.31	45.5
T700/4901/MDA	65	682	8.8	1.25	59.9
T700/4901/mPDA	65	709	4.6	1.34	58.7

[a]Eight specimens tested for T300/ and T700/BP907. Ten specimens tested for the other laminates.
[b]Two specimens tested.

The average compressive strengths of quasi-isotropic laminates are plotted against the resin tensile moduli in Fig. 8. As for unidirectional composites [*7*], the compressive strength increases with increasing resin modulus.

Although in tension the T700 fiber is stronger than the T300 fiber, the corresponding laminates show the opposite trend in compression. Both failure stresses and failure strains for T700 laminates are lower than for T300 laminates probably because of the smaller diameter of the T700 fiber. According to the nonlinear microbuckling model presented in Ref *7*, the compressive strength of a 0° ply is inversely related to the initial fiber curvature. Since a larger diameter fiber is likely to have a smaller initial curvature, compressive strength of composites will increase with fiber diameter. Although initial fiber deflections have significant effect on compressive strength, they are very small (0.005 times the deflection wavelength [*7,8*]) and difficult to measure accurately.

The T300/5208 laminate shows the highest compressive failure strain despite the fact that the 5208 resin has the lowest tensile failure strain of all four resins used. The failure strains of quasi-isotropic laminates are compared with those of unidirectional laminates of the same material systems [*7*] in Fig. 9. The quasi-isotropic laminates are seen to have considerably higher failure strains than the corresponding unidirectional laminates. This is due to a better lateral support provided to the 0° plies by adjacent off-axis plies in quasi-isotropic laminates. The in-plane kinking in the quasi-isotropic laminate is influenced by the higher stiffness of off-axis plies normal to the loading. The out-of-plane kinking also is retarded by the off-axis fibers bridging over the kink band.

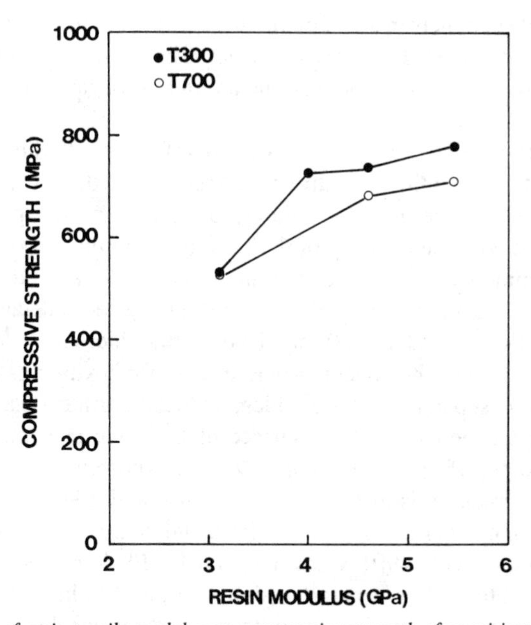

FIG. 8—*Effect of resin tensile modulus on compressive strength of quasi-isotropic laminates.*

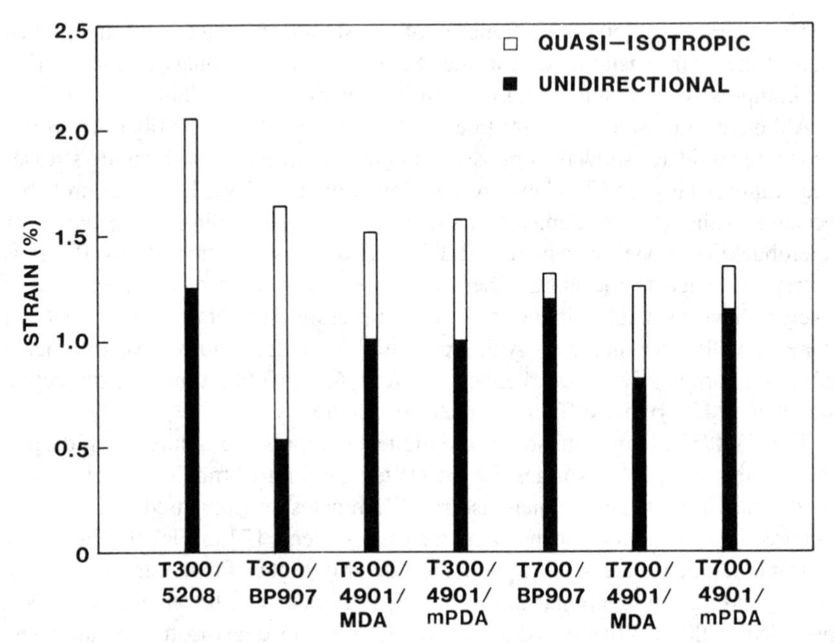

FIG. 9—*Comparison in failure strain between quasi-isotropic laminates and unidirectional laminates of the same material systems.*

Specimens With Holes

The failure of specimens with a circular hole was also initiated by fiber kinking in the 0° plies, as reported in Refs *4* and *10*. Similar to the findings of Ref *4*, the initial failure started at the hole boundary and propagated towards the specimen edges.

The sequence of failure events suggested in Refs *4* and *10* did not include any comparison between laminates with brittle resin and with soft resin. As in the unnotched laminates, the mode of failure propagation following the initial fiber kinking changed with the ductility of the resin used. In the T300/5208 laminate, delamination immediately followed the initiation of a kink band (Fig. 10). Two outer 0° plies failed in the form of kinking, causing local delamination of the outside 45° ply from the adjacent 0° ply. Upon further increase in load, the failure band propagated along the specimen width. The fiber kinking in the 0° plies, coupled with the separation of 45° plies, induced further delamination. Figure 10*b* shows the polished lateral surface of the same specimen as shown in Fig. 10*a*. The outer 45° ply was ground away to expose the fiber kinking in the 0° ply. Intraply cracking is also observed in a 45° and a 90° ply (Fig. 10).

A combination of fiber kinking in 0° plies and delamination was also observed in laminates made with 4901/MDA and 4901/mPDA resins. Similar to the T300/5208 laminate, failure was initiated in the outer 0° plies (Fig. 11). In these laminates, however, the kink band had propagated farther through the width of

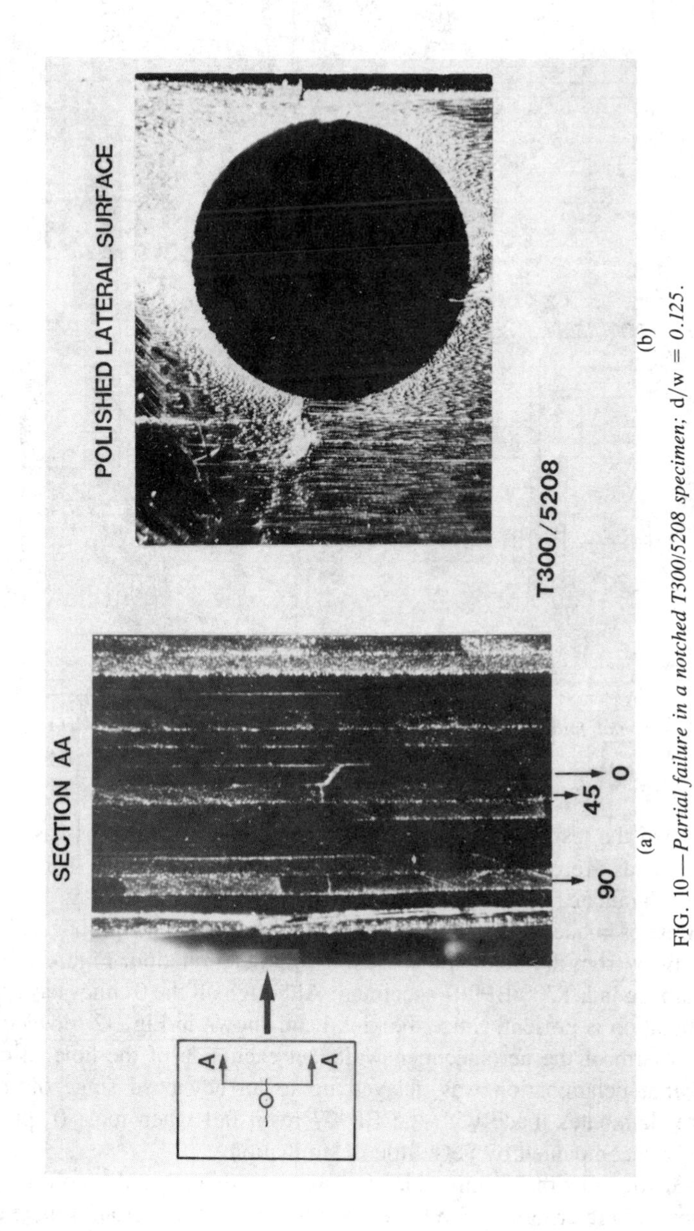

FIG. 10—*Partial failure in a notched T300/5208 specimen;* d/w = *0.125.*

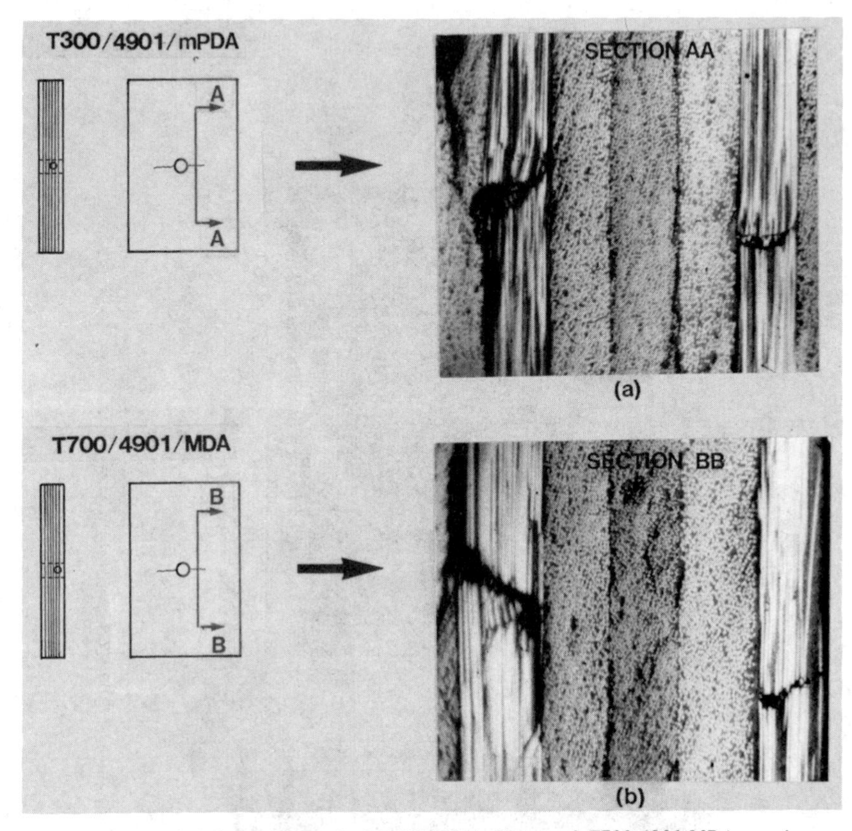

FIG. 11—*Partial failures in notched T300/4901/mPDA and T700/4901/MDA specimens;* d/w = *0.125.*

the ply when the test was stopped. It was therefore difficult to assess whether these laminates were any more tolerant of the initial damage than was the T300/5208 laminate.

In laminates made with the BP907 resin, the kink band propagated some distance away from the hole before causing any delamination. Figure 12 shows partial damage in a T700/BP907 specimen. Although all the 0° plies have failed, no delamination is present. Since the kink band shown in Fig. 12 traveled more than one fourth of the net specimen width on each side of the hole, it can be deduced that delamination was delayed up to an advanced stage of failure. Therefore, laminates made with the BP907 resin fail when most 0° plies fail completely accompanied by very little delamination.

Notched strengths of 25-mm-wide specimens were analyzed according to the point-stress failure criterion of Whitney and Nuismer [*11*]. For an infinite quasi-isotropic plate with a hole of radius R, the point-stress failure criterion predicts the notched strength $\sigma_N{}^\infty$ to be

$$\sigma_N{}^\infty/\sigma_0 = 2/(2 + \zeta^2 + 3\zeta^4) \qquad (1)$$

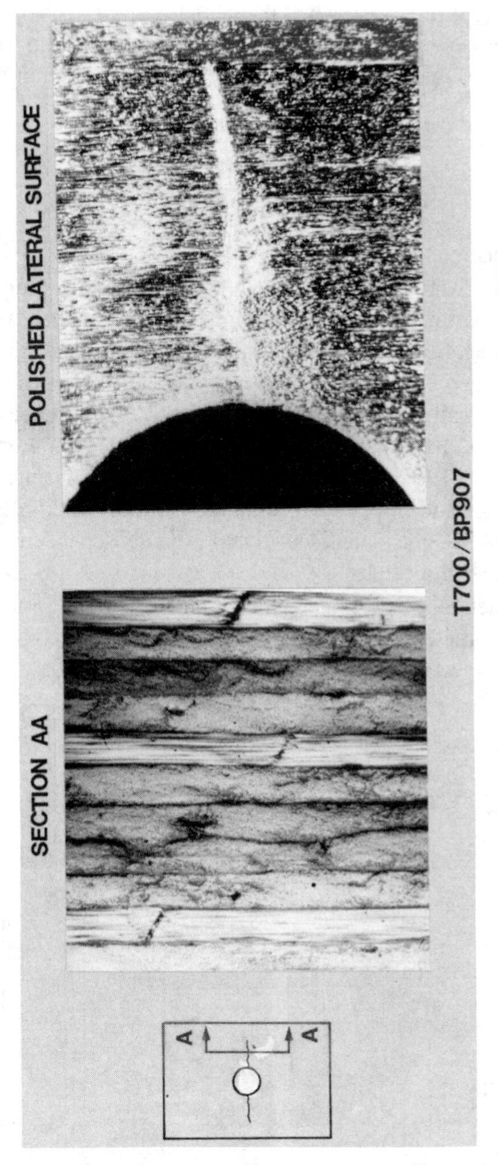

FIG. 12—*Partial failure in a notched T700/BP907 specimen;* d/w = 0.250.

where σ_0 is the unnotched strength and

$$\zeta = R/(R + d_0) \tag{2}$$

Here, d_0 defines the distance from the hole boundary to a point where the normal stress in the loading direction reaches σ_0 at final failure.

The experimental notched strength σ_N obtained from a specimen of finite width can be converted to $\sigma_N{}^\infty$ using the finite width correction factor Y:

$$\sigma_N{}^\infty = Y\,\sigma_N \tag{3}$$

$$Y = \frac{2 + (1 - 2R/W)^3}{3(1 - 2R/W)} \tag{4}$$

where W is the specimen width.

Equations 1 to 4 were used together with the unnotched strengths (given in Table 2) and the experimental notched strengths to calculate the average value of d_0 for each composite system tested, and the results are shown in Fig. 13. As can be seen from Eqs 1 and 2, a higher d_0 indicates less notch sensitivity, that is, less reduction in the normalized notched strength $\sigma_N{}^\infty/\sigma_0$ with increasing hole size.

As expected, T300/BP907 and T700/BP907 laminates show the least notch sensitivity while the T300/5208 shows the most. The two curing agents methylenedianiline (MDA) and *m*-phenylenediamine (mPDA) yield almost the same notch sensitivity. Also, the notch sensitivity is more dependent on the matrix material type than on the fiber.

With d_0 experimentally determined, the change of strength ratio σ_N/σ_0 with normalized hole diameter $2R/W$ as calculated by Eqs 1 through 4 is shown together with the experimental data for each laminate in Figs. 14 and 15. The

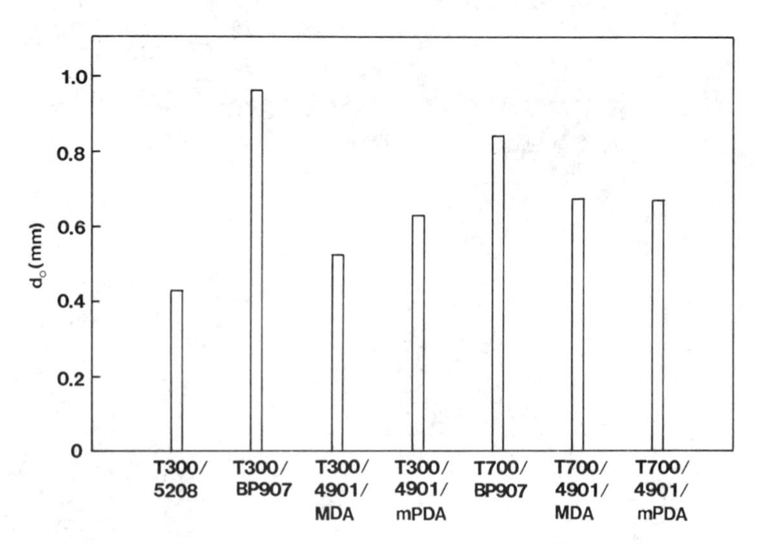

FIG. 13—*Average values of characteristic length* d_0 *for different laminates.*

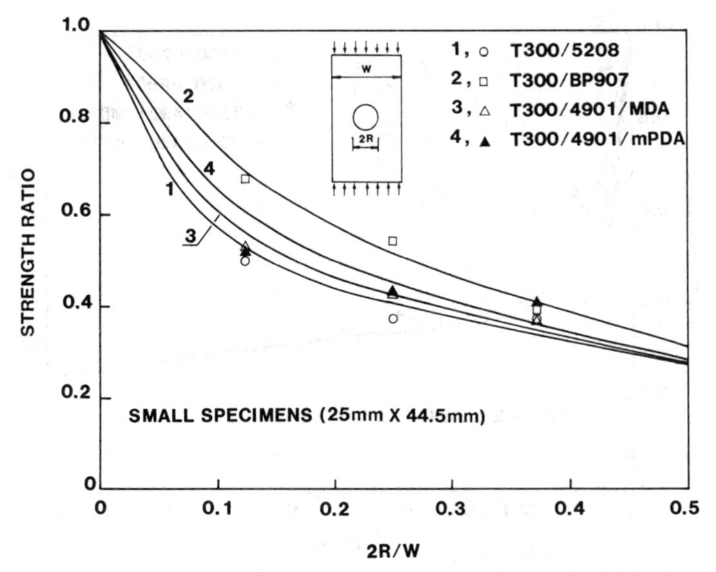

FIG. 14—*Change of unnotched-to-notched strength ratio with normalized hole diameter: T300 laminates 25 mm wide by 44.5 mm long.*

experimental data for $[45/0/-45/90]_{6S}$ specimens 127 mm wide and 254 mm long are shown in Figs. 16 and 17. Also shown in the latter figures are the predictions for the large specimens using the d_0 values obtained from the small

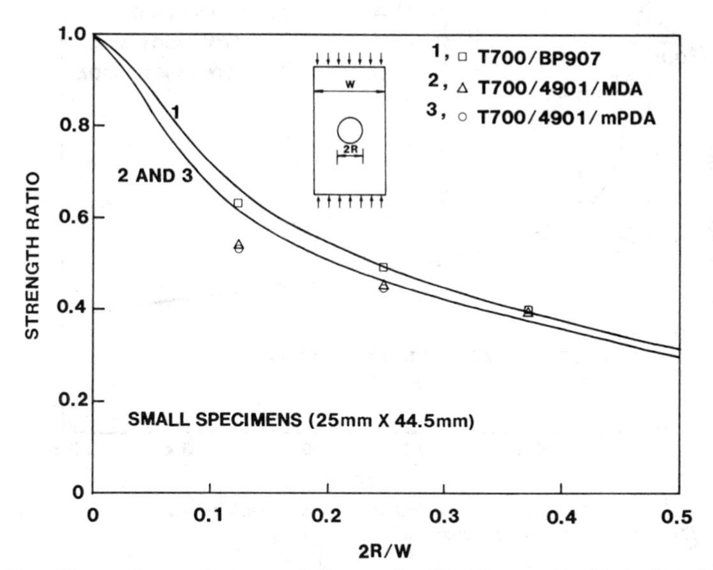

FIG. 15—*Change of unnotched-to-notched strength ratio with normalized hole diameter: T700 laminates 25 mm wide by 44.5 mm long.*

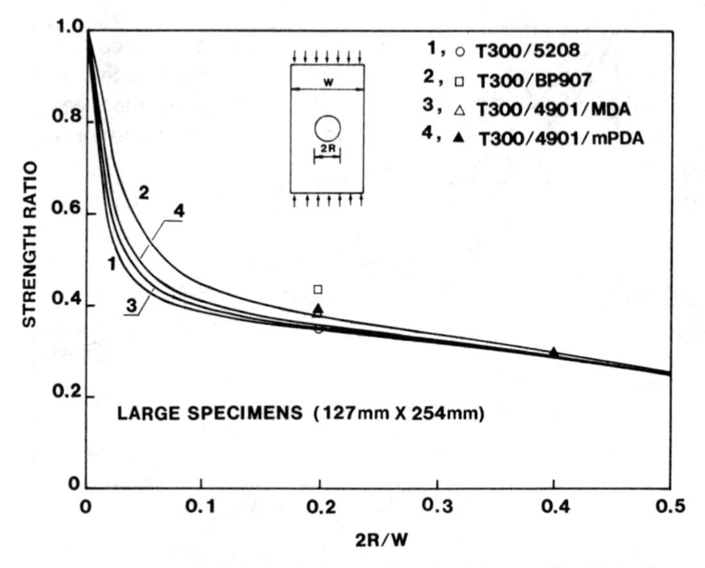

FIG. 16—*Change of unnotched-to-notched strength ratio with normalized hole diameter: T300 laminates 127 mm wide by 254 mm long.*

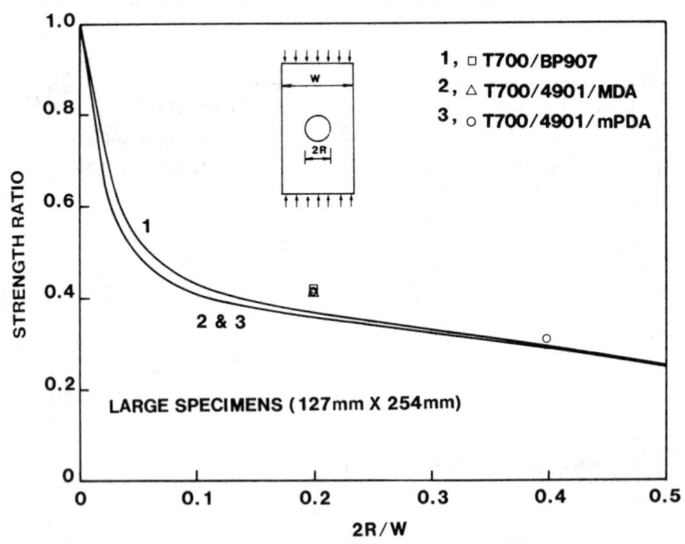

FIG. 17—*Change of unnotched-to-notched strength ratio with normalized hole diameter: T700 laminates 127 mm wide by 254 mm long.*

specimens. The agreement between the experimental results and the predictions is quite reasonable.

Impact Damage

The extent of damage as a result of impact is shown in Fig. 18, where the damage areas measured on ultrasonic C-scan records are plotted against the impact energies. The figure also includes the data for other composite systems reported in Ref *5*. Whereas two different impact energies were used for the laminates made with 4901/mPDA resin, only the lower impact energy was used for the laminates made with 4901/MDA resin. Of all resins shown, BP907 gives the best impact resistance.

The T700 fiber shows better impact resistance than the T300 fiber. During impact, large local deformation gradients occur which cause high tensile stresses to develop on the back surface. The T700 fiber with its higher strain capability can sustain higher back surface tensile strains before failure than the lower strain capability T300 fiber.

The 4901 resin yields laminates that, in impact damage size, rank somewhere between the BP907 laminates and the T300/5208 laminate, the former being the most resistant to impact.

The failure strength for specimens loaded in compression following impact damage at selected impact energies is presented in Fig. 19. Laminates constructed using BP907 resin and the higher strain T700 fiber recorded a higher failure strength than did BP907 laminates with T300 fiber.

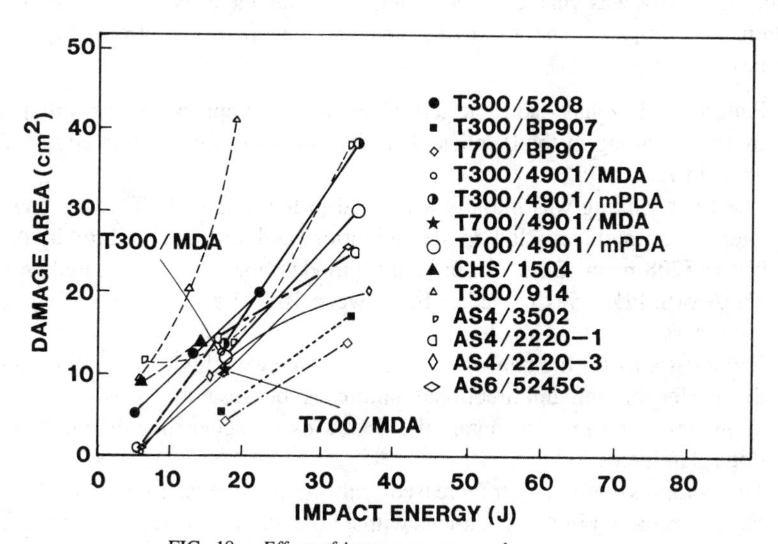

FIG. 18—*Effect of impact energy on damage area.*

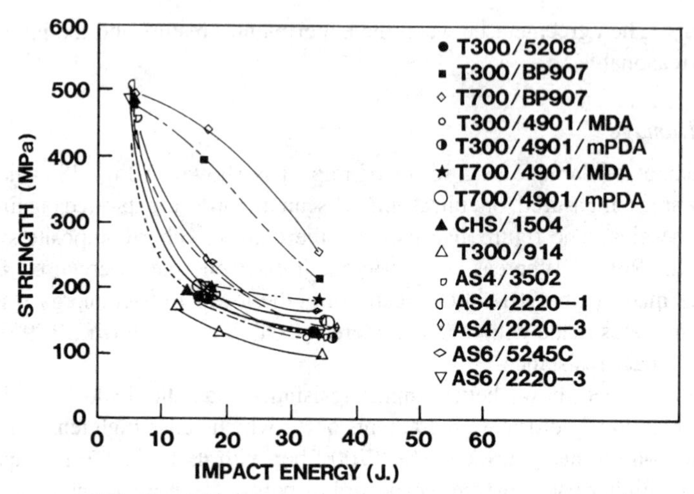

FIG. 19—*Effect of impact energy on compressive strength.*

Conclusions

Compressive behavior of quasi-isotropic graphite/epoxy laminates was studied using two different fibers (T300 and T700) and four different resins (5208, BP907, 4901/MDA, and 4901/mPDA). The IITRI compression fixture was used to study the unnotched behavior, while specimens with holes of varying diameters were used to study the notch sensitivity. The effect of low velocity impact on structural integrity was characterized in terms of damage area and the strength retention after impact. The following conclusions are drawn from the present study.

1. Failure of the quasi-isotropic laminates under compression was initiated by fiber kinking in the 0° plies. The fiber kinking started at an edge and grew inward.
2. The fiber kinking was initially contained in the tough BP907 resin; however, it immediately led to delamination and ultimate failure in the brittle 5208 resin. Based on the failure modes, the 4901 resin cured with MDA or mPDA was judged to be between BP907 and 5208 resins in its toughness.
3. Failure strains for the quasi-isotropic laminates were higher than those for the corresponding unidirectional laminates because the quasi-isotropic laminates could locally contain the fiber kinking better than the unidirectional laminates.
4. The compressive strength increased with the resin tensile modulus.
5. Failure initiation in the specimens with a hole was also fiber kinking in the 0° plies. Fiber kinking occurred at the point of maximum compressive stress and grew inward. Growth of the kink band was most gradual in the BP907 resin and the least stable in the 5208 resin. The notch sensitivity was the lowest for the laminates made with the BP907 resin and the highest

for those made with the 5208 resin. The fiber type (T300 versus T700) did not seem to affect significantly the notch sensitivity. Previous investigations (Ref *4*) have shown higher strain fibers (AS4 versus T300) to improve open-hole compression strength. The smaller diameter of the T700 fiber (Table 1) may be a factor that counteracts the improvement expected from a higher strain fiber.

6. The BP907 resin allowed the least impact damage while the 4901 resins were comparable to the 5208 resin. The T700 fiber resulted in better resistance to impact damage than the T300 fiber probably because of its higher strain capability. The T700/BP907 laminate showed the highest residual strength, followed by the T300/BP907 laminate. The remaining laminates were comparable to one another in their residual strengths.

Acknowledgments

The first and second authors' work was supported by NASA Langley Research Center through Grant NAG-1-295.

APPENDIX

Calculation of the Critical Buckling Stress

Notation — Figure 1A shows the coordinate system, the laminate orientation, and the stress notation used.

Buckling Formula — If the transverse shearing forces induced by deformation are taken into account, the critical buckling force in a column is given by [*12*]:

$$P_{cr} = \frac{P_e}{1 + (nP_e)/(AG_{13})} \quad \text{(A-1)}$$

where

$$P_e = (\pi/L_e)^2 D_{11} \quad \text{(A-2)}$$

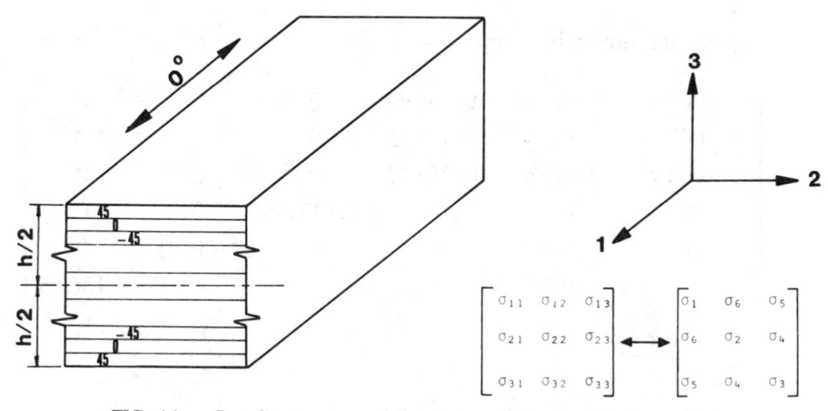

FIG. 1A—*Coordinate axes and the stress notation used in Appendix A.*

and

D_{11} = flexural stiffness,
L_e = effective length (= L for simply supported ends),
A = cross-sectional area,
n = a numerical factor depending on the cross section (= 1.2 for rectangular shape), and
G_{13} = transverse shear modulus.

Unidirectional T300/5208 Properties Used in The Calculations —

E_1 = 131.0 GPa ν_{21} = 0.380
E_2 = 13.0 GPa ν_{12} = 0.038
$E_3 = E_2$ ν_{13} = 12
G_{12} = 6.4 GPa ν_{23} = 0.492
$G_{31} = G_{12}$ ν_{32} = 23
G_{23} = 3.02 GPa (via the modified rule of mixtures assuming $V_f = 0.6$, $E_m = 4.0$ GPa, $\nu_m = 0.35$, and $G_{f2} = 5.9$ GPa).

where

E_m = Young's modulus of matrix,
ν_m = Poisson's ratio of matrix,
V_f = fiber volume fraction, and
G_{f2} = transverse shear modulus of fiber.

The On-Axis Compliance Matrix—

$$[S] = \begin{bmatrix} 1/E_1 & -\nu_{12}/E_2 & -\nu_{13}/E_3 & 0 & 0 & 0 \\ -\nu_{21}/E_1 & 1/E_2 & -\nu_{23}/E_3 & 0 & 0 & 0 \\ -\nu_{31}/E_1 & -{}_{32}/E_2 & 1/E_3 & 0 & 0 & 0 \\ 0 & 0 & 0 & 1/G_{23} & 0 & 0 \\ 0 & 0 & 0 & 0 & 1/G_{31} & 0 \\ 0 & 0 & 0 & 0 & 0 & 1/G_{12} \end{bmatrix} \quad \text{(A-3)}$$

substituting the unidirectional properties,

$$[S] = \begin{bmatrix} 0.00763 & -0.00290 & -0.00290 & 0 & 0 & 0 \\ -0.00290 & 0.07692 & -0.03785 & 0 & 0 & 0 \\ -0.00290 & -0.03785 & 0.07692 & 0 & 0 & 0 \\ 0 & 0 & 0 & 0.33069 & 0 & 0 \\ 0 & 0 & 0 & 0 & 0.15625 & 0 \\ 0 & 0 & 0 & 0 & 0 & 0.15625 \end{bmatrix} \quad \text{(A-4)}$$

On-Axis Stiffness Matrix —

$$[C] = \begin{bmatrix} 138.9 & 10.3 & 10.3 & 0 & 0 & 0 \\ 10.3 & 17.9 & 9.2 & 0 & 0 & 0 \\ 10.3 & 9.2 & 17.9 & 0 & 0 & 0 \\ 0 & 0 & 0 & 3.02 & 0 & 0 \\ 0 & 0 & 0 & 0 & 6.4 & 0 \\ 0 & 0 & 0 & 0 & 0 & 6.4 \end{bmatrix} \tag{A-5}$$

Calculation of D_{11} — D_{11} was calculated using the following formula

$$D_{ij} = \int_{-h/2}^{+h/2} C_{ij} z^2 \, dz \tag{A-6}$$

where z is the coordinate in the thickness direction. Assuming a $[45/0/-45/90]_{3S}$ stacking sequence and a ply thickness of 0.14 mm yields

$$D_{11} = 203\,500 \text{ Nt mm}$$

Calculation of G_{13} — Using contracted notation, stress-strain relation of a constituent ply can be written as

$$\sigma_i = C_{ij}\varepsilon_j + C_{iA}\varepsilon_A \tag{A-7}$$

$$\sigma_A = C_{Aj}\varepsilon_j + C_{AB}\varepsilon_B \tag{A-8}$$

where $i,j = 1, 2, 6$ are associated with the in-plane coordinates and $A,B = 3, 4, 5$ with the out-of-plane coordinates. According to the classical laminated plate theory, σ_A are assumed to be constant throughout the thickness. If the total strains ε_i are displayed as $\varepsilon_i = \varepsilon_i^\circ + kz$, then one can show that [*13*],

$$\overline{\varepsilon_A} = \overline{C_{AB}}^{-1}\sigma_B + \overline{C_{AB}{}^{-1}C_{Bj}}\varepsilon_j^\circ - \overline{C_{AB}{}^{-1}C_{Bj}z}k_j \tag{A-9}$$

where ε_i° and k_j are the in-plane strains and curvatures, respectively. An overbar stands for average through the thickness h, that is,

$$\overline{(\ )} = \frac{1}{h}\int_{-h/2}^{h/2} (\ )\, dz$$

Assuming $\sigma_3 = \sigma_4 = \varepsilon_j = k_j = 0$, Eq A-9 yields

$$\overline{\varepsilon_5} = \overline{C_{55}}^{-1}\sigma_5 \tag{A-10}$$

where $C_{55}{}^{-1}$ is the element in the fifth row and fifth column of $[C_{ij}]^{-1}$. The transverse shear modulus is then given by

$$G_{13} = \frac{1}{\overline{C_{55}{}^{-1}}} \tag{A-11}$$

Using the on-axis stiffness matrix given in Eq A-5, a $[45/01\text{-}45/90]_{3S}$ stacking sequence and a ply thickness of 0.14 mm results in

$$G_{13} = 4.1 \text{ GPa}$$

The Critical Buckling Stress—Substituting $D_{11} = 203\,500\ Nt$ mm, $G_{13} =$ 4.1 GPa, $n = 1.2$, $A = 21.3$ mm^2, and $L_e = 12.7$ mm in Eqs A-2 and A-1, gives

$$P_{cr} = 37\,828 \text{ Nt}$$

or

$$\sigma_{cr} = P_{cr}/A = 1774 \text{ MPa}$$

References

[1] Rhodes, M. D., Williams, J. G., and Starnes, J. H., "Effect of Low Velocity Impact Damage on the Compressive Strength of Graphite/Epoxy Hat-Stiffened Panels," NASA TN D-8411, National Aeronautics and Space Administration, 1977.

[2] Starnes, J. H., Jr., Rhodes, M. D., and Williams, J. G., "Effect of Impact Damage and Holes on the Compressive Strength of a Graphite/Epoxy Laminate," in *Nondestructive Evaluation and Flaw Criticality for Composite Materials, ASTM STP 696,* 1979, American Society for Testing and Materials, Philadelphia, pp. 145–171.

[3] Williams, J. G. and Rhodes, M. D., "Effect of Resin on Impact Damage Tolerance of Graphite/Epoxy Laminates," *Composite Materials: Testing and Design, ASTM STP 787,* I. M. Daniel, Ed., American Society for Testing and Materials, Philadelphia, 1982, pp. 450–480.

[4] Williams, J. G., "Effect of Impact Damage and Open Holes on the Compression Strength of Tough Resin/High Strain Fiber Laminates," NASA TM-85756, National Aeronautics and Space Administration, 1984.

[5] Williams, J. G., O'Brien, T. K., and Chapman, A. J., III, "Comparison of Toughened Composite Laminates Using NASA Standard Damage Tolerance Tests," NASA CP-2321, ACEE Composite Structures Technology Conference, Seattle, WA, Aug. 1984.

[6] Palmer, R. J., "Investigation of the Effect of Resin Materials on Impact Damage to Graphite/Epoxy Composites," NASA CR 165677, National Aeronautics and Space Administration, March 1981.

[7] Hahn, H. T. and Williams, J. G., "Compression Failure Mechanisms in Unidirectional Composites," NASA TM 85834, National Aeronautics and Space Administration, Aug. 1984.

[8] Sohi, M. M., "Compressive Behavior of Graphite/Epoxy Composites," M. S. thesis, Washington University, St. Louis, MO, Dec. 1984.

[9] Hahn, H. T., "Effect of Constituent Properties on Compressive Failure Mechanisms. Tough Composite Materials Workshop," NASA CP 2334, May 1983.

[10] Rhodes, M. D., Mikulas, M. M., Jr., and McGowan, P. E., "Effects of Orthotropy and Width on the Compression Strength of Graphite-Epoxy Panels with Holes," *AAIA Journal,* Vol. 22, No. 9, Sept. 1984, pp. 1283–1292.

[11] Whitney, J. M. and Nuismer, R. J., "Stress Fracture Criteria for Laminated Composites Containing Stress Concentrations," *Journal of Composite Materials,* Vol. 8, 1974, pp. 253–265.

[12] Timoshenko, S. P. and Gere, J. M., *Theory of Elastic Stability,* McGraw Hill Book Co., Inc., New York, 1961, p. 132.

[13] Hahn, H. T. and Kim, R. Y., "Swelling of Composite Laminates," in *Advanced Composite Materials—Environmental Effects, ASTM STP 658,* American Society for Testing and Materials, Philadelphia, 1978, pp. 98–120.

Kevin R. Hirschbuehler[1]

A Comparison of Several Mechanical Tests Used to Evaluate the Toughness of Composites

REFERENCE: Hirschbuehler, K. R., **"A Comparison of Several Mechanical Tests Used to Evaluate the Toughness of Composites,"** *Toughened Composites, ASTM STP 937,* Normal J. Johnston, Ed., American Society for Testing and Materials, Philadelphia, 1987, pp. 61–73.

ABSTRACT: A variety of mechanical tests are in use to evaluate the "toughness" of finished composite parts. These include compressive strength after impact, critical Mode I strain energy release rate (G_{Ic}), edge delamination G_{Ic}, and damage area generated after impact. While much data has been generated on these, and other "toughness" tests, it has usually been gathered at a variety of locations using different variations on a given test, with few materials being subjected to more than one or two of these tests. All of these factors combine to make it difficult to compare different composite matrices on an equal basis, or to predict how neat resin mechanical properties will translate to a composite structure.

This paper discusses the results of a study comparing the neat resin and composite mechanical properties of a variety of thermoset matrices. Both single and dual matrix ("interleaved") composite systems were studied. Comparisons of composite test data have shown large differences between the various toughness tests, with improvements in one test not necessarily being reflected in the others. For example, the effect of interleafing two resin matrices of vastly different strain capabilities results in significant increases in impact related tests, with much smaller changes in G_{Ic}.

This study involved mechanical testing of several thousand coupons. Analysis of the results has shown that a strong relationship exists, albeit not a perfect one, between neat resin and composite properties.

KEY WORDS: advanced composites, aerospace composites, toughened composites, epoxy matrix composites, hybrid composites, composite epoxy resins, toughened epoxy resins, dual matrix epoxy resins, toughness mechanical tests, test correlations, impact resistance, crack propagation, neat resin mechanical tests, shear mechanical tests

The use of high performance composite materials has increased to the point that they are being actively studied for primary structure applications. A fundamental prerequisite to their usage is an understanding of their toughness, and

[1]Product development engineer, American Cyanamid Co., Old Post Rd., Havre de Grace, MD 21078.

most importantly, their ability to carry load after impact damage. As part of the effort to understand composite toughness characteristics, a number of tests have been devised that measure various aspects of a composite material's resistance to damage and subsequent damage spread. Because of the large variety of tests, as well as the many variations on each test, few materials are subjected to more than one or two such tests, and the results are seldom compared to neat resin properties.

As part of the composite research effort at American Cyanamid Company, a number of these toughness tests have been evaluated and correlated to neat resin properties to aid in the development of tough composite matrices. Neat resins were examined using flexural strength, modulus, and strain to failure. Promising candidates were then impregnated onto graphite tape, and their toughness evaluated by one or more of the following properties: damage size and compressive strength after impact, critical Mode I strain energy release rate (G_{Ic}) by double cantilever beam, open hole compressive strength, and short beam shear strength after impact. Relationships between these test properties and neat resin properties form the basis of this report. A small amount of work was done using open hole tension and edge delamination tests, but not enough to allow conclusions to be drawn.

Work has also centered on interleafing graphite epoxy prepregs with lower modulus, high elongation resin layers to enhance impact resistance. The effect of this hybrid construction on impact-related properties will also be discussed.

Experimental Procedure

Neat resin data were generated from 0.16-cm-thick resin plaques cured 2 h at 135°C and 3 h at 180°C, either in an oven or pressclave as required to provide void-free test specimens. As all the resin systems studied here were single-phase epoxies, the void content of the plaques was determined using micrographic examination of the clear resins. All test coupons had less than 1% voids. Flexural modulus, strength, and strain to failure were determined per ASTM Tests for Flexural Properties of Unreinforced and Reinforced Plastics and Electrical Insulating Materials (D 790). The test coupons were 1.27 by 3.18 cm long, and tested in three-point loading at a span to depth ratio of 16/1.

All laminates were laid up from machine made unidirectional graphite/epoxy tape. All the tape used in the study was 145-g/m^2 graphite areal weight, 34 ± 2% resin content. All graphite fiber employed was 32-Msi (221-GPa) modulus aerospace grade material, about ⅔ of all samples employing Celanese Celion 6000 of 1.5 to 1.7% strain to failure. The remaining ⅓ of the systems were coated on Hercules AS-6 (1.7 to 1.9% strain) or AS-4 (1.5 to 1.6% strain) 12 000 filament bundles. The prepreg laminates were bagged using a zero bleed, net resin content technique, then cured for 2 h at 177°C and 100 psi (690 kPa) in an autoclave. Cured laminates were inspected gravimetrically and ultrasonically to measure cured resin content and void content. All coupons used in the study were between 31 and 35% cured resin content, and less than 1% voids.

Compressive strength after impact and open hole compressive strength were determined using the test technique first described by Byer [1]. For both tests, the tape was laid up in a 36-ply, near quasi-isotropic stacking sequence:

$$[\pm45/0/90/0/90/\pm45/0/90/0/90/\pm45/0/90/\pm45]_s$$

Coupons were cut 10.2 by 15.2 cm and end ground to provide aligned test surfaces. Open hole coupons were prepared by drilling 2.22-cm diameter holes with a water-cooled diamond drill bit. Impact coupons were damaged by dropping a 4.86-kg dart into a 0.79-kg impact tip with a 0.79-cm radius tip resting on the specimen surface. The drop height was adjusted to provide 6670 J/m of impact energy per laminate thickness. To prevent impact energy dissipation as a result of test specimen vibrations, the specimen was mounted on a steel base, 30.5 by 30.5 by 5.1 cm thick, centered over a 7.6- by 12.7-cm cutout to permit deflection, and restrained with Destaco clamps. After impact, the damage size was determined using an ultrasonic inspection system in reflector plate mode. Average damage diameter was determined by measuring diameter at eight locations. Both open hole and impact damaged specimens were then tested in edgewise compression at 0.13 cm/min using a test fixture as reported earlier [1].

G_{Ic} was determined using a width tapered double cantilever beam, 4.45 by 15.2 cm long, 16 ply all 0° layup. A 2.5-cm-long fluoropolymer crack starter was used between Plies 8 and 9, and aluminum tape was used to pull the coupon open. Short beam shear strength after impact was tested as described by Susman [2]. For this test, 16-ply quasi-isotropic laminates (stacking sequence: $[\pm45/0/90/0/90/\pm45]_s$) were cut into 1.27- by 1.9-cm long coupons, impacted at 0-, 1.13-, 2.26-, or 3.30-J energy in a Gardner model IG1120 impact tester, then tested in short beam shear per ASTM Test for Apparent Interlaminar Shear Strength of Parallel Fiber Composites by Short Beam Method (D 2344) at a test span of 1.02 cm. For all tests, a minimum of 4 coupons were tested, with the average reported as the plotted data point.

Results and Discussion

Neat Resin Properties as Prediction of Composite Properties

In the search for promising composite matrices, the most important resin properties are those that accurately predict the final composite performance. Our expectations were that neat resin modulus, strain to failure, or work to failure (the integrated area under the stress-strain curve) would be the most significant predictors of compressive strength after impact. The results for these three neat resin properties are shown in Figs. 1, 2, and 3.

Figure 1 is a plot of compression after 6670 J/m of impact ("impact strength") as a function of neat resin flexural modulus (21°C, dry). Although a general trend of lower impact strengths at higher resin moduli seems apparent, the data are well scattered and no correlation seems to be evident. The effect of resin flexural strain to failure on impact strength is shown in Fig. 2. The relationship here is quite

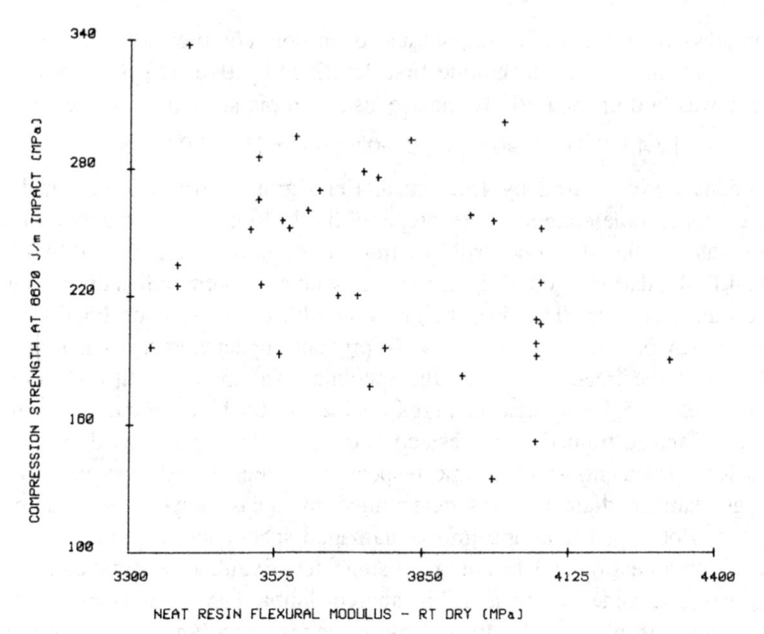

FIG. 1.—*Compression after impact versus resin flexural modulus.*

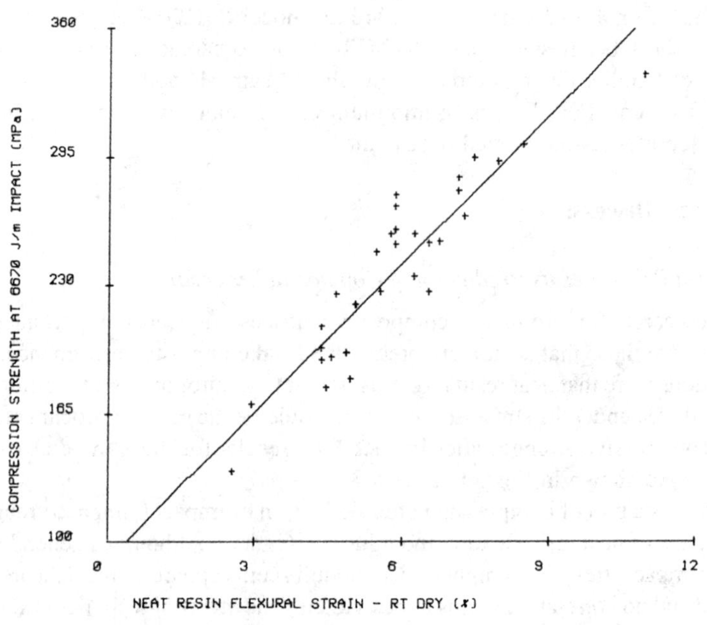

FIG. 2—*Compression after impact versus resin flexural strain.*

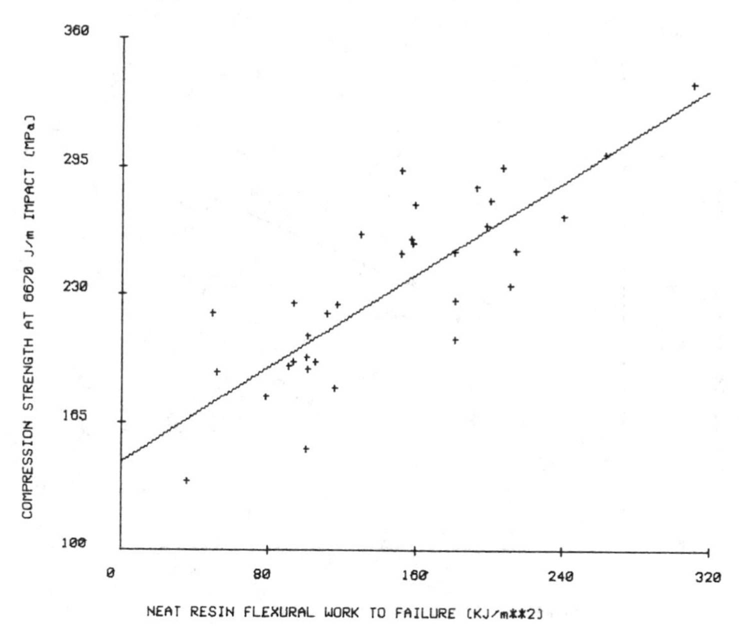

FIG. 3—*Compression strength after impact versus resin work.*

strong, with the least squares fitted line plotted: compression after impact (MPa) = 25.17 × (elongation, %) + 89.03 MPa having a correlation coefficient of 0.918. While the correlation is not perfect, the scatter band is tight, and for virtually every resin studied here, an increase in resin strain to failure results in an increase in impact strength.

Figure 3 shows impact strength as a function of neat resin work to failure. There is obviously a correlation here, as the plotted least squares line has a correlation coefficient of 0.82. Equally as obvious, the correlation is not as good as that for resin elongation versus impact strength. This is logical, as the resin work to failure is the area under the stress strain curve. Since the range of resin moduli under study here is relatively small (about 3310 to 4275 MPa), work to failure will be approximately linear in resin elongation, with "noise" superimposed by the scatter in initial moduli. Thus, the correlation is decreased as compared to the impact strength versus elongation case.

Figures 4 and 5 are plots of G_{Ic} by double cantilever beam as a function of resin flexural elongation and work to failure, respectively. In both cases, a definite trend of increasing G_{Ic} with both increasing elongation and increasing work to failure is apparent. However, the data are highly scattered, with linear regression analysis correlations of only about 0.55 to 0.60. Thus, the neat resin properties do not appear to allow a prediction of composite G_{Ic} values.

The final neat resin correlation is shown in Fig. 6, where neat resin strain to failure is plotted against the ratio of short beam shear strength at 1.13-J impact

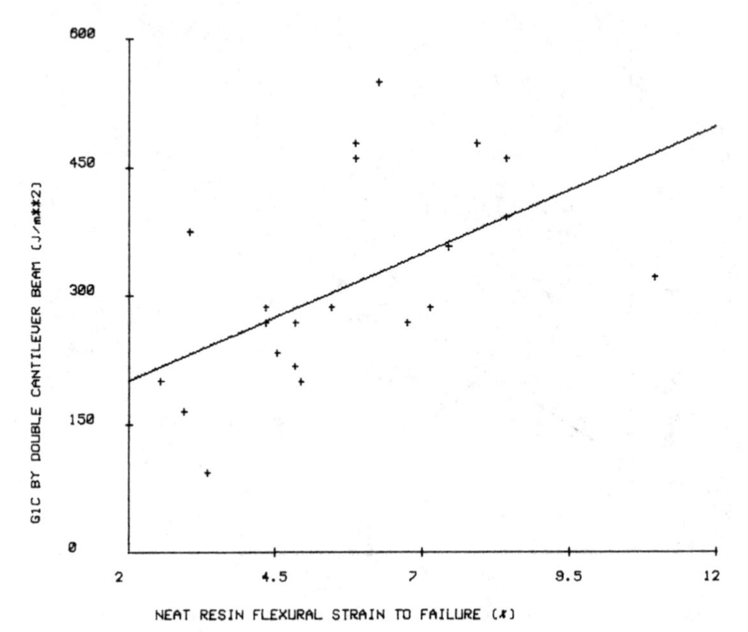

FIG. 4—*Composite* G_{Ic} *versus neat resin strain to failure.*

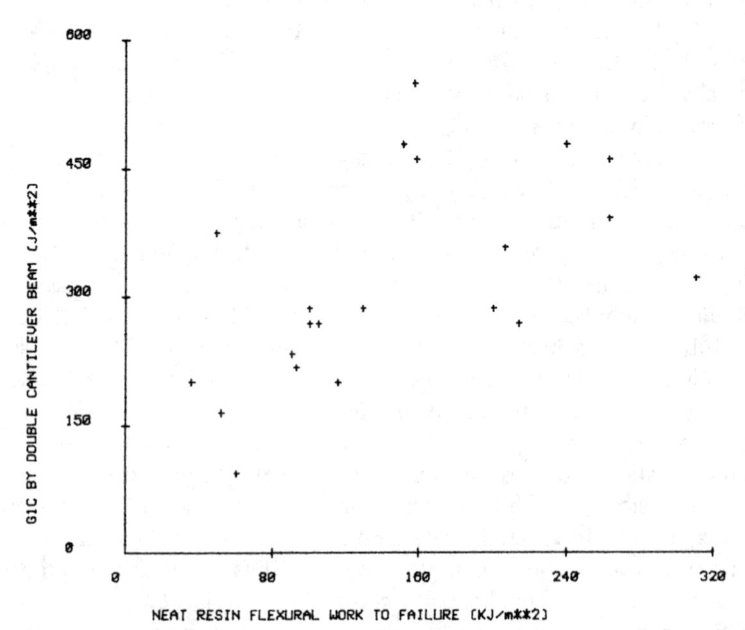

FIG. 5—*Composite* G_{Ic} *versus neat resin work to failure.*

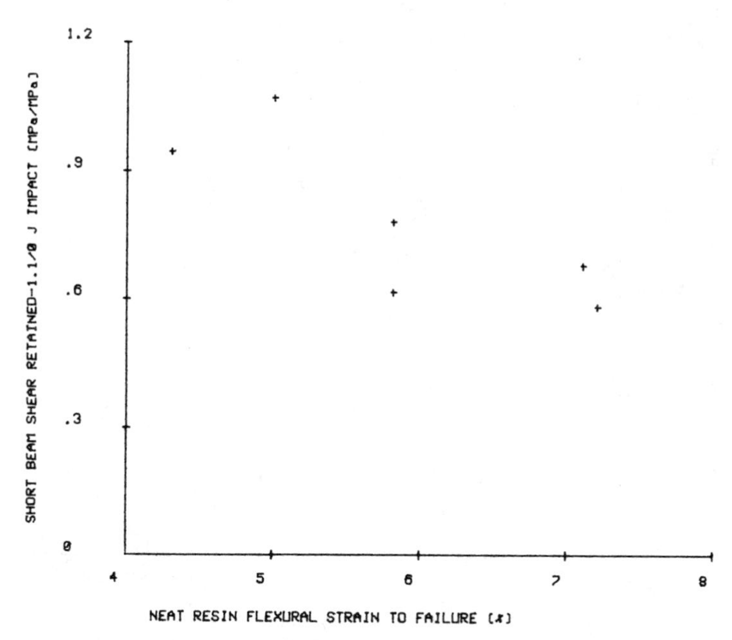

FIG. 6—*Impact short beam shear strength versus resin strain.*

over shear strength at 0-J impact. Although not many data points were available for this plot, increasing resin strain to failure resulted in lower short beam shear strength retentions after impact. The above effect opposes the one seen with compressive strength after impact, and would seem to eliminate the small impact short beam shear test as a screen for the larger compression test.

Comparisons Between Composite Tests

The original driving force behind comparisons between composite toughness tests was to find a cheaper and faster alternative to the compressive strength after impact test. For experimental resin evaluation, the amount of material and labor involved to prepare these 10.2- by 15.2-cm coupons can be prohibitive. Accordingly, the first test studied was the short beam shear strength after impact shown as a plot in Fig. 7. It was conceptually similar to the compression test, measuring property retention after impact damage while requiring only a small fraction of the materials and labor. The data in Fig. 7 show a relationship existed between the two, with higher retention from 0 to 1.13 J of impact in shear generally resulting in higher compressive strengths after impact. A linear least squares regression yielded a correlation of only 0.613, so the relationship was not as strong as was hoped. It has been a useful predictor of general trends, with only one material with an impact shear retention above 0.9 yielding a compressive strength after impact below 276 MPa. The data of Fig. 6 as discussed above

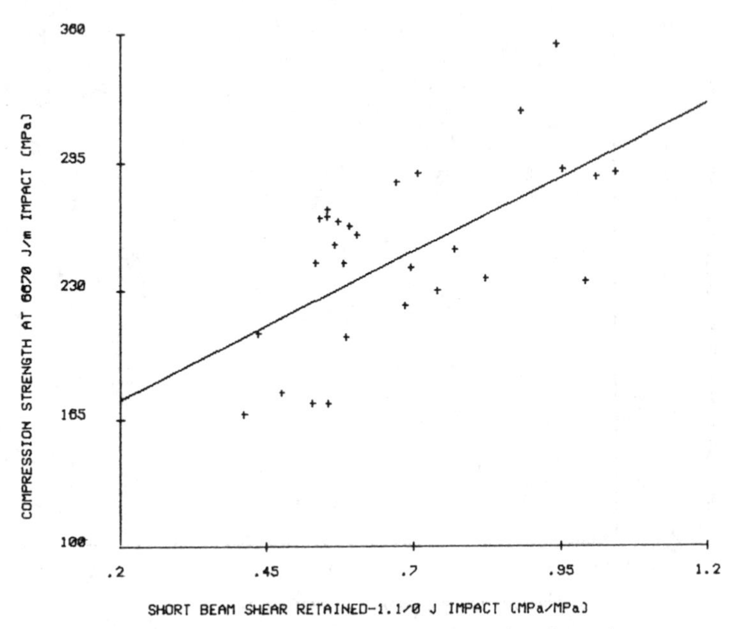

FIG. 7—*Compressive strength versus impact short beam shear.*

indicate the opposite conclusion. This was most probably because insufficient data points were available for the impact shear versus resin flexural strain plot to establish a valid correlation.

Figure 8 shows compressive strength after impact as a function of G_{Ic} by double cantilever beam. The situation here is similar to that seen with the impact shear data, that is, a strong general trend of increasing impact strength with increasing G_{Ic}, but with a large amount of scatter. A linear regression yields a correlation of 0.725. Note the large amount of scatter in the region of G_{Ic} between 262 and 450 J/m^2. This may be due to variations in tape quality. While hard to quantify, it has been observed that the poorer the visual appearance of a batch of tape (puckers, ripples, and so forth), the higher the recorded G_{Ic} for that batch. As an example, the range of G_{Ic} for CYCOM® 1806 advanced composites production batches was 245 to 480 J/m^2, while compression after impact varied from 136 to 276 MPa.

While not useful to a formulator as a predictive tool, the amount of damage measured in a panel (by ultrasonic inspection) after impact may give an indication of the expected residual strength. The data for impact strength as a function of average damage diameter are plotted in Fig. 9. As expected, a general trend of lower residual compressive strengths for larger impact damage diameters is revealed. However, the scatter is even greater than the previous two correlations (linear correlation coefficient of 0.52). This large scatter from one fixed impact instrument configuration effectively eliminates impact damage area as a predictor

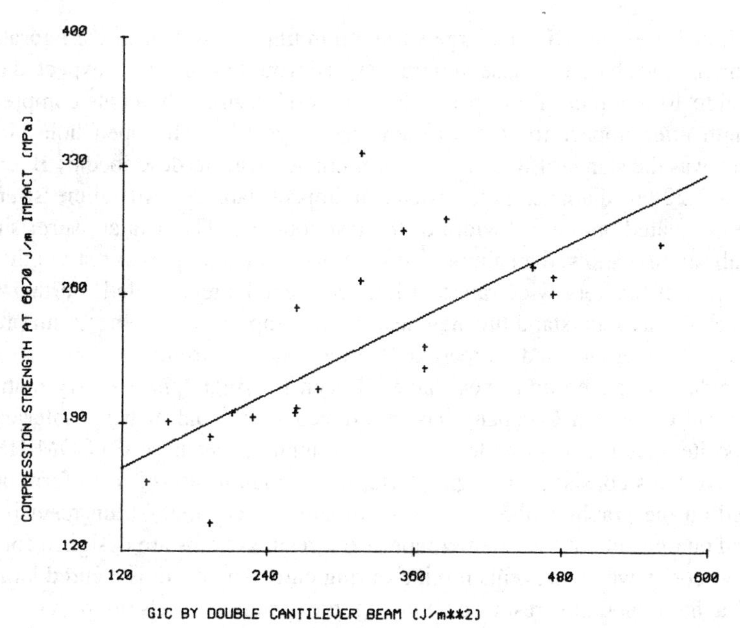

FIG. 8—*Compressive strength after impact versus composite* G_{Ic}.

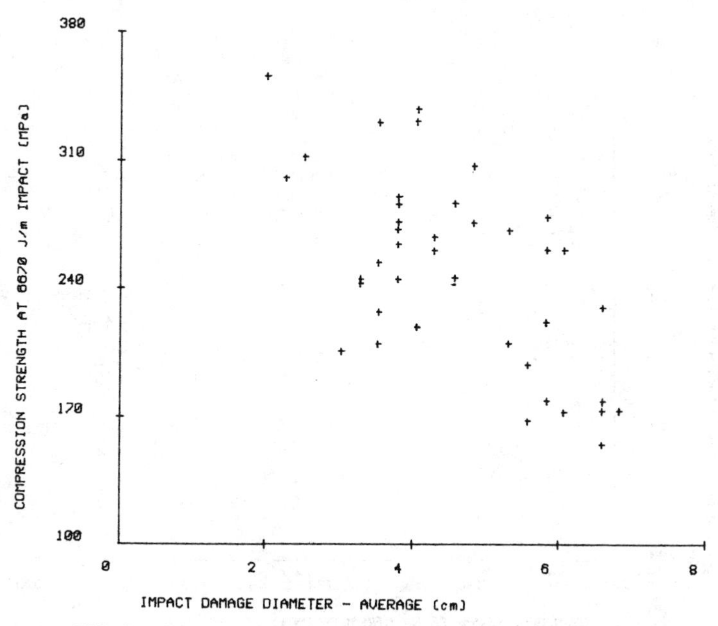

FIG. 9—*Compression after impact versus impact damage diameter.*

of residual strength. Given a large variation in impact instrument configurations, any predictions based on observed impact size would have a large expected error.

Figure 10 is a plot of an open hole compressive strength versus compressive strength after impact for ten different resin systems. The open hole coupon chosen was the standard 10.2- by 15.2-cm impact panel as described by Byers [1] with a 2.22-cm diameter hole instead of impact damage. All failure strengths were calculated on the full width of the test coupons. The data are surprising in that all samples showed an almost constant open hole compressive strength. This is in spite of the very wide range of impact strengths represented. Further work is necessary to understand this new test, with comparisons to other resin properties, and tests at elevated temperatures after water exposure.

The data presented up to now have all been for single-phase epoxy matrices. American Cyanamid Company has introduced some dual matrix, "interleafed" composite systems to provide enhanced toughness, such as CYCOM HST-7. These systems consist of a single-phase, high-modulus epoxy resin formulation coated on the graphite fibers. A lower modulus, very high strain resin is then coated on one side of the prepreg tape. Both resin systems are designed for high melt viscosity which prevents mixing during cure. The result is a cured laminate with a high modulus resin surrounding the graphite fibers to provide good stability at elevated temperatures, and a high strain resin layer isolated (or "interleafed") between the plies of graphite to prevent delamination during impact. For the systems discussed here, the total resin content is about 40%. Of

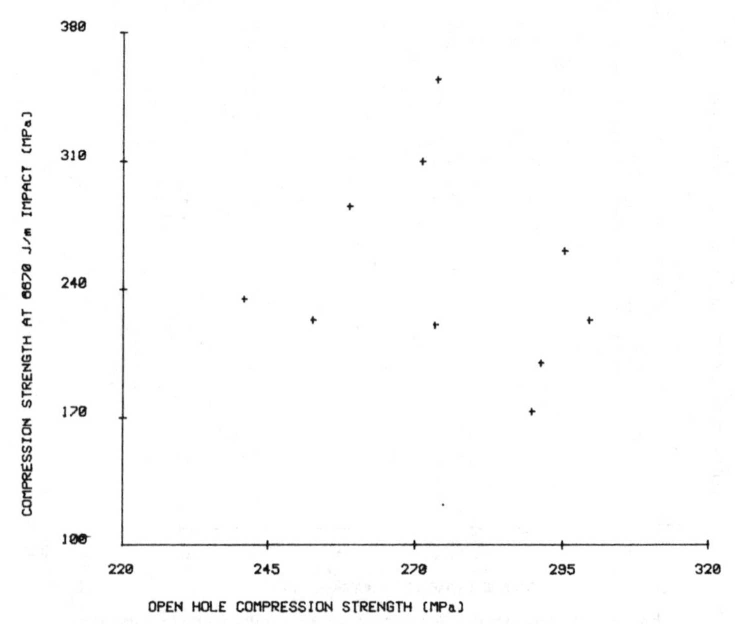

FIG. 10—*Compression after impact versus open hole compression.*

the two resins involved, about 70% is the high modulus resin on the fiber, and the remaining 30% is the high strain interleaf.

Figures 11 and 12 show the effect of interleafing on three commercial systems, CYCOM HST-7, 985, and 1808 advanced composites. For all three systems, the noninterleafed G_{Ic} values are about 262 J/m^2 (despite large differences in both chemistry and other mechanical performance). Addition of interleaves (again, different chemistry in all three cases) resulted in about a 20% improvement in G_{Ic} to 315 J/m^2. However, the increase in compressive strength after impact was much larger, from 40 to 55%. A similar effect would also be seen if short beam shear strength after impact (retention at 1.13 J/0 J) had been plotted. In all three cases, impact shear retention improved from about 0.55 to 0.60 to above 0.9. This lack of significant change in G_{Ic} may be part of the explanation of the large scatter seen in G_{Ic} versus impact data seen in Fig. 8. At least in the case of interleaved systems, G_{Ic} does not appear to be strongly linked to impact damage resistance.

Conclusions

The most important conclusions in this data, from a resin developer's viewpoint, are first, neat resin strain to failure is a strong predictor of composite

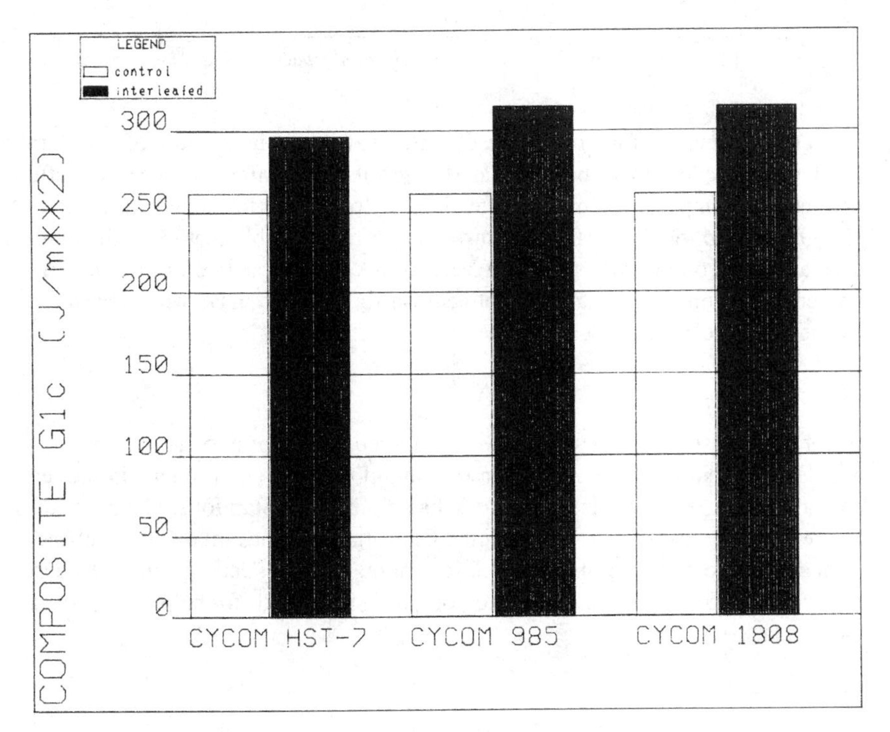

FIG. 11—*Effect of interleaf on* G_{Ic}.

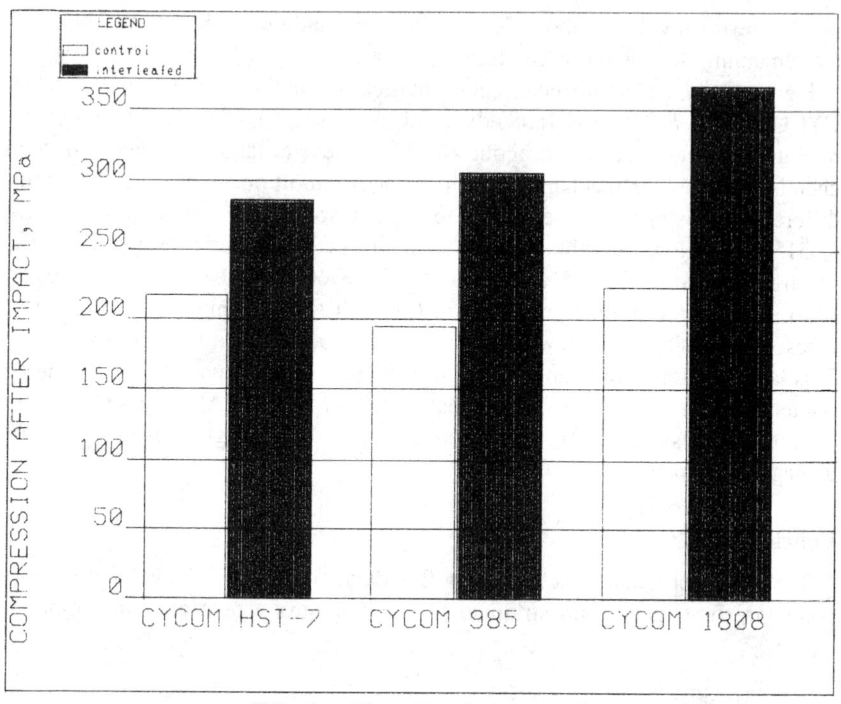

FIG. 12—*Effect of interleaf on impact.*

impact performance for single-phase systems with resin moduli of 3450 to 4140 MPa. Second, short beam shear strength retention after impact is a useful predictor of compressive strength after impact for a resin formulator, especially at high strength retention. Third, interleafing offers a viable method to increase dramatically compressive strength retention after impact; and fourth, improvements in impact strength are not necessarily reflected in G_{Ic} when measured by the double cantilever beam.

Acknowledgments

The data presented in this paper were gathered as part of a long-term research and development effort by American Cyanamid Company to produce improved composite matrix materials. This work has gone on at Stamford, Connecticut, Havre de Grace, Maryland, and Saugus, California, and has involved the efforts of many people. I would like to thank Raymond Krieger, Gene Susman, Robert Evans, David Bullock, Jeanne Courter, and Stan Kaminski, for help in preparing these data.

References

[1] Byers, B. A., "Behavior of Damaged Graphite/Epoxy Laminates Under Compression Loading," NASA Contractor Report 159293, National Aeronautics and Space Administration, Aug. 1980.

[2] Susman, S. E., "Graphite Epoxy Toughness Studies," presented at 12th National SAMPE Technical Conference, Oct. 1980, sponsored by the Society for the Advancement of Material and Process Engineering, Corina, CA.

DISCUSSION

Golam M. Newaz[1] *(written discussion)*—Can you elaborate on the "compressive strength after impact" testing? Doesn't impact cause fiber damage which would contribute to local instabilities leading to premature buckling and fracture of fiber under subsequent compressive loads? If this is true, then fiber strength can partially influence compressive strength. Are we then clearly determining the role of the matrix through this type of testing? I would appreciate your view on this matter.

K. R. Hirschbuehler (author's closure)—Impact damage in a composite coupon includes both resin and fiber damage. The work presented in this paper is primarily concerned with resin properties and resin development, so fiber properties were held constant as much as possible. As explained in the Experimental Section, all fibers studied here were 32-Msi (221-GPa) modulus, with strengths in the range of 480 to 600 ksi (3310 to 4140 MPa) (strains of 1.5 to 2.0% at failure). By comparison, resins strains studied range from 2.5 to 11%. Thus, the variations in performance shown in this paper should reflect mostly resin variations, as the fiber properties were held almost constant.

Looking beyond the scope of the data presented here, I agree that variations in fiber properties would affect impact performance. I would expect that for any given impact energy, a stronger fiber would show decreased damage and increase retained compressive strength. It would require much larger variations in fiber properties than those studied here to see those effects, but fiber effects were not the point of this study.

[1]Owens-Corning Fiberglas Technical Center, Granville, OH.

Donald L. Hunston,[1] *Richard J. Moulton,*[2] *Norman J. Johnston,*[3] *and Willard D. Bascom*[4]

Matrix Resin Effects in Composite Delamination: Mode I Fracture Aspects

REFERENCE: Hunston, D. L., Moulton, R. J., Johnston, N. J., and Bascom, W. D., **"Matrix Resin Effects in Composite Delamination: Mode I Fracture Aspects,"** *Toughened Composites, ASTM STP 937,* Norman J. Johnston, Ed., American Society for Testing and Materials, Philadelphia, 1987, pp. 74–94.

ABSTRACT: A variety of thermoset, toughened-thermoset, and thermoplastic polymers were characterized for Mode I critical strain energy release rates, G_{Ic}, and their composites were tested for interlaminar G_{Ic}s using the double-cantilever beam specimen. A clear correlation between the data from the two types of experiments was found. With brittle polymers (resin G_{Ic}s less than 200 J/m^2), the composite G_{Ic}s varied from slightly greater than to three times greater than the resin values. Although the resin toughness may represent the lower limit for the composite, the increased G_{Ic} value usually found in the composite was attributed to the fiber breakage and pullout that generally accompany composite crack growth. For tough matrix resins, an increase of 3 J/m^2 in resin G_{Ic} resulted in approximately a 1-J/m^2 increase for the composite. This less than complete transfer of toughness was attributed to the fibers restricting the crack-tip deformation zone that is associated with the high G_{Ic}s of the polymers. The transition between these two types of behavior occurred at a point where the size of the deformation zone was roughly comparable to the fiber-fiber spacing between plies. Other factors that tended to increase the interlaminar G_{Ic} included fiber nesting and bridging, while weak fiber-matrix bonding tended to decrease the crack growth resistance when the poor bonding occurs over extensive areas. This factor is particularly common in thermoplastic composites which exhibited lower interlaminar G_{Ic}s than the corresponding thermosets. Scanning electron microscope pictures of the fracture surfaces showed significant regions of failure at or near the interface in the thermoplastic composites.

KEY WORDS: adhesion, composites, delamination, epoxy, fracture, polyamideimide, polycarbonate, polyethyerimide, polysulfone, toughened epoxy, thermoplastic, thermoset

[1]Polymer composites group leader, National Bureau of Standards, Polymers Division, Gaithersburg, MD 20899.

[2]Manager of Composite Research and Development, Hexcel Aerospace, 11711 Dublin Blvd., Dublin, CA 94566, and consultant, 18 Westwind Rd., Lafayette, CA 94549.

[3]Senior scientist, NASA-Langley Research Center, Materials Division, M/S 226, Hampton, VA 23665.

[4]Research associate, Hercules Aerospace Division, Hercules Inc., Bacchus Works, Magna, UT 84044; present address, research professor, University of Utah, Salt Lake City, UT 84112.

In the last several years, considerable concern has been expressed over possible problems resulting from delamination in fiber-reinforced composites. This concern has generated numerous efforts to find ways of minimizing the initiation or growth or both of delaminations [*1–20*]. Since one of the factors thought to contribute to the susceptibility of composites to delamination is the brittle nature of the matrix resins generally employed, a number of groups are investigating the use of tougher matrix materials [*4–7,11–20*]. One difficulty in such studies is the selection of a test method or methods to characterize the resistance of a composite to the initiation or growth or both of interlaminar cracks. Many of the studies have used the double cantilever beam (DCB) geometry for this purpose. Obviously, this test focuses on Mode I loading and damage growth rather than initiation. Nevertheless, the simplicity of the DCB test makes such studies a useful first step.

Experiments using this approach have shown that a large improvement in the Mode I critical strain energy release rate, G_{Ic}, for the resin itself (that is, a 20-fold increase) produce an important but only moderate improvement in composite interlaminar G_{Ic} (that is, a 4- to 8-fold increase) [*5,6,12–19*]. Moreover, important trade-offs were found. For example, a matrix material with increased toughness often has inferior properties in a hot humid environment.

These results raise an interesting question. Since a large improvement in G_{Ic} for the resin translates to only a modest increase in interlaminar G_{Ic}, could the same improvement in composite fracture behavior be obtained by using a moderately tough matrix resin? If so, the sacrifices that must be made in other properties could be minimized. To state this question in a more general way, what is the detailed relationship between the toughnesses of the matrix resins and the interlaminar G_{Ic}s of their corresponding composites? This study seeks to address this question.

To develop such a relationship, the influence of parameters other than matrix toughness must be minimized, or the matrix materials must span a sufficiently wide range of G_{Ic}s so that the effects of other parameters are overshadowed. In addition, the influence of different fabrication procedures must be controlled or factored into the analysis. In general, with the exception of a preliminary note [*19*] based on the work reported in detail here, the data that presently exist in the literature [*3,6,12,16,20*] do not meet these criteria. The purpose of this paper is to provide such data and to investigate their implications.

Experimental Procedure[5]

Fracture experiments were performed using several different resins and their graphite fiber composites. These materials represent three of the common classes of matrix resins.

[5]The use of trade names and the identification of manufacturers in this paper are for the purpose of adequately specifying the Experimental Procedure only. In no case does such identification express or imply an official recommendation or endorsement by the Government, nor does it imply necessarily that they are the best available for the purpose.

121°C (250°F) Cure Epoxies

Three commercial 121°C cure epoxies were evaluated. Designated as HX205, HX206, and F185 (Hexcel Corporation), their chemical compositions have been reported elsewhere [*5*]. What is important is that they contained 0, 8, and 13% elastomer, respectively, as a toughening agent, and this gave them a wide range of resin G_{Ic}s (see Table 1).

Neat resin plates approximately 15.24 by 30.48 by 0.635 to 0.953 cm (6 by 12 by ¼ to ⅜ in.) from which test specimens could be cut were fabricated as follows. The liquid resin was heated in a vacuum oven at 66 to 93°C (150 to 200°F) to remove solvent and air bubbles. The deareated material was cured in molds using an autoclave with the same cure cycle employed for the composites, namely, heat to 121°C at 1 to 4°C (2 to 8°F) per minute and 0.59-MPa (85-psi) pressure, hold 90 min, then cool under pressure.

Two sets of composite specimens were fabricated with these resins: one set by Hexcel Corporation using standard-sized Thorne® 300 fibers and the other by NASA-Langley using polyimide-sized Celion® 6000 fibers. Composites were fabricated at Hexcel from commercial production prepreg tape in an autoclave

TABLE 1—*Experimental fracture data.*

Resin	Fiber[a]	Resin[b] Type	Fiber Volume, %	G_{Ic}, J/m² (Reference) Neat Resin	Composite
5208	T-300	BE	65	80	100[d]
3501-6	AS-4	BE	67	95	175
HX205	T-300	BE	58		380
				230 [*5,21*][c]	
HX205	C-6000	BE	56		790[d]
Ultem[e]	T-300	TP	65		935[d]
				3300 (footnote 6)[c]	
Ultem[e]	T-300	TP	60		1060[d]
P-1700[f]	T-300	TP	52		1200[d]
				2500 [*22,23*] (footnote 7)[c,g]	
P-1700[f]	T-300	TP	52		1340[d]
HX206	T-300	TE	55		830
				2200 [*5,21*][c]	
HX206	C-6000	TE	49		1550[d]
F185	T-300	TE	59		1960
				5830 [*5,21*][c,h]	
F185	C-6000	TE	57		2250[d]

[a] Standard epoxy size for T-300 and AS-4 fibers, polyimide size for C-6000 fibers.
[b] BE = brittle epoxy, TP = thermoplastic, TE = toughened epoxy.
[c] Data from present study as well as from literature.
[d] Composites fabricated at NASA-Langley Research Center.
[e] Ultem 1000 polyetherimide, first entry for composite with insert for precrack, second entry had no insert.
[f] Udel P1700 polysulfone, first entry for composite with insert for precrack, second entry had no insert.
[g] Data for neat polysulfone fracture exhibited unusually large variations.
[h] Significantly higher value recently obtained by Chakachery and Bradley [*32*].

using a vacuum bag (56 cm [22 in.] Hg), and the recommended cure cycle described above. At NASA-Langley, wet prepreg was made by drum winding a methyl ethyl ketone solution of the resin. The prepreg was dried at 55°C (130°F) in a preheated oven and press molded as follows. F185 and HX206 prepregs were placed in a cold mold, 0.52-MPa (75-psi) pressure applied, and the mold heated at 5°C (10°F) to 121°C and held 1 h. The HX205 prepreg was placed in a mold preheated to 121°C, contact pressure applied and held 15 min, 0.52-MPa pressure applied, and the mold held at 121°C for 1 h. All panels were postcured unrestrained in an air oven at 121°C for 2 h.

177°C (350°F) Cure Epoxies

For comparison with the three tough epoxy systems, two additional sets of neat resin and composite specimens were fabricated from brittle 177°C cure epoxies: 3501-6 (Hercules, Inc.) on standard-sized AS-4 fiber and 5208 (Narmco Materials, Inc.) on standard-sized Thornel 300 fiber. Both net resin plates and composites were fabricated following recommended manufacturer, vacuum-bag autoclave procedures for low viscosity resins, the 3501-6 at Hercules and the 5208 at NASA-Langley.

Thermoplastic Composites

In addition to the five epoxies, neat resin and composite specimens were also fabricated from three commercial thermoplastics: Udel® P1700 polysulfone (Union Carbide Corporation), Ultem® 1000 polyetherimide (General Electric Company), and Lexan® 101 polycarbonate (General Electric Company). The first two were fabricated with standard-sized Thornel 300 fiber, the third with standard-sized AS-4 fiber. The three thermoplastic matrices are amorphous materials and were chosen so no complications would be introduced by variations in the type and amount of crystallinity between the bulk resin and the corresponding composite.

Polysulfone prepreg was available commercially (U.S. Polymeric) as 30.5-cm (12-in.)-wide tape on a 25.4-cm (10-in.)-diameter spool and contained several percent cyclohexanone solvent. The polyetherimide prepreg was commercially available (McCann Manufacturing Company) as 14-cm (5.5-in.)-wide tape in 122-cm (48-in.) strips containing traces of methylene chloride. It was necessary to predry both prepregs at 288°C (550°F) for 1 h in a forced air oven to remove all solvent and moisture. Polysulfone and polyethyerimide composites were fabricated at NASA-Langley from this boardy prepreg by standard press mold procedures at a maximum temperature/pressure of 343°C/4.14 MPa (650°F/600 psi) and 427°C/5.17 MPa (800°F/750 psi), respectively, then cooled below glass transition temperature (T_g) under pressure before removing from the mold. Neat resin specimens of both were molded at 288°C/1.72 MPa (550°F/250 psi).

Polycarbonate prepreg was fabricated at NASA-Langley by running fiber through a dip tank containing an 18% w/w solution of the polymer in methylene chloride. The wet tow was wound onto a drum coated with release paper, then air-dried. The boardy prepreg was further B-staged at 205°C (400°F) for 1 h in a forced air oven and press molded at 288°C (550°F) for 30 min at 0.69-MPa (100-psi) pressure. Although this composite was prepared by NASA-Langley, the fracture tests were not performed as part of this work. Nevertheless, the results, which were obtained elsewhere (see Table 2) will be included in the comparison of literature data made later in this paper.

Composite Characterization

All composite panels were ultrasonically C-scanned on an Ultrasonic Systems, Inc., instrument. Only panels showing no porosity under NASA-Langley's high sensitivity tests preestablished for T300/5208 composites were machined into specimens.

Composite fiber volume fractions are listed in Table 1. They were calculated from the product of fiber areal weight and number of plies divided by the product of fiber density and carefully determined panel thickness, assuming (on the basis of C-scan measurements) void free laminates. This method of calculating fiber volume fractions was used because the values agreed within 2% with those calculated from fiber areal weights and composite and resin densities experimentally determined from a density gradient column.

To help characterize interlaminar crack growth behavior in the composites, DCB fracture surfaces were examined using an AMR Model 1000 scanning electron microscope (SEM). SEM micrographs of selected fracture surfaces at several magnifications are shown in Figs. 1 through 4.

TABLE 2—*Literature fracture data.*

Resin	Fiber[a]	Resin[b] Type	G_{Ic}, J/m^2 (Reference)	
			Neat Resin	Composite
2220-3	AS-4	BE	95 [*18*]	250 [*18*]
U-Resin-1[c]	AS-4	ER	90 [*18*]	300 [*18*]
U-Resin-2	AS-4	ER	200 [*18*]	310 [*18*]
U-Resin-3	AS-4	ER	245 [*18*]	410 [*18*]
U-Resin-4	AS-4	ER	754 [*18*]	821 [*18*]
U-Resin-5	AS-4	ER	950 [*18*]	810 [*18*]
U-Resin-6	AS-4	ER	1100 [*18*]	763 [*18*]
3502	AS-1	BE	69 [*16*]	155 [*16*]
3502	AS-4	BE	69 [*16*]	255 [*16*]
F-155	C-6000	TE	730 [*16*]	600 [*16*]
F-185	C-6000	TE	6040 [*16*][e]	2600 [*16*]
Lexan[d]	AS-4	TP	5500 [*23,27,28*]	1300 (footnote 8)

[a] Standard-sized fibers.
[b] BE = brittle epoxy, ER = experimental resin, TP = thermoplastic, TE = toughened epoxy.
[c] U = resin composition unknown.
[d] Lexan 101 polycarbonate.
[e] Significantly higher value recently obtained by Chakachery and Bradley [*32*].

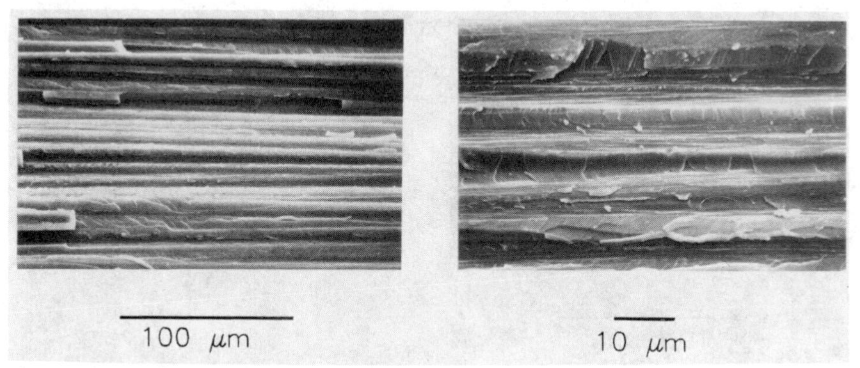

FIG. 1—*DCB fracture surface for a brittle epoxy (5208) composite at two magnifications shows fiber breakage and pullout, and good fiber-matrix bonding.*

Neat Resin Fracture Toughness Testing

The Mode I critical strain energy release rates for neat resin samples were obtained either from test runs using compact tension specimens or, in a few cases, from the literature (see Table 1). Average values for the G_{Ic}s of HX205, HX206, and F185 were taken from the literature [*5,21*] and a few tests conducted in the present study. The G_{Ic} data for polyetherimide were obtained in this work, and they were in good agreement with values reported by the manufacturer.[6] The result used for polysulfone was an average of data from this work and the literature [*22,23*].[7] Both the results of this study and the literature values showed a wide variation and, consequently, this average is subject to a good deal of

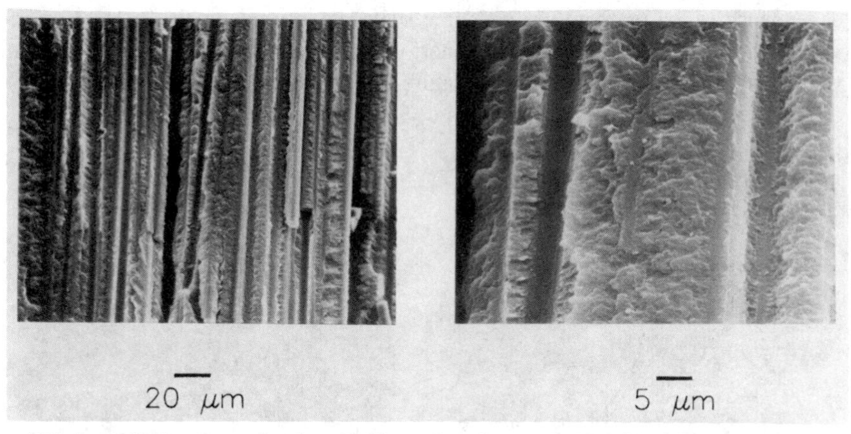

FIG. 2—*DCB fracture surface for a highly toughened epoxy (F185-C6000) at two magnifications shows deformation in the polymer, fiber breakage and pullout, and good fiber-matrix bonding.*

[6]A. Yee, General Electric Co., R&D Laboratory, personal communication, Jan. 1983.
[7]J. M. Whitney, Air Force Wright Aeronautical Laboratory, personal communication, March 1982.

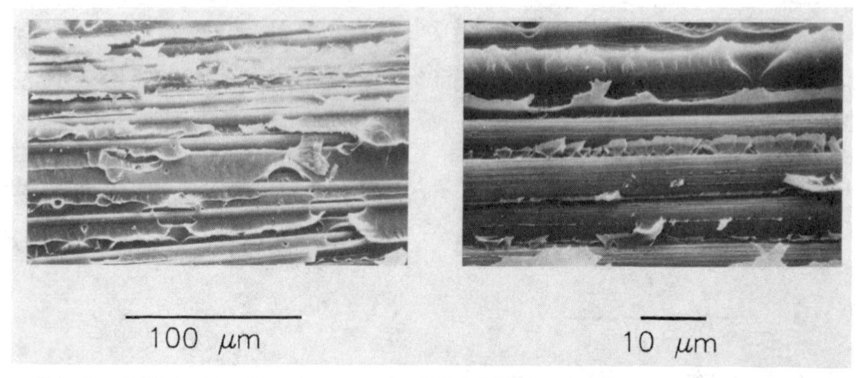

FIG. 3—*DCB fracture for a thermoplastic (Udel P1700) polysulfone composite at two magnifications shows deformation of the polymer, fiber breakage and pullout, and significant regions of poor fiber-matrix bonding.*

uncertainty. The data for 5208 and 3501-6 were determined in this study and are in general agreement with values determined elsewhere [*8,15*].

The compact tension specimens used for the fracture tests were rectangular or circular in shape with thicknesses up to 1.25 cm for the tougher materials so that plane strain values could be obtained. The ASTM Test for Plane-Strain Fracture Toughness of Metallic Materials (E 399) guidelines were used to analyze the data and assure valid G_{Ic} determinations in plane-strain conditions. Modulus values for the calculations were obtained from three-point bonding tests or from the manufacturer's data sheets. For the experiments performed in this work, the time scales for the modulus measurements were matched to those of the neat resin fracture tests, although even the toughest material showed relatively little time dependence in the temperature and time scale range used for the tests.

It is known that the critical strain energy release rates for tough polymers can vary considerably with the test temperature and crosshead speed (loading history)

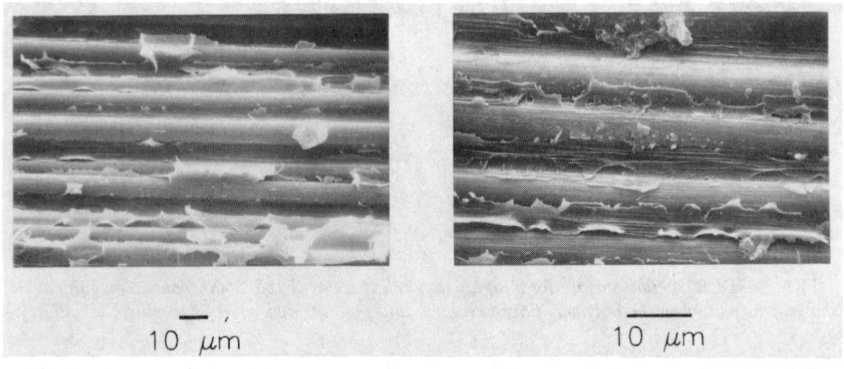

FIG. 4—*Fracture surface for polyetherimide (Ultem 1000) composite at two magnifications shows deformation of the polymer and significant regions of poor fiber-matrix bonding.*

[*12,17,24,25*]. A thorough comparison of neat and composite fracture results therefore requires a detailed knowledge of these effects so that the comparison can be made over a wide range of conditions with the correct correspondence between the two geometries. Such knowledge, however, is available only for a few resins and none of the composites used in this study. Thus such a comparison cannot be made at the present time. As a first step toward such a goal, this study attempts to make the comparison for one particular set of conditions. Moreover, the goal in this work was to investigate major trends rather than make detailed comparisons, so it was adequate to have similar, if not identical, conditions for the neat resin and composite tests. Nevertheless, the experiments on neat resin were constrained by requiring that they involve a slow crosshead speed (0.05 cm/min) and, to the extent possible, identical conditions for all materials (temperature = 22°C). Since the composites failed by slow crack growth, it was hoped that this produced test conditions for the resins and the composites that were in the same general range. Where literature values were used, the references were chosen so that the samples were tested under roughly these same conditions.

Another cautionary note should be made about comparisons between bulk resins and their composites. Although an effort was made to cure each resin and its composite in the same way, even identically cure conditions may not give exactly the same material in the two different cases. Moreover, the differences that could be present—for example, variations in network structure, morphology, degree of cure, and so forth—will affect the G_{Ic} values. In most cases, however, such effects are known to be small compared to the range of toughnesses examined here. There is one potential exception. The G_{Ic}s for rubber-modified epoxies can vary significantly with changes in morphology and cross-link density. For the materials tested here, F185 is of some concern because evidence does exist that suggests there are differences in morphology between the resin and its composite. This will be discussed in more detail later in this paper.

Composite Interlaminar Fracture Testing

Most of the composites examined in this work were 24-ply unidirectional lay-ups with the fiber direction down the length of the sample. A few NASA-fabricated HX205, HX206, and F185 specimens had a more complex $[(0_2/\pm45)_3 0]_s$ lay-up. However, this difference did not appear to have any major effect on the interlaminar G_{Ic} values. Midplane inserts were not used as crack starters for the epoxy composites tested in this study. Cracks were initiated with a razor blade as close to the midplane as could be achieved. For the thermoplastic composites, a 0.002 54-cm (0.001-in.)-thick Telfon® or Kapton® film was inserted during fabrication at one end of the specimen plate between Plies 12 and 13 for use as a crack starter.

Interlaminar G_{Ic} values were measured using double cantilever beam (DCB) specimens (2.54 cm by 14.24 cm by 24 plies) cut from the plates. The results were analyzed using both the compliance and area methods for determining fracture energies [*8,13,15*], and in all cases, the two methods gave similar results.

For the tests involving the toughest matrix resins, a detectable amount of nonlinearity and creep that could be attributed to the resin was observed. This added some uncertainty to the exact values of the G_{Ic}s obtained. Here again, however, the purpose in the work was to investigate major trends, and the uncertainties involved were much less than the effects that were of interest.

The G_{Ic} values obtained with the DCB specimens correspond to steady crack growth at a velocity determined by the crosshead speed and the crack length. Although the velocity decreased as the crack length increased during the test, the range of values was not very large (0.04 to 0.02 cm/s, for example), so a velocity of about 0.03 cm/s was a good average. Studies by Russell [*26*] have found that specimens with inserts for precracks give G_{Ic} values that increase as the cracks grow away from the insert but then reach a steady value for growth through the remainder of the specimen. He attributed this to fiber nesting and bridging which will be discussed later in this paper. Since the work here involves specimens both with and without inserts, however, only data away from the insert will be discussed.

The composite fracture experiments are performed by repeatedly loading and unloading the specimen to produce short segments of crack growth (1 to 2 cm) each time. Consequently, there is an opportunity to observe both the steady crack growth and how the growth starts (that is, the onset) during each cycle. For brittle matrix materials, the first detectable crack growth occurred at almost the same strain energy release rate as that for steady growth. With tough matrix resins, however, the onset of crack growth occurred at a significantly lower strain energy release rate than that for steady crack growth. When such differences were observed, both the onset and the steady growth values were determined. The value for the onset may not be a true material property, however, since it may depend on other factors such as the sensitivity of the experiment; that is, the closer one looks, the earlier one may detect the crack growth. Nevertheless, reporting both values was useful in providing an indication of the onset process for interlaminar crack growth in the various materials.

Literature Fracture Data

In addition to the seven resins and ten composites described above, a further comparison can be made by considering three sets of selected results from the literature. These results and the references are summarized in Table 2. The first set of data involves a commercial resin, Hercules 2220-3, and a series of six experimental Hercules resins of unknown composition but with a variety of measured G_{Ic} values. The composites were made with AS-4 fiber, processed by standard 177°C cure autoclave procedures, and tested for interlaminar fracture behavior at Hercules. The results were reported elsewhere [*18*].

The second set of data involves three commercial resins: Hercules 3502 on AS-1 and AS-4 fibers, Hexcel F155 on T-300 fibers, and F185 on Celion 6000 fibers [*16*]. The G_{Ic} values are for steady crack growth at a lower velocity than that measured in the present work; however, this velocity difference is not large enough to influence the comparisons made in this paper.

The final set of data is for a commercial amorphous thermoplastic resins: polycarbonate (General Electric Lexan 101). The composite was made at NASA-Langley as described above; its interlaminar G_{Ic} value was measured by Bradley.[8] The polycarbonate neat resin G_{Ic} reported in Table 2 is an average taken from the literature [*23,27,28*], and agrees closely with recent values determined experimentally on minaturized specimens [*23*].

On the 13 composite systems reported in Table 2, only data for steady crack growth were available for all systems. Consequently, comparisons could only be made on that basis.

Results and Discussion

Influence of Fiber Volume and Type

It is obvious that the samples used in this work provided a wide range of critical strain energy release rates. However, to achieve this, a variety of fabrication procedures were involved; for example, some prepregs were drum wound while others were made on commercial machines. As a result, factors such as prepreg uniformity and fiber volume fraction varied considerably. The significant influence of these factors on G_{Ic} values will be discussed later.

Prepreg resin content and fiber areal weight also varied, and this led to some scatter in composite fiber volume fractions (see Table 1). For example, composites made with the two brittle epoxies (5208, 3501-6) had very high fiber volumes. The tougher composites had consistently lower fiber volumes. In fact, fiber volume fractions for composites made with the three tougher epoxies (HX205, HX206, and F185) and two fiber types exhibited a very narrow range—55 to 59%. Thus, variation in fiber volume fraction was not a major contributing factor to the significant variations observed in G_{Ic} values among these six composites. The one exception was the HX206/C6000 composite which was extremely resin rich. It also had a much higher G_{Ic} value than the HX206/T300 composite with the lower resin content. This indicates that interlaminar G_{Ic} can vary with composite resin content, especially when the latter is extremely high. Similar results were found by Jordan and Bradley [*18*]. They went on to examine the influence of the distribution of the resin in the composites. They found that significant differences in this distribution from one sample to the next could have an important effect because, as might be expected, the critical parameter is the resin content near the crack tip and not the average resin content.

In contrast to the epoxy results, the two Ultem composites tested here gave reasonably similar G_{Ic} values, 1060 and 935 J/m^2, although a larger difference might have been expected based on the fact that their fiber volume fractions varied by 5%. However, a complicating feature in this case was that the composite with the higher G_{Ic} value (and the higher resin content) did not have a midplane insert for a starter crack while the other did. The absence of such a crack starter may lead to more fiber bridging which also helps contribute to a higher G_{Ic}

[8]W. L. Bradley, Texas A&M University, unpublished data.

value. In this case, a 5% variation in fiber volume fraction did not appear to affect appreciably the interlaminar G_{Ic}. The identical observation concerning the effect of an insert on G_{Ic} values was made for the two P1700 polysulfone composites. Comparisons between the P1700 and the Ultem data are complicated by the fact that their fiber volumes differ by 8 to 13% and both yielded interfacial failures. The effect of the latter will be discussed later. None of these composites were characterized for the distribution of their resin, and hence the effects discussed by Jordan and Bradley [*18*] may also play a role here.

To summarize, small variations in fiber volumes within any one class of matrices do not seem to influence significantly the interlaminar fracture behavior so long as the distribution of the resin does not vary greatly. Large variations, as expected, do produce differences in interlaminar fracture behavior. The influence on G_{Ic} of fiber volume fraction variations among the different classes of matrices is beyond the scope of this work. What is important here, however, is that even the largest variations attributed to fiber volume fraction changes are small compared to the range of G_{Ic} values covered by the different resins examined. Consequently, in the resin-versus-composite comparisons that will be made later, the variations in fiber volume fraction contribute to the data scatter but do not obscure the general trends.

The composites examined here also involve several different types of fibers, but all of the fibers have mechanical properties which fall within a narrow range. Studies [*16*] that investigated this type of variation in fiber have found differences in interlaminar fracture behavior, but these differences are quite small and would not affect the overall trends determined in this study.

Onset Versus Steady Growth

The tough resin composites show a clear difference between the strain energy release rates associated with the first detectable interlaminar crack growth and those measured for steady crack growth. This can be shown by plotting the onset strain energy release rate values as a percentage of the steady growth G_{Ic}s against the steady growth G_{Ic}s for the ten composite systems evaluated in this study (Fig. 5). The graph indicates that although the differences between the onset and steady growth strain energy release rates are small (16% at most), a clear trend exists with tougher systems showing larger differences. It is tempting to speculate that this effect is related to the highly viscoelastic fracture behavior seen in the neat resin specimens for tough polymers [12,24,29]. Whatever the explanation, however, this feature is an important topic for future study.

Correlation of Neat Resin and Composite Experimental Fracture Behavior

The neat resin and composite fracture results for all ten materials evaluated in this study are plotted in Fig. 6. The data for the brittle epoxies, 5208 and 3501-6, are represented by X and +, respectively. For the other systems the open symbols represent the composites fabricated by Hexcel while the filled symbols are for the NASA fabricated composites. In those situations where the onset and propagation

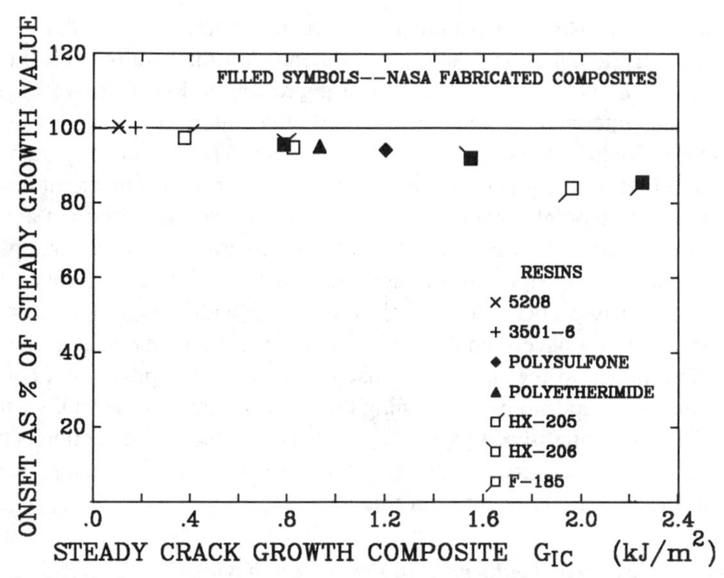

FIG. 5—*Interlaminar strain energy release rates for the onset of crack growth as a percentage of the steady growth values as a function of the latter.*

fracture strain energy release rates are significantly different, both points are shown, and they are connected by vertical lines.

Based on these results it is possible to offer a number of hypotheses and draw some important conclusions. The data as a whole show a general correlation between neat resin and composite fracture results. As suggested by some results in previous studies [*3,18,19*], this correlation can be divided into two regions. At low G_{Ic} values the points fall somewhat above a line of Slope 1 (Fig. 6). This can be explained by asserting that the resin behavior is generally transferred fully to the composite, but, in addition, the composite fracture involves toughening mechanisms [*3*] not present in the neat resin. For example, some fibers are always broken during crack propagation, and this clearly absorbs some energy. Moreover, the crack must often go around fibers so that the fracture surface in the composite is quite rough, and this additional surface area is not taken into account in the strain energy release rate calculation. Of course, some differences may also exist in the crack-tip stress fields in the neat resin and the composite. These and other similar factors can be used to rationalize why the G_{Ic}s in the composites are higher than those for the resins. It might be speculated that the resin toughnesses represent the lower limiting values for the composites, but more study is needed before this idea can be critically assessed.

With the tougher resins, however, this relationships breaks down; the neat fracture behavior is not fully transferred into the composite. It has been suggested [*2,5,12,17,30*] that this can be explained by noting that the high G_{Ic}s of these resin materials are associated with large crack-tip deformation zones, and in the composites, the fibers restrict the size of these zones and thereby limit the interlaminar G_{Ic}s. The data obtained here for the extreme cases, that is, very brittle and very tough resins, are in agreement with the results from previous studies [*2,5,6,12,14,16,17*]. What is most interesting in Fig. 6, however, is the evidence that intermediate toughness resins give intermediate values for the interlaminar fracture behavior of their composites. This is an important new result, but in retrospect it is probably not surprising since the constraint imposed by the fibers on the deformation zone size is known to be more than a simple limiting value. Such a limiting value would produce not only a horizontal plateau in Fig. 6, but also, as noted in previous studies [*2,5 12,16,17*], a much lower interlaminar fracture energy for composites made with very tough resins. As these earlier studies have pointed out, if the deformation zone size in the composite were limited to the fiber-fiber spacing, very little improvement could be obtained in the composite by toughening the resin. In fact, observations with the scanning electron microscope [*5,16,30*] have shown that the interaction between the fibers and the deformation zone is a more complex relationship, and the results in Fig. 6 are consistent with this idea.

Influence of Composite Uniformity on Fracture Behavior

To examine this correlation in more detail, it is useful to consider the results for the three Hexcel epoxy materials fabricated by Hexcel and NASA. For these samples, only a general correlation exists between the neat resin and the composite fracture behavior. Although there is no obvious reason to expect this relationship to be linear, a good general feeling for the relative variations is given

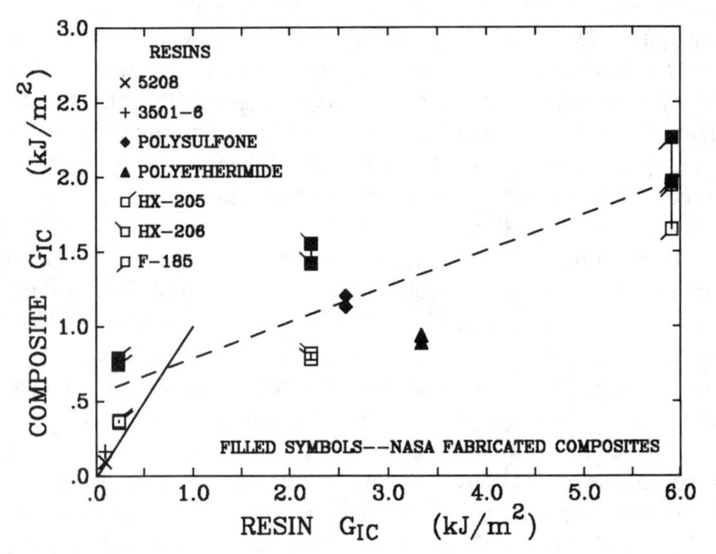

FIG. 6—*Composite interlaminar strain energy release rates for onset and steady crack growth as a function of neat resin* G_{Ic}*s for same systems as in Fig. 5.*

by the dotted line in Fig. 6 which has a slope of 0.24 (best fit line for the onset and steady crack growth data points for the three epoxies, that is, the squares). If, on the one hand, the composite fabricated by NASA and those made by Hexcel are considered separately, the correlation seems to be much better, although the data are obviously very limited. The trends for the two sets of data are similar except that the NASA-fabricated composites (filled squares) have significantly higher G_{Ic} values in all cases than the composites made by Hexcel (open squares). This suggests that the fabrication procedure is introducing another parameter into the comparison.

There are a number of possible factors that might contribute to this difference in behavior. In observing the experiments, however, one factor is strongly suggested. In the Hexcel samples, which are fabricated in a way that is more representative of commercial fabrication procedures, the fibers are generally confined to uniform layers, and thus the fracture produces a relatively flat fracture surface. With the NASA samples, on the other hand, the fiber layers are not well-defined, and the fibers from the various layers tend to intermingle. This feature, which has been termed nesting, makes it necessary to pull fibers from one layer out of the adjacent layer to propagate the crack. The propagation can occur in two ways. First the crack tip may bend around the fibers as it advances, thus loosing its planarity. This complicates the crack-tip stress field and leads to an increase in the fracture surface area that is not accounted for in the calculations. Secondly, some of the fibers may extend at an angle across the crack opening behind the crack tip. As the crack surfaces separate, these fibers must be either broken or pulled out of the fiber layers on one or both sides of the failure surface. This effect has been called fiber bridging. Both nesting and fiber bridging lead to a rough fracture surface and higher values for the apparent interlaminar G_{Ic}. If the results in this study are typical, it would seem that these effects can play a major role in determining the measured interlaminar fracture behavior.

Influence of Interfacial Failure on Fracture Behavior

The final observation that can be made from the data in Fig. 6 concerns the thermoplastic composites. They seem to follow the same general relationship as the epoxy based materials, that is, intermediate toughness resins gave composites with intermediate interlaminar G_{Ic}s. This is interesting because the toughening mechanisms in these materials are quite different than those in the rubber-toughened epoxies [*31*].

A more detailed comparison between the epoxies and the thermoplastics should probably be restricted to the NASA fabricated composites since all of these samples had similar fiber nesting and bridging effects. If this comparison is made (that is, comparing the filled symbols in Fig. 6), the thermoplastics would fall considerably below the relationship for the epoxies. One possible explanation for this can be gained from observations of the fracture surfaces. Scanning electron microscope pictures of the fracture surfaces in the epoxy composites (Figs. 1 and 2) show that virtually all of the fibers are coated with a thick layer of polymer;

that is, the failure occurred cohesively and clearly in the polymer rather than adhesively or cohesively but very near the interface. Even in the toughest epoxy composite systems, where the samples were subjected to the highest loads, the fracture path (Fig. 2) was found to be predominantly through the matrix, although occasionally some bare or lightly coated fibers were observed.

In contrast, the fracture surfaces in the thermoplastic composites show a large fraction of area where the failure had occurred either at or very near the fiber-matrix interface. Fracture surfaces, such as those shown in Figs. 3 and 4 for polysulfone and polyetherimide, respectively, are typical. This suggests that with the thermoplastics, there is poor fiber-matrix bonding, and thus significant regions where the crack can propagate along a weak path either at or near the fiber-matrix interface.

It can be argued that small, dispersed regions of poor bonding might be advantageous and increase the interlaminar G_{Ic} since they could divert the crack or lower the constraint in the matrix (that is, reduce the triaxiality of the stress field) or both. The latter effect would facilitate matrix yielding. In fact, the fracture surfaces of the thermoplastics do exhibit evidence of extensive yielding. When these poorly bonded regions become pervasive, however, the possible advantages may be outweighed by the presence of a weak path along which the crack can propagate. Consequently, if, in fact, the thermoplastics do exhibit a lower interlaminar G_{Ic} than they should, these observations of extensive failure at or near the fiber-matrix interface may provide a possible explanation.

The existence of less than ideal bonding in the thermoplastics is not unreasonable. A great deal of study has been devoted to the design of fiber surface treatments for epoxy based composites, and thus good bonding would be expected for these systems. With thermoplastics, however, relatively little work on fiber surface treatments has been performed. The two thermoplastics evaluated here were made using fibers with a standard epoxy sizing which is unlikely to promote optimum bonding with a thermoplastic matrix. If this explanation is correct, it implies a need to improve fiber surface treatment methods for thermoplastic matrix composites to achieve their true interlaminar fracture potential. The interlaminar fracture energies of the thermoplastic composites tested here are already quite high. Consequently, the differences under discussion here are not between poor and good performance but between good and very good performance.

Further Correlations—Experimental and Literature Fracture Data

All of the observations and hypotheses made above can be further examined by adding data points from the literature to the results obtained here. However, only interlaminar G_{Ic} values for steady crack growth can be considered. Figure 7 shows such a graph. This new plot is consistent with all of the results discussed above. The lines in this figure are drawn to represent general trends in the data and are not meant to imply that the relationship should necessarily be linear. The

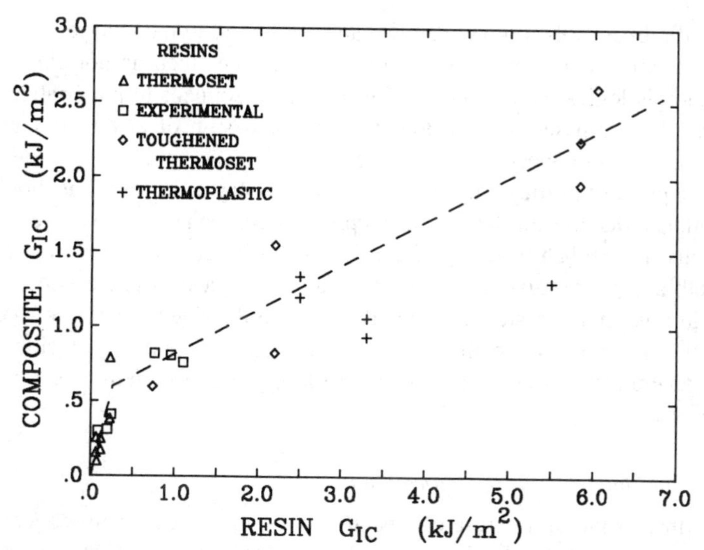

FIG. 7—*Composite interlaminar strain energy release rates for steady crack growth as a function of the neat resin* G_{Ic}*s for the systems in Fig. 6 and selected literature data.*

new data for brittle polymers again indicate that the G_{Ic} values of the composites are greater than those for the corresponding resins. In this case, the slope of the best fit line through the data points in this region is slightly greater than 2.

For the tough resins, on the other hand, the trend line has a slope of 0.31 which is quite similar to what was found in Fig. 6 where both onset and steady growth were considered. One observation worth noting is that this slope is very dependent on the fracture energies for F185 since all of the points at the extreme right involve this system. In recent work, Chakachery and Bradley [*32*] found neat resin G_{Ic} for F185 using a J integral approach that was higher than that reported previously [*5,21*]. This difference may be due to variations in neat resin morphologies from lab to lab for this complex two-phase system, or the newer numbers may represent a more realistic measure of the true value. Clearly, more work needs to be conducted with these very tough materials. Regardless of the outcome, however, it appears that, although the slope of the line in Fig. 7 may change somewhat, the general trend remains the same.

As mentioned earlier, evidence also exists indicating that the morphology in the F185 composites is not the same as that observed in the bulk samples used to measure neat resin G_{Ic}. This evidence includes small angle X-ray scattering data on neat resin moldings and composites [*33*], SEM photographs of fracture surfaces for neat resin moldings and composites, and mechanical properties for unidirectional tape and woven fabric composites.[9] The F185 composite matrix apparently contains a greater fraction of small elastomeric domains and more

[9]N. J. Johnston, NASA-Langley Research Center and R. M. Moulton, Hexcel Corp., unpublished data.

dissolved rubber in the epoxy matrix than is observed in the neat resin castings. One speculation is that the constrictive effect of the fibers is not allowing the larger morphologies seen in the F185 neat resin specimens to materialize. If so, this must be given serious consideration in the design of any matrix material containing a cure-induced second phase.

For the purpose of the comparison made in this paper, however, it should also be mentioned that the similarities in morphology for the neat resin and composite specimens are probably as significant as the differences which may explain why both specimens give very high G_{Ic} values. As a result, the general trend shown in Figs. 5 and 6 should be valid, although the exact position of the F185 points is uncertain. Since these points are at the extreme right side of the graph, generalizations about the behavior of very tough systems must be made with great care.

Constraint of the Composite Deformation Zone

One final point of interest can be drawn from Fig. 7. This concerns the transition between the behavior seen for the brittle resins and that found for the tough polymers. This transition occurs at a neat resin G_{Ic} value of about 200 J/m^2. By using this value, an interesting comparison can be made between the fiber-fiber spacing in the composite and the size of the crack-tip deformation zone in the resin. If a simple elastic-plastic model is used to estimate the size of this zone in polymers with G_{Ic} values near 200 J/m^2, a result of roughly 20 μm is obtained. In the composite, the fiber-fiber spacing varies considerably, and although the average spacing is roughly 9 μm, it is significantly larger than that between plies. Consequently, the results suggest an interesting similarity between the zone size and the fiber-fiber spacing between plies. Although this comparison is rather tentative, it is nevertheless reminiscent of the result obtained by Bascom and Cottington for adhesive bond fracture [*25*]. They found a change in the failure behavior when the deformation zone size was similar to the bond thickness and suggested that the adherends acted as a constraint to limit the G_{Ic} value of very tough polymer adhesives. The results here suggest a qualitatively similar picture. As would be expected, however, the details of the constraint and its quantitative effect on the G_{Ic} behavior is somewhat different in the composite.

The data in Fig. 7 also suggest another result. The point at which composite interlaminar G_{Ic} changes from greater than to less than that of neat resin occurs at a value of approximately 700 J/m^2. Consequently, a G_{Ic} somewhere between 200 and 700 J/m^2 may represent the useful trade-off point for toughening a neat resin. Below this value, improvements in the resin toughness have a sufficient benefit in the composite so that sacrifices in other properties may be justified. Above this value, a point of diminishing returns is reached. Consequently, it is necessary to consider the loss of features such as hot/wet properties with increases in resin-toughness to be sure they do not outweigh the relatively small gains in the interlaminar G_{Ic} that are obtained.

Conclusions

Fracture experiments conducted on neat and composite specimens with a series of resin systems suggests a strong correlation between the failure behavior exhibited in the two geometries, in particular, the Mode I critical strain energy release rates of the neat resins and the interlaminar G_{Ic}s of their composites as measured by the double cantilever beam. For brittle resins, the interlaminar G_{Ic}s are somewhat larger than the neat resin G_{Ic} values because a full transfer of the neat resin toughness to the composite occurs and, in addition, crack growth in the composite involves other toughening mechanisms, such as fiber pullout and breakage that are not present in the bulk resin tests. Although the lower limiting values of the interlaminar G_{Ic}s in this range may be the neat resin G_{Ic}, the specimens tested in this study gave interlaminar G_{Ic}s that varied from just slightly above the neat resin values to more than three times these numbers. On average, the interlaminar G_{Ic}s of the composites were slightly more than twice those of the neat resins.

With tougher matrices, the higher G_{Ic}s of the polymers are only partially transferred to the composites, presumably because the fibers restrict the crack-tip deformation zones. All other factors being equal, however, the results suggest that for tough resins an improvement in neat resin G_{Ic} of 3 J/m^2 may be expected to improve the interlaminar G_{Ic} by approximately 1 J/m^2.

The transition between the brittle resin and tough resin regions occurred in a range where the size of the crack-tip deformation zone is roughly similar to what might be expected for the fiber-fiber spacing between plies in the composite. This general range (200 to 700 J/m^2 resin G_{Ic}) may also mark a useful transition for resin design. Below this region, improvements in the resin have a significant benefit in the composite, and thus, sacrifices in other properties can often be justified. Above this region, increases in resin toughness must be examined more closely to be sure that the sacrifices in other properties do not overshadow the modest benefits for interlaminar fracture resistance.

The results also suggest that other factors play a major role in determining the interlaminar fracture behavior for composites. Fiber nesting, bridging, pull-out, and breakage can produce an increase in the apparent interlaminar G_{Ic}, while poor fiber-matrix bonding produces a decrease. This latter factor may be particularly important for thermoplastic composites since they appear to exhibit large regions of poor bonding and interlaminar G_{Ic} values that are lower than those found for toughened thermosets with similar resin G_{Ic}s.

References

[1] McKenna, G. B., Mandell, J. F., and McGarry, F. J., "Interlaminar Strength and Toughness of Fiberglass Laminates," in *Society of Plastics Industries Annual Technical Conference RPD*, Society of Plastics Industries, New York, 1974, Section 13-C.

[2] Scott, J. M. and Phillips, D. C., "Carbon Fibre Composites with Rubber-Toughened Matrices," *Journal of Materials Science*, Vol. 10, No. 4, April 1975, pp. 551–562.

[3] McKenna, G. B., "Interlaminar Effects in Fiber-Reinforced Plastics—A Review," *Polymer and Plastics Technology Engineering,* Vol. 5, No. 1, 1975, pp. 23–53.
[4] de Charentenary, F. X. and Benzeggagh, M., "Fracture Mechanics of Mode I Delamination in Composite Materials," *Proceedings of ICCM 3,* Vol. 1, Paris, France, Aug. 1980, pp. 186–197; *Advances in Composite Materials,* Pergamon Press, Oxford, 1980.
[5] Bascom, W. D., Bitner, J. L., Moulton, R. J., and Siebert, A. R., "The Interlaminar Fracture of Organic-Matrix, Woven Reinforcement Composites," *Composites,* Vol. 11, No. 1, Jan. 1980, pp. 9–18.
[6] Miller, A. G., Hertzberg, P. E., and Ranatal, V. W., "Toughness Testing of Composite Materials," in *Proceedings of the Twelfth National SAMPE Technical Conference,* SAMPE, Covina, CA, 1980, p. 279.
[7] Devitt, D. F., Schapery, R. A., and Bradley, W. L., "A Method for Determining the Mode I Delamination Fracture Toughness of Elastic and Viscoelastic Composite Materials," *Journal of Composite Materials,* Vol. 14, Oct. 1980, pp. 270–285.
[8] Wilkins, D. J., Eisenmann, J. R., Camin, R. A., Margolis, W. S., and Benson, R. A., "Characterizing Delamination Growth in Graphite-Epoxy," in Damage in Composite Materials, *ASTM STP 775,* K. L. Reifsnider, Ed., American Society for Testing and Materials, Philadelphia, 1981, pp. 168–183.
[9] Ashizawa, M., "Fast Interlaminar Fracture of a Compressively Loaded Composite Containing a Defect," in *Fifth DOD/NASA Conference on Fibrous Composites in Structural Design,* New Orleans, LA, Jan. 1981; NADC Report 81-096-60, Naval Air Development Center, Jan. 1981.
[10] Wilkins, D. J., "A Comparison of the Delamination and Environmental Resistance of a Graphite-Epoxy and a Graphite-Bismaleimide," *Report NAV-GD-0037,* Naval Air Systems Command, Aug. 1981.
[11] Williams, J. G. and Rhodes, M. D., "The Effects of Resin on the Impact Damage Tolerance of Graphite-Epoxy Laminates," *NASA Technical Memorandum 83213,* Oct. 1981.
[12] Hunston, D. L., Bascom, W. D., and Bitner, J. L., "Comparison Between the Fracture Behavior of Polymer Resins in Adhesive Bonds and Fibrous Composites," in *Proceedings of the Annual Meeting of the Adhesion Society,* Adhesion Society, AL, 1982.
[13] Whitney, J. M., Browning, C. E., and Hoogsteden, W., "A Double Cantilever Beam Test for Characterizing Mode I Delamination of Composite Materials," *Journal of Reinforced Plastic Composites,* Vol. 1, No. 4, 1982, pp. 297–313.
[14] O'Brien, T. K., Johnston, N. J., Morris, D. H., and Simonds, R. A., "A Simple Test for the Interlaminar Fracture Toughness of Composites," *SAMPE Journal,* July/Aug. 1982, pp. 8–16.
[15] Ashizawa, M., "Improved Damage Tolerance of Laminated Composites Through the Use of New Tough Resins," in *Sixth DOD/NASA Conference on Fibrous Composites in Structural Design,* New Orleans, LA, Jan. 1983, AMMRC MS83-2, Army Materials and Mechanics Research Center, Nov. 1983.
[16] Bradley, W. L. and Cohen, R. N., "Delamination and Transverse Fracture in Graphite/Epoxy Materials," in *Mechanical Behavior of Materials-IV; Proceedings of the 4th International Conference,* J. Carlsson and N. G. Ohlson, Eds., Pergamon Press, New York, 1984, pp. 595–601.
[17] Hunston, D. L. and Bascom, W. D., "Effects of Lay-up, Temperature, and Loading Rate in Double Cantilever Beam Tests of Interlaminar Crack Growth," *Composites Technology Review,* Vol. 5, No. 4, Winter 1983, pp. 118–119.
[18] Bascom, W. D., Bullman, G. W., Hunston, D. L., and Jensen, R. M., "The Width Tapered DCB for Interlaminar Fracture Testing," in *Proceedings of the 29th National SAMPE Symposium,* SAMPE, Covina, CA, April 1984.
[19] Hunston, D. L., "Composite Interlaminar Fracture: Effect of Matrix Fracture Energy," *Composites Technology Review,* Vol. 6, No. 4, Winter 1984, pp. 176–180.
[20] DiSalvo, G. D. M. and Lee, S. M., "Fracture Tough Composites—The Effects of Toughened Matrices on the Mechanical Performance of Carbon Fiber Reinforced Laminates," *SAMPE Quarterly,* Vol. 14, No. 1, Jan. 1983, pp. 14–17.
[21] Ting, R. Y. and Cottington, R. L., "Comparison of Laboratory Techniques for Evaluating the Fracture Toughness of Glassy Polymers, *Journal of Applied Polymer Science,* Vol. 25, No. 9, Sept. 1980, pp. 1815–1823.
[22] Ting, R. Y. and Cottington, R. L., "The Rate-Dependent Fracture Behavior of High Performance Sulfone Polymers," in *Proceedings of the 8th International Congress on Rheology,* G. Astarita, and G. Marruci, Eds., Plenum Press, New York, 1980.

[23] Hinkley, J. A., "Small Compact Tension Specimens for Polymer Toughness Screening," *Journal of Applied Polymer Science*, Vol. 31, 1986, in press.
[24] Bitner, J. L., Rushford, J. L., Ross, W. S., Hunston, D. L., and Riew, C. K., "Viscoelastic Fracture of Structural Adhesives," *Journal of Adhesion*, Vol. 13, No. 1, 1981, pp. 3–28.
[25] Bascom, W. D. and Cottington, R. L., "Effects of Temperature on the Adhesive Fracture Behavior of Elastomer-Epoxy Resins," *Journal of Adhesion*, Vol. 7, No. 4, 1976, pp. 333–346.
[26] Russell, A. J., "Factors Affecting the Opening Mode Delamination of Graphite/Epoxy Laminates," Materials Report 82-Q, Defence Research Establishment Pacific, Victoria, B.C., Canada, Dec. 1982.
[27] Petrie, S. P., Diebenedetto, A. T., and Miltz, J., "The Effects of Stress Cracking on the Fracture Toughness of Polycarbonate," *Polymer Engineering and Science*, Vol. 20, No. 6, April 1980, p. 385.
[28] Gerberich, W. W. and Martin, G. C., "Fracture Toughness Predictions in Polycarbonate," *Journal of Polymer Science, Polymer Physics*, Vol. 14, No. 5, May 1976, p. 897.
[29] Hunston, D. L. and Bullman, G. W., "Viscoelastic Fracture Behavior for Different Rubber-Modified Epoxy Adhesive Formulations," *International Journal of Adhesion and Adhesives*, Vol. 5, No. 2, April 1985, pp. 69–74.
[30] Bradley, W. L. and Cohen, R. N., "Matrix Deformation and Fracture in Graphite-Reinforced Epoxies," in *Delamination and Debonding of Materials, ASTM STP 876*, W. S. Johnson, Ed., American Society for Testing and Materials, Philadelphia, 1985, pp. 389–410.
[31] Kinloch, A. J. and Young, *Fracture Behavior of Polymers*, Applied Science, London, 1977.
[32] Chakachery, E. and Bradley, W. L., "Fracture of Hexcel F-185 Neat Resin and T6T145/F185 Composite," *Polymer Science and Engineering*, in press.
[33] Hong, Su-D., Chung, S. Y., Fedors, R. F., Moacanin, J., and Gupta, A., "Morphology and Dynamic Mechanical Properties of Diglycidyl Ether of Bisphenol-A Toughened with Carboxyl-Terminated Butadiene-Acrylonitrile," NASA CP 2334, 1983, p. 285.

DISCUSSION

G. M. Newaz[1] (written discussion)—In correlating matrix G_{Ic} with composite G_{Ic}, are you ensuring that G_{Ic} of the matrix is evaluated under plane-strain condition? My concern is whether you are using a property that is specimen geometry dependent or a matrix material property (plane-strain G_{Ic}). In the former case, there may even be discrepancy in results among researchers based on the choice of test method and selected specimen geometry. Should we not attempt to correlate matrix plane-strain toughness unless thickness of the specimen required for valid G_{Ic} is a concern? What is your opinion?

D. L. Hunston, R. J. Moulton, N. J. Johnston, and W. D. Bascom (authors' response)—This is an important point since the comparisons that are the major objective in this work can be meaningful only if the nature of the measurement does not vary from one material to another. Consequently, we make every effort to assure that the specimen dimensions in the resin tests produce conditions where plane strain dominates. Our experience over the years has been that the ASTM guidelines in E 399 represent an adequate (although perhaps conservative) criteria. This is based on tests with a number of different tough materials conducted

[1]Owens-Corning Fiberglas, Technical Center, Granville, OH 43056.

as a function of specimen width. For the materials tested here, the one place where we do have some concern about the application of simple fracture testing methods is the F-185 resin. This is particularly worth noting because F-185 is the toughest material in the comparison, and consequently, it plays an important role in estimating the trends. As a result, the manuscript does try to place the uncertainties with regard to F-185 in some prospective.

D.L. Hunston, R.J. Moulton, N.J. Johnston, and W.D. Bascom (authors' closure)—The task of defining the relationship between the fracture behavior of neat resins and that of their composites is complex but very important. We hope that this paper provides a useful first step in addressing that question. It should be clear, however, that much more remains to be done both theoretically and experimentally. Consequently, an important objective of the paper is to encourage additional studies to examine and refine the many uncertainties that remain.

William M. Jordan[1] *and Walter L. Bradley*[1]

Micromechanisms of Fracture in Toughened Graphite-Epoxy Laminates

REFERENCE: Jordan, W. M. and Bradley, W. L., "**Micromechanisms of Fracture in Toughened Graphite-Epoxy Laminates,**" *Toughened Composites, ASTM STP 937,* Norman J. Johnston, Ed., American Society for Testing and Materials, Philadelphia, 1987, pp. 95–114.

ABSTRACT: The combination of low cross-link density and elastomer additions has been seen to give the most potent toughening for neat resins. A relatively small increment of the additional neat resin fracture toughening above 800 J/m^2 is actually reflected in the delamination fracture toughness of a composite. Mode II delamination toughness of brittle systems may be as much as three times the Mode I delamination fracture toughness, while a ductile system may have a Mode II delamination fracture toughness that is similar to the Mode I value. The energy absorbed per unit area of crack extension for delamination seems to be independent of ply orientation if proper accounting of the near and far field energy dissipation is made.

KEY WORDS: composite materials, delamination, Mode I, Mode II, mixed mode, toughened epoxy

Nomenclature

- B Specimen
- E Elastic modulus in the fiber direction
- G_{Ic} Critical energy release rate for stable crack growth for Mode I loading
- G_{IIc} Critical energy release rate for stable crack growth for Mode II loading
- $G_{Tot,c}$ Total critical energy release rate for mixed mode loading
- I Moment of inertia for cracked portion of split laminate
- L_c Crack length in split laminate
- P_s Asymmetric load component $(P_u + P_L)/2$ (see Fig. 1)
- P_a Symmetric load component $(P_u - P_L)/2$ (see Fig. 1)
- P_u Load applied to upper half of split laminate (see Fig. 1)

This work performed at Texas A&M University was supported by the Air Force Office of Scientific Research with Major David Glasgow as project monitor.

[1]Research assistant and professor of mechanical engineering, Mechanical Engineering Department, Texas A&M University, College Station, TX 77843.

P_L Load applied to lower half of split laminate (see Fig. 1)
Δ_s Total opening displacement (see Fig. 1)

Introduction

Graphite/epoxy composite materials are very attractive for a number of aerospace applications because of their high strength and stiffness to weight ratios. Furthermore, the ability to tailor the stiffness and thermal expansion coefficients of a laminate by using an appropriate layup for the composite laminate is a very attractive feature. The first generation of graphite/epoxy composites were developed to maximize stiffness and glass transition temperature T_g by utilizing a high cross-link density. However, these systems were found to delaminate rather easily when out-of-plane stresses were applied. Subsequently, attempts have been made to improve the composite toughness by improving the toughness of the resin systems. This has had somewhat disappointing results in that a large increase in resin toughness has not been found to give a proportionate increase in composite toughness. Scott and Phillips [*1*] found that a tenfold increase in resin toughness increased composite toughness by a factor of two. Similar results on different systems were found by Bascom et al. [*2,3*], Vanderkley [*4*], and Bradley and Cohen [*5,6*].

In this paper, we will consider the question of whether some types of resin toughening mechanisms are more effective than others in enhancing delamination toughness. In particular, toughening by reductions in cross-link density and elastomer additions will be considered. The efficacy of the toughening mechanisms for Mode I, Mode II, and mixed mode loading will be considered.

Experimental Procedures and Analysis

Materials

Four graphite epoxy composite systems have been studied in this research effort. Hercules AS4/3502 composite which utilizes a highly cross-linked, and therefore, relatively brittle, resin system was chosen to be compared with three tougher systems, namely, Hexcel T6T145/F155, T6T145/F185, and T6T145/HX205. The Hexcel F155 resin has a lower cross-link density (280 atomic mass units between cross-links[2]) than the Hercules 3502 resin (AMU between cross-links not available). The Hexcel F185 and HX205 resins have even lower cross-link densities (430 AMU between cross-links[2]). Thus, the three Hexcel resins differ from the Hercules 3502 resin in that they have lower cross-link densities with the associated lower T_g values and higher resin ductility.

A second difference between the Hexcel F155 and the Hexcel F185 resins and the Hercules 3502 resin is that the two Hexcel resins have elastomeric material added to enhance their respective toughnesses. Approximately 6% carboxy-

[2]R. Moulton, Hexcel Corporation, Dublin, CA, personal communication.

terminated butadiene acrylonitride (CTBN) is included in the F155 resin. This CTBN rubber precipitates as a second phase with a diameter which varies from 0.1 to 1.0 μm [2,3]. The F185 resin includes both a 6% addition of the CTBN rubber which precipitates and approximately 8% of prereacted rubber which has been mechanically blended into the resin [2,3]. These prereacted rubber particles have a bimodal distribution of diameters with peaks at 2 and 8 μm. The Hexcel HX205 has had no elastomer additions. Whatever toughness it manifests is a result of its relatively low cross-link density, which is the same as the Hexcel F185 resin, as previously noted.

Unidirectional, 24-ply-thick panels of the AS4/3502 and the T6T145/F155 were laid up using prepreg from the respective manufacturers and then cured in an autoclave/press at Texas A&M University. Additional panels of T6T145/F185 containing some plus and minus 45° plies were prepared to study the effect of ply orientation across the delaminating plane on the delamination fracture toughness. Unidirectional, 24-ply-thick panels of T6T145/F185 and T6T145/HX205 were prepared at NASA Langley for this study.

All panels contained a 0.025-mm-thick strip of Teflon® laid to a depth of 3 cm from one edge of the panel between the center two plies to provide a crack starter. Since the Teflon strip introduces a relatively blunt notched "crack," a natural crack extension of at least 3 cm was made before the measurement of any critical energy release rates. Thus, the G_c values reported in this study are for crack growth rather than crack initiation.

Split laminate specimens 2.5 cm wide by 30 cm in length were cut from the composite panels (fibers running the length of the specimens for the unidirectional panels) for macroscopic testing, while much smaller specimens (3 cm long by 0.6 cm wide) were cut to be fractured in the scanning electron microscope.

Castings of neat resin were provided by Hexcel for the F155, F185, and HX205 systems. Rectangular specimens 1.27 cm wide by 0.318 cm thick by 3.56 cm long were machined from the neat resin castings to be tested on a dynamic mechanical spectrometer manufactured by Rheometrics, Inc. Standard tension specimens were also machined from the neat resin castings, as were compact tension specimens for fracture toughness testing. Standard compact tension specimens were prepared according to the ASTM Test for Plane-Strain Fracture Toughness of Metallic Materials (E 399) with $W = 5.08$ cm, but with a thickness of approximately 0.5 cm, which was the thickness of the castings.

Mechanical Properties Testing

Delamination Fracture Toughness Tests—Delamination fracture toughness tests were run at ambient temperature (24°C) in opening mode (Mode I), shear mode (Mode II), and mixed mode conditions on split laminate test specimens using an MTS materials testing system operated in stroke control at 0.0085 cm/s. Load and displacement were continuously measured, while the crack length was measured visually at the surface at discrete intervals (about once every 1 cm of crack extension). Unload compliance measurements were made at intervals of about 2 to 3 cm.

Mixed mode and Mode II tests were also performed using split laminate specimens. This was accomplished by asymmetrically loading the cracked end of the split laminate while restricting vertical displacement of the uncracked end, which was still free to translate horizontally, as seen in Fig. 1. This procedure was first developed by Vanderkley [*4*] at Texas A&M University. The mixed mode loading can be analyzed by the superposition principle and noting that an asymmetrically loaded split laminate can be treated as the sum of pure bending and pure Mode I loading, with the pure bending giving essentially a pure Mode II state of stress (see Fig. 1). When the specimen is loaded as shown in Fig. 1 with the uncracked end and the upper arm of the split laminate held stationary and the lower arm of the split laminate displaced (through actuater displacement), the percentage of Mode II loading will vary continuously throughout the test from pure Mode I to a Mode II energy release rate that is approximately 40% of the total energy release rate.

Two special cases of the loading configuration seen in Fig. 1 may be noted. First, if the upper arm of the cracked end of the split laminate is unconstrainted ($P_u = 0$), then the specimen will experience a constant fraction of Mode II

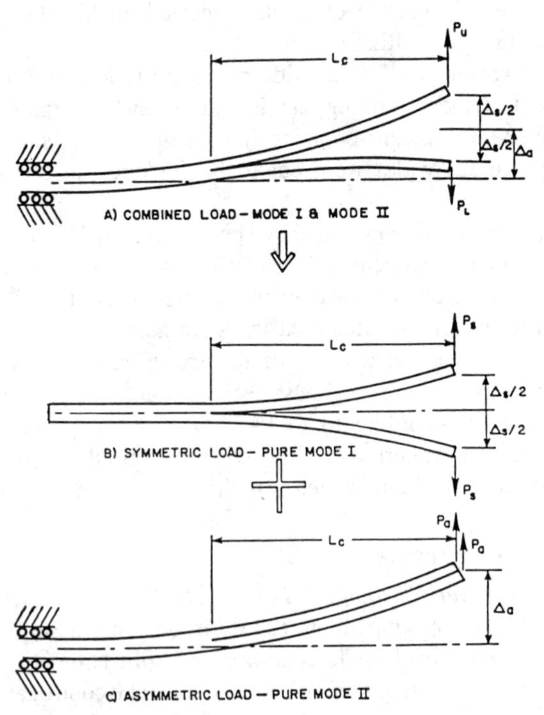

FIG. 1—*Schematic showing how asymmetric loading of split laminate can introduce a mixed Mode I/Mode II state of stress at the crack tip.*

loading throughout the test, which gives a Mode II energy release rate that is 43% of the total energy release rate. Second, if both upper and lower arms of the split laminate are pulled down with equal force using the actuater ($P_u = -P_L$), then a pure Mode II loading condition results. This loading arrangement is, in effect, one half of a three-point bend test, and thus, is very similar to the end notch flexural test developed by Russell and Street [7].

For the Mode II tests, a 0.79-mm Teflon spacer was placed between the crack faces of the split laminate specimen to minimize any frictional effects which might occur by the rubbing of the two surfaces together. This spacer would constitute a superimposed Mode I loading, but because the spacer was very thin, the Mode I energy release rate it produces is trivially small. A scanning electron microscopy (SEM) examination of the fractured surfaces after Mode II testing gave no indication of rubbing between the two fractured surfaces.

Neat Resin Fracture Toughness Tests—Fracture toughness tests were run on several of the neat resins where G_{Ic} values were either unavailable or were regarded with some suspicion (literature values on F185 were calculated assuming linear elastic fracture mechanics, which is questionable for such a ductile system). The procedures specified in either ASTM E 399 or ASTM Test for J_{IC}, a Measure of Fracture Toughness (E 813) were followed, with a single-specimen, multiple compliance approach used for ASTM E 813 on the Hexcel F185 system. Fatigue precracking was used to introduce cracks in the compact tension specimens.

Rheometrics Tests—Rheometric tests were run on neat resin specimens. These specimens were subjected to dynamics torsional cycling over a wide range of temperatures which bracketed the glass transition temperature T_g. The loss and storage modulus were recorded as a function of temperature. The glass transition temperature was assumed to correspond to the temperature where the loss modulus value was a maximum.

Tension Tests—Tension tests were conducted on the neat resin systems according to ASTM Test for Tensile Properties of Plastics (D 638). Strain was measured using strain gages which were mounted on the front and back face of each specimen and were capable of measuring strain of up to 10%. Complete stress-strain behavior was determined for each specimen from which tensile strength, yield strength, and elongation were noted.

Analysis of Delamination Fracture Toughness Tests

Mode I Critical Energy Release Rate Analysis G_{Ic} *For Delamination Fracture*—Generally, the term critical energy release rate is used to refer to initial crack extension from some preexisting flaw or fatigue precrack. In this work it is used to refer to the work required per unit area of new crack surface created for stable crack propagation, rather than for initiation. The Mode I critical energy release rate was calculated using linear beam theory as described by Vanderkley [4] where

$$G_{Ic} = (P_s L_c)^2 / BEI \quad (1)$$

where P_s is the symmetric loading ($[P_u + P_L]/2$, see Fig. 1), L_c is the crack length, B is the specimen width, and EI is the flexural stiffness of the arms of the cracked portion of the specimen.

The flexural stiffness was calculated from measured values of load P_s, displacement Δ_s, and crack length L_c using a relationship also derived from linear beam theory, namely,

$$EI = 2P_s L_c^3 / 3\Delta_s \quad (2)$$

A value for EI was calculated at each of the approximately twenty points per specimen where crack length was measured. An average value of EI for the specimen was then determined and used in Eq 1 to calculate G_{Ic} values, also at approximately twenty points for each specimen. Three split laminate specimens of each composite were tested in Mode I with the reported G_{Ic} values being the average of approximately sixty calculated values.

The results from all Mode I tests were also analyzed using the unload compliance measurements analyzed using the method suggested by Wilkins et al. [*8*], with the calculated values of G_{Ic} typically falling within 3% of the values calculated using linear beam theory (that is, Eqs 1 and 2). The only Mode I tests for which it was not possible to use the linear beam theory analysis (or the Wilkins analysis) were the T6T145/F155 specimens containing all plus and minus 45° plies. Because of the lower compliance, a significant degree of geometric nonlinearity was noted in the load-displacement records for these specimens. Furthermore, some far field damage in the split laminate arms was evident in view of the fact that the load-displacement record did not return to the origin on unloading. As a result, the nonlinear beam theory analysis first developed by Devitt et al. [*9*] was used to analyze these results. However, this approach is only good for nonlinear elastic behavior. A J integral analysis is currently being developed to use in analyzing these results which include not only geometric nonlinearities but also nonlinear viscoelastic behavior of the resin.

Mixed Mode and Mode II Critical Energy Release Rate Analysis G_{IIc} *For Delamination Fracture*—For Mode II delamination fracture toughness analysis, the following relationship for Mode II critical energy release rate was derived by Vanderkley [*4*], again assuming linear beam theory (see Fig. 1):

$$G_{IIc} = 3(P_a L_c)^2 / 4BEI \quad (3)$$

where P_a is the asymmetric load $(P_u - P_L)/2$ and the other terms are as previously defined.

For all pure Mode II tests, the load-displacement records were found to be significantly nonlinear. Calculation of G_{IIc} values using Eq 3 based on linear beam theory gave artificially high values of G_{IIc} in comparison to the results of Russell and Street [*7*] on similar material. A critical energy release rate calculation for split laminate specimens based on nonlinear elastic beam theory has previously been published for Mode I loading [*9*], but not for Mode II loading.

Thus, we chose to use the area method approximation, whereby the area under a load-displacement record for loading, crack advance from L_1 to L_2, and unloading is determined and assumed to be equal to the work required to increase the crack area by $B[L_2 - L_1]$, where B is the specimen width. This method assumes all of the work represented by the area bounded by the load, crack advance, unload-displacement curve goes into crack advance. Any far field energy dissipation would erroneously be lumped into the energy absorbed per unit area of crack extension relationship. Thus, the calculated values of G_{IIc} using this approach may be considered upper bound estimates. Nevertheless, our results for G_{IIc} using the area method were in reasonable agreement with Russell and Street's results [7] (which implies far field damage is minimal for the unidirectional, split laminate specimens studied in this work) and were much smaller than the values one would obtain by naively using linear beam theory on the obviously nonlinear load-displacement records obtained on the Mode II specimens.

Mixed Mode Critical Energy Release Rate Analysis $G_{Tot,c}$ *For Delamination Fracture*—Most of the load-displacement records for the mixed mode delamination tests were quite linear so that G_{Ic} and G_{IIc} could be calculated using Eqs 1 and 3 with $G_{Tot,c}$ calculated as the sum of G_{Ic} and G_{IIc}. Where this was not the case, no analysis was attempted since neither superposition nor linear beam theory would be appropriate.

Real-Time Observations of Delamination Fracture In Scanning Electron Microscope

Real-time observation of fracture in the scanning electron microscope (SEM) is a relatively new technology. Theocaris and Stassinakis, [10] have fractured composite specimens loaded in tension in the SEM while Beaumont [11] has fractured composite and polymeric systems using torsional loading in the SEM. In this study, small split laminate specimens were fractured in a JEOL 35 scanning electron microscope specially equipped with a loading stage. Delamination is obtained by pushing a delaminating specimen over a blunt stationary wedge. A blunt wedge was used so that crack growth would result from pushing apart the two crack surfaces in essentially Mode I loading, without any direct pressure applied at the crack tip by the wedge. The wedge tip remained well away from the crack tip so that Mode I conditions would dominate at the crack tip. The fracture process was recorded on video tape and standard sheet film. A post-mortem fractographic examination was made of the fractured macroscopic split laminate specimens and the miniature split laminate specimens fractured.

Experimental Results

Results from Mechanical Properties Tests

The results of the various mechanical properties tests are summarized in Tables 1 to 4. The glass transition temperature results are presented in Table I,

TABLE 1—*Glass transition temperature of resins.*

Resin System	Glass Transition Temperature, °C
3502	191[a]
F155	118
HX205	118
F185	109

[a] Obtained from literature from Hercules, Inc., Magna, UT 84044.

TABLE 2—*Resin tensile properties.*

Resin System	Yield Strength, MPA	Tensile Strength, MPA	Elongation, %
3502	...	...	...
F155	58	69	3.10
F185	38	46	8.87
HX205	61	73	3.24

TABLE 3—*Critical energy release rates* G_c, J/m^2.

Material	AS4/3502	T6T145/F155	T6T145/HX205	T6T145/F185
Mode I (neat resin)	70 (13)	730 (14)	460	8100
Mode I (composite)	189	520	455	2205
20% Mode II	264	525	796	...
43% Mode II	...	548	789	...
Mode II	570	1270	1050	2440

TABLE 4—*Effect of fiber orientation upon Mode I* G_{Ic}.

Material	Layup	G_{Ic}
F155	0 (24)	520
F155	+45/−45/0 (8)/−45/+45/ −45/+45/0 (8)/+45/−45/	600
F155	+45 (2)/−45 (4)/+45 (2)/ −45 (2)/+45 (4)/−45 (2)/	1400[a]

[a] Note this is not a true G_{Ic} value because of inadvertent inclusion of far field damage, as explained in text.

where the 3502 is seen to have the highest T_g, F185 has the lowest, and HX205 and F155 have an identical value of T_g. The three more ductile systems, HX205, F185, and F155, all have T_g values well below the relatively brittle 3502. The tension tests results are summarized in Table 2. No as-cast 3502 resin was obtained. Thus, a tension test was not run on this resin. The F155 and the HX205 have nearly identical tensile properties, which is not surprising since they have the same T_g. The F185 resin is much more soft and ductile than the F155 or the HX205, as expected. Though tensile properties were not obtained for the 3502 resin, it would be expected to be somewhat stronger with a smaller elongation to failure than the other three resins.

The neat resin fracture toughness values G_{Ic} are summarized in Table 3 and correlate nicely with the T_g results and the tension test results presented in Tables 1 and 2. The ratio of composite delamination toughness G_{Ic} to neat resin fracture toughness G_{Ic} is seen to decrease as the resin toughness increases, as previously noted by Bradley and Cohen [6] and Hunston [12]. A second important trend to note from Table 3 is that the Mode II delamination toughness is always higher than the Mode I delamination toughness, with a ratio of Mode II delamination toughness to Mode I delamination toughness of approximately $\times 3$ for the brittle system decreasing to a value just barely larger than $\times 1.0$ for the most ductile system.

The effect of fiber orientation across the plane of delamination on delamination fracture toughness G_{Ic} is seen in Table 4. For a specimen with plus or minus 45° plies across the interface that is debonding, but with a stiffness similar to a unidirectional composite, the delamination critical energy release rate was found to be very similar to the results for unidirectional laminates (600 compared to 520 J/m^2). Chai [13] has previously noted a similar result; namely, that ply orientation across the plane of delamination does not significantly affect the delamination fracture toughness for laminates with similar stiffness. The much larger value for G_{Ic} indicated for the second multiaxial layup is a result of far field damage as a result of nonlinear viscoelastic behavior by the resin. As will be seen presently, the resin carries a significant load in the axial direction for a composite laminate with all ±45° plies. This significant resin loading not only in the crack tip region but at other locations removed from the crack tip will cause the resin to undergo nonlinear viscoelastic deformation. This energy dissipation in the far field should not be counted in the crack tip energy dissipation per unit area of crack extension. However, our G_I calculation using the area method does not distinguish between crack tip and far field energy dissipation. Thus, the G_{Ic} value of 1400 J/m^2 indicated in Table 4 is not really a G_{Ic}, but includes G_{Ic} plus far field damage.

Results From Real-Time Observations of Delamination Fracture in Scanning Electron Microscope SEM

The observations of fracture in the SEM had as its purpose the determination of the details of the delamination fracture process for the various systems to gain a better understanding of how resin toughness can be translated into delamination fracture toughness. The damage zone size around the crack tip as well as the critical event (that is, resin fracture, interfacial debond, and so forth) can be determined from such observations.

In situ fractography for the four composite systems are shown in Figs. 2 through 5. The damage zone as evidenced by fine microcracking is seen to be

FIG. 2—*In situ delamination of AS4/3502 showing debonding with very little resin damage (*top, *at ×1000).*

very small for the AS4/3502 and the critical fracture event is usually interfacial debonding (see Fig. 2). In Fig. 2, the cracking appears to be through the resin, but is in fact along a fiber just beneath the surface, as indicated by the charging (light-colored region) adjacent to the fracture plane. Crack advance in the AS4/3502 was usually discontinuous. As the wedge was inserted further into the specimen, crack advance would not occur continuously but in a burst of interfacial debonding which resulted in significant crack advance. This unstable mode of crack growth suggests that the interfacial bonding in this system is quite heterogeneous. Crack arrest probably occurs at some locally better bonded region. The energy release rate must be increased to initiate crack growth. However, once the locally better interfacial bonding is overcome, crack advance occurs until the energy release rate decreases (which it does for crack extension under displacement controlled loading) to a value lower than the local resistance G_{Ic}, which is probably at the next region of better interfacial bonding.

A much more extensive damage zone is seen around the crack tip for T6T145/F155 (see Fig. 3). The microcracked zone is much larger in extent than for the AS4/3502 and has a much greater density of microcracks. In spite of this fact, a significant amount of crack extension occurred by interfacial failure. It did not appear that the interfacial failure in the T6T145/F155 was always a result of debonding. Rather, it sometimes appeared that the microcrack density was greater adjacent to fibers, and therefore, made coalescence more likely in that region. Thus, it was rather an exception to see a bare fiber in the postmortem fractography of the T6T145/F155 while bare fibers were quite common in the AS4/3502.

The in situ fractography for the T6T145/HX205 is presented in Fig. 4 where the damage zone is again seen to be characterized by significant microcracking. The extent of the damage zone is somewhat greater than for the T6T145/F155, but the density of microcracking is much less. Failure seemed to proceed in the

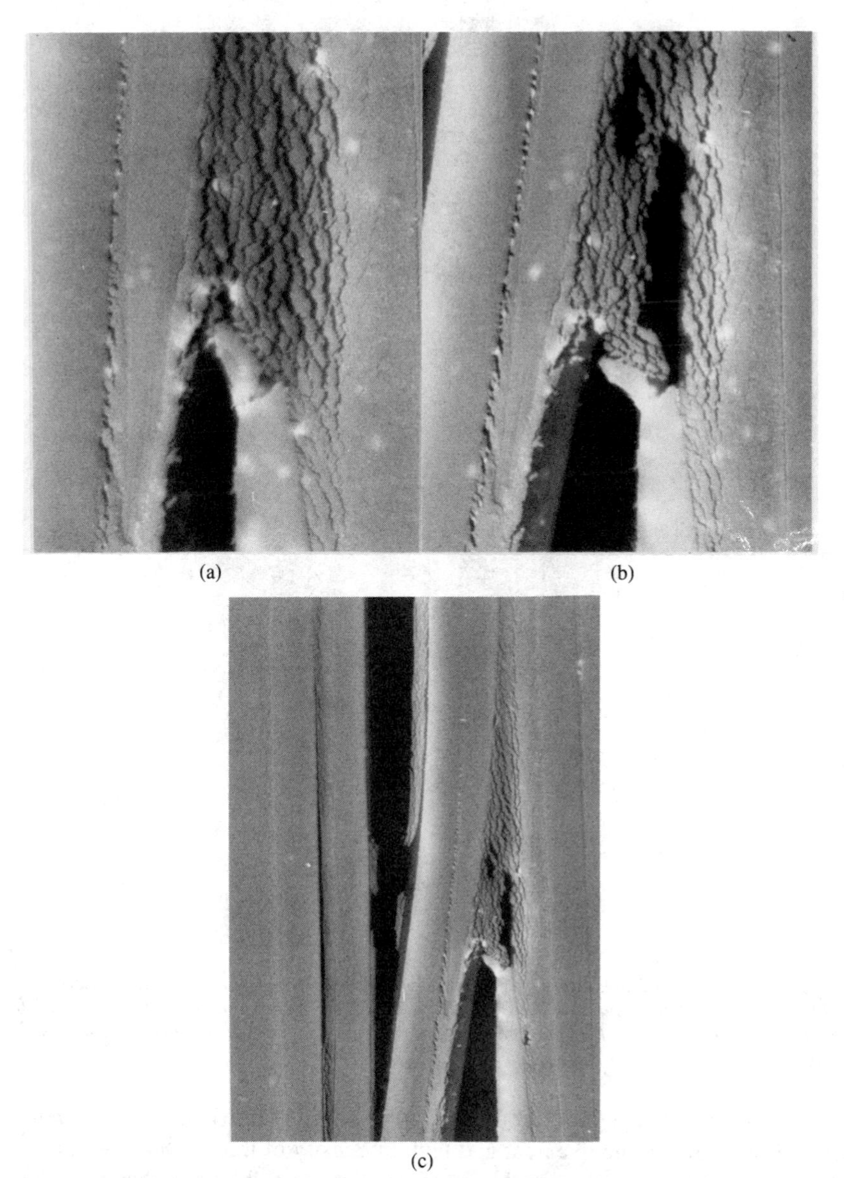

(a) (b)

(c)

FIG 3—*(a) In situ delamination of T60045/F155 showing extensive microcracking around the crack tip (left, at ×3900). (b) In situ delamination of T6T145/F155 showing coalescing of microcracks to form macroscopic crack growth (right, at ×3000). (c) In situ delamination of T6T145/F155 system. Debonding as well as microcracking is visible (×1000).*

interfacial area between resin and fiber, not by debonding but by coalescence of microcracks which seemed to exist in greater number in the interfacial region in the T6T145/HX205.

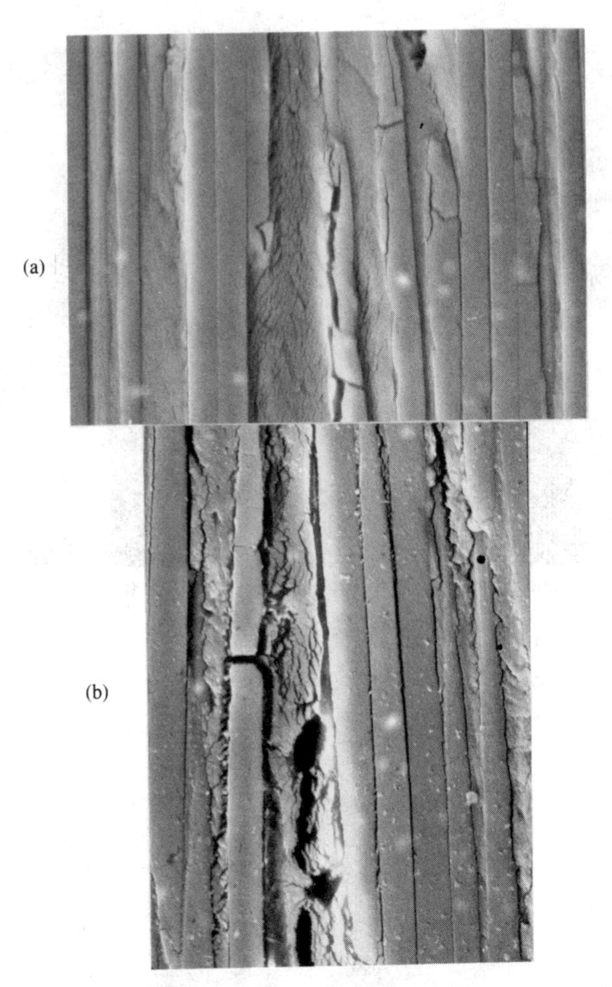

FIG. 4—*(a) In situ delamination of HX205 system. Main crack grows is preceeded by developing of large microcrack zone ahead of the crack tip (*top, *at ×800). (b) In situ delamination of HX205 system. Extensive microcracking that coalesces into macroscopic crack growth (*bottom, *at ×1000).*

Finally, the damage zone around the crack tip in the T6T145/F185 is seen in Fig. 5 to be both large in extent and high in density of microcracks. Again, the failure is still frequently near the interface, but a result of a higher density of microcracks in this region rather than debonding. Very few regions of bare fiber are noted in the postmortem fractography.

The in situ fracture behavior of the T6T145/F155 composite with off-angle plies is seen in Fig. 6. Because the fibers are at an angle of 45° to the length of the specimen, Mode I loading of the split laminate gives significant loading of the resin along the length of the specimen as well as in the Mode I opening direction. Thus, while microcracking similar to that observed in the specimens made from

(a) (b)

FIG. 5—*(a) In situ delamination of T6T145/F185 system. Large microcrack zone ahead of crack tip, with a significant extent above and below plane of crack (*left, *at ×1000). (b) In situ delamination of T6T145/F185 system. Resin tearing as well as microcracking is evident (*right, *at ×1000).*

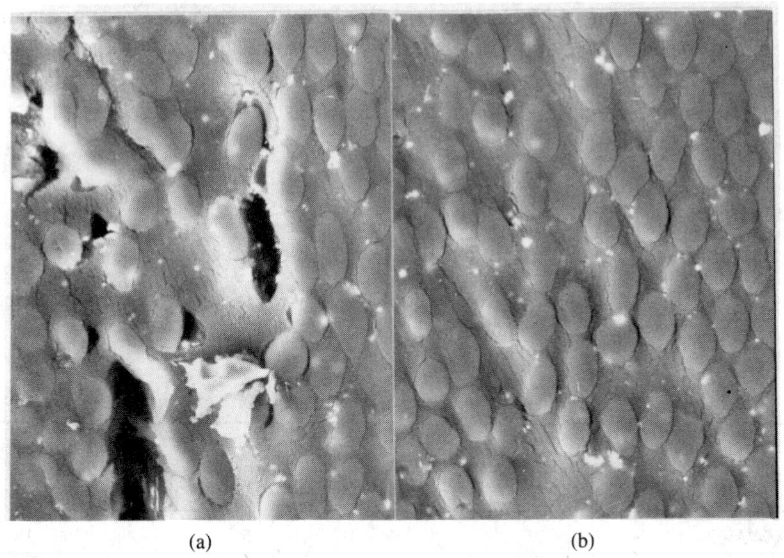

(a) (b)

FIG. 6—*(a) In situ delamination of T6T145/F155 system with fibers at ±45. (*left, *at ×1000). (b) In situ delamination of F155 system with fibers at ±45. Note microcracks that point back to macroscopic crack tip (*right, *at ×1000).*

a unidirectional laminate is still evident, the orientation is different and voiding at the resin/fiber interface is noted. In fact, coalescence of these voids seems to play a significant role in the delamination fracture process in a split laminate with a large number of off-axis plies.

Typical damage zone sizes around the crack tip of the various composites tested have been quantified and are summarized in Table 5.

Results of Postmortem Fractographic Examination of Fractured Surface of Delaminated Specimens

The postmortem fractographic examination included specimens fractured for Mode I, Mode II, and mixed mode loading conditions. The results for the AS4/3502 consistently indicated bare fibers (except possible sizing) and interfacial debonding, whereas the three ductile systems gave only occasional indication of fiber debonding. Postmortem fractographic results for T6T145/F155 and T6T145/HX205 are seen in Figs. 7 and 8. The duplex appearance to the fracture surface in the T6T145/F155 appears to be the result of a variable thickness of the resin rich region between plies. Only a few bare fibers are noted.

Figures 9 through 11 present highlights of the results from the postmortem fractographic examination of the specimens fractured in Mode II and mixed mode. At the time this work was done, our stage was not yet adapted for Mode II testing in the SEM. Thus, the postmortem fractographic results on the Mode II and the mixed mode delamination fractures constitute all of the fractographic information obtained for these loading conditions. The distinctive features on the fractured surface of the three more ductile composite systems loaded with a significant percentage of Mode II loading are leaf-like artifacts whose orientation relative to the delamination plane increase monotonically with increasing percentage of Mode II loading. The more brittle AS4/3502 had a very regular array of what appear to be sigmodial shaped microcracks which have coalesced to give macrocrack advance.

Discussion

The significant results to be discussed in this section are as follows:

(1) the efficacy of rubber particle additions and lower cross-link density in enhancing neat resin fracture toughness;

TABLE 5—*Damage zone size and corresponding delamination fracture toughness.*

Material	Size of Damaged Zone, Mm		G_{Ic} J/m^2
	Ahead of Crack Tip	Above or Below Delamination Plane	
AS4/3502	20	5	189
F155	20	10	520
F155 ± 45	50	10	...
HX205	75	35	455
F185	200	35	2205

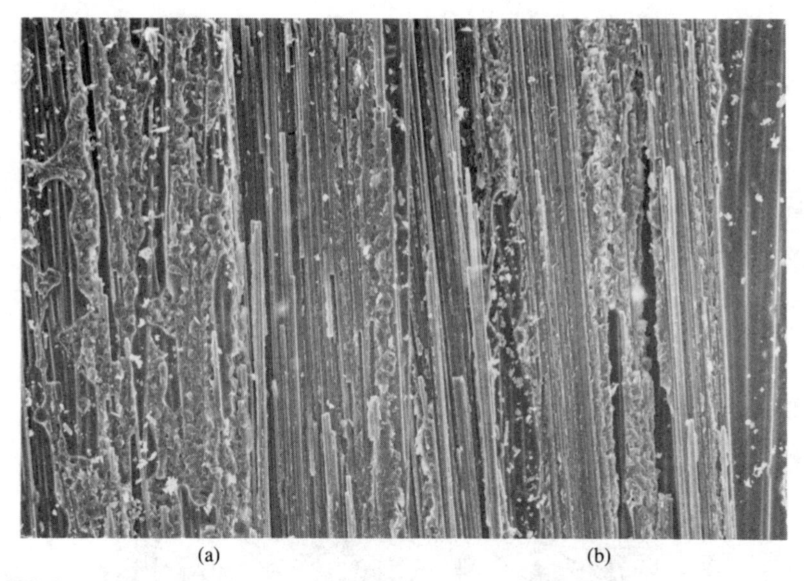

FIG. 7—*(a) Postmortem fractography of F155 composite delaminated in Mode I conditions (near center of specimen) (*upper left, *at ×100). (b) Postmortem fractography of F155 composite delaminated in Mode I conditions (near edge of sample) (*upper right, *at ×100).*

(2) the efficacy of rubber particle additions and lower cross-link density in enhancing composite delamination fracture toughness;
(3) the change in fracture toughness with increasing fraction of Mode II shear loading; and
(4) the effect of ply orientation and laminate stiffness on delamination fracture toughness.

These macroscopic fracture toughness results will be discussed in light of the in situ and postmortem fractographic observations.

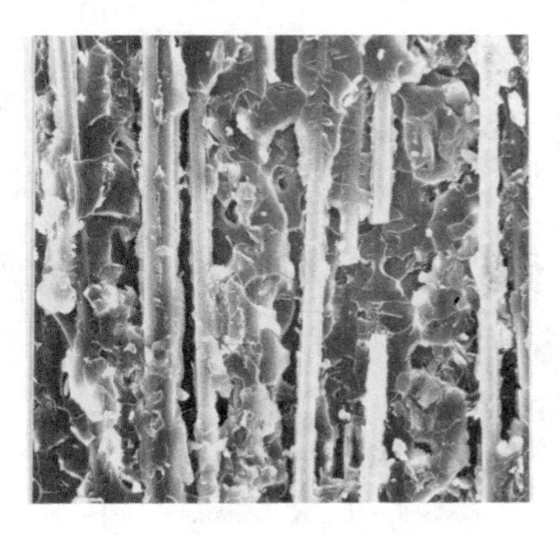

FIG. 8—*Postmortem fractography of HX205 composite delaminated in Mode I conditions (*lower center, *at ×300).*

(a)

(b)

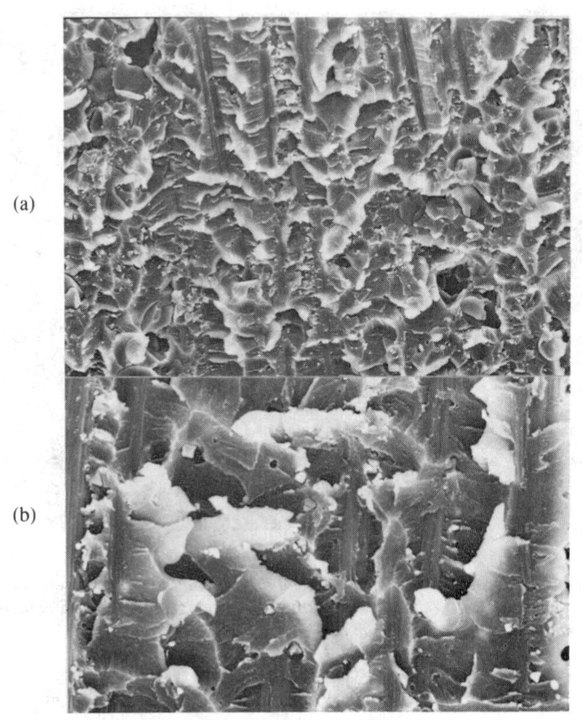

FIG. 9—*(a) Postmortem fractography of F155 composite delaminated in 43% Mode II conditions (top surface of fractured specimen)* (top, *at* ×*450). (b) Postmortem fractography of F155 composite delaminated in 43% Mode II conditions (bottom surface of fractured specimen). Note leaf-like artifacts are oriented in opposite direction to those on the top surface* (center, *at* ×*1500).*

FIG. 10—*Postmortem fractography of HX205 composite delaminated in 43% Mode II conditions* (bottom, ×*300).*

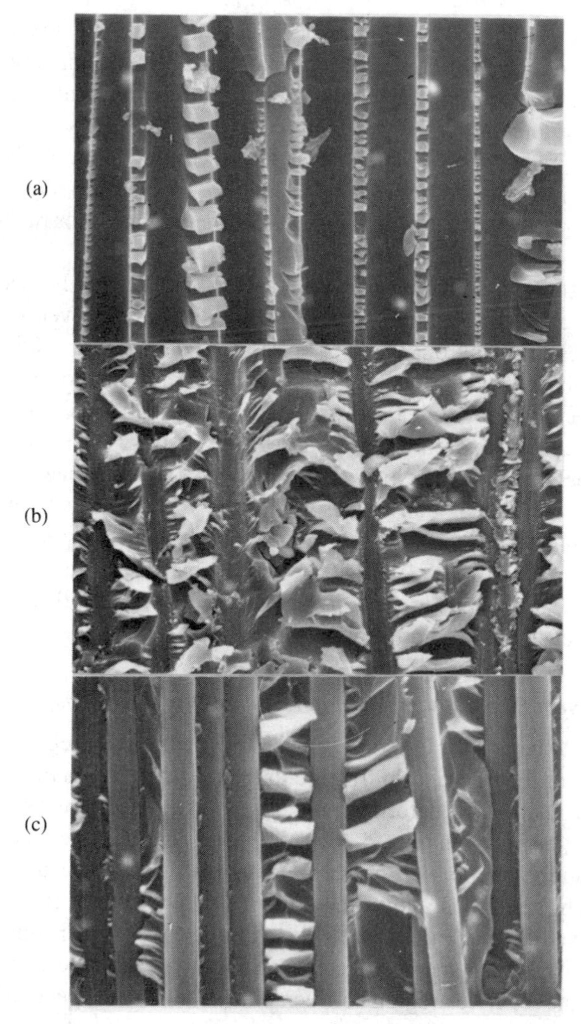

FIG. 11—*(a) Postmortem fractography of AS4/3502 composite delaminated in Mode II conditions. Zipper-like artifacts appear to be formed by coalescence of series of sigmodial shaped microcracks (*top, ×*1500). (b) Postmortem fractography of F155 composite delaminated in Mode II conditions. Note extensive resin deformation, (*center, at ×*1000). (c) Postmortem fractography of HX205 composite delaminated in Mode II conditions. Note extensive resin deformation (*bottom, at ×*1000).*

Neat Resin Fracture Toughness

The neat resin toughness was found to correlate with T_g and with elongation in a tension test for the four systems studied. Note that the F155 and the HX205 had quite similar tensile properties and T_g, though they were toughened somewhat differently, that is, HX205 relied entirely on decreasing the cross-link density while the F155 used a combination of reduction in cross-link density (compared to an epoxy like Hercules 3502) and rubber particle additions. The F155 had a better fracture toughness for both neat resin and composite delamination than did the HX205. Materials with similar tensile elongations may have significantly different fracture toughnesses since fracture toughness depends on the ductility under a triaxial rather than a uniaxial states of stress. Since rubber particle additions can relieve this triaxial state of tension by voiding, the same elongation in a tension test may correlate with a higher fracture toughness for rubber particle toughened resins.

The beneficial effect of rubber particle additions is particularly evident in the F185, whose neat resin toughness is ×18 that of the HX205 even though their tensile ductilities differ by only a factor of ×3. The elastomer additions while giving some enhancement to tensile elongation give a dramatic increase to the fracture toughness. Yee and Pearson [*14*] have noted that a resin needs to have some intrinsic capacity to deform in response to shear stress if it is to be benefited by rubber particle additions, which principally increase the shear stress in the crack tip region by relaxing constraint. Again, the very large fracture toughness of the F185 compared to the other systems suggests a strong synergistic effect between lowering the cross-link density, which increases the freedom to deform in response to shear stress and rubber particle additions, which allow larger shear stresses to be developed at the tip of a crack.

Translation of Neat Resin Toughness into Delamination Toughness

There appear to be at least two reasons why tougher resin systems give a much smaller fraction of their neat resin fracture toughness in delamination toughness. First, a tougher resin has a deformation zone that is much more extensive than the resin rich region between plies. In the composite, the fibers act like rigid filler in this deformation zone, which reduces the degree of load redistribution away from the crack tip, and thus allows the critical strain or critical stress condition for local failure at the crack tip to be achieved more easily. For resins where the deformation/damage zone ahead of the crack tip is on a scale of less than or equal to the height of the resin rich region between plies, one might expect the neat resin toughness and the delamination toughness to be similar if interfacial failures do not dominate the delamination fracture behavior. Only in the case of the F185 is there extensive damage outside of the resin rich region between plies, and only in this system is there a dramatic difference in the neat resin fracture toughness and the composite delamination fracture toughness.

A second way the fibers may prevent delamination fracture toughness from achieving neat resin fracture toughness is that they allow heterogeneous nucleation sites for fracture. For Mode I loading, all four of the composite systems studied had failure primarily in the interfacial region. Even where fiber bonding is good, the interfacial region seems to have a greater density of microcracks (or deformation), which leads to premature failure of the composite before extraction of the full toughness from the resin system.

Mode II Delamination Fracture Toughness

The more brittle the system, the greater is the ratio of G_{IIc} to G_{Ic}. This is the generalization from this work and other unpublished work at Texas A&M University. The increase in total energy release rate for increasing Mode II in brittle systems can be understood to be the result of a whole series of brittle sigmodial shaped microcracks, impeded in their growth on their respective principle normal stress plane by the presence of the fibers. This both increases the area of fracture surface created and constitutes a more torturous path for crack propagation, each of which would require greater energy dissipation per unit area of crack extension.

Where the fracture is more ductile, the energy dissipation does not seem to be such a sensitive function of the imposed state of stress. For example, the fracture mode and the appearance of the fracture surface were similar for pure Mode I and pure Mode II loading of T6T145/F185. For such systems, one would expect $G_{Ic} \approx G_{IIc}$, which was the case for T6T145/F185.

The Effect of Off-Angle Plies

The in situ fractography seen in Fig. 6 clearly indicates that significant resin deformation is occurring along the axis of the specimen. This specimen with all $\pm 45°$ plies would certainly experience significant resin loading in the direction of the specimen axis as a result of bending stresses. While similar stresses would also be experienced by a unidirectional laminate specimen, the fibers would carry essentially all of the loading in the axial direction. It is this additional deformation, not only in the crack tip region, but presumably all along the specimen that is the far field damage previously mentioned as being responsible for giving an artificially high value for G_{Ic} of 1400 J/m^2. It is worth noting that the specimen with $\pm 45°$ plies at the interface where delamination is occurring, but mainly unidirectional plies otherwise had a delamination fracture toughness similar to that for the unidirectional laminate. A J integral analysis being developed at Texas A&M University has given approximately 600 J/m^2 for G_{Ic} for the laminate with all $\pm 45°$ plies. If this proves to be a reliable result, then it suggests that the delamination fracture toughness in composite materials may be a material property independent of stacking sequence if the near and far field damage are properly separated.

Summary

The combination of low cross-link density and elastomer additions has been seen to give the most potent toughening for neat resins. A relatively small increment of the additional neat resin fracture toughening above 800 J/m^2 is actually reflected in the delamination fracture toughness of a composite. Mode II delamination toughness of brittle systems may be as much as three times the Mode I delamination fracture toughness while a ductile system may have a

Mode II delamination fracture toughness that is similar to the Mode I value. The energy absorbed per unit area of crack extension for delamination seems to be independent of ply orientation if proper accounting of the near and far field energy dissipation is made.

Acknowledgment

The generous support of this work by the Air Force Office of Scientific Research (Major David Glasgow, project monitor) is acknowledged. Rich Moulton's (Hexcel Corporation) assistance in providing the materials studied is also recognized. The authors also acknowledge the technical support given by the Electron Microscopy Center of Texas A&M University.

References

[1] Scott, J. M. and Phillips, D. C., "Carbon Fibre Composites with Rubber Toughened Matrices," *Journal of Materials Science,* Vol. 10, 1975, pp. 551–562.

[2] Bascom, W. D., Bitner, J. L., Moulton, R. J., and Siebart, A. R., "The Interlaminar Fracture of Organic-Matrix, Woven Reinforcement Composites," *Composites,* Jan. 1980, pp. 9–18.

[3] Bascom, W. D., Ting, R. Y., Moulton, R. J., Riew, C. K., and Siebert, A. R., "The Fracture of an Epoxy Polymer Containing Elastomeric Modifiers," *Journal of Materials Science,* Vol. 16, 1981, pp. 2657–2664.

[4] Vanderkley, P. S., "Mode I-Mode II Delamination Fracture Toughness of a Unidirectional Graphite/Epoxy Composite," Master's thesis, Texas A&M University, College Station, TX, Dec. 1981.

[5] Cohen, R. N., "Effect of Resin Toughness on Fracture Behavior of Graphite/Epoxy Composites," Master's thesis, Texas A&M University, College Station, TX, Dec. 1982.

[6] Bradley, W. L. and Cohen R. N., "Matrix Deformation and Fracture in Graphite Reinforced Epoxies," presented at ASTM Symposium on Delamination and Debonding of Materials, Pittsburgh, PA, 8–10 Nov. 1983.

[7] Russell, A. J. and Street, K. N., "Moisture and Temperature Effects on the Mixed-Mode Delamination Fracture of Unidirectional Graphite/Epoxy," *Delamination and Debonding of Materials, ASTM STP 876,* S. Johnson, Ed., American Society for Testing and Materials, Philadelphia, 1985.

[8] Wilkins, D. J., Eisemann, J. R., Cumin, R. A., and Margolis, W. S., "Characterizing Delamination Growth in Graphite-Epoxy," in *Damage in Composite Materials: Basic Mechanisms, Accumulation, Tolerance and Characterization, ASTM STP 775,* American Society for Testing and Materials, Philadelphia, 1982.

[9] Devitt, D. F., Schapery, R. A., and Bradley, W. L., "A Method for Determining the Mode I Delamination Fracture Toughness of Elastic snf Biscoelastic Composite Materials," *Journal of Composite Materials,* Vol. 14, Oct. 1980.

[10] Theocaris, P. S. and Stassinakis, C. A., "Crack Propagation in Fibrous Composite Materials Studied by SEM," *Journal of Composite Materials,* Vol. 15, March 1981, p. 133–141.

[11] Mao, T. H., Beaumont, P. W. R., and Nixon, W. C., "Direct Observations of Crack Propagation in Brittle Materials," *Journal of Materials Science Letters,* Vol. 2, 1983.

[12] Hunston, D. A., "Composite Interlaminar Fracture: Effect of Matrix Fracture Energy," presented at ASTM Symposium on Toughened Composites, Houston, TX, March 1985.

[13] Chai, H., "The Characterization of Mode I Delamination Failure in Non-Woven, Multi-Directional Laminates," *Composites,* Vol. 15, No. 4, Oct. 1984, pp. 277–290.

[14] Yee, A. F. and Pearson, R. A., "Toughening Mechanism in Elastomer-Modified Epoxy Resins—Part 1," NASA Contractor Report 3719, 1983, experimental work completed.

Mike F. Hibbs, [1] *Ming Kwan Tse,* [2] *and Walter L. Bradley* [1]

Interlaminar Fracture Toughness and Real-Time Fracture Mechanism of Some Toughened Graphite/Epoxy Composites

REFERENCE: Hibbs, M. F., Tse, M. K., and Bradley, W. L., **"Interlaminar Fracture Toughness and Real-Time Fracture Mechanism of Some Toughened Graphite/Epoxy Composites,"** *Toughened Composites, ASTM STP 937,* Norman J. Johnston, Ed., American Society for Testing and Materials, Philadelphia, 1987, pp. 115–130.

ABSTRACT: Five graphite/epoxy composites containing toughened epoxies prepared at Dow Chemical and AS-4 graphite fibers from Hercules have been studied in Mode I, mixed mode, and Mode II to determine their delamination fracture toughnesses G_c and the controlling micromechanism of fracture. The G_c values were determined using split laminate specimens. The delamination fracture process was observed in real time in the scanning electron microscope. To increase delamination fracture toughness, both an improved interface as well as higher resin toughness were found to be required. Increasing Mode II loading, particularly of the more brittle systems, gives a significantly greater resistance to crack propagation as measured by the total energy release rate required to propagate the crack.

KEY WORDS: composite material, delamination, mixed mode, Mode I, graphite/epoxy, toughened, in situ fracture

Nomenclature

- B Specimen thickness
- C Compliance of test coupon
- E Elastic modulus in the fiber direction
- G_{I} Energy release rate for Mode I loading (opening mode)
- G_{II} Energy release rate for Mode II loading (in plane shear)
- G_{Ic} Critical energy release rate for stable crack growth for Mode I loading
- G_{IIc} Critical energy release rate for stable crack growth for Mode II loading

[1]Research assistant and professor, Mechanical Engineering Department, Texas A&M University, Engineering Research Center, College Station, TX 77843.

[2]Senior research engineer, The Dow Chemical Company, Resins Research, B-1215 Bldg., Freeport, TX 77541.

$G_{\mathrm{Tot,c}}$ Total critical energy release rate for stable crack growth
I Moment of inertia for cracked portion of split laminate
L_c Crack length in split laminate
P_s Symmetric load component $(P_u + P_L)/2$ (see Fig. 1)
P_a Asymmetric load component $(P_u - P_L)/2$ (see Fig. 1)
P_u Load applied to upper half of split laminate (see Fig. 1)
P_L Load applied to lower half of split laminate (see Fig. 1)
Δ_s Total opening displacement (see Fig. 1)

Introduction

Graphite/epoxy composite materials are beginning to be used in many new aerospace applications, as a result of their very high strength-to-weight and stiffness-to-weight ratios. However, design strains used in the aircraft industry are sometimes limited by concerns about delamination. The first generation resins developed for use in graphite/epoxy composite materials optimized stiffness and high glass transition temperature (T_g) by using a very high cross-link density. Unfortunately, such resins are quite brittle. Recent developments have centered on how to increase resin toughness with a minimum penalty in stiffness and T_g. As tougher resins systems are developed, however, a commensurate improvement in interfacial bonding will be required if the full benefit of the increased toughness is to be realized in practice.

The purpose of this study has been to evaluate the delamination toughness of five graphite/epoxy composite materials systems where the resin toughness and interfacial bonding strength have been systematically increased. The delamination fracture toughness for Mode I, Mode II, and mixed mode loading has been determined using split laminate specimens. The fracture process has been studied using real-time in situ fracture observations in a scanning electron microscope (SEM) with a specially designed stage which allows specimen fracture in the SEM. Postmortem fractography has been used to supplement the micromechanistic portion of the study.

Experimental and Analytical Procedures

Sample Preparation

The five graphite/epoxy composites studied are toughened high temperature epoxies combined with AS-4 fibers from Hercules. Systems P4 and P5 are composites which consist of brittle resins and graphite fibers without the sizing which is often added to enhance interfacial bonding and protect the fibers during processing. Resin P4 is Dow Chemical XD7342 and P5 is Hercules 3501-6. Systems P6 and P7 are composed of the somewhat tougher Novalac epoxy, used with sized graphite fibers. The P7 system is a modified version of P6, altered to enhance both resin toughness and adhesion. A fifth system (Q6) is a much tougher cross-linkable thermoplastic epoxy (Dow XU71788) with sized fibers to enhance interfacial bonding.

Mechanical Testing

Delamination fracture toughness measurements for opening mode (G_{Ic}), shear mode (G_{IIc}), and mixed mode were made using split laminate specimens, sometimes called double cantilevered beam specimens (DCB), as shown in Fig. 1. These specimens were tested at room temperature (24°C) in stroke control at a rate of 0.0085 cm/s. A partial unload compliance measurement was made after each approximate 1 cm of crack growth. Stroke, displacement, and load were measured continuously as a function of time while crack length was measured discretely using visual measurements on the edge of the specimen. A check of the accuracy of the measurement of crack length by visual observation on the edge of the specimen was made by performing the final several centimetres of delamination at liquid nitrogen temperatures, which gives a distinctly different fracture surface appearance. This allowed the estimation of the degree of crack tunneling

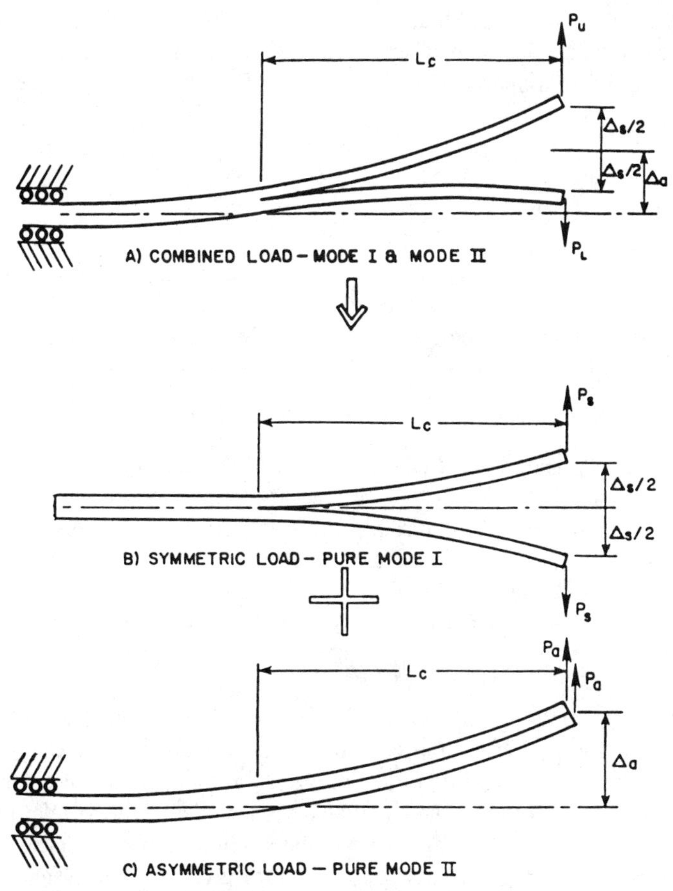

FIG. 1—*Schematic showing how asymmetric loading of split laminate can introduce a mixed Mode I/Mode II state of stress at the crack tip.*

(or thumb nail formation) to determine whether surface measurements of crack length needed to be corrected. Typically, the correction to the crack measured on the edge of the specimen was +1.0 mm.

The mixed mode and Mode II tests were also performed using split laminate specimens. The uncracked end was supported against vertical displacements, but was free to move horizontally. The cracked end was then asymmetrically loaded, as shown in Fig. 1, and described more fully elsewhere by Vanderkley [1]. A 0.8-mm-thick Teflon® spacer was used between the ends of the cracked surfaces for the pure Mode II tests to avoid frictional loading between the two fractured surfaces. Confirmation that no significant frictional loading occurred was found in the fractographic examinations subsequent to mechanical testing. Using superposition, one can then analyze the Mode I and Mode II contributions to the total energy release rate using linear beam theory.

Analysis of Results

Mode I Analysis—The Mode I critical energy release rate for delamination G_{Ic} was calculated from the measured data in three ways. The linear beam theory was used in the first approach as described by Devitt et al. [2] where

$$G_{Ic} = P_s^2 L_c^2 / BEI \tag{1}$$

and

$$EI = 2P_s L_c^3 / 3\Delta_s \tag{2}$$

The area method was used in the second approach as suggested by Whitney [3] and can be shown to be identical to linear beam theory when all the unload compliances are linear and pass through the origin. For this approach, the following relationship is used for linear load displacement behavior:

$$G_{Ic} = (P_{s1}\Delta_{s2} - P_{s2}\Delta_{s1}) / 2B\,\Delta L_c \tag{3}$$

where P_{s1} and P_{s2} and Δ_{s1} and Δ_{s2} are consecutive loads and opening displacements for a crack extension ΔL_c. The area under the load-displacement record which represents the work required to grow the crack a finite distance is seen in Fig. 2.

The third approach used in this work was the compliance change method based on data analysis as suggested first by Wilkins [4]. In this approach

$$G_{Ic} = (nH^{1/n}/2B)(P_s^{1+1/n})(\Delta_s^{1-1/n}) \tag{4}$$

where the constants n and H are determined from a least squares fit of $\log C = n \log L_c + \log H$, and C is measured at each unload. Using each of these approaches, it is possible to determine G_{Ic} at many points of crack growth along the specimen. Ideally, the calculated values of G_{Ic} will be relatively constant for various crack lengths during delamination. Some variation is occasionally observed during the early stages of crack growth. This variation may be associated with the development of a fiber bridging zone behind the crack tip, the

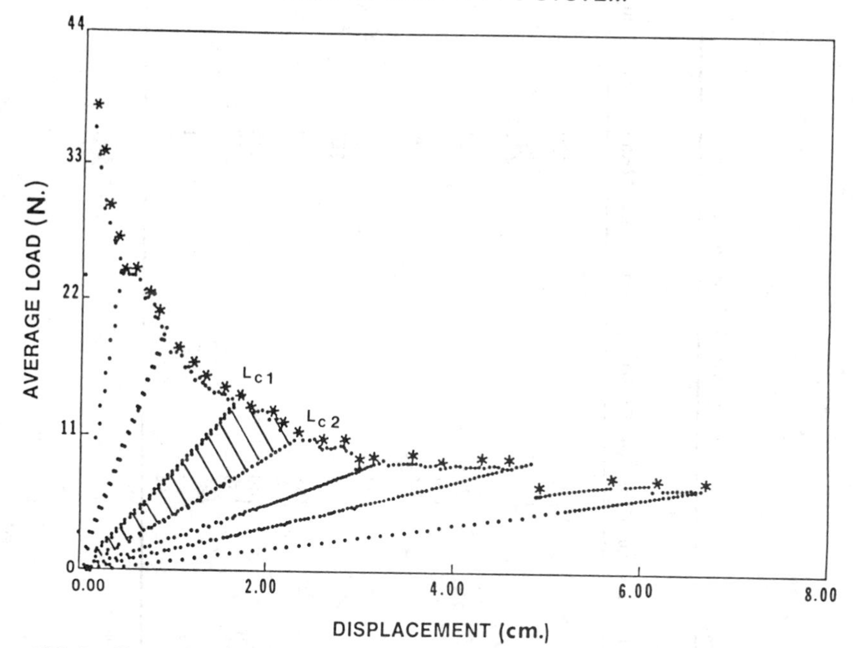

FIG. 2—*Typical load/opening displacement curve for the P5 system. Cross-hatched area represents the energy expended for the crack extension.* ($L_{c2} - L_{c1}$).

development of a steadystate damage zone ahead of the natural crack as it grows from the artifical crack produced by the Teflon, or a result of the fact that beam theory is strictly applicable only when the beam length (and crack length) is several times the beam width. The latter seemed to be the case in this study since no significant fiber bridging was noted and the calculated values of G_{Ic} actually decreased during the transient. Thus, only the "steadystate" values of G_{Ic} were considered when an average value of G_{Ic} was calculated. A typical load-displacement record is shown in Fig. 2 with the reduced results given in Table 1.

Mode II Analysis—The Mode II critical energy release rate for delamination fracture was initially calculated from measured quantities using a relationship based on linear beam theory; namely,

$$G_{IIc} = 3P_a^2L_c^2/4BEI \tag{5}$$

However, it was noted that the load-displacement record for the Mode II tests were somewhat nonlinear and the calculated G_{IIc} values seemed artificially high. This suspicion was confirmed by comparison of our G_{IIc} values for System P5, obtained using Eq. 5 (based on linear beam theory), to results by Street and Russell [*5*] using an end notch flex test on a similar composite material. Reanalysis of the same measured load-displacement records for Mode II delami-

TABLE 1—*Sample calculation of composite mode I energy release rate* (G_{Ic}) *for a typical DCB specimen.*

Input Data						
P_s, N	Δ, cm	L_c, cm	C, cm/N	Area Method G_{Ic}(A), J/m^2	Linear Beam Theory G_{Ic}(B), J/m^2	Compliance Method G_{Ic}(C), J/m^2
			SYSTEM P5			
13.3	1.8	9.3	...	122	134	139
12.9	2.1	9.8	...	166	139	141
11.9	2.2	10.2	...	141	131	131
11.1	2.3	10.8	0.205	130	125	122
10.7	2.6	11.3	...	156	128	123
10.7	2.9	11.8	...	120	139	131
9.0	3.1	12.3	...	240	107	99
9.2	3.2	12.8	0.339	31	121	111
9.4	3.6	13.3	...	137	136	122
9.0	4.0	13.8	...	160	136	121
9.1	4.3	14.3	...	120	149	131
9.0	4.7	14.8	0.482	140	155	134
6.8	5.0	16.5	...	152	110	91
7.5	5.8	16.9	...	96	140	115
7.2	6.3	17.4	...	196	138	112
7.3	6.8	17.9	0.845	125	151	121
			Ave. G_{Ic}:	139	134	121
			Standard Deviation:	43	12	17

nation using the area method gave results for our P5 system that were similar to Street and Russell's results. The energy release rate was calculated as the area under the load-displacement record for a load-crack extension—unload divided by the increase in crack area $(L_{c2} - L_{c1}) \times B$ (see Fig. 2). It is worth noting that the area method implicitly implies that all energy dissipation occurs in the crack tip rather than in the far field. Because the nonlinear load-displacement records returned to the origin, this implied minimal energy dissipation outside the crack tip volume. In general, the reanalysis of the measured load-displacement records gave G_{IIc} values that were 60 to 70% lower than those calculated using linear beam theory. Thus, it was concluded that only the area method or a nonlinear beam analysis could be used to analyze the Mode II results. The area method was used for this work.

Relative good agreement between linear beam theory and the area method calculations of G_{Ic} were obtained, as seen in Table 1. However, Fig. 2 indicates that the load-displacement records for the Mode I tests were quite linear. Thus, good agreement would be expected. The higher G_{IIc} values necessitated much higher loads and deflections to give delamination for Mode II loading, and these higher loads gave the observed geometrically induced nonlinearity in the load-displacement behavior.

Mixed Mode Analysis—For the specimens loaded asymmetrically as shown in Fig. 1, the total energy release rate was calculated by combining the Mode I and Mode II energy release rates calculated from linear beam theory (Eqs 1 and 5) to give

$$G_{Tot,c} = G_{Ic} + G_{IIc} = (4P_s^2 + 3P_a^2)L_c^2/4BEI \tag{6}$$

The fraction of Mode II energy release rate was calculated by combining Eqs 5 and 6 to give

$$G_{IIc}/G_{Tot,c} = \frac{3P_a^2}{4P_s^2 + 3P_a^2} \tag{7}$$

Finally, the fracture toughness of the neat material was determined using compact tension specimens tested according to ASTM Test for Plane-Strain Fracture Toughness of Metallic Materials (E 399). Because of the relative brittleness of the resins studied, it was not possible to precrack the specimens. Thus, a razor blade was used to pop in very sharp cracks.

Fractographic Examination

All fractured surfaces were subsequently examined in a JEOL-25 scanning electron microscope to determine the degree of resin fracture versus interfacial debonding. The nature of the resin fracture (brittle versus ductile) and the degree of microcracking which precedes fracture, as evidenced by scallops on the fracture surface, was also noted.

Real-Time Observation of Delamination Fracture in SEM

The details of the delamination fracture process were determined using real-time observations of the fracture process in the scanning electron microscope, which displays the image on a cathodic ray tube (CRT). Delamination was achieved by pushing a wedge into the precracked portion of a small DCB-type specimen. The wedge tip was sufficiently blunt to ensure that the wedge remained well away from the crack tip, giving essentially pure Mode I loading. The crack propagation was then observed by viewing the edge of the specimen near the region of the crack tip. The experimental results were recorded on both video tape and on standard photographic film.

Experimental Results

Fracture Toughness Results

The Mode I (G_{Ic}), mixed mode ($G_{I,IIc}$), and Mode II (G_{IIc}) fracture toughness results are summarized in Table 2. As has been previously mentioned, the G_{Ic} values were calculated using three different approaches, but the results were generally quite similar, with the average values from each analysis method following within one standard deviation of each other (see Table 1). The average values of G_{Ic} obtained for each of the three analyses on each Mode I specimen were in turn averaged to obtain the values reported in Table 2. Mixed mode results analyzed using linear beam theory are also presented in Table 2 along with Mode II values analyzed using the area method. Fracture toughness values for the neat resin obtained from compact tension specimens are also included in Table 2.

The delamination fracture toughness for Mode I loading was found to be greater than the neat resin toughness for the most brittle systems (P4 and P5) and similar to the neat resin toughness for the tougher systems (P6, P7, and Q6). Similar behavior has previously been reported by Bradley and Cohen [*6*].

Increasing the fraction of Mode II loading increases the total energy release rate for crack growth. Most of this increase was observed as the percentage of Mode II energy release rate increased from 40 to 100%.

Fractography

The fractured surfaces for specimens delaminated in Mode I loading are found in Fig. 3. As evident by the smooth, "clean" surfaces of the fibers, the fracture micromechanism in Systems P4, P5, and P6 is seen to be primarily interfacial debonding. Where a larger resin rich region is observed, a very brittle cleavage fracture is noted. Ductile resin failure is seen to be a fracture micromechanism in Q6, with significant resin deformation preceding fracture. Resin is seen to be completely coating the fibers indicating good interfacial adhesion. System P7 is seen to have a combination of resin fracture and interfacial debonding. Fiber pullout is evident in both P4 and P5. The fracture surface in the systems with failure by interfacial debonding (P4, P5, and P6) has the appearance of a cor-

TABLE 2—*Resin and composite energy release rates.*[a]

Resin System	Resin G_{Ic}, kJ/m^2	Mode I, kJ/m^2	20% Shear, kJ/m^2	43% Shear, kJ/m^2	Mode II, kJ/m^2
P4	0.08	0.165 (45)[b]	0.373 (2)	0.532 (14)	0.769 (5)
P5	0.07	0.137 (64)	0.532 (1)	0.695 (14)	1.267 (4)
P6	0.14	0.165 (26)	...	0.334 (6)	1.532 (3)
P7	0.32	0.340 (29)	0.522 (1)	0.629 (9)	1.724 (2)
Q6	0.73	0.848 (21)	1.533 (1)	0.969 (10)	2.836 (1)

[a] A summary of the energy release rates for the five resin systems. Mode I G_{Ic} was calculated by averaging the results from the area method, linear beam theory, and change in compliance analysis. The mixed modes, 20 and 43% shear energy release rates were calculated using linear beam theory. Mode II G_{IIc} was calculated using the area method.

[b] ( ) indicates the number of data points used to determine G_{Ic}.

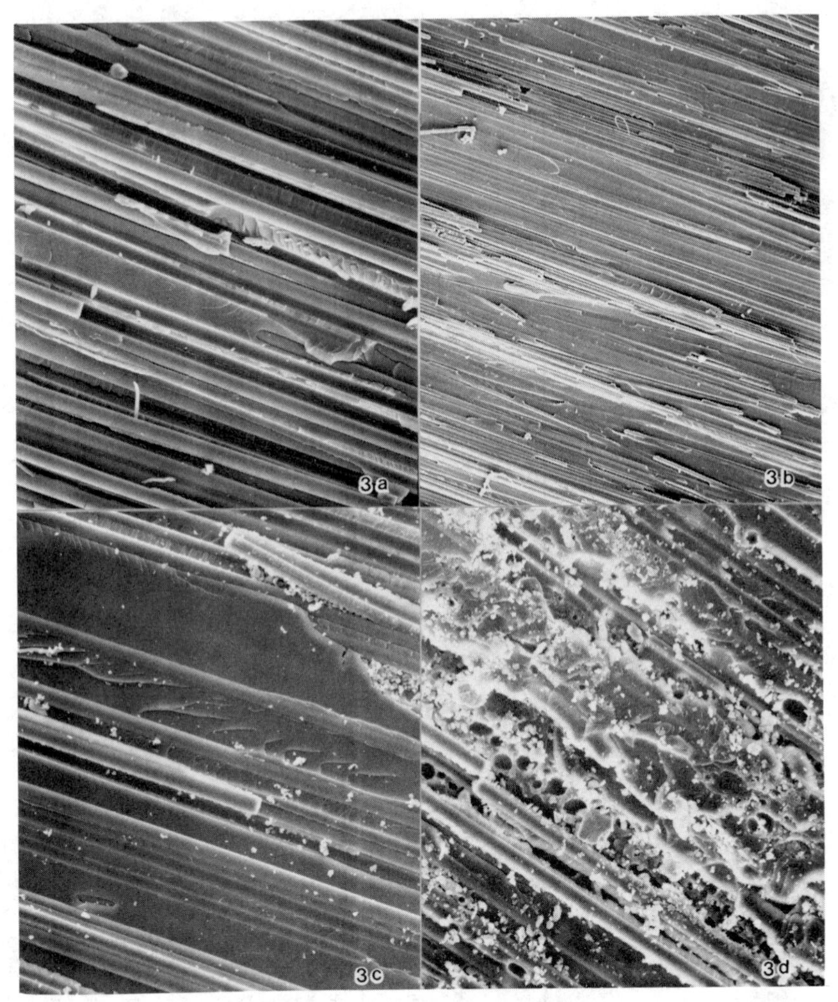

FIG. 3—*Postmortem fractography of Mode I delamination.* (a) *P4, ×450;* (b) *P5, ×70;* (c) *P6, ×450;* (d) *P7, ×450;* (e) *Q6, ×450; and* (f) *Q6, ×1500. P4, P5, and P6 are characterized by a smooth corrugated surface and fiber debonding. P7 displays some resin deformation with improved resin adhesion. Q6 shows extensive resin adhesion and at higher magnifications* (f), *failure can be clearly seen to be due to resin cracking and deformation.*

rugated roof (or a plowed field), which has the effect of increasing the area of the fractured surface. This is typical of delamination fracture in composite systems with relative brittle resin.

The fractured surfaces for the various specimens with mixed mode loading up to 43% Mode II were essentially the same as for Mode I, except for a somewhat greater incidence of individual or multiple fiber pullout. A significantly different fracture surface for all of the systems fractured in Mode II was found, which correlates with the dramatic increase in total energy release rate (see Fig. 4). The

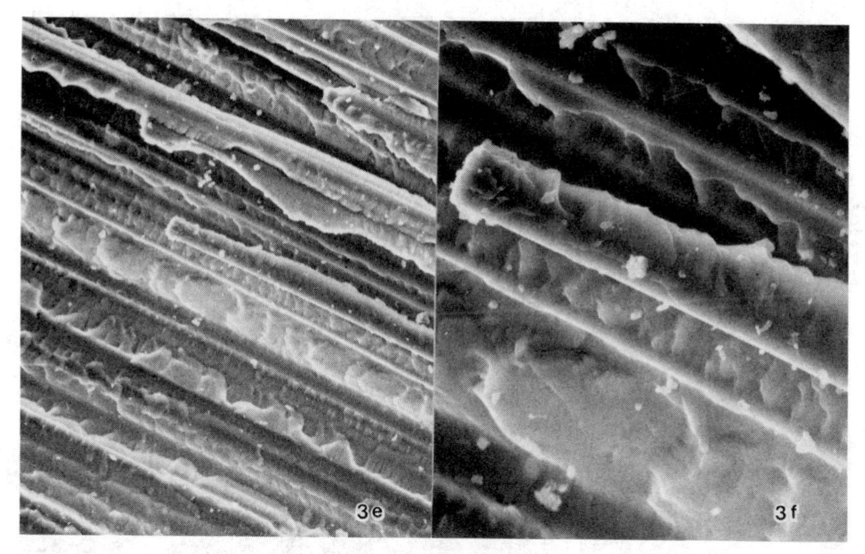

FIG. 3—*Continued.*

scalloped appearance of the fractured surfaces seen in Fig. 4 is typical of high Mode II delaminations of composite materials made from brittle resin systems and constitutes resin cracking.

Real-Time Observations of Delamination in SEM

In situ observations of fracture in the SEM of the P4 system are seen in Fig. 5. The primary crack is seen to proceed by interfacial debond. Some microcracking is seen to occur in the resin behind the primary crack tip and appears to occur concurrently with fiber bridging and eventual pullout.

Fracture of the Q6 system in the SEM is seen in Fig. 6. In contrast to P4, crack propagation occurs primarily by resin deformation and fracture, with only occasional interfacial debonding. Considerable resin deformation and microcracking is seen in regions outside the resin rich region between plies. This larger deformation and damage zone is undoubtedly the result of the better interfacial bonding which allows a greater stress buildup to occur in the crack tip region before crack advance occurs.

Discussion

Increasing the delamination toughness requires an increase in both the resin toughness and the interfacial bond strength. Otherwise, the resin toughness will not be extracted in delamination fracture as a result of "short circuiting" via fiber debonding. The two best composite systems were seen to be the ones with the best interfacial bonding (P7 and Q6) and with the greatest resin deformation before fracture. In this regard, Q6 was particularly good, showing fractographic evidence of considerable resin yielding and deformation.

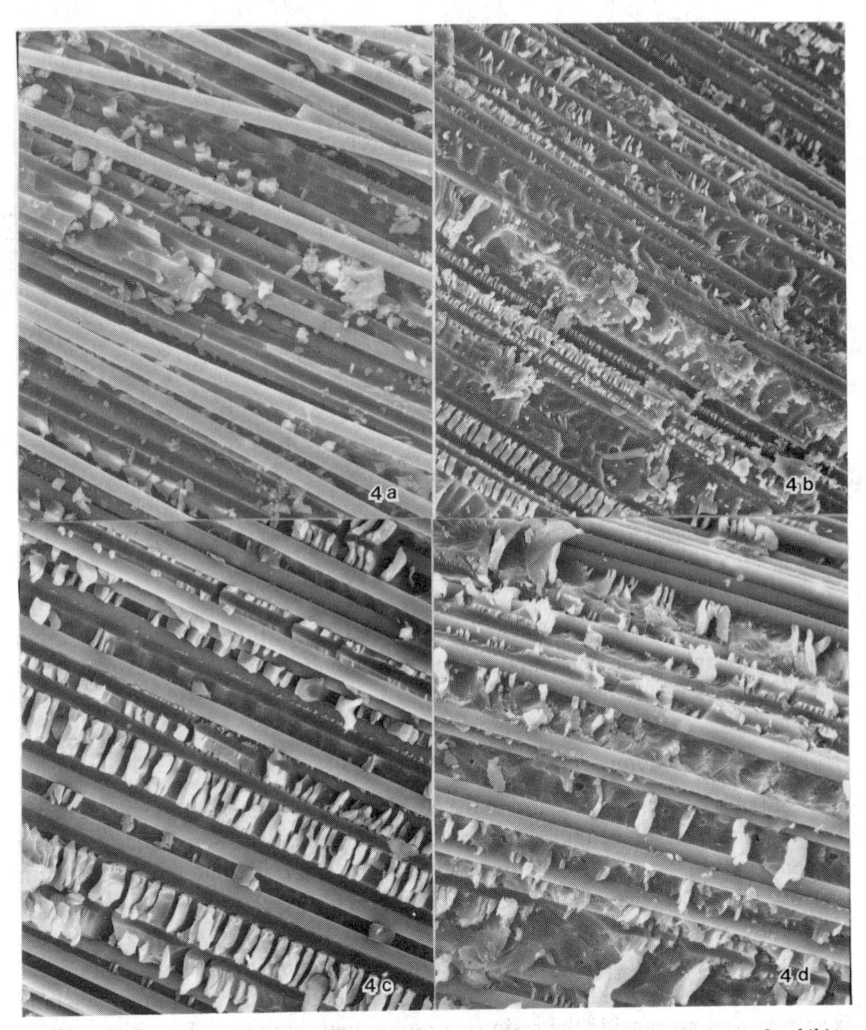

FIG. 4—*Postmortem fractography of Mode II delamination. All the systems tested exhibit a rougher or more scalloped fracture surface than seen in Mode I failure:* (a) *P4,* (b) *P7,* (c) *P5, and* (d) *P6. All at ×450.*

The delamination toughness for Mode I loading was found to be greater than the neat resin toughness for the most brittle systems, namely, P4 and P5. This type of behavior has previously been reported by Bradley and Cohen [*6*] and is explained in terms of a greater fracture area as a result of a "corrugated roof" fracture surface in the composite (see Fig. 3) as compared to a mirror smooth fracture surface in the neat resin. This greater fracture area along with fiber bridging and fiber breakage apparently compensates for the premature failure by fiber debonding before resin cracking, giving a net increase in toughness.

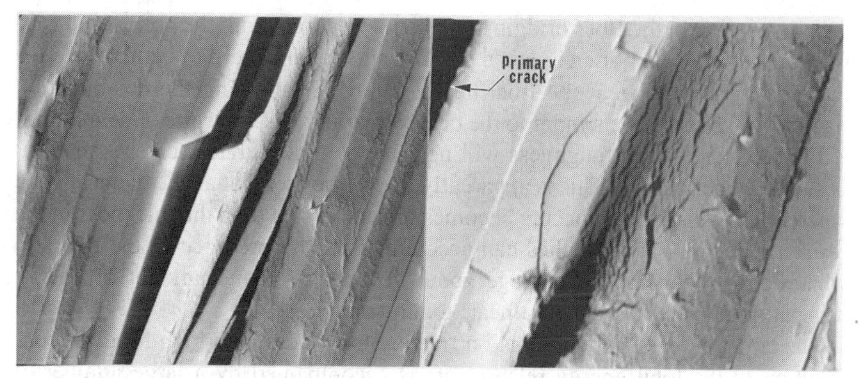

FIG. 5—*In situ delamination fracture in P4. Primary cracking along an interfacial boundary and fiber bridging,* ×*1000* (left). *Resin deformation and microcracking resulting from fiber bridging. Note the primary crack seen in the upper left corner,* ×*3000* (right).

At higher resin toughness values, the Mode I delamination fracture toughness is seen to be similar to the neat resin toughness (see P6, P7, and Q6 in Table 2). This again agrees with earlier results by Bradley and Cohen [*6*] in which it was noted that the damage zone in the neat resin may be similar in height to the thickness of the resin rich region between plies, and thus, the fracture process will be similar in neat material and composite. This does appear to be the case for P7 and Q6. On the other hand, P6 seems to have failed primarily by fiber debonding.

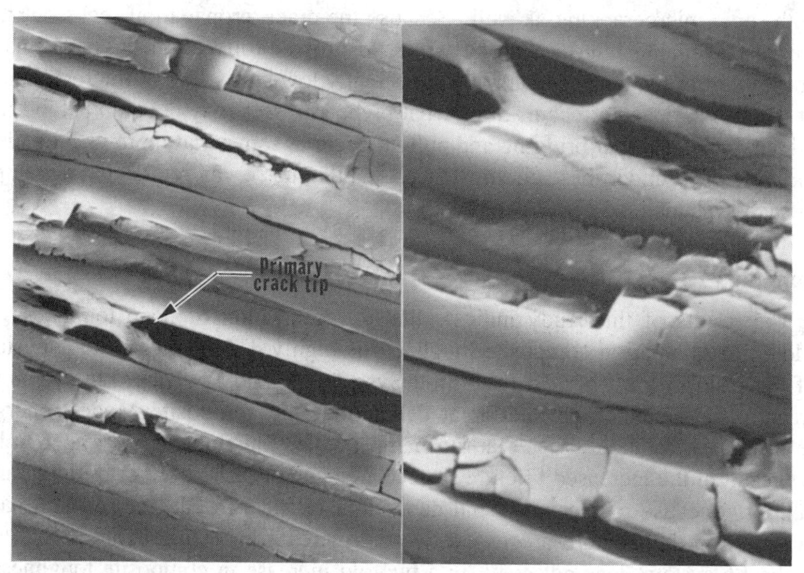

FIG. 6—*In situ delamination fracture in Q6. Resin deformation and bridging near the primary crack tip. Numerous secondary cracks are seen adjacent to the primary crack,* ×*1100* (left). *Resin drawing and microcracking seen within the resin bridging,* ×*2000* (right).

Again, apparently the fiber bridging and breakage along with the corrugated roof fracture surface which increases the fracture surface area help to compensate for the premature fracture at the fiber/matrix interface, allowing the delamination fracture toughness to be similar to the neat resin toughness, in spite of debonding.

Further increases in toughness will not give commensurate increases in composite toughness [*6,7*]. This is apparently due to the fact that as the damage zone where energy dissipation occurs becomes greater in size than the thickness in the resin rich region between plies can accommodate, the presence of the fibers in what would otherwise be damage zone prevents the same amount of energy dissipation as would occur in a neat, tough resin system.

The large increase in delamination fracture toughness with high Mode II contribution to the total energy release rate is accompanied by a large number of scallops appearing on the fractured surface in the systems studied. These scallops apparently result from the attempt of the primary crack to propagate in a brittle fashion on the principal normal stress plane, which for increasing Mode II is no longer parallel to the resin rich region between plies. As a result, a whole series of microcracks form and grow in the resin rich region between plies, terminating at the fibers which bound this region. The coalescence of these cracks constitutes growth of the primary crack. This much more tortuous path for the crack is responsible for the increased resistance to crack growth evidenced in the increase in total energy release rate with increasing Mode II loading.

Delamination crack propagation in P4 as observed in the SEM occurs with growth along one interface with occasional jumps to an adjacent interface on the same fiber or an adjoining fiber. When this occurs, fiber bridging usually results with some microcracking as well. Because the crack propagation seeks out the weakest available interface along which to grow, propagation of the primary crack in P4 is always occurring at a sufficiently low stress so that microcracking of the matrix does not occur. Somewhat higher stresses may develop behind the primary crack as fiber bridging and pullout occur at somewhat more well bonded locations. It is under these circumstances that the matrix microcracking develops in such systems (see Fig. 5).

Summary

The results of this study indicate that the translation of resin toughness to Mode I delamination composite toughness is highly dependent on the adhesion of the resin to the fiber. In the brittle Systems P4 and P5, where the primary fracture mechanism was interfacial debonding, the composite Mode I energy release rate was about twice that for the neat resin. This increase was seen to be the result of the increased fracture surface area and occasional fiber bridging and pullout. Because of the interfacial debonding restricts the degree of resin participation in the fracture process, a twofold increase in resin toughness such as in the P6 system does not result in a twofold increase in composite toughness. Instead, the P6 resin system gave a delamination composite toughness similar to those of the brittle resin systems.

When the interfacial adhesion is enhanced, composite toughness can approach that of the neat resin toughness for moderately tough systems ($G_{Ic} \cong 600$ J/m^2), since resin deformation and fracture is not preempted by interfacial failure. Alternatively, a combination of increased surface area or fiber bridging and breaking or both, and resin deformation can also give composite delamination toughness values which equal the neat resin toughness (for example, the P7 system).

In all of the graphite/epoxy systems of this study, the delamination toughness is seen to increase with higher percentages of Mode II shear loading. In the systems where Mode I delamination is dominated by interfacial debonding (P4, P5, and P6), this increased toughness in the mixed mode and Mode II loading conditions is the result of the microcracking in the resin rich region between plies on planes perpendicular to the principal normal stress plane. Such cracks can only grow a short distance before being obstructed by the fibers. Macrocrack advance requires the coalescence of these microcracks. This microcracking increases the fracture surface area and results in a more tortuous path for delamination crack growth. The increase from Mode I to Mode II delamination toughness was less dramatic in the P7 system where there was less interfacial debonding for Mode I fracture. This is expected since some of the resin toughness was already extracted in the Mode I delamination through resin deformation. In the Q6 system where resin adhesion was seen to be very good, the out of plane stress resulting from the shear loading was seen to increase only moderately the amount of resin deformation and resin cracking, resulting in only a threefold increase in delamination toughness as one changes from Mode I to Mode II loading. Finally, the more ductile resins do not tend to give brittle microcracking and scallops formation for Mode II loading. Thus, the fracture micromechanism is less effected by a change in loading mode.

Acknowledgments

This study was made possible by the material and financial support of the Dow Chemical Company.

References

[1] Vanderkley, P. S., "Mode I-Mode II Delamination Fracture Toughness of a Unidirectional Graphite/Epoxy Composite," Master's thesis, Texas A&M University, College Station, TX, Dec. 1981.

[2] Devitt, D. F., Schapery, R. A., and Bradley, W. L., "A Method for Determining the Mode I Delamination Fracture Toughness of Elastic and Viscoelastic Composite Materials," *Journal of Composite Materials,* Vol. 14, Oct. 1980.

[3] Whitney, J. M., Browning, C. E., and Hoogsteden, W., "A Double Cantilever Beam Test for Characterizing Mode I Delamination of Composite Materials," *Journal of Reinforced Plastics and Composites,* Vol. 1, Oct. 1982.

[4] Wilkins, D. J., Eisenmann, J. R., Camin, R. A., Margolis, W. S., and Benson, R. A., "Characterizing Delamination Growth in Graphite-Epoxy," in *Damage in Composite Materials: Basic Mechanisms, Accumulation, Tolerance, and Characterization, ASTM STP 775,* K. L. Reifsnider, Ed., American Society for Testing and Materials, Philadelphia, 1982, pp. 168–183.

[5] Russell, A. J. and Street, K. N., "Moisture and Temperature Effects on the Mixed-Mode Delamination Fracture of Unidirectional Graphite/Epoxy," in *Delamination and Debonding of Materials, ASTM STP 876,* W. S. Johnson, Ed., American Society for Testing and Materials, Philadelphia, 1985, pp. 349–370.
[6] Bradley, W. L. and Cohen, R. N., "Matrix Deformation and Fracture in Graphite-Reinforced Epoxies," in *Delamination and Debonding of Materials, ASTM STP 876,* W. S. Johnson, Ed., American Society for Testing and Materials, Philadelphia, 1985, pp. 389–410.
[7] Hunston, D. L. and Bullman, G. W., "Characterization of Interlaminar Crack Growth in Composites: Double Cantilever Beam Studies," in *1985 Grant and Contract Review, NASA Langley Research Center Materials Division, Fatigue and Fracture Branch,* Vol. II, 13–14 Feb. 1985.

Willard D. Bascom, [1] *D. J. Boll,* [1] *D. L. Hunston,* [2] *Bret Fuller,* [3] *and P. J. Phillips* [4]

Fractographic Analysis of Interlaminar Fracture

REFERENCE: Bascom, W. D., Boll, D. J., Hunston, D. L., Fuller, B., and Phillips, P. J., **"Fractographic Analysis of Interlaminar Fracture,"** *Toughened Composites, ASTM STP 937,* Norman J. Johnston, Ed., American Society for Testing and Materials, Philadelphia, 1987, pp. 131–149.

ABSTRACT: The failed surfaces of interlaminar fracture (Mode I) specimens were examined using scanning electron microscopy. The matrix resins were Hercules 3501-6 and 2502 epoxys and Phillips Petroleum polyphenylene sulfide (PPS) and the reinforcing carbon fibers were Hercules AS4 and AS6G. The epoxy matrix composites exhibited fiber pull-out, hackle markings, and regions of smooth resin fracture. Considerable (30 to 50%) relaxation of the deformed resin occurred when the epoxy matrix specimens were heated above the matrix glass transition temperature T_g. Some of the fractography features are discussed in terms of transverse tensile stresses and peeling stresses acting on the fibers. The PPS/AS4 fracture surfaces were characterized by large amounts of apparent interfacial failure at the fiber-resin boundary and massive deformation of the PPS matrix.

KEY WORDS: composites, carbon fiber (reinforcement), epoxy (matrix), polyphenylene sulfide (matrix), fractography, interlaminar fracture, resin yielding, crack propagation

The fractography of continuous carbon fiber reinforced polymers (CFRP) has been the subject of numerous investigations [*1–5*]. Some of the features most often cited have been broken fibers, fiber pullout, apparent fiber/matrix adhesive failure, and markings variously referred to as hackle, serrations, lacerations, and chevrons. The origin of these hackle markings has been discussed by Richards-

Certain commercial materials and equipment are identified in this paper to specify adequately the experimental procedure. In no case does such identification imply recommendation or endorsement by the National Bureau of Standards, nor does it imply an assertion by the National Bureau of Standards that these items are necessarily the best available for the purpose.

[1]Research associates, Hercules Aerospace Co., P.O. Box 98, Bldg. X143A, Magna, UT 84092.

[2]Group leader, Composite Materials, Polymer Science Division, National Bureau of Standards, Gaithersburg, MD 20899.

[3]IBM, Endicott, NY.

[4]Professor, Department of Material Science and Engineering, University of Tennessee, Knoxville, TN 37996-2200.

Frandsen and Naerheim [*5*], Russell and Street, [*6*], and most recently by Robertson [*7*].

Much of the fractography of CFRP has been done in connection with the interlaminar fracture of double cantilever beam specimens which, from a continuum point of view, measures an opening mode (Mode I) interlaminar fracture energy G_{Ic}. However, the micro deformation processes in the crack tip damage zone are undoubtedly more complex and involve combined shear and tensile components. Bradley and Cohen [*8*] observed the development of the damaged zone in situ in a scanning electron microscope (SEM) and noted matrix cracking and apparent interfacial failure. They observed a much larger region of damage above and below the main crack front for composites with "tough" matrix resins than for "brittle" matrix resins. Another important feature of CFRP delamination is fiber bridging across the crack opening to create a tied zone [*6*]. Russell and Street have demonstrated the direct influence of this tied zone on interlaminar G_{Ic} and that the effect of the bridging increases with fiber volume.[5] Finally, there is substantial evidence of damage occurring well ahead of the apparent crack front as shown by de Charentenay et al. [*9*] using acoustic emission techniques.

In the work reported here, the fractographic features associated with interlaminar fracture are described for CFRP with a state-of-the-art epoxy matrix, a somewhat tougher epoxy matrix, and a thermoplastic matrix. Interpretation of these features gives some insight as to the deformation processes in the crack front damage zone. Consequently, double cantilever beam specimens that had been tested for interlaminar fracture energy were examined using high resolution SEM. To determine if any of the features were the result of plastic deformation, the specimens were heated above the resin T_g and then examined for changes in appearance.

Experimental Procedure

Width tapered double cantilever beam (WTDCB) specimens were fabricated from Hercules AS4/3501-6 and AS4/2502 epoxy matrix prepreg by hand layup of 24-ply (~3-mm) unidirectional panels and autoclave cured at 177°C (350°F) for 2 h. The 3501-6 matrix material was postcured at 177°C for 4 h. The AS4/polyphenylene sulfide (PPS) panels were obtained from Phillips Petroleum Company as 24-ply unidirectional panels. The WTDCB specimens were cut from the cured panels with the 0° fiber direction along the length of the specimen in all cases. The interlaminar fracture energy of the three composite materials is listed in Table 1 along with the G_{Ic} of the matrix resins. With PPS, the comparison between the resin and composite has a degree of uncertainty since PPS is partially crystalline and the amount and type of crystallinity may not be the same in both cases. All tests were conducted using standard methods as described previously [*10–13*]. For electron microscope examinations of the fracture sur-

[5]A. J. Russell, Defense Research Establishment Pacific Victoria, BC, Canada, personal communication, 1985.

TABLE 1—*Fracture energies.*

Material	Resin Fracture Energy G_{Ic}, J/m^2	Interlaminar Fracture Energy G_{Ic}, J/m^2
3501-6/AS-4	95	180
2502/AS6G	250	497
PPS/AS-4	1365 [*12*]	729
3501-6/AS6G	95	300

faces in the composite, sections were cut from the tapered region as shown in Fig. 1. The fracture load (and thus the fracture energy) were essentially constant through the tapered section. Specimens measuring 4 by 6 mm in cross section and 1.5 mm thick (12 plys) were cut and mounted for SEM examination. A thin (10 to 20-μm) gold coating was deposited to reduce charging. The SEM was a JOEL 200CX scanning transmission electron microscope used in the normal electron scanning mode. Because of the relatively large specimen size, the incident angle had to be fixed at 90°. In those instances where it was necessary to vary the tilt angle, a Cambridge Stereoscan II was used but with some loss of resolution compared to the JOEL microscope.

Results

Three major features that characterize much of the fracture surface of 3501-6/AS4 are shown in Fig. 2. These features include fiber pullout, matrix fracture, and hackle markings. Fiber pullout and breakoff are shown in Figs. 2*a* and *b*. The right-hand side of 2*c* illustrates a region of smooth resin fracture and 2*d* illustrates the hackle markings associated with the pullout of two adjacent

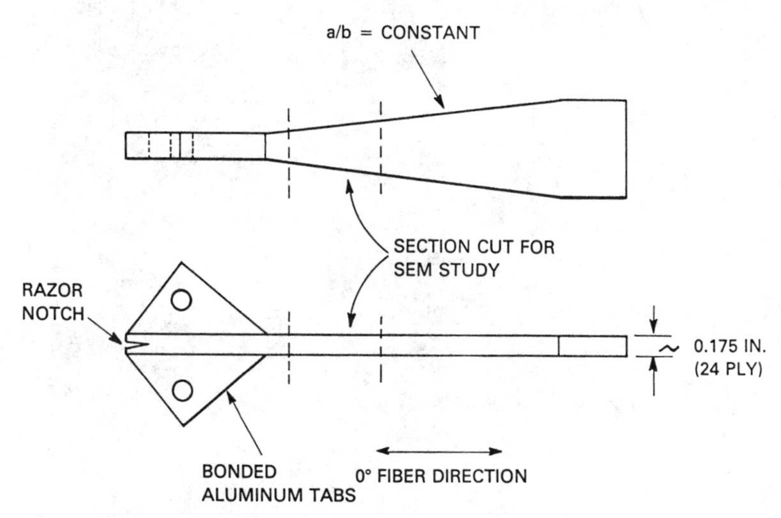

FIG. 1—*Schematic of width tapered double cantilever beam specimen showing section cut for SEM examination.*

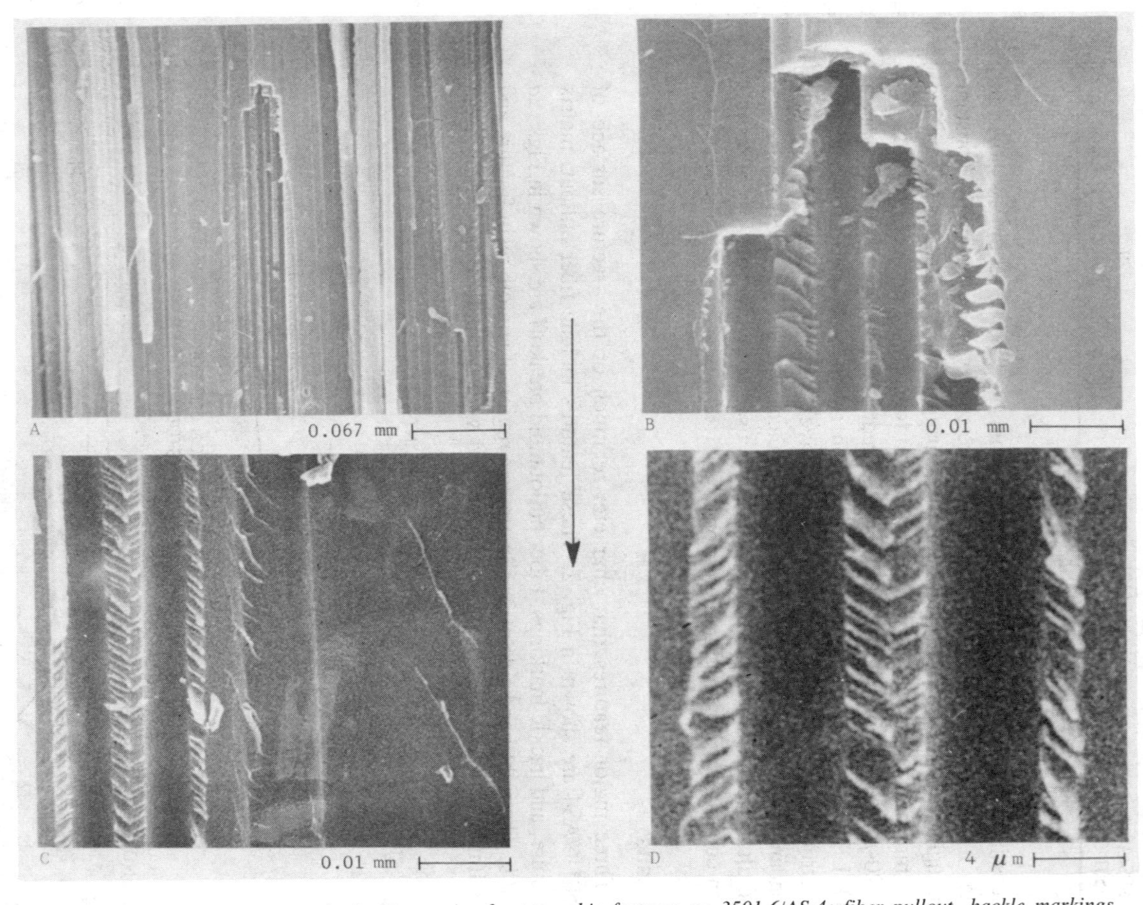

FIG. 2—*SEM photomicrograph showing major fractographic features on 3501-6/AS-4; fiber pullout, hackle markings, smooth resin fracture, and channels left by fiber pullout.*

fibers. In general, pullout involved fiber bundles; in Fig. 2*b* at least five and possibly eight fibers were drawn from two adjacent layers. The hackle markings in 2*c* and *b* appear to initiate from the periphery of the fibers and form "river markings" at a slight angle toward the direction of crack propagation. Note that the channels left by the fiber between the rows of hackle markings are essentially featureless which suggests relatively little resin deformation.

Figure 3 also shows fiber pullout, hackle markings, and areas of smooth resin fracture. In addition, a single fiber is shown that extends out of the general plane of fracture at one end but is embedded in the resin at the other. A relatively high magnification view of the loose end (Fig. 3*b*) suggests that there are "flanges" of resin extending from the sides of the fiber. These flanged (or winged) fibers were a common feature although the placement and size of the "wings" were not always as symmetric as in Fig. 3*b*.

The effect of thermal treatment on the 3501-6 deformation features is shown in Figs. 4 and 5. Specific locations on the fracture surface were selected and their coordinate positions on the microscope stage recorded. The specimen was then removed from the microscope and held at 50° above the resin T_g (225°C) for 10 min, cooled, and repositioned in the microscope and examined for changes in the deformation features. A comparison of Fig. 4*a* and *b* indicates considerable reduction in the size and shape of the deformation after thermal treatment. Notably, a reduction in the size and number of hackle markings and the width of the flange structure on the fiber in Fig. 4*c* and *d*. Some features were unaffected by heating, for example, the pebble-like fragment in the upper right-hand corner of Fig. 4*c* and *d*. In Fig. 5, a channel left by fiber pullout appears to have been "straightened" by the thermal treatment. As in Fig. 4 there was a distinct reduction in the number and size of the hackle markings.

A comparison of specific features before and after thermal treatment suggests dimensional changes in local regions of the matrix that are from 30 to 50% and even greater after 10 min at 50°C above T_g. These changes correspond to retractions along the direction in which the matrix was extended during fracture. Since the photographs presented in Figs. 4 and 5 were all taken with the electron beam directed perpendicular to the specimen, some of the apparent dimensional changes could be due to movement of the feature rather than shrinkage. However, examinations of these specimens at a tilt angle of 55° established that size changes of 30 to 50% had in fact occurred. Heat treatment temperatures below T_g but 10°C above the temperature at which the elastic modulus of 3501-6 begins to fall abruptly from the glassy to rubbery state ($T\Delta G'$) produced less dramatic relaxation effects than shown in Figs. 4 and 5. Nonetheless, dimensional changes did occur at 10° above $T\Delta G'$ for the more severely deformed regions.

The fracture surfaces of AS6G/2502 composites exhibited essentially the same features as observed on AS4/3501-6. In Fig. 6*a* there is evidence of fiber pullout, hackle markings, resin fracture surfaces, and resin attached to the fibers. The major difference between the fracture surfaces of the composites with 3501-6 and 2502 were that the resin deformation features were more "massive" for the 2502

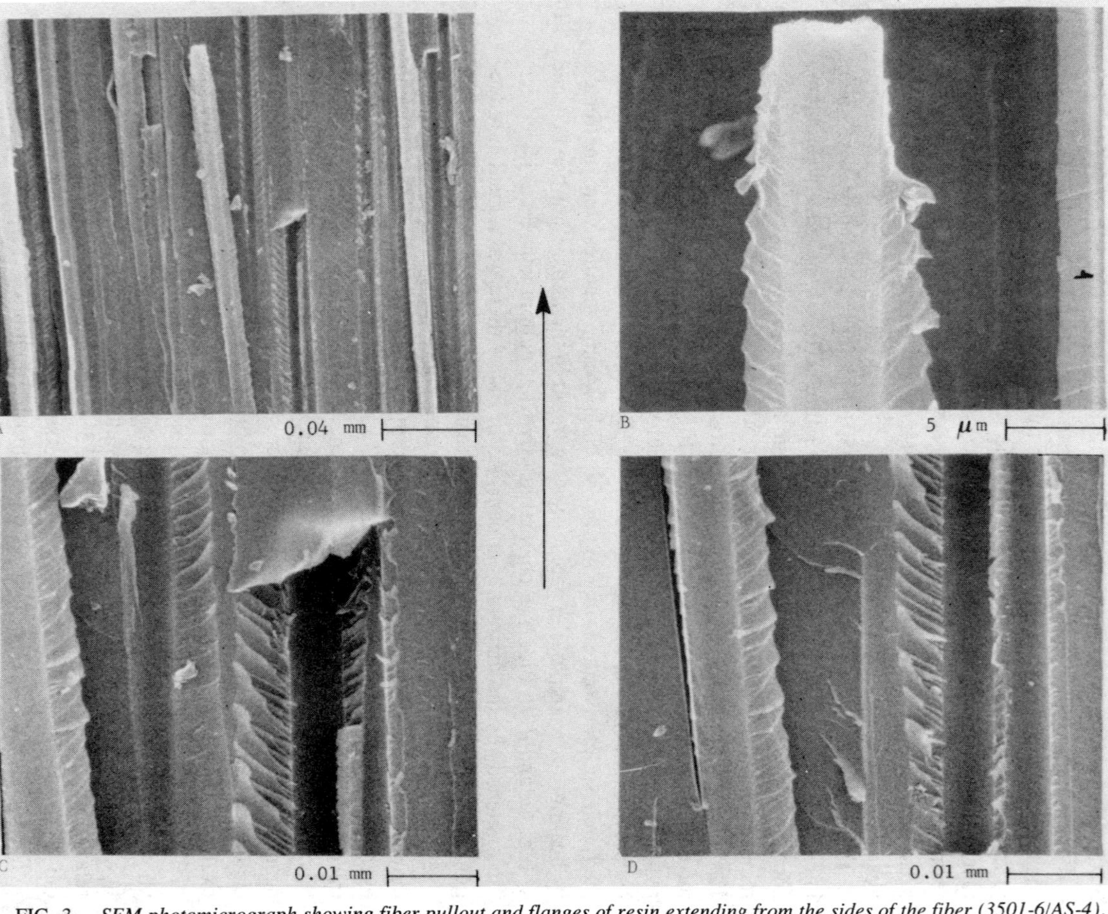

FIG. 3—*SEM photomicrograph showing fiber pullout and flanges of resin extending from the sides of the fiber (3501-6/AS-4).*

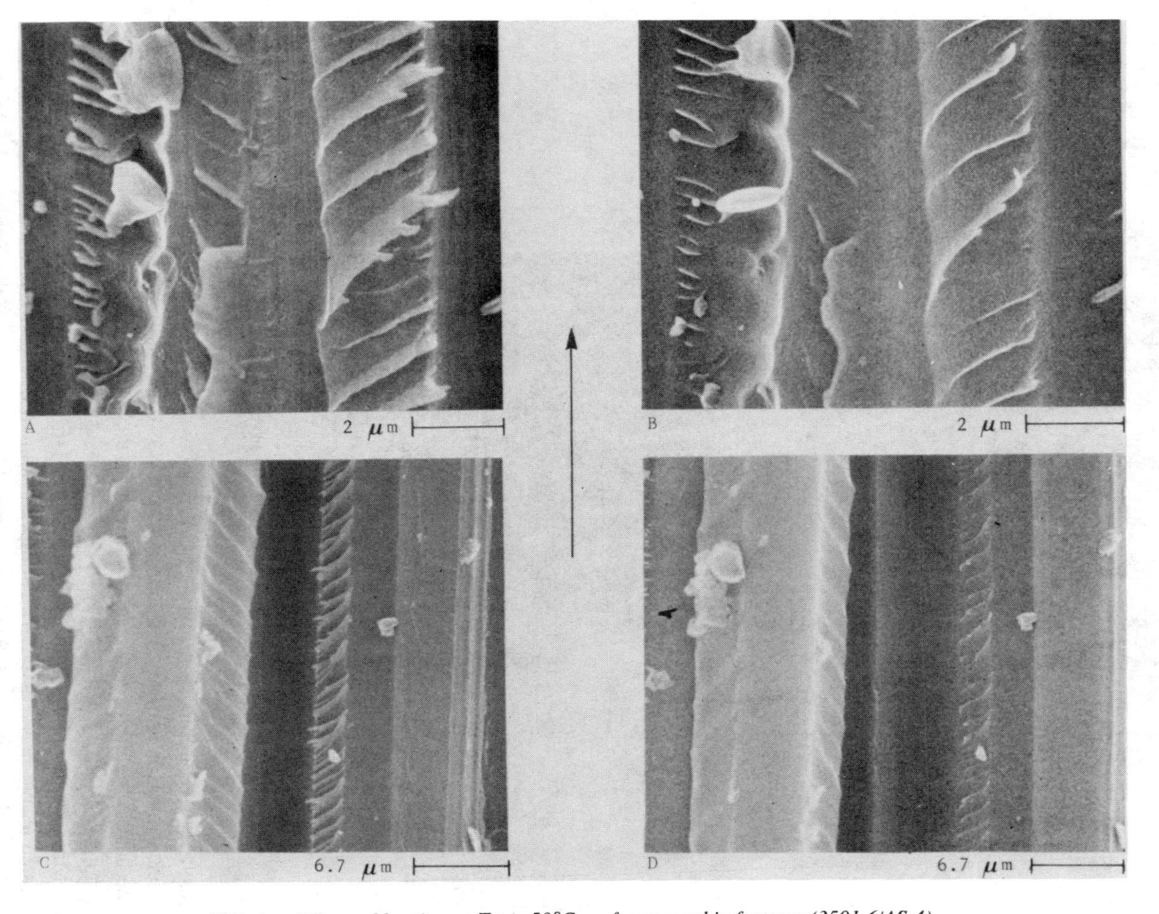

FIG. 4—*Effects of heating at* T_g + *50°C on fractographic features (3501-6/AS-4).*

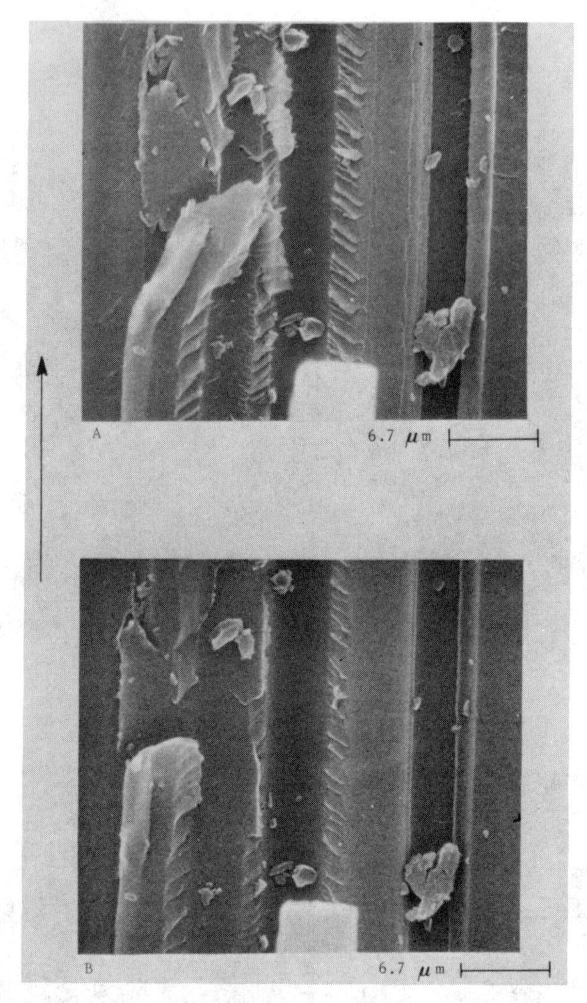

FIG. 5—*Effects of heating at* T_g + *50°C on fractographic features (3501-6/AS4).*

resin. For example, the layer of 2502 left on pulled-out fiber was generally thicker than for 3501-6. Another example was the hackle markings; for the 2502 the hackle features were larger and more widely spaced as indicated by the bar graphs in Fig. 7.

Many of the 2502 features, like the 3501-6, relaxed when the specimen was heated above T_g (170°C). An example is shown in Fig. 8. The hackle markings in 8*a* have essentially disappeared in 8*b*. Also, the edge of the right-hand side of 8*a* became less irregular after heating (8*b*).

The essential differences in properties between the AS4 and the AS6G fibers are presented in Table 2. The AS6G (G refers to an epoxy compatible sizing, the AS4 was unsized) has a higher tensile strength and elongation and a smaller diameter than the AS4.

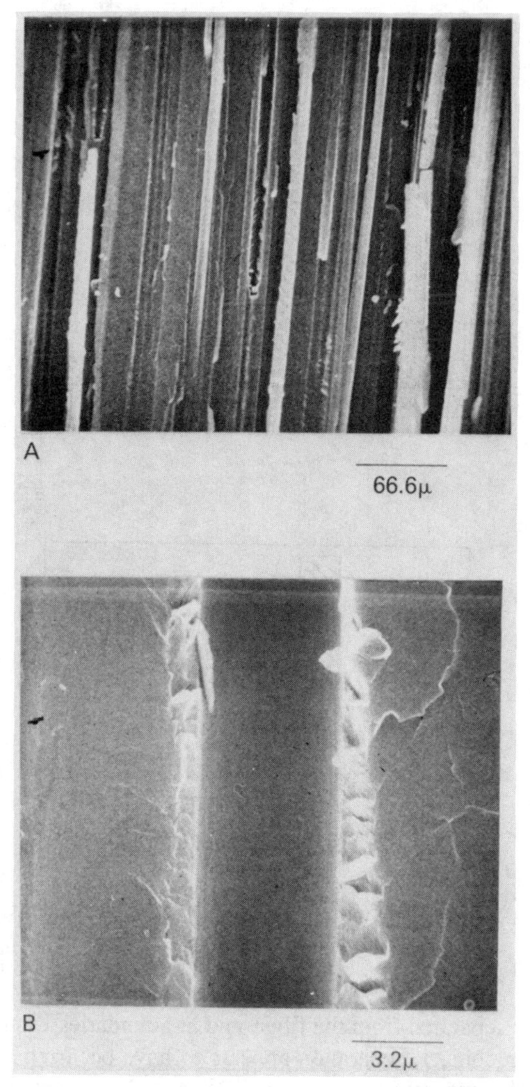

FIG. 6—*Fractographic features characteristic of 2502/AS6G delamination.*

To determine any effect of the fiber on the fracture markings, specimens of AS6G/3501-6 were examined for fracture features and found to exhibit the same markings as on the AS4 reinforced material except that "resin threads" were observed in the AS6/3501-6. These were rarely observed on the AS4/3501-6 or AS6G/2502 delamination surfaces. Examples of these threads are shown in Figs. 9 and 10.

The polyphenylene sulfide/AS4 fracture surfaces shown in Figs. 11 and 12 were characterized by what appeared to be many regions of essentially interfacial failure; the fibers appeared denuded of resin. The resin appears as ruptured tubes

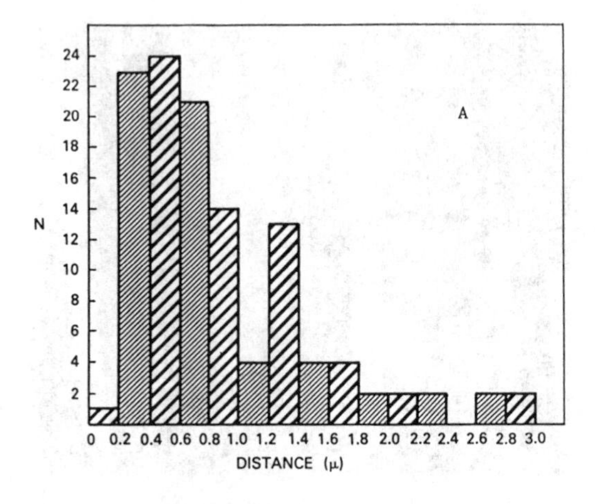

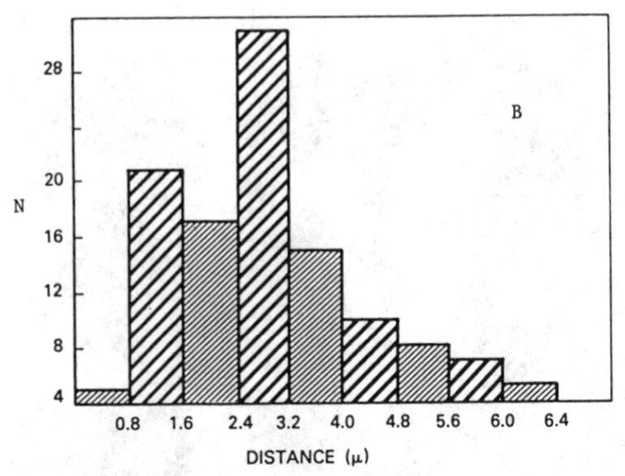

FIG. 7—*Typical distribution of spacing between hackle marking for 3501-6/AS-4(A) and 2502/AS6G(B).*

as if it had freely separated from the fiber, and at boundaries where the fiber and resin were still in contact, separation appears to have occurred along the resin-fiber interface. Heating above the matrix T_g (Fig. 12) indicated relaxation of the deformed resin.

It should be stressed that actual interfacial failure cannot be determined unambiguously from these SEM photomicrographs. However, the photographs suggest a weak boundary layer between the PPS and AS4, especially compared to the 3501-6 and 2502 composites.

Discussion

Deformations of 30 to 50% seem rather large for an epoxy such as 3501-6. This polymer is a highly crossed-linked thermoset based on tetraglycidylether

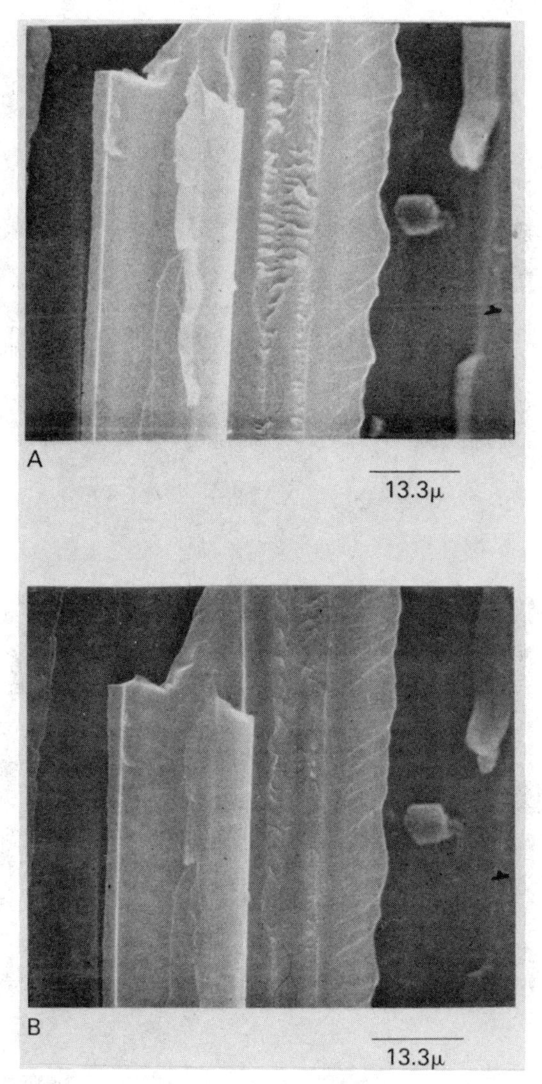

FIG. 8—*Effect of heating at* T_g + *50°C on fractographic features (2502/AS6G).*

TABLE 2—*Carbon fiber properties.*

	Tensile Laminate Properties			
Fiber	Strength, GPa	Modulus, GPa	Elongation, %	Diameter, μm
AS-4	3.59	235	1.53	6.96
AS6G	4.17	243	1.65	5.52

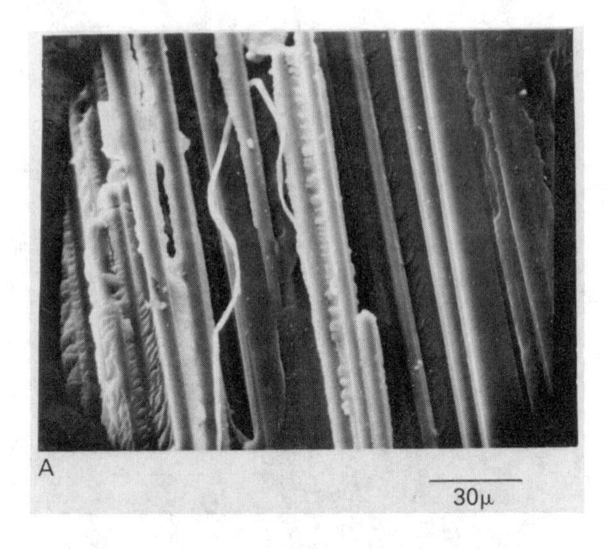

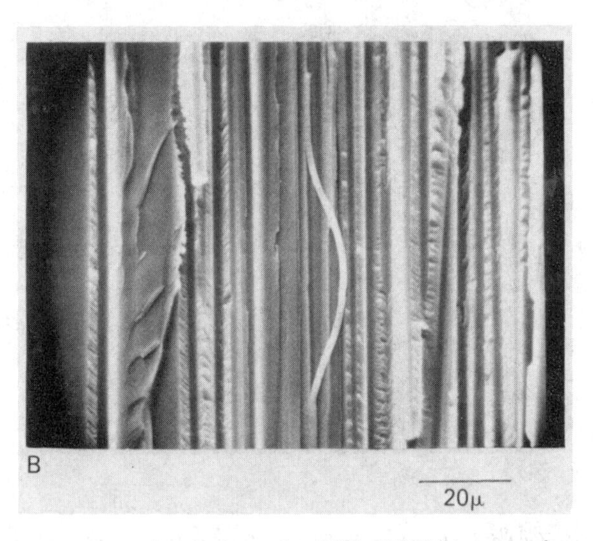

FIG. 9—*Resin threads observed on 3501-6/AS6G fracture surfaces.*

methylene dianiline and diaminodiphenylsulfone with minor amounts of epoxy modifiers and catalysts. In conventional tension tests (ASTM Test for Tensile Properties of Plastics [D 638]), the elongation is of the order of 2 to 3% or less. On the other hand, studies of epoxy resin fracture [*13,14*] indicate extensive yielding of the resin at the tip of the precrack, particularly in regions where slow crack growth precedes the usual unstable crack growth. For unstable growth, the crack speed accelerates, and the fracture surface is essentially featureless. Indeed, the hackle markings observed on the 3501-6 and 2502 composites look very

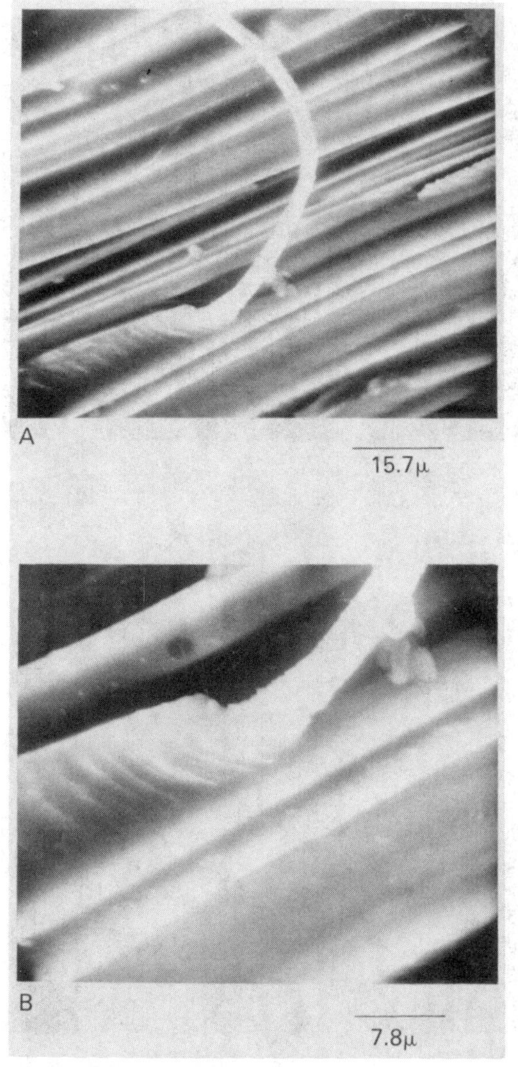

FIG. 10—*Development of a resin thread on a 3501-6/AS6G fracture surface.*

similar to the deformation associated with a slow crack region in neat epoxy resins (see especially Ref *14,* Fig. 1).

Actually the conventional tension testing of matrix resins may be misleading since it is dominated by flaws in the specimen that lead to failure by unstable crack growth. The composites, however, fail by stable crack growth so that the entire fracture surface is subjected to the high crack tip stress field. Moreover, recent work by Bradley and Cohen [*8*] as well as the examination of the fracture surfaces conducted here suggest that microcracking occurs ahead of the crack tip.

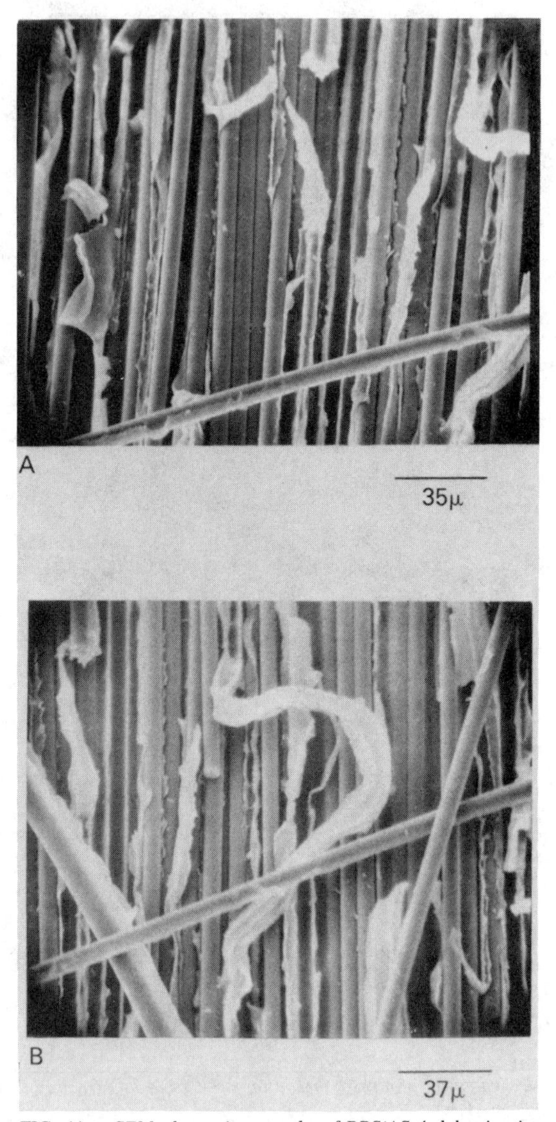

FIG. 11—*SEM photomicrographs of PPS/AS-4 delamination.*

This may relieve the triaxiality of the stress field and thus permit drawing of local regions of the matrix under a stress field that is more uniaxial or biaxial in character. This greatly enhances the potential of the matrix to flow. Recent experiments by Odom and Adams[6] have found that tension tests conducted on specimens of 3501-6 with a cross section of less than 0.1 cm^2 gave substantially

[6]E. M. Odom and D. F. Adams, College of Engineering, University of Wyoming, personal communication, 1984.

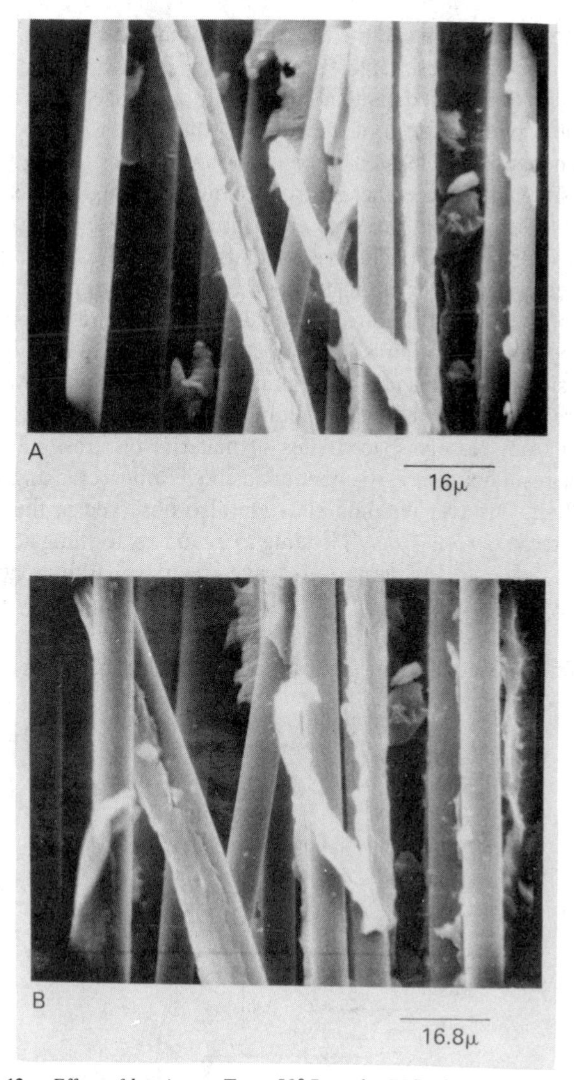

FIG. 12—*Effect of heating at* T_g + *50°C on the PPS/AS-4 fracture surfaces.*

larger extensions to failure than the conventional specimens with cross sections of 0.5 cm^2. This may be the result of the factors mentioned above and if so the composite may represent an even more extreme case since the resin regions between fibers in the composite are the order of 10^{-5} cm^2. If microcracking occurs, very thin resin regions may be subjected to drawing forces.

Most of the fractography features of the 3501-6 and 2502 matrix materials can be understood if delamination is considered in a simplistic fashion as a two-step process. Ahead of the main crack front, the fiber can be viewed as a high modulus inclusion in a lower modulus matrix subject to essentially tensile stresses normal to the plane of the specimen and the fiber longitudinal axis. In Fig. 13, the stress distribution of the fiber-resin interface is shown for the circumference of an isolated fiber [*15,16*]. It predicts that failure will be focused near the fiber-resin interface at the top and bottom of the fiber. A shear stress will exist where the radial stress changes from tension to compression; Point a in Fig. 13.

For composites with good fiber-matrix bonding, the high stress levels at the fiber-matrix interface would promote crack growth near the interface. As seen previously [*17*] for failure in adhesive bonds, however, the failure occurs in the matrix thus leaving a layer of resin on the fiber. In some cases a fiber is "peeled-out" of the resin. If failure occurs along the top and bottom of the fiber as suggested by Fig. 13, for an isolated fiber it could lead to a flange or wing of resin on both sides of the fiber, a configuration similar to that shown in Fig. 3*b*. On the fracture surface, resin yielding along the fiber is not uniform but appears to be initiated from closely spaced sites of material or stress inhomogeneities along the fiber-resin boundary. As mentioned above, microcracking would facilitate this yielding. Discreet yielding sites are also observed in the slow growth yield zone of cracks in resin [*10,11*] leading to ligaments forming across the crack opening which subsequently rupture to leave fracture markings not unlike the hackle markings observed here. The hackle markings, especially near isolated fibers, are bounded by areas of featureless resin fracture which correspond to slow growth failure becoming unstable and the onset of catastrophic fast fracture.

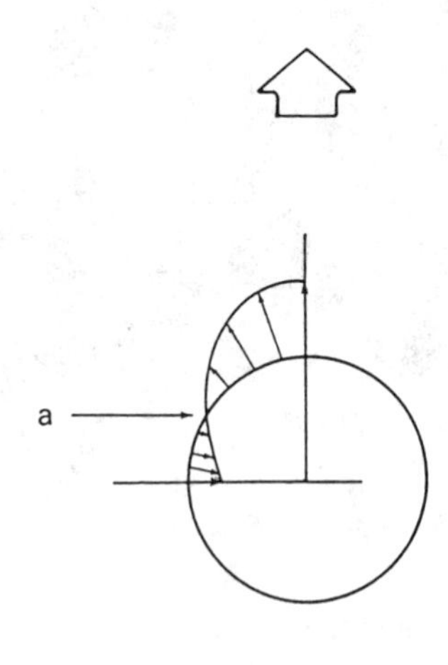

FIG. 13—*Stress distribution (one quandrant) at the resin/fiber boundary for a fiber in transverse loading.*

There were no significant differences in the fracture markings on the 2502 matrix laminates compared to the 3501-6 markings except that the deformation for the tougher resin was more massive, for example, the hackle markings were larger and more widely separated, and loose broken fibers had a heavier coating of resin. These differences are not unexpected considering the fracture energy of the 2502. A high toughness implies a large degree of deformation and yielding whether it is in the damage zone of a single crack of a resin specimen or in the complex stress field of the damage zone of a laminate. For example, in the pullout or peelout of a fiber the stress field focuses failure into the interfacial region, but studies [*17*] of adhesive bond fracture suggest that the larger the damage zone associated with the crack tip the further will be the locus of separation from the interface. This effect results in the apparent interfacial failure when the damage zone is small, that is, 3501-6, but there is an easily detectable coating on the fiber when the damage zone is larger, that is, 2502.

The origin of the resin threads, which were most prominent on the 3501-6/AS6G fracture surfaces is not entirely clear. It may be related to the smaller diameter of the AS6G fiber compared to AS-4 fiber. The cross-sectional area of the resin between fibers is less for AS6G compared to AS-4, by almost ×2. The effective diameter of the resin between AS6G fibers is 2 to 3 μm which is approaching the crack tip yield zone size for resin fracture, 1 to 2 μm for 3501-6. The yield zone diameter δ was calculated from [*18*],

$$\delta = \frac{G_{Ic}}{\sigma_y}$$

where σ_y is the tensile yield strength of 3501-6, about 70 MPa (10 ksi). It is possible that during the fiber pullout process the interfiber resin yields rather than fractures as would appear to be the case in Fig. 10.

The interlaminar fracture of AS-6/3501-6 is almost two thirds greater than for AS-4 in 3501-6 (Table 1). One possible explanation for this difference is the smaller diameter of the AS-6 fiber (Table 2). Since the fiber volume is the same in both composites (~65%), the area of deformation involved in separating fiber from resin (fiber pullout) is greater by a factor of about ×1.6 for the smaller diameter fiber which of itself would explain the difference in fracture energy. Also, as already mentioned, the smaller interfiber distance in the AS-6/3501-6 appears to promote more resin yielding as evidenced by resin threads, on the fracture surface.

Polyphenylene sulfide matrix composites exhibited a totally different behavior than either of the epoxy-matrix materials in that there was extensive apparent interfacial failure. What is interesting is that despite the poor bonding between fiber and resin, the interlaminar fracture energy reflects the high fracture energy of the polymer. It is problematical whether an increase in the apparent adhesion will significantly improve delamination resistance since this resistance is largely due to matrix deformation which for the PPS composites was quite extensive as shown in Figs. 11 and 12. Work is now in progress at Hercules on the adhesion of thermoplastics to carbon fiber to determine the differences in behavior compared to thermosetting resins and the extent to which adhesion actually influences laminate properties. As with the epoxies, a high level of shear yielding is revealed by the relaxation of the fractographic features when heated above T_g. In the PPS composites, however, the deformation was more widespread as reflected by the higher interlaminar fracture energy. In the case of a thermoplastic high, extension to failure may not be so surprising. Nevertheless, the presence of long strands and ligaments of highly drawn material suggests that here too microcracking may relieve the triaxial stress field at the crack tip thus facilitating yielding. In this connection the poor bonding to the fibers may have some benefit (unless so extensive as to provide an easy path for the crack propagation).

Unquestionably, the overall deformation processes suggested here for the epoxy and thermoplastic matrix materials in Mode I delamination are more complex than the schematics would suggest. A more detailed analysis is necessary to understand fiber bundle pullout, fiber fracture, fiber crossover effects, and crack front branching.

Conclusions

The SEM fractography of two epoxy matrix composites after Mode I delamination revealed extensive localized yielding of the resin in the form of hackle markings and other features. Heating the SEM specimens 50° above the resin T_g caused most of the matrix deformations to relax and indicated local resin regions where the yielding may involve 30 to 50% elongation. The origin of hackle markings is attributed to yielding at the periphery of individual fibers or fiber bundles during pullout. Further evidence of plastic deformation was the development of elongated resin threads on the delaminated surfaces of 3501-6/AS6G. The 2502 matrix composites exhibited more extensive yielding, for example, larger hackle markings and a thicker resin coating on the pulled-out fibers, compared to the 3501-6. This difference is consistent with the higher toughness and larger yield zone of the 2502 resin.

The PPS/AS4 composite exhibited significant amounts of apparent interfacial failure, but still showed massive deformation of the matrix polymer.

Acknowledgments

This work was supported by Hercules IR&D funding and by the National Aeronautics and Space Administration. We wish to thank Dr. Norman Johnston, NASA Langley, for support and helpful comments during the progress of this work; to Dr. Tim Murtha, Phillips Petroleum, Bartlesville, for providing the PPS laminates; and Mr. Timo Hoggart, University of Utah, for the PPS SEM studies.

References

[1] Sinclair, J. H. and Chamis, C. C., "Mechanical Behavior and Fracture Characteristics of Off Axis Fiber Composites I, Experimental Investigation," NASA Technical Paper 1081, National Aeronautics and Space Administration, Washington, DC, Dec. 1977.

[2] Morris, G. E., *Nondestructive Evaluation and Flaw Criticality for Composite Materials, ASTM STP 696,* R. B. Pipes, Ed., American Society for Testing and Materials, Philadelphia, 1979, pp. 274–297.
[3] Miller, A. G. and Wingert, A. L., *Nondestructive Evaluation and Flaw Criticality for Composite Materials, ASTM STP 696,* R. B. Pipes, Ed., American Society for Testing and Materials, Philadelphia, 1979, pp. 223–273.
[4] Johannesson, T., Sjoblom, P., and Selden, R., *Journal of Materials Science,* Vol. 19, 1984, p. 1171.
[5] Richards-Frandsen, R. and Naerheim, Y., *Journal of Composite Materials,* Vol. 17, 1983, p. 105.
[6] Russell, A. J. and Street, K. N., *Proceedings of the 4th International Conference on Composite Materials,* 1982, Japan Society for Composites, Tokyo, p. 279.
[7] Robertson, R. E., Mindroiu, V. E., and Cheung, M.-F., *Composite Science & Technology,* in press.
[8] Bradley, W. L. and Cohen, R. N., *Delamination and Debonding of Materials, ASTM STP 876,* W. S. Johnson, Ed., American Society for Testing and Materials, Philadelphia, 1985, pp. 389–410.
[9] deCharentenay, F. X. and Benzeggagh, M., *Advances in Composite Materials, Proceedings,* ICCM-3, Vol. 1, Pergammon Press, New York, 1980, p. 186.
[10] Bascom, W. D., Bitner, J. L., Moulton, R. J., and Siebert, A. R., *Composites,* 11, 1980, p. 9.
[11] Bascom, W. D., Bullman, G. W., Hunston, D. L., and Jensen, R. M., *SAMPE National Meeting Proceedings,* Vol. 29, Society for the Advancement of Material and Process Engineering, Covina, CA, 1984, p. 970.
[12] Ma, M. C. C., O'Connor, J. E., and Lou, A. Y., *SAMPE National Meeting Proceedings,* Vol. 29, Society for the Advancement of Material and Process Engineering, Covina, CA, 1984, p. 755.
[13] Bascom, W. D., Ting, R. Y., Moulton, R. J., Riew, C. K., and Siebert, A. R., *Journal of Materials Science,* Vol. 16, 1981, p. 2657.
[14] Moloney, A. C. and Kausch, H. H., *Journal of Materials Science,* Vol. 4, 1985, p. 289.
[15] Schuiech, H., "Mechanics of Composite Materials," in *Proceedings of the 5th Symposium on Naval Structural Mechanics, Phil.,* F. W. Wendt, H. Liebowitz, and N. Perrone, Eds., Pergammon Press, New York, 1967.
[16] Newag, G. H., *SAMPE Quarterly,* Vol. 15, 1984, p. 20.
[17] Bascom, W. D., Timmons, C. O., and Jones, R. L., *Journal of Materials Science,* Vol. 10, 1975, p. 1037.
[18] Knott, J. F., *Fundamentals of Fracture Mechanics,* Butterworths, London, 1973, p. 153.

Herbert S. Schwartz[1] *and John Timothy Hartness*[2]

Effect of Fiber Coatings on Interlaminar Fracture Toughness of Composites

REFERENCE: Schwartz, H. S. and Hartness, J. T., **"Effect of Fiber Coatings on Interlaminar Fracture Toughness of Composites,"** *Toughened Composites, ASTM STP 937,* Norman J. Johnston, Ed., American Society for Testing and Materials, Philadelphia, 1987, pp. 150–178.

ABSTRACT: A candidate approach for improving the fracture toughness of graphite fiber epoxy composites was investigated in which coatings of tough epoxy based adhesives were applied to the graphite fibers and subsequently overcoated with a 350°F (177°C) service epoxy laminating resin. Composites were fabricated from the coated fiber/epoxy matrix prepreg and from prepreg made with uncoated fibers and the same matrix. Tests on the composites included Mode I interlaminar fracture G_{Ic} by the double cantilever beam method, 0° compression, and 90° tension. Scanning electron microscope photomicrographs of fracture surfaces were made. Comparison of G_{Ic} values for adhesive coated fiber composites with those of uncoated fiber composites was inconclusive because of anomalies in specimen behavior during testing. However, photomicrographs of fracture surfaces of the G_{Ic} specimens indicated that the coated fiber composites had much greater fracture surface area and deformation in the polymeric constituents than did the uncoated fiber composites. Although there were no major differences in 0° compressive strength and 90° tensile strength between the coated fiber composites and the uncoated fiber composites, in some cases the coated fiber composites were slightly superior.

KEY WORDS: graphite fiber composites, coated fibers, adhesives, toughness, delamination resistance, fractography

Graphite fiber plastic composites are being used to an increasing extent in flight critical structures on aircraft. One important goal for these structures is that they have the capability to maintain structural integrity (for example, withstand flight loads) in the presence of initial manufacturing defects or after damage has occurred in service. Defects that occur between laminations and damage that causes delaminations can, under flight loads, propagate more readily than defects or cracks that are perpendicular to the plane of laminations. Therefore, there is

[1]Materials research engineer, Air Force Wright Aeronautical Laboratories, AFWAL/MLBC, Wright-Patterson AFB, OH 45433.
[2]Associate research chemist, University of Dayton Research Institute, 300 College Park, Dayton, OH 45469.

an increasing interest in, and need for graphite fiber composites that have improved delamination resistance (or more conveniently termed improved toughness) over contemporary graphite fiber composites.

There are several possible approaches for developing improved toughness graphite fiber composites. One could be to develop a higher toughness matrix. Another could be to develop special physical constructions such as tough resin layers sandwiched between conventional graphite fiber prepreg [1]. A third could be to coat the graphite fibers with a tough resin before prepregging with a conventional matrix resin. Previous analytical and experimental work has indicated that the use of tough (that is, low modulus, high strain) polymeric coatings on glass, asbestos, and boron fibers improves the impact strength, transverse strength, and total strain energy of composites made with the coated fibers compared to those made with uncoated fibers [2–6].

In all of these approaches, however, there are constraints or limits on the degree to which toughness can be increased without adversely affecting other required properties. For example, one important structural property is in-plane compressive strength. If increased toughness of the matrix or other polymeric constituent is achieved by excessive reduction of modulus of elasticity, this could result in a decrease of in-plane compressive strength and other matrix sensitive structural design properties.

The use of tough coatings on graphite fibers represents an extrapolation of current prepregging technology in which a matrix compatible size (coating) is applied to the fibers. Therefore, the application of tough coatings to the fibers could be implemented in a straightforward and expeditious manner, if warranted.

Our ultimate objective is to develop improved toughness composites for use in the 350 to 450°F (177 to 232°C) temperature range, and to investigate the merits of high toughness polymeric coatings to achieve this. Since the current availability of high temperature resistant, high toughness polymers is limited, we used a model system approach in which adhesive formulations having very high toughness but moderate temperature capability were used as coatings in conjunction with a state-of-the-art 350°F (177°C) service epoxy matrix. In this manner, we hoped to assess the merits of the coated fiber approach for improving composite toughness.

Material Types, Processing, and Fabrication

The adhesives selected for use as coatings were American Cyanamid's FM-1000 nylon epoxy and 3M Co.'s AF 163-2 elastomer modified epoxy. In the cured state, the FM-1000 is a single phase material, and the AF 163-2 is a two-phase material, containing spherical particles of elastomer in an epoxy continuum. Both nylon-epoxy and two-phase elastomer modified epoxies have been found to have very high toughness in adhesive bonded joints [7]. The two-phase elastomer modified epoxies have also been found to have improved toughness over conventional single-phase epoxies in neat form and in composites [8,9]. Both of the adhesives were received from the manufacturer as "unsupported" (no fabric or mat carrier) film materials. Unsized Hercules/AS-4 12 K tow graphite fiber was used to make prepreg for composites fabrication. The laminating resin used as being representative of a 350°F (177°C) service epoxy was Fiberite 976.

All composite panels fabricated were unidirectional. Thickness of the composites ranged from 0.040 in. (0.102 cm) for tensile test panels to 0.215 in. (0.546 cm) for double cantilever beam and compressive test panels.

The following material systems were used to fabricate panels for testing and characterization:

(1) sheets of "neat" unreinforced cured adhesive, 0.020 to 0.040 in. (0.051 to 0.102 cm) thick;
(2) composites made with the following matrices:
  (a) Fiberite 976,
  (b) FM-1000, and
  (c) AF 163-2; and
(3) composites made with AS-4 fiber coated with either FM-1000 or AF 163-2 and subsequently impregnated with Fiberite 976 as the matrix.

The prepreg made with Fiberite 976 matrix was made by pulling the graphite tow through a heated steel crucible (a rectangular cross-section die) containing the resin. The crucible was kept at a temperature of about 200°F (93.3°C) which kept the resin at the proper viscosity for good impregnation of the tow. The impregnated tow was wound on a drum having a diameter of 24 in. (61 cm) and a length of 12 in. (30.5 cm). The impregnated tow was slightly over 0.2 in. (0.51 cm) wide and about 0.015 in. (0.038 cm) thick. After fabrication into composites, the "per ply" thickness was nominally 0.008 in. (0.020 cm).

The preliminary preparation for making the coated fiber prepreg and the prepreg using the adhesive as the matrix consisted of making solutions of the adhesives in organic solvents. The solvent for FM-1000 was a mixture of 60% methanol and 40% methylene chloride by weight. The solvent for AF 163-2 was toluene. For impregnating the fiber with the adhesive as the matrix, a solution containing 30% solids by weight was used. For impregnating the fiber with the adhesive as a coating, a solution containing 5% solids by weight was used.

In making prepreg with an adhesive as the matrix, the tow was passed through a tray containing the resin solution, then past a hot air dryer, and then on to the takeup drum. In making prepreg using the adhesive as a coating, the procedure was similar except that after the tow passed the hot air dryer, it went through the heated die containing the Fiberite 976 resin and then on to the takeup drum. A schematic diagram of the apparatus is shown in Fig. 1.

The prepreg made with the adhesive solutions was dried for an additional 1 h at 120°F (48.9°C) in a vacuum. It was found that these drying conditions reduced the residual solvent to a negligible amount and did not cause any significant advancement in cure of the adhesives.

For the coated fiber prepreg, the weight of the coating was about 5% of the weight of the fiber. For a fiber diameter of 2.76 by 10^{-4} in. (7 by 10^{-4} cm), a fiber density of 0.065 lbm/in.^{3} (1.8 g/cm^3) and a coating density of 0.045 lbm/in.^{3}

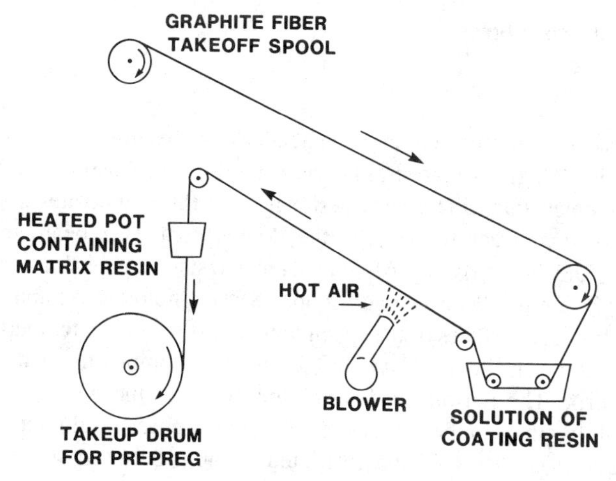

FIG. 1—*Schematic of apparatus used to make coated fiber prepreg.*

(1.25 g/cm³), then calculated coating thickness was 4.7 by 10^{-6} in. (1.2 by 10^{-5} cm). The typical average resin content by weight in the prepreg was about 35%.

All of the test panels were fabricated in hydraulic presses with electrically heated platens. The neat adhesive panels were fabricated from multiple plies of film adhesive. Both the neat adhesive panels and the composites were fabricated in steel molds with provision for flowout of excess resin.

The FM-1000 neat adhesive panel was cured between stops at 350°F (177°C) for 1 h. The AF 163-2 neat adhesive panel was cured at 300°F (149°C) for 1 h at a pressure of 15 to 30 psi (0.103 to 0.207 MPa).

The composites were cured under the following conditions:

1. FM-1000 matrix—350°F (177°C) for 1 h at 200 psi (1.38 MPa).
2. AF 163-2 matrix—300°F (149°C) for 1 h at 100 psi (0.689 MPa).
3. Fiberite 976 matrix—270°F (132°C) for 1 h; 350°F (177°C) for 2 h; apply 85 psi (0.586 MPa) after 15 min at 270°F (132°C) and hold for entire cure cycle.
4. FM-1000 coating on AS-4 fibers with Fiberite 976 matrix—Same cure cycle as for Fiberite 976 matrix.
5. AF 163-2 coating on AS-4 fibers with Fiberite 976 matrix—3 h at 250°F (121°C) contact pressure; then apply 50 psi (0.344 MPa) and raise temperature to 350°F (177°C); hold at 350°F (177°C) for 2 h.

Note that the AF 163-2 coating on AS-4 fibers permitted excessive flow-out of the Fiberite 976 matrix at the fabrication pressures normally used for this matrix. It was therefore necessary to use contact pressure initially until the Fiberite 976 had gelled sufficiently to avoid excessive flow-out.

Tests and Test Procedures

Overview

Tension tests were performed on the neat adhesive specimens from which their ultimate tensile strength, stress-strain curves, modulus of elasticity, and strain at fracture were determined. The tests performed on the composites consisted of Mode I interlaminar fracture (G_{Ic}) by the double cantilever beam method, 0° compression, and 90° tension. All mechanical tests were performed at room temperature. Scanning electron microscope (SEM) photomicrographs were obtained of the fracture surfaces of the specimens. Glass transition temperatures (T_g) were determined for the neat AF 163-2 and the composite made with the AF 163-2 matrix. The reason for determining the T_g of the AF 163-2 materials was the possibility that, in the composites, the rubber phase might not completely precipitate out as a result of the restricted volume between the fibers. If a significant percentage of the rubber remained in the epoxy phase, then it should be indicated by a lower T_g than that for the neat AF 163-2.

Specific Procedures

Tension tests on the neat adhesives were performed on tensile dogbones using head travel rates in accordance with Federal Specification L-P-406b, Method 1011. The AF 163-2 specimens were nominally 0.040 in. (0.102 cm) thick with a gage length 1 in. (2.5 cm) long and 0.180 in. (0.457 cm) wide. Strain was measured with an extensometer. The FM-1000 specimens were nominally 0.020 in. (0.051 cm) thick with a gage length 1 in. (2.5 cm) long and 0.255 or 0.375 in. (0.255 or 0.952 cm) wide. Because of the high elongation of this material, strain was not measured with an extensometer or strain gages. Instead, ink marks were made to define the gage length and elongation measurements were made with dial calipers at predetermined increments of elongation for which the load on the specimen was noted.

The specimens used for G_{Ic} determinations were the double cantilever beam type with a Teflon® film crack starter insert at the midplane and adhesively bonded "T" cross-section aluminum loading tabs. Details of the specimen configuration and construction have been previously described [*10*]. The specimens were made of 24 plies of prepreg and the per ply thickness in the cured composite was 0.007 to 0.0085 in. (0.018 to 0.022 cm). Width was 0.250 to 0.300 in. (0.635 to 0.762 cm) and total length was 6 in. (15.2 cm), of which the central 3 in. (7.72 cm) was used for G_{Ic} determinations. G_{Ic} was determined for crack length increments of 0.25 or 0.50 in. (0.635 or 1.27 cm). Head travel rate during testing was 0.05 in. (0.127 cm)/min.

The 0° compressive strength specimens were made from the same 24-ply panels from which the G_{Ic} specimens were machined. Length was 0.475 in. (1.21 cm) and width was 0.30 in. (0.762 cm). The specimens were tested unsupported between spherical seat loading plates at a head travel rate of 0.02 in.

(0.051 cm)/min. Compressive strength was determined, but stress-strain data were not obtained.

The 90° tension specimens were made from four plies of prepreg (nominally 0.035 in. [0.089 cm] cured thickness), 4 in. (10.2 cm) long and 1 in. (2.54 cm) wide. Strain was measured with adhesively bonded strain gages. Head travel rate was 0.01 in. (0.025 cm)/min.

SEM photomicrographs of fracture surfaces of specimens were made with an ETEC scanning electron microscope. The fracture surfaces were coated with carbon before making the photomicrographs.

Glass transition temperatures were determined from shear modulus and loss modulus data on 0.040-in. (0.102-cm) thick rectangular specimens subjected to oscillating torsion in a Rheometrics Dynamic Spectrometer.

It was noticed that the texture of the laboratory prepared prepregs was much coarser than that of production prepreg. Since the coarser texture could enhance nesting of prepreg plies and thereby affect measured G_{Ic} values, it was decided to measure surface roughness on AS4/Fiberite 976 laboratory prepared prepreg and production prepreg. Surface roughness was measured with a Surfanalyzer System 4000 from Federal Products Corporation, Providence, Rhode Island.

Discussion of Results

Tension Tests on Neat Adhesives

The tensile stress-strain curves for the neat AF 163-2 and FM-1000 materials are shown in Figs. 2 and 3, respectively. The AF 163-2 had a classic "elastic-plastic" stress-strain curve, whereas the FM-1000 stress-strain curve

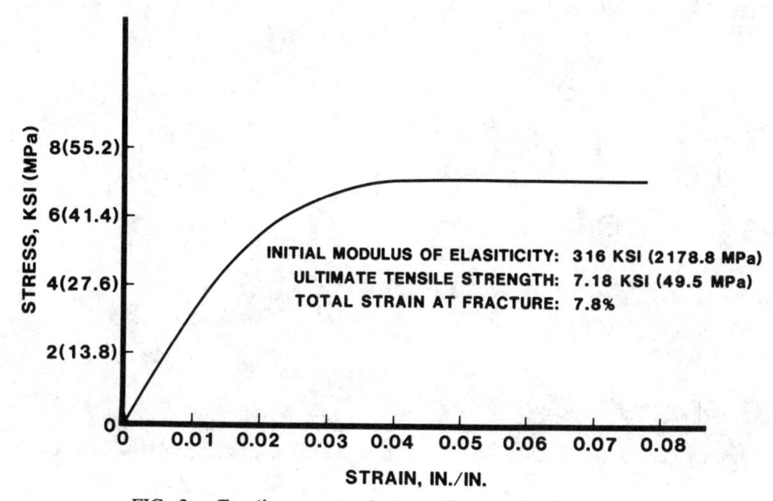

FIG. 2—*Tensile stress-strain curve for AF 163-2 neat resin.*

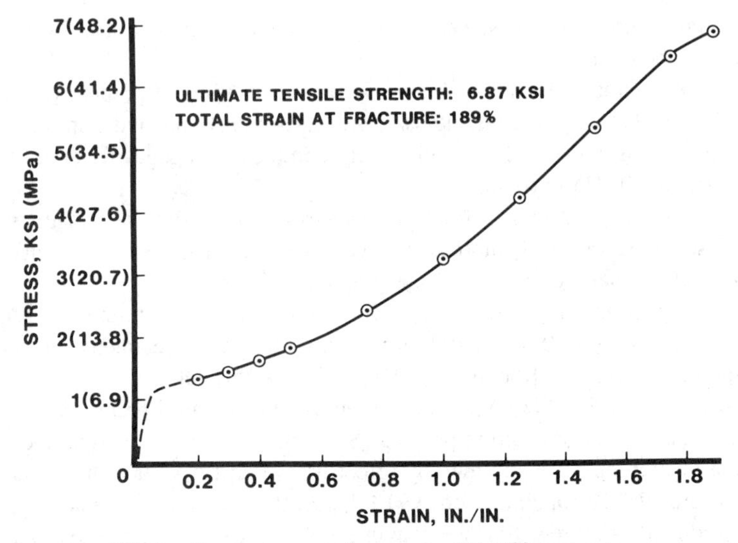

FIG. 3—*Tensile stress-strain curve for FM-1000 neat resin.*

indicated some strain hardening in the 0.4 to 1.4 strain levels. The tensile strengths of the two adhesives were fairly close, 7.18 ksi (49.50 MPa) for AF 163-2 and 6.87 ksi (47.37 MPa) for FM-1000. However, the strain at fracture of the FM-1000 was much greater, being 189% compared to 7.8% for the AF 163-2. The fracture surface of the neat AF 163-2 tension specimen, shown in Fig. 4, exhibits the classical two-phase morphology for this type of material, with

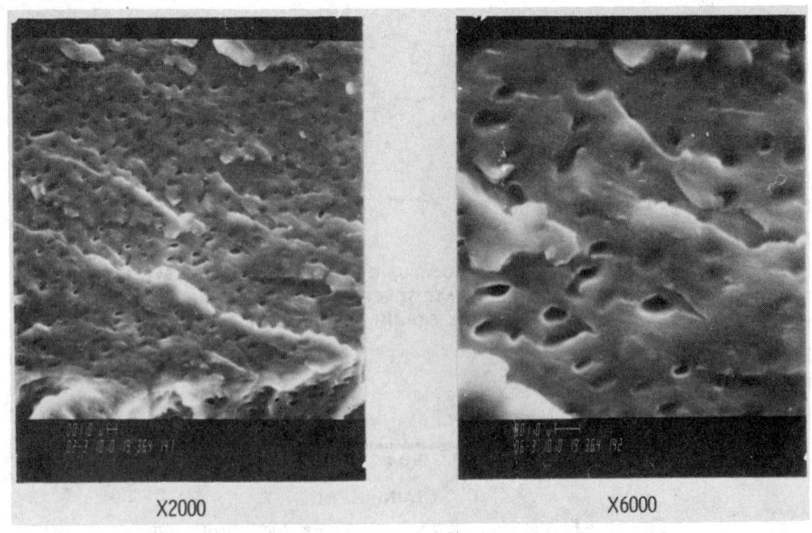

FIG. 4—*Fracture surface of AF 163-2 neat tension specimen.*

the rubber particles having a diameter from several tenths of a micrometer to about 1 μm.

G_{Ic} *Tests on Composites*

G_{Ic} data on the "control reference composites," that is, those with Fiberite 976, FM-1000, and AF 163-2 matrices, are given in Table 1.

The G_{Ic} tests on the AS4/Fiberite 976 were intended to provide baseline data for comparison with corresponding data on the coated fiber composites. However, the measured G_{Ic} value for this material of slightly over 2.3 in. · lb/in.^{2} (403 J/m^2) is much greater than the value of 0.5 in. · lb/in.$^{-2}$ (87.6 J/m^2) obtained in prior investigations using the same test method on composites made with production prepreg of AS4/Fiberite 976. The principal reason for this was that during the G_{Ic} tests, the crack propagated not only along the midplane, but also along planes above and below the midplane. This was observed as "fiber bridging" between the upper and lower halves of the specimen as shown in Fig. 5. Therefore, the additional fracture surface that was created gave artificially high G_{Ic} values for the AS4/Fiberite 976. This type of behavior has been previously reported by other investigators and has been attributed to "fiber nesting" which prevents a discrete layer of resin from forming between adjacent plies [*11*]. The laboratory prepreg had a much coarser texture than corresponding production prepreg. The surface roughness of the laboratory prepreg, shown in Table 2, was about two and one-half times that of the production prepreg. It is believed that this coarse texture promoted fiber nesting which contributed to the artificially high G_{Ic} values.

The G_{Ic} tests on the composites made with the adhesives as matrix materials were intended to determine whether the high toughness of the adhesives in bonded joints would similarly demonstrate high toughness in graphite fiber composites. In particular, one previously reported investigation for a two-phase rubber toughened epoxy similar to AF 163-2 showed greatly reduced toughness in the composite compared to the bulk material [*12*]. However, another investigation showed a much greater percentage of bulk material toughness was achieved in the composite [*13*].

As seen in Table 1, the G_{Ic} value of 16.2 in. · lb/in.^{2} (2840 J/m^2) for AF 163-2 matrix composite is a very high value, as is the value of 23.5 in. · lb/in.^{2} (4110 J/m^2) for the FM-1000 matrix composite. These values are considered valid because the crack propagated along the midplane without fiber bridging. In our laboratory, it has been observed in prior tests on composites made with matrices of varying toughness, that the higher the toughness of the matrix, the less tendency there is for multiple crack formation and fiber bridging [*14*].

G_{Ic} data on coated fiber composites are given in Table 3. Of the eight specimens tested, one had FM-1000 coating and the other seven AF 163-2 coating. A greater number of the latter materials was tested because it was believed that the two-phase coating concept had greater potential for future development of high toughness composites.

TABLE 1 — G_{Ic} *of control reference composites.*

Composite Material	Fiber Volume, %	G_{Ic}, in. · lb/in.^{2} (J/m^2)		Number of Specimens	Number of Data Points
		Average	Range		
Commercial 1 prepreg with resin similar to Fiberite 976	...	...	0.5–1.2 (87.6–210)	...	...
LAB PREPREGS					
Fiberite 976/AS4					
Panel 1	65.5	2.32 (406)	1.88–2.88 (329–504)	2	16
Panel 2	69.4	2.31 (404)	1.79–2.84 (313–497)	2	16
FM-1000/AS4	...	23.5 (4110)	16.3–19.4 (2850–3400)	1	4
AF 163-2/AS4	...	16.2 (2840)	13.9–18.2 (2430–3190)	3	14

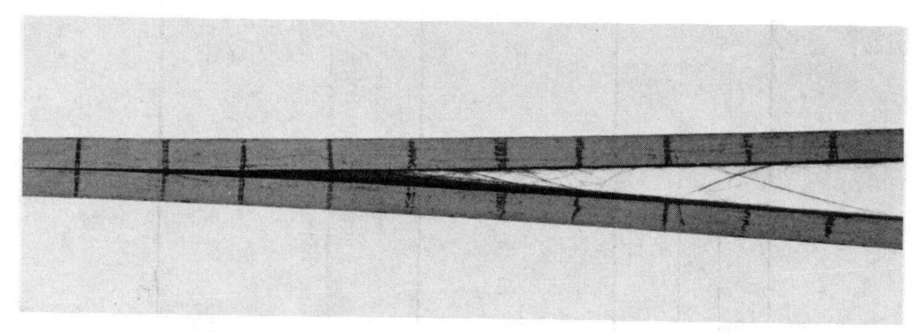

FIG. 5—*Example of fiber bridging in AS-4/976 double cantilever beam specimen.*

The composite with FM-1000 coated fibers had a G_{Ic} of 1.75 in.·lb/in.$^{-2}$ (306 J/m^2) which is considered valid based on visual observations of specimen behavior during test.

Of the composites made with AF 163-2 coated fibers, Panels 3 and 4 having G_{Ic} of 1.88 and 1.77 in.·lb/in.$^{-2}$ (329 and 310 J/m^2), respectively, are considered to be more representative of the intrinsic toughness of this material system than Panels 1 and 2 having G_{Ic} of 2.96 and 4.02 in.·lb/in.^{2} (518 and 704 J/m^2), respectively. This is because of the lower fiber volume content (conversely, higher resin contact) of Panels 3 and 4 which tends to minimize multiple plane crack propagation.

The contention that the intrinsic fracture toughness of the AF 163-2 coated fiber composites is greater than that of the baseline AS-4/Fiberite 976 composites is borne out by SEM photomicrographs of fracture surfaces of the double cantilever beam (DCB) specimens of these materials. Figure 6 is a pair of SEM photomicrographs of the fracture surface of an AS-4/Fiberite 976 DCB specimen. The relevant features are a considerable amount of debonding of the matrix from the fibers, and fracture within the matrix that is of a brittle nature, with little deformation and low fracture surface. Figure 7 is a pair of SEM photomicrographs of the fracture surface of a DCB specimen made with AF 163-2 coated AS-4 fibers with Fiberite 976 matrix. The coating is the material containing the holes and the Fiberite 976 matrix is the material having the parallel blade-like projections. Deformation of the coating and a large amount of fracture surface within the matrix are evident. Figure 8, showing the fracture surface of another DCB specimen of the same material system at a magnification of ×6000, gives (in the right-hand photomicrograph) a more detailed picture of the deformation and fracture within the coating. The nature and extent of these fracture processes indicate much greater energy dissipation in fracturing the AF 163-2 coated fiber composites compared to that for AS-4/Fiberite 976 composites.

Compressive Strength of Composites

The 0° compressive strengths of the composites are given in Table 4. All of the specimens except those made with AF 163-2 matrix failed by delamination.

TABLE 2—*Surface roughness of AS4/Fiberite 976 prepreg.*[a]

Prepreg Material	Surface Roughness Parameter, In.	
	R_A[b]	R_Y[c]
Production	203	984
Lab	550	2330

[a] Stylus traversed perpendicular to fiber direction for 1 to 2 in.
[b] R_A is arithmetic average height of roughness irregularities measured from a mean line within the sampling length.
[c] R_Y is the maximum peak to valley height measured from the mean line within the sampling length.

TABLE 3—G_{Ic} *of coated fiber composites (all with fiberite 976 matrix and AS4 fiber).*

Coating	Fiber Volume, %	G_{Ic}, in. · lb/in.^{2} (J/m^2)		Number of Specimens	Number of Data Points
		Average	Range		
FM-1000	. . .	1.75 (306)	1.14–2.29 (200–401)	1	6
AF 163-2					
Panel 1	71.6	2.96 (518)	2.58–3.29 (452–576)	2	10
Panel 2	70.2	4.02 (704)	3.21–6.71 (562–1170)	3	17
Panel 3	68.1	1.88 (329)	1.34–2.28 (235–399)	1	9
Panel 4	57.6	1.77 (310)	1.54–2.40 (270–420)	1	12

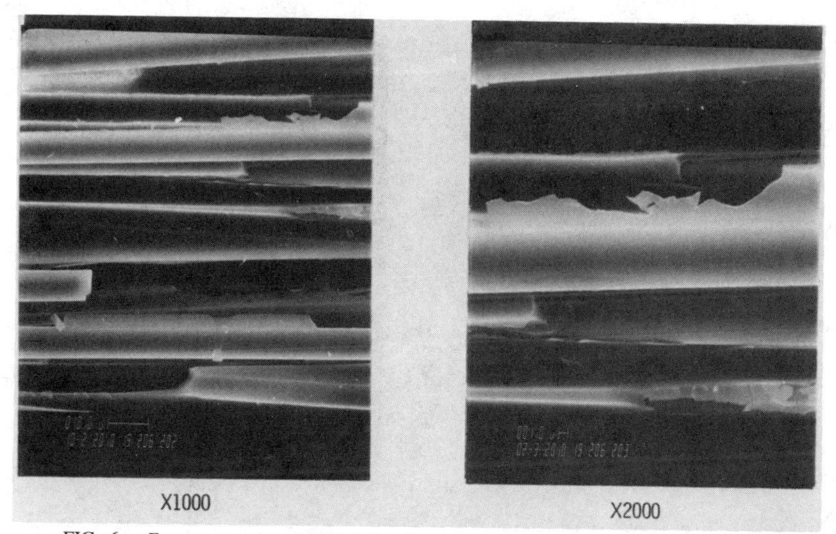

FIG. 6—*Fracture surface of Mode I DCB specimen Fiberite 976 matrix with AS-4.*

Specimens of the latter material failed by local buckling although there were delamination cracks within the specimens when failure occurred. Although the AF 163-2 matrix composites had 36% lower strength than that of the Fiberite 976 matrix baseline, the AF 163-2 coated fiber composites had 17% higher strength than that of the baseline. Photomicrographs of the fracture surface of the AF 163-2 coated fiber composites (Fig. 9) show that fracture occurred primarily within the Fiberite 976 matrix.

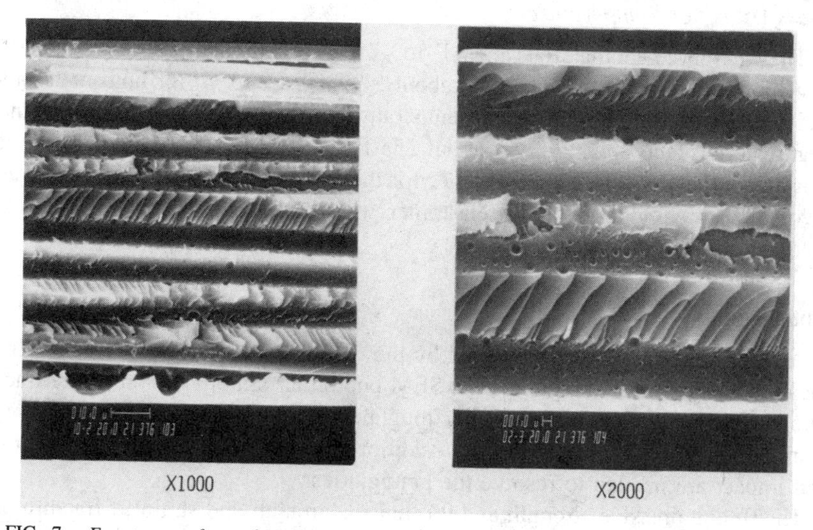

FIG. 7—*Fracture surface of Mode I DCB specimen AF 163-2 coating on AS-4, Fiberite 976 matrix, Panel 3.*

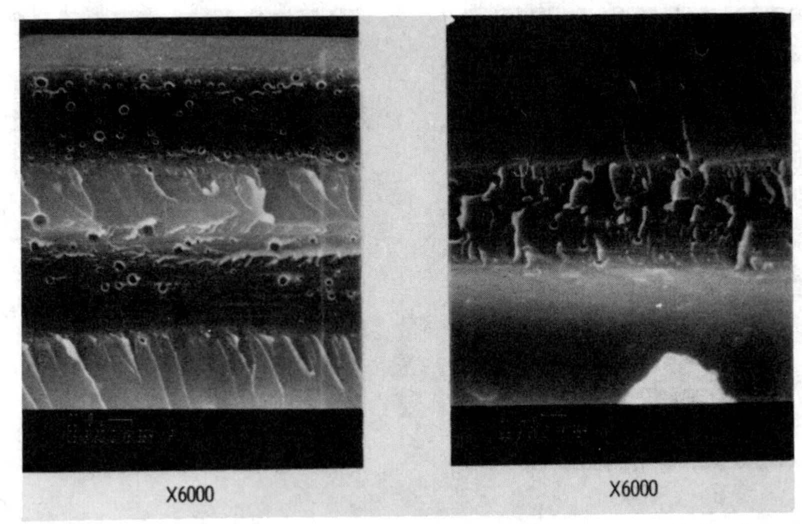

FIG. 8—*Fracture surface of Mode I DCB specimen AF 163-2 coating on AS-4, Fiberite 976 matrix.*

Tensile Properties of Composites

The 90° tensile strengths and strain-to-fracture of the composites are given in Table 5. Only the FM-1000 matrix composites and FM-1000 coated fiber composites had significantly higher tensile strength and strain-to-fracture than that of the Fiberite 976 matrix composites.

Glass Transition Temperature

The AF 163-2 neat materials and AF 163-2 matrix composite showed two glass transition temperatures. One was at about −99.4°F (−73°C) for both materials and represents the glass transition temperature of the elastomer phase. For the neat materials, the upper T_g was about 253°F (123°C), and for the composite it was about 194°F (90°C). The lower T_g for the composite implies that there was incomplete precipitation of the elastomer when the AF 163-2 was used as a matrix in the composite.

Conclusions

Although the G_{Ic} values measured by the double cantilever beam method for the composites were ambiguous, the SEM photomicrographs of composite fracture surfaces indicated much greater toughness of the coated fiber composites than the uncoated fiber composites. Additional tests such as edge delamination and impact are needed to resolve the ambiguities.

The 0° compressive strength and 90° tensile strength and strain-to-fracture of the coated fiber composites were either equal to or slightly greater than those of the uncoated fiber composites.

TABLE 4—*0° compressive strength of composites (specimens tested as unsupported short columns).*

Composite Material	Ultimate Compressive Strength, ksi (MPa)		Number of Specimens Tested
	Average	Range	
Fiberite 976 matrix w/AS4 fibers	127 (876)	114–139 (786–958)	4
FM-1000 matrix w/AS4 fibers	135 (933)	107–152 (738–1050)	10
AF 163-2 matrix w/AS4 fibers	81.7 (563)	68–95 (469–655)	5
Fiberite 976 matrix w/FM-1000 coated AS4 fibers	139 (959)	124–145 (855–1000)	4
Fiberite 976 matrix w/AF 163-2 coated AS4 fibers	149 (1030)	137–159 (945–1100)	5

TABLE 5—*90° tensile properties of composites (all with AS4 fibers).*

Composite Material	Ultimate Tensile Strength, ksi (MPa)		Strain at Fracture, %		Number of Specimens Tested
	Average	Range	Average	Range	
Fiberite 976 matrix	3.34 (23.0)	2.63–3.78 (18.1–26.1)	0.22	0.18–0.25	3
FM-1000 matrix	5.30 (36.5)	...	0.48	...	1
AF 163-2 matrix	3.04 (21.0)	1.77–4.57 (12.2–31.5)	0.38	0.18–0.64	3
Fiberite 976 matrix w/FM-1000 coated AS4 fibers	6.62 (45.6)	5.60–7.74 (38.6–53.4)	0.46	0.40–0.51	3
Fiberite 976 matrix w/AF 163-2 coated AS4 fibers	3.02 (20.8)	1.83–3.90 (12.6–26.9)	0.24	0.15–0.32	3

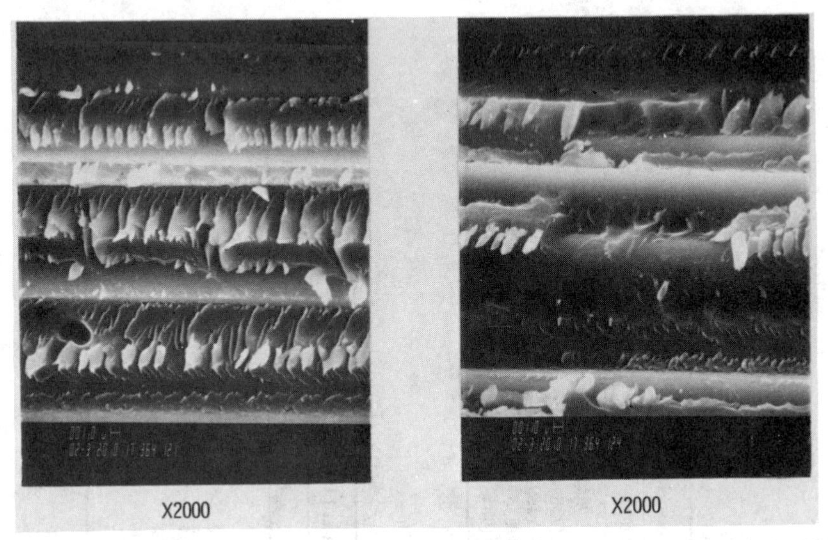

FIG. 9—*Fracture surface of 0° compression specimen AF 163-2 coating of AS-4, Fiberite 976 matrix.*

The results obtained in this investigation warrant further evaluation of the merits of coated fiber composites for improved toughness.

References

[*1*] Browning, C. E. and Schwartz, H. S., "Delamination Resistant Composite Concepts," in *Composite Materials: Testing and Design (Seventh Conference), ASTM STP 893,* J. M. Whitney, Ed., American Society for Testing and Materials, Philadelphia, 1986, pp. 256–265.

[*2*] Broutman, L. J. and Agarwal, B. D., "A Theoretical Study of the Effect of an Interfacial Layer on the Properties of Composites," *Polymer Engineering and Science,* Vol. 14, No. 8, Aug. 1974, pp. 581–588.

[*3*] Lavengood, R. E., Michno, M. J., and Fairing, J. D., "The Effects of Thick Interfaces on the Mechanical Performance of Fiber Reinforced Composites," DTIC Report AD-776592, March 1974.

[*4*] Plueddeman, E. P., "Bonding Rigid Polymers to Mineral Surfaces Through Rubbery Interface," *Proceedings of the Reinforced Plastics Division,* Society of the Plastics Industry, Inc., Feb. 1974.

[*5*] Nielsen, L. E. and Peiffer, D. G., "Preparation and Mechanical Properties of Thick Interlayer Composites," *Journal of Applied Polymer Science,* Vol. 23, April 1979, pp. 2253–2264.

[*6*] Woodhams, R. T. and Xanthos, M., "Polymer Encapsulation of Colloidal Asbestos Fibrils," *Journal of Applied Polymer Science,* Vol. 16, Feb. 1972, pp. 381–394.

[*7*] Mostovoy, S. and Ripling, E. J., "Fracturing Characteristics of Adhesive Joints," Final Report under Naval Air Systems Command Contract N00019-73-C-0163, Jan. 1974.

[*8*] Drake, R. and Siebert, A., "Elastomer Modified Epoxy Resins for Structural Applications," *SAMPE Quarterly,* Vol. 6, No. 4, July 1975.

[*9*] Diamant, J. and Moulton, R. J., "Development of Resins for Damage Tolerant Composites—a Systematic Approach," *SAMPE Quarterly,* Vol. 16, No. 1, Oct. 1984.

[*10*] Whitney, J. M., Browning, C. E., and Hoogsteden, W., "A Double Cantilever Beam Test for Characterizing Mode I Delamination of Composite Materials," *Journal of Reinforced Plastics and Composites,* Oct. 1982, pp. 297–313.

[11] Hunston, D. L., "Characterization of Interlaminar Crack Growth in Composites with the Double Cantilever Beam Specimen," in *Tough Composite Materials,* NASA Conference Publication 2334, Hampton, VA.

[12] Scott, J. M. and Phillips, D. C., "Carbon Fiber Composites with Rubber Toughened Matrices," *Journal of Materials Science,* Vol. 10, 1975, pp. 551–562.

[13] Bascom, W. D., Bitner, J. L., Moulton, R. J., and Siebert, A. R., "The Interlaminar Fracture of Organic-Matrix Woven Reinforcement Composites," *Composites,* Vol. 11, No. 1, Jan. 1980.

[14] Abrams, F., Riechman, J., and Browning, C. E., "Determining and Improving the Toughness of a Brittle Matrix Composite," presented at the Symposium on Toughened Composites, Houston, TX, sponsored by American Society for Testing and Materials, Philadelphia, 1985.

Mark Weinberg [1]

Surface Energy Measurements of Graphite and Glass Filaments

REFERENCE: Weinberg, M., "**Surface Energy Measurements of Graphite and Glass Filaments,**" *Toughened Composites, ASTM STP 937,* Norman J. Johnston, Ed., American Society for Testing and Materials, Philadelphia, 1987, pp. 166–178.

ABSTRACT: The surface energies of graphite and glass filaments were determined by the Wilhelmy balance technique and two semi-empirical expressions. The work of adhesion of various thermoplastics on these filaments was then predicted. The technique successfully distinguished surface changes caused by processing variations. Polar matrix materials such as nylon 6,6 were predicted to adhere significantly better than nonpolar resins such as polyethylene. The notch sensitivity of composite plaques correlated with these predictions. However, there was no relationship in silane systems where wetting is not the primary mechanism for improved adhesion and tensile properties.

KEY WORDS: interfaces, wettability, thermoplastic resins, composite materials, glass fibers, graphite, mechanical tests, adhesion

A fundamental understanding of the wetting of the individual filaments by the matrix resin is crucial to the production of quality composite parts. This is because wetting directly affects the fiber-matrix adhesion. Incomplete wetting can produce interfacial defects and thereby lower both the tensile and compressive strengths [*1,2*]. Thus, measurements to quantify these phenomena are needed to choose the optimum fiber, fiber sizing, and matrix resin for the particular application.

Unfortunately, direct measurements of the wetting of single filaments by molten thermoplastics are very difficult to obtain. In this study, the wetting of graphite and glass filaments by room temperature liquids was determined. The solid surface energies were found from these measurements and two semi-empirical expressions—the harmonic mean and the geometric mean equations. These equations were then used with literature values of polymer surface tensions to predict the polymer wetting in terms of work of adhesion. Finally, the im-

[1]Research engineer, E. I. du Pont de Nemours and Company, formerly with Central Research and Development Department, presently with Polymer Products Department, Experimental Station, E323/302, Wilmington, DE 19898.

portance and limitations of these predictions were demonstrated by measurements of the notch sensitivity and transverse tensile properties of composite plaques.

Basic Theory

The work of adhesion, W_A, is the decrease of Gibbs free energy per unit area when an interface is formed from two individual surfaces [*3a*]. The greater the work of adhesion, the greater the interfacial attraction. The work of adhesion is related to the contact angle θ and liquid surface tension γ by:

$$W_A = (1 + \cos\theta)\gamma \tag{1}$$

The various molecular attractive forces are assumed to be linearly additive. Both the work of adhesion and the surface tension therefore can be separated into two terms [*3b*]:

$$W_A = W_A{}^d + W_A{}^p$$

$$\gamma = \gamma^d + \gamma^p$$

where the superscript d refers to dispersive (nonpolar) contributions and the superscript p refers to polar interactions, including dipole energy, induction energy, and hydrogen bonding.

Two semi-empirical expressions relate the surface properties of each individual material to the work of adhesion [*3b*]. These are the geometric mean equation, preferred between a low energy and a high energy material:

$$W_A = 2[(\gamma_i{}^d\gamma_s{}^d)^{1/2} + (\gamma_i{}^p\gamma_s{}^p)^{1/2}] \tag{2}$$

and the harmonic mean equation, preferred between low energy materials:

$$W_A = 4\left[\frac{\gamma_i{}^d\gamma_s{}^d}{\gamma_i{}^d + \gamma_s{}^d} + \frac{\gamma_i{}^p\gamma_s{}^p}{\gamma_i{}^p + \gamma_s{}^p}\right] \tag{3}$$

where the subscripts i and s refer to any given liquid and solid, respectively.

In this work, the surface energy components of the filaments, $\gamma_s{}^d$ and $\gamma_s{}^p$, are the unknown quantities that need to be determined. The work of adhesion can be experimentally found by measuring the contact angle on the solid (described below) and combining this in Eq 1 with the known liquid surface tension. If this procedure is performed for two dissimilar liquids, i and j, whose surface energy components are known, then Eq 2 becomes a set of two simultaneous equations with two unknowns. Thus, the polar and dispersive parts of the filament surface energy can be calculated. Once these are known, Eq 2 can again be used to find the work of adhesion for any polymer or other material whose energy components are available. This procedure can also be followed with Eq 3, if desired.

The Wilhelmy Technique [*4–7*] is a versatile method of measuring contact angles on filaments. A single filament is suspended from a microbalance and then immersed in the test fluid. The force on the filament is monitored as a function of time and immersion depth. This force can be expressed as the sum of the capillary force, fiber weight, and buoyancy force, or

$$F = \gamma \pi d \text{ COS } \theta + W_f + B \tag{4}$$

where

F = force as measured by microbalance,
d = fiber diameter,
γ = liquid surface tension,
θ = contact angle,
W_f = weight of the fiber, and
B = buoyancy force.

The buoyancy is insignificant when compared with the capillary force for the ~7 to 15 μm diameter filaments used in composites. The filament weight can be nulled out in air before measurement. Therefore, Eq 4 reduces to:

$$\text{COS } \theta = \frac{\text{F}}{\gamma \pi d} \tag{5}$$

Thus, measurements of the capillary force and the fiber diameter and literature values for the liquid surface tension will allow calculation of the contact angle through Eq 5. The work of adhesion can then be found from Eq 1.

A suitable technique to determine directly fiber-matrix stress transfer for ductile resins involves the determination of the fiber "critical length" [*7–9*]. A single filament is embedded axially in a tension specimen of a matrix with an elongation to break much greater than the filament. A tensile force is applied to the specimen and is transferred to the fiber by shear at the fiber-matrix interface. Since the volume fraction of fiber will be below the minimum loading required for reinforcement, the fiber axial stress will rise until the fiber breaks at some point, depending upon defects and probability. Continued application of stress will continue this breaking until all remaining fiber lengths are shorter than the "critical length," below which sufficient stress can not be transferred to the filament to break it. The Kelly-Tyson shear-lag analysis [*10*] relates this critical length, L_c, to the effective interfacial shear strength τ by:

$$\tau = \frac{\sigma_f d}{2L_c} \tag{6}$$

where

σ_f = fiber tensile strength and
d = fiber diameter, both of which are assumed to be constant. Note that τ may be affected by many simultaneous processes occurring in the fiber, matrix, and interfacial regions.

The data are quantitatively analyzed by fitting the length/diameter ratios of the experimentally found segments to the two-parameter Weibull distribution. The average effective interfacial shear stress was found from the cumulative distribution function. These analysis procedures with their limitations are described in detail in Refs *8* and *9*, respectively.

Experimental Procedures

Surface Energy Measurements

The filaments were attached to a metal hook using adhesive tape and then suspended from the loop of a Cahn Model 25 microbalance. A beaker of test fluid was raised until the fluid surface was close to the filament. The balance was tared and an "inch-worm" precision elevator was activated to further raise the breaker and immerse the filament. The elevator speed was constant at 0.005 mm/s; this was sufficiently slow so that the dynamic contact angle was independent of velocity and identical to the static angle [*3c*]. The force was monitored for at least 300 s to give the advancing force. The elevator direction was then reversed and the receding force levels were measured. The apparatus was placed within a metal enclosure to eliminate the effects of air convection during operation. At least ten runs were made with each filament/liquid combination; a different filament was used for each run. All tests were performed at room temperature (23°C).

The AS-4 filaments used in this study were taken from a Hercules A-193P graphite fabric. Two variations of these filaments were examined. The "as-received" samples had a proprietary epoxy sizing on them. The "400°C" samples were held in an oven at 400°C in air for 2 h to modify the surface layer. Both filament varieties had an average diameter of 7.56 μm as determined from microscope observations on at least twenty of each type.

Two kinds of E-glass filaments (Owens-Corning Fiberglas) were examined—a "bare" glass and one with 1.5% by weight A-1100 γ-aminopropylsilane (Union Carbide). The average diameters for the "bare" and silane-treated filaments were found to be 13.5 and 14.2 μm, respectively.

As described above, contact angle measurements on at least two dissimilar liquids are needed to solve for the components of the solid surface energy. However, this method can be applied with any larger number of liquids. Three test materials were used in this study; they are listed in Table 1 with their surface energy components. They were chosen to cover a wide range of polar and nonpolar (dispersive) characteristics. The total surface tensions were experimentally verified using the Du Nouy Ring technique. By using these three test liquids, it is possible to get three independent solutions for the solid surface energies. These solutions were averaged to give more reliable values.

TABLE 1—*Liquids used for surface energy measurements* [6].

	Surface Tension, mN/m			
Liquids	Dispersive	Polar	Total	% Dispersive
Ethylene glycol	29.3	19.0	48.3	61
Glycerol	34.0	30.0	64.3	53
Water	22.1	50.7	72.8	30

Notch Sensitivity Study

All samples were made with Hercules A-193P 0/90° plain weave graphite fabric. The AS-4 filaments were washed to remove the sizing, once in methyl ethyl ketone and then three times in methylene chloride. The resins were commercial grade thermoplastics—Alathon® 7050 (polyethylene (PE), Du Pont), Hytrel® 5526 (thermoplastic elastomer, Du Pont), PETG 6763 (amorphous polyester, Eastman Kodak), and Fostarene® 817 (polystyrene (PS), American Hoechst). All resin was dried overnight at 60°C before molding. The plaques were made by layering fabric and ground resin in a mold kept within a vacuum bag with a nitrogen bleed. The mold was preheated for at least 15 min before increasing the pressure to 3.8 MPa (555 psi) for at least 30 min. The temperatures were adjusted within the range of 200 to 280°C to give the approximately same viscosity for all the resins. This was based upon previously obtained rheological data. The plaques were 50 to 55% by volume fibers and were shown by C scans to be well consolidated. Test specimens (1.27 by 15.2 cm) were cut from the plaques while they were sandwiched between Lucite® sheets. Holes of either 0.635, 0.476, and 0.318 cm (¼, 3⁄16, and ⅛ in.) were cut in the centers. The specimens were pulled on an Instron tester at 4.2×10^{-3} cm/s (0.1 in./min) with a 2.54-cm (1-in.) gage length.

Transverse Tests

Glass tow that had been wrapped around a plate was impregnated with a 10% solution of Lexan® polycarbonate (General Electric) in methylene chloride. Sheets of this were plied up in a 7.6- by 15-cm (3- by 6-in.) mold with Lexan film to achieve a 55% volume fraction of fibers. The mold was placed in a vacuum bag and heated to 260°C for 15 min before pressing at 7.7 MPa (1100 psi) for 30 min. Two kinds of glass (Owens-Corning Fiberglas) were used—"bare" and coated with 0.25% by weight A-1100 aminosilane. Specimen cut to dimensions of 7.6 by 1.3 cm (3 by ½ in.) were pulled on an Instron tester at 4.2×10^{-3} cm/s (0.1 in./min) with a 2.54-cm (1-in.) gage length.

Critical Length Experiments

A single glass filament—either "bare" or with 1.5% by weight A-1100 amino-silane (Owens-Corning)—was taped with Kapton® between two pieces of 7.62 by 10^{-3}-cm (3-mil) thick sheets of PETG film. This "sandwich" was heated in a 180°C press at minimum pressure for 10 min to fuse the polymer films. A dogbone per ASTM Test for Tensile Properties of Plastics by Use of Microtensile Specimens (D 1708) was cut so that the filament was aligned axially in the center of the gage section. The specimens were pulled on an Instron tension testing machine at a crosshead speed of 1.12×10^{-3} cm/s (0.05 in./min) until they began yielding. The segment lengths were measured from photographs taken through a Nikon polarizing microscope at ×50.

Results and Discussion

Surface Energy Measurements

Graphite—A summary of the Wilhelmy balance measurements is given in Table 2. This shows that for each test liquid there is significant hysteresis in the contact angles between the advancing and receding directions. This indicates considerable surface nonhomogeneity. Two very low surface tensile materials, hexane (γ = 18.4 mN/m) and silicone oil (γ = 18.7 mN/m), showed no hysteresis. This is consistent with the conclusion of Penn and Miller [*11*] that wetting hysteresis would disappear below a critical surface tension of the wetting liquid.

The average cosines were then used to calculate the components of the graphite surface energy that are given in Table 3. Since the advancing angles reflect the lower surface energy regions of the surface, the results for the epoxy-sized, as-received graphite show that this portion is considerably more polar than dis-

TABLE 2—*Summary of Wilhelmy balance measurements.*[a]

	COS θ	
	Advancing	Receding
	GRAPHITE—AS RECEIVED	
Water	0.383 (0.058)	0.840 (0.053)
Ethylene glycol	0.700 (0.016)	0.935 (0.038)
Glycerol	0.330 (0.057)	0.836 (0.058)
	400°C	
Water	0.235 (0.060)	0.698 (0.044)
Ethylene glycol	0.602 (0.038)	0.878 (0.065)
Glycerol	0.271 (0.037)	0.855 (0.067)
	GLASS—1.5% SILANE	
Water	0.171 (0.056)	0.745 (0.190)
Ethylene glycol	0.512 (0.031)	0.850 (0.064)
Glycerol	0.298 (0.030)	0.900 (0.148)
	BARE	
Water	0.525 (0.086)	0.810 (0.140)
Ethylene glycol	0.793 (0.010)	0.932 (0.011)
Glycerol	0.568 (0.082)	1.000 (0.070)

[a] Numbers in parentheses are standard deviations.

TABLE 3—*AS-4 graphite surface energy components, mN/m.*

	As-Received			400°C in Air			% Dispersive	
	Dispersive	Polar	Total	Dispersive	Polar	Total	As-Received	400°C
				ADVANCING				
Harmonic	10.8	28.6	39.4	12.2	20.9	33.1	27	37
Geometric	9.6	27.3	36.9	13.5	15.9	29.4	26	46
				RECEDING				
Harmonic	14.8	51.7	66.5	20.3	34.6	54.9	22	37
Geometric	15.9	46.5	62.4	18.9	35.8	54.7	26	35

persive ($\gamma_s^p = 29$ and $\gamma_s^d = 11$ mN/m for the harmonic mean and $\gamma_s^p = 27$ and $\gamma_s^d = 10$ for the geometric mean equations). These values are somewhat different than those found by Hammer and Drzal [*12*] ($\gamma_s^d = 26$, $\gamma_s^p = 30$) and Kaelble et al. [*13*] ($\gamma_s^d = 27$, $\gamma_s^p = 30$) for the AS filament and Benatar [*6*] ($\gamma_s^d = 25$, $\gamma_s^p = 28$) for the same AS-4 filament used in this study. Penn and Miller [*11*] did not calculate surface energy but their contact angle data for an unknown type of carbon fiber are much closer to the results reported here than are those of Hammer and Drzal and Kaelble et al. (Benatar's contact data is not available). Heating the filaments in air for 2 h at 400°C significantly increases the dispersive component from 27 to 37% (harmonic mean) while decreasing the total energy from 39 to 33 mN/m. These may be due to a low energy carbon layer deposited on the surface as a result of the heat treatment.

Table 3 also shows that with both filaments, the surface energy values computed from receding contact angles are higher than those computed from advancing angles. This is because the receding angles reflect the higher energy regions of the heterogeneous surface. There is apparently no difference in polarity (that is, percent dispersive) between the high and low energy regions.

In addition, these data point out that there are no significant differences when using either the harmonic mean or the geometric mean equations. Therefore, all subsequent results reported were calculated with only the harmonic mean equation.

The components of the graphite surface energy were then used to predict the work of adhesion between various polymers and the AS-4 filaments. The components of the polymeric surface energies used were taken from Ref *3d*. Table 4 lists the predicted work of adhesion between the graphite and the polymers. This indicates that polar resins, such as nylon 6,6, are predicted to adhere significantly better than nonpolar materials such as polystyrene or polyethylene. The graphite heated in air at 400°C shows the same general ranking, but with adhesion predicted to be slightly poorer for the polar polymers and slightly better with the more dispersive polymers. This obviously reflects the change in polarity shown in Table 3. Thus, this technique can show the effect of selective processing variations on the predicted work of adhesion between filaments and various thermoplastics. The works of adhesion calculated from the receding angles also

TABLE 4—*Graphite predicted work of adhesion, mN/m.*

	As-Received	400°C in Air
ADVANCING—HARMONIC MEAN		
Nylon 6,6	72	70
Poly(vinylidene fluoride)—PVDF	63	62
Poly(vinyl fluoride)—PVF	63	63
Poly(methyl methacrylate)—PMMA	62	62
Poly(ethylene terephthalate)—PET	61	61
Poly(vinylidene chloride)—PVDC	61	62
Poly(vinyl chloride)—PVC	53	55
Poly(butyl methacrylate)—PBMA	52	53
Polystyrene—PS	48	51
Polyethylene—PE	36	39
Poly(tetra fluoroethylene)—PTFE	36	38
RECEDING		
Nylon 6,6	87	89
Poly(vinyl fluoride)—PVF	75	77
Poly(methyl methacrylate)—PMMA	74	80
Poly(vinylidene chloride)—PVDC	73	81
Poly(ethylene terephthalate)—PET	72	80
Poly(vinylidene fluoride)—PVDF	66	66
Poly(vinyl chloride)—PVC	64	73
Poly(butyl methacrylate)—PBMA	61	68
Polystyrene—PS	58	68
Polyethylene—PE	45	55
Poly(tetrafluoroethylene)—PTFE	42	45

have the same general ranking as those from the advancing angles. The values are all higher, reflecting the better adhesion that occurs on higher energy surfaces.

Glass—The Wilhelmy Balance data are given in Table 2. As with the graphite, the surfaces are energetically heterogeneous. The surface energy components for the glass are given in Table 5. "Bare" glass was 33% dispersive with a total energy of 45 mN/m. The addition of the aminopropylsilane coating increased the dispersive component to 43%, but reduced the total energy to 31 mN/m. This is consistent with the addition of a low energy polymer to the surface. The receding components indicate the higher energy regions of the filament are more polar than the low energy regions.

The effect of the silane coating is shown in Table 6 in terms of the predicted work of adhesion with polymeric materials. The work of adhesion decreases 10 to 15% for the more polar resins and 5 to 10% for the more dispersive materials. The ranking is generally similar to that of the graphite.

TABLE 5—*Glass filament surface energy components, mN/m.*

	Amino—Silane			"Bare"			% Dispersive	
	Dispersive	Polar	Total	Dispersive	Polar	Total	Silane	No Silane
Advancing	13.3	17.8	31.1	14.6	30.3	44.9	43	33
Receding	17.8	41.9	59.7	23.1	39.4	62.5	30	27

TABLE 6—*Glass predicted work of adhesion, mN/m.*[a]

	Advancing—Harmonic Mean	
	"Bare"	1.5% Silane
Nylon 6,6	80	69
PVF	71	63
PMMA	70	63
PVDC	69	63
PET	69	62
PVDF	66	62
PVC	62	57
PBMA	59	55
PS	57	53
PE	44	42
PTFE	42	40

[a] See Table 4 for polymer code.

Notch Sensitivity Study

One example of the importance of these predictions is in the notch sensitivity of composite plaques. Figure 1 shows the normalized tensile strength (notched strength/unnotched strength) as a function of relative hole size (hole diameter/specimen width) for the four different resins. Each point represents the average of eight pulls taken from four separate plaques. As indicated by the

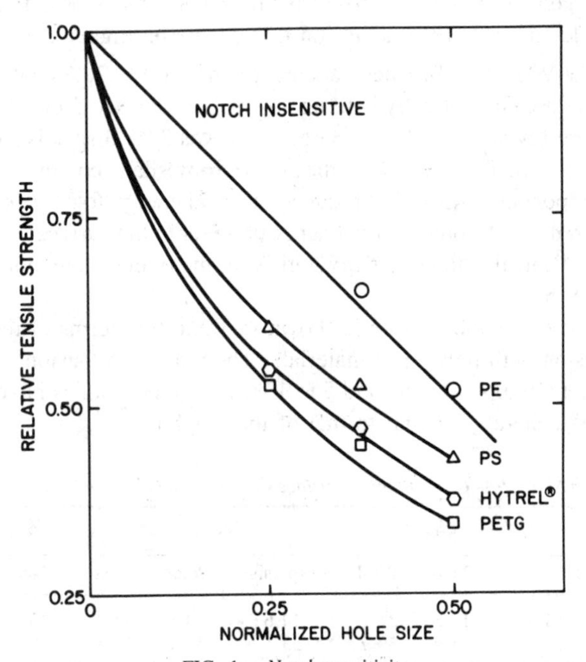

FIG. 1—*Notch sensitivity.*

notch-insensitive line, the normalized strength will decrease because of the area reduction that results from the presence of the hole. However, the magnitude of the decrease is generally greater than that merely caused by area reduction. The stress concentration caused by the hole causes earlier failure. The PETG has the largest decrease, while the polyethylene is essentially on the notch insensitive line.

Several parameters influence notched strength in composite materials. Some of these are fiber-matrix interfacial bonding, matrix properties, fiber properties, fiber orientation, and lamina stacking sequences [*14*]. In this work, the fiber, the orientation, and the stacking were kept constant, thereby eliminating these factors from causing variations in results. The ranking does not correlate with matrix strength, modulus, or energy to break as determined from the areas under the tensile stress-strain curve. However, the predictions made above do correlate. The works of adhesion predicted were 61, 51, and 39 mN/m for PET, polystyrene, and polyethylene, respectively. Though a prediction was not made for Hytrel®, it is reasonable to expect its work of adhesion with the fibers to be similar to that of the PET because of its butylene terephthalate hard segments, but somewhat poorer (that is, lower work of adhesion) as a result of the dispersive poly(tetramethyleneoxy) terephthalate soft segments. PETG should be very similar to PET, since they differ only in the presence of 30 mol % of a secondary glycol. Thus, the ranking of the predictions is inversely proportional to the experimental ranking of the relative strength at any hole size. PETG, with the highest predicted work of adhesion, has the largest notch sensitivity and lowest relative strength, while polyethylene, with the lowest predicted value is essentially notch insensitive. The Hytrel® and polystyrene are appropriately positioned between these extremes. Mechanistically, the poorer adhesion between the fiber and matrix may cause these filaments to debond under stress and thereby dissipate the energy that may otherwise break the filaments. The stress concentration caused by the hole would therefore have a smaller effect on the poorer adhering resins and result in a smaller decrease in relative strength for any hole size.

Transverse Tests

The limitations of these predictions are exemplified by the analysis of transverse tension tests of unidirectional glass/polycarbonate composites. The average results of twelve pulls from three different plaques for both "bare" and silane-coated glass are given in Table 7. These clearly show that the silane treatment doubles both the transverse elongation and strength, contrary to the work of adhesion predictions. Though the modulus is an elastic property, the fact that the two moduli are identical is proof that this test is examining a material property rather than differences in sample quality.

This property improvement appears to be due to an increase in fiber-matrix adhesion through the reaction between the matrix and the silane's organofunctional group. This was demonstrated by direct observation of the stress transfer by the "critical length" technique. This shear method can be applied to a

TABLE 7—*Average transverse tensile properties of unidirectional glass/polycarbonate.*[a]

Glass Type	Strength (MPa)	Elongation (%)	Modulus (GPa)
Silane-coated	22.3	0.21	10.1
	(6.1)	(0.09)	(2.5)
"Bare"	10.0	0.13	9.78
	(2.5)	(0.03)	(3.03)

[a] Note: numbers in parentheses are standard deviations.

transverse tension test since there is a large modulus mismatch between the fiber and the matrix. This results in a significant shear force because of Poisson's ratio effects. The relative length of the filaments compared to the average distance between them yields a region of pure shear [*3e*]. Shear failure is consistent with the visual observation of the failure surfaces. Experiments with "bare" and silane-treated glass fibers in PETG show that the mean critical length is significantly *shorter* with silane. These results are shown in the histograms of Fig. 2 which plot the fraction of the total number of segments found as a function of length. Since the average effective interfacial shear strength τ is inversely proportional to the critical length L_c (Eq 6), these figures demonstrate that the shear adhesion strength between the fiber and matrix is *increased* by the addition of the silane. Quantitatively, the average effective interfacial strengths were 12 and 23 MPa (1.8 and 3.3 kpsi) for the "bare" and silane treated glass, respectively. This appears to be the source of the improved transverse tensile properties described above, though these studies were done with polycarbonate as the matrix resin. In fact, the PETG results here are probably underestimates since polycarbonate would have reacted better with the particular silane used [*15*].

Contrary to these adhesion results, the studies described earlier predicted that the silane would *decrease* the work of adhesion by 5 to 15%. This is because absorbed organo-silanes have low surface energies [*3f*]. Wettability is apparently not the dominant factor in determining properties in reactive systems such as this. While good wetting is always necessary to avoid the formation of interfacial voids, it is not sufficient for good adhesion and properties. The use of work of adhesion predictions should therefore be limited to nonreactive systems where there are no chemical bonds that dominate the adhesion across the interface.

Conclusions

The surface energy technique described here can distinguish surface changes caused by processing variations and predict their effect on wettability in terms of the work of adhesion. The results of a notch sensitivity study in composite plaques are consistent with these predictions. However, since good wetting is apparently not the primary mechanism for improved adhesion and tensile properties in silane systems, the predictions do not apply.

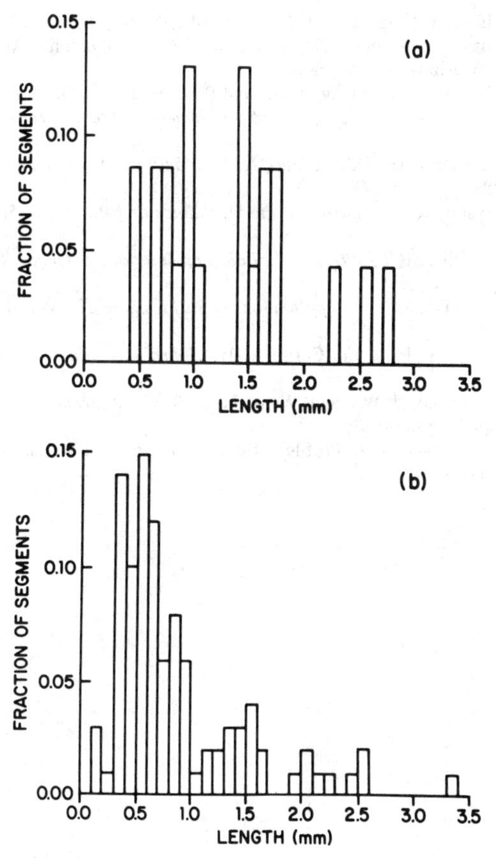

FIG. 2—*Critical length determination where* (a) *is "bare" glass and* (b) *is silane-coated glass.*

Acknowledgments

The author would like to thank Dr. S. Wu for his many insightful comments and suggestions.

References

[1] Martinez, G. M., Piggott, M. R., Bainbridge, D. M. R., and Harris, B., *Journal of Materials Science,* Vol. 16, No. 10, Oct. 1981, pp. 2831–2836.

[2] Hancox, N. L., *Journal of Materials Science,* Vol. 10, No. 2, Feb. 1975, pp. 234–242.

[3] Wu, S., *Polymer Interface and Adhesion,* Marcel Dekker, New York, 1982 (*a*) p. 4, (*b*) pp. 98–104, (*c*) p. 248, (*d*) p. 180, (*e*) p. 488, (*f*) pp. 594–600.

[4] Miller, B. and Young, R. A., *Textile Research Journal,* Vol. 45, No. 5, May 1975, pp. 359–365.

[5] Wesson, S. P. and Tarantino, A., *Journal of Non-Crystalline Solids,* Vols. 38 and 39, Part II, May/June 1980, pp. 619–624.

[6] Benatar, A., "Effects of Surface Modification of Graphite Fibers by Plasma Treatment on the Hygral Behavior of Composites," Master's thesis, MIT—Industry Polymer Processing Program, Massachusetts Institute of Technology, Cambridge, MA, May 1983.

[7] McMahon, P. E. and Ying, L., "Effects of Fiber/Matrix Interactions on the Properties of Graphite/Epoxy Composites," NASA Report CR-3607, National Aeronautics and Space Administration, Washington, DC, Sept. 1982.
[8] Drzal, L. T., Rich, M. J., Camping, J. D., and Park, W. J., in *35th Annual Conference of the Reinforced Plastics/Composites Institute of the Society of Plastics Industries,* Section 20-C, 1980, p. 1.
[9] Fraser, W. A., Ancker, F. H., Dibenedetto, A. T., and Elbirli, B., *Polymer Composites,* Vol. 4, No. 4, Oct. 1983, pp. 238–248.
[10] Kelly, A. and Tyson, W. R., *Journal of the Mechanics and Physics of Solids,* Vol. 13, 1965, pp. 329–350.
[11] Penn, L. S. and Miller, B., *Journal of Colloid and Interface Science,* Vol. 78, No. 1, Nov. 1980, pp. 238–241.
[12] Hammer, G. E. and Drzal, L. T., *Applications of Surface Science,* Vol. 4, Nos. 3 and 4, April 1980, pp. 340–355.
[13] Kaelble, D. H., Dynes, P. J., and Curlin, E. H., *Journal of Adhesion,* Vol. 6, No. 1, 1974, pp. 23–48.
[14] Pipes, R. B., Gillespie, J. W., and Wetherhold, R. C., *Journal of Composite Materials,* Vol. 13, April 1979, pp. 148–160.
[15] "Organofunctional Silanes—A Profile," Bulletin SUI-6, Union Carbide Corp., Silicones, Danbury, CT, May 1981.

James M. Whitney[1] *and Lawrence T. Drzal*[2]

Axisymmetric Stress Distribution Around an Isolated Fiber Fragment

REFERENCE: Whitney, J. M. and Drzal, L. T., **"Axisymmetric Stress Distribution Around an Isolated Fiber Fragment,"** *Toughened Composites, ASTM STP 937,* Norman J. Johnston, Ed., American Society for Testing and Materials, Philadelphia, 1987, pp. 179–196.

ABSTRACT: An analytical model is presented for predicting the stresses in a system consisting of a broken fiber surrounded by an unbounded matrix. The model is based on classical theory of elasticity and is applicable to the stress analysis of a single-fiber interfacial shear strength test specimen. Stresses in the vicinity of the broken fiber are approximated by a decaying exponential function multiplied by a polynomial. An exact solution is obtained for the far field stresses away from the broken end of the fiber. The fiber is assumed to be transversely isotropic. The model also includes the effect of expansional strains as a result of moisture and temperature. Numerical results are presented for AS-4 graphite fiber and Kevlar® 49 aramid fiber embedded in an epoxy matrix. Predicted values of critical fiber lengths are compared to experimentally measured values.

KEY WORDS: interface mechanics, single fiber test, shear lag analysis, broken fiber

A single-fiber interfacial shear strength test has been shown to provide a reproducibly accurate and sensitive means of determining the level of fiber and matrix adhesion in shear. However, this approach provides a single parameter which is the average shear strength of the fiber fragment which characterizes the fiber-matrix interface. To evaluate the interface and its role in the composite fracture process, or in determining composite toughness, the three-dimensional stress distribution around the fiber is desirable.

Stress distributions around discontinuous fibers have been studied by a number of researchers. Analysis of the class often referred to as "shear lag" have been performed by Cox [*1*], Rosen [*2*], and Kelly and Tyson [*3*]. In these models, only the fiber axial stress distribution and the fiber matrix interfacial shear stress

[1]Materials research engineer, Materials Laboratory, Air Force Wright Aeronautical Laboratories, AFWAL/MLBM, Wright-Patterson Air Force Base, OH 45433-6533; also adjunct associate professor, University of Dayton, Dayton, OH.

[2]Associate professor of chemical engineering, Composite Materials and Structures Center, Michigan State University, East Lansing, MI 48824-1226.

distribution are determined. In addition, the models are applicable to composites with multiple fibers, that is, fiber volume fraction is an input parameter. Amirbayat and Hearle [*4*] included the distribution of fiber matrix interfacial pressure in their model. As in the previous models, this solution also contained fiber volume fraction as an input parameter.

In the present paper, an approximate closed form solution is developed which predicts the axisymmetric stress distribution in a system consisting of a single broken fiber surrounded by an unbounded matrix. The approximate solution is based on a knowledge of the basic nature of the stress distribution near the end of a broken fiber. These stresses are represented by a decaying exponential function multiplied by a polynomial. Equilibrium equations and the boundary conditions of classical theory of elasticity are exactly satisfied throughout the fiber and matrix, while compatibility is only approximately satisfied. The far field solution away from the broken end exactly satisfies all of the equations of elasticity. The model also includes the effects of expansional strains as a result of moisture and temperature. Axisymmetric behavior is assumed in the development of the model.

Current reinforcements such as graphite and aramid fibers are orthotropic in nature. Previous "shear lag" type models have assumed isotropic fibers. The model developed in the present paper considers orthotropic fibers of the transversely isotropic class. In particular, the plane of the fiber cross section is assumed to be isotropic.

For purposes of continuity, a brief description of the single fiber test is presented before the development of the analytical model.

Single-Fiber Test

A single-fiber specimen is illustrated in Fig. 1. The gage section of this specimen is nominally 25 mm (1 in.) in length, while the cross section dimensions are 3 mm (0.125 in.) wide and 1.5 mm (0.0625 in.) thick. Fiber diameters are in the order of 0.01 mm (0.0004 in.) with the fiber centered axially in the coupon.

The specimen is loaded under a microscope and fiber breakage monitored. In the single-fiber tests under consideration in the present paper, the matrix consists of Epon® 828 cured with the stoichiometric amount of metaphenylenediamine (mPDA). This material provides an ideal model resin system for interface studies. In particular, Epon 828 has a high strain to failure which allows for multiple fiber breaks before matrix failure, an initial modulus which is comparable to epoxy resins utilized in high performance structural composites, good photoelastic properties, and processes at a relatively low temperature which minimizes residual stresses as a result of cure. Typical single-fiber specimen behavior is illustrated in Fig. 2. Although complete specimen fracture does not occur until a strain level of approximately 6%, fiber breakage is observed in the range of 1 to 3%.

Before the first fiber break, the axial strain in the gage section of a single-fiber specimen is assumed to be uniform. After the final loading, it is assumed that all

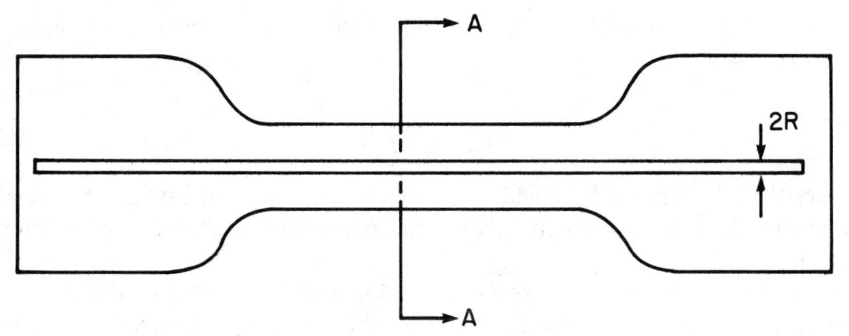

FIG. 1—*Single-fiber test specimen.*

of the fiber fragments have been reduced to their critical length $2L_c$ (see Fig. 2), or less. A theoretical value of L_c can be determined from a stress analysis and is associated with the distance along the broken fiber length required to dissipate end effects. Thus, when the fiber reaches the critical length of $2L_c$, the original unbroken fiber axial stress is only attained at the center cross section. Although further fiber fracture is possible, the probability is low, as it is unlikely that a defect will occur at the center cross section of the fiber. If fiber matrix interfacial fracture occurs before the fiber reaching its critical length, then the final fragment length will be larger than the theoretical critical length.

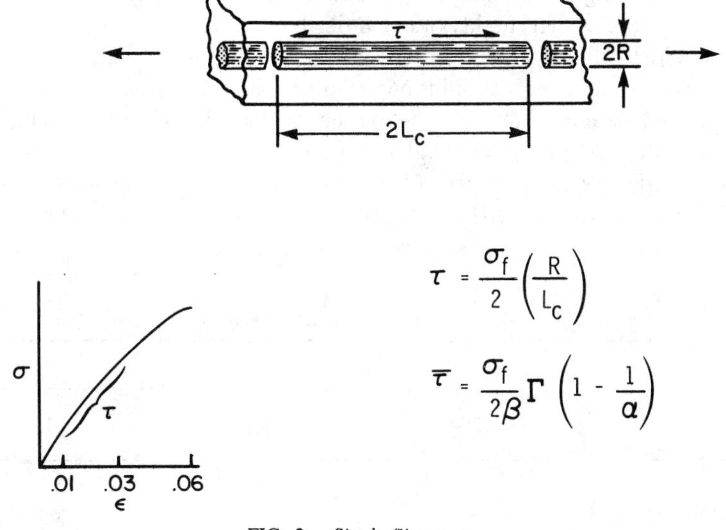

FIG. 2—*Single-fiber test.*

An average shear strength τ can be determined from a simple force balance with the result

$$\tau = \frac{\sigma_f}{2}\left(\frac{R}{L_c}\right) \tag{1}$$

where R is the fiber radius, and σ_f is the independently measured fiber axial strength at the critical length as determined from tension tests performed on fiber segments.

As a result of the variation of material properties and failure mode, (for example, interfacial failure versus the fiber reaching its critical length), a distribution of critical lengths will be obtained from a single-fiber test specimen. If the resulting values of L_c/R are fit to a two-parameter Weibull distribution, then a mean value of the interfacial shear strength can be obtained from the relationship [5]

$$\bar{\tau} = \frac{\sigma_f}{2\beta}\Gamma\left(1 - \frac{1}{\alpha}\right) \tag{2}$$

where $\bar{\tau}$ denotes the mean value of τ, while β and α are maximum likelihood estimates of the scale and shape parameters, respectively, for the Weibull distribution. In addition, Γ is the gamma function.

Typical experimental results are shown in Table 1 for AS-4 graphite fibers and Kevlar® 49 fibers. Both the critical length and the shear strength are listed.

The interfacial shear strength is very low and the critical length very high for the Kevlar 49 fiber compared to the AS-4 fiber. The failure modes for these two fibers are quite different as illustrated in Fig. 3. In particular, these fiber failures reveal a brittle-type fracture for the AS-4 fiber, while the Kevlar 49 fiber displays longitudinal splitting. A comparison of the substructures [6,7] of these two fibers is illustrated in Fig. 4. The highly oriented fibrils of the Kevlar 49 fibers promote the longitudinal splitting. Thus, in the case of the single-fiber interfacial shear strength test, the longitudinal splitting failure leads to apparent low values of interfacial shear strength, thus preventing the Kevlar 49 fiber from reaching its theoretical critical length in the single-fiber test.

With the structure of Kevlar 49 fibers, the radial stress components take on particular significance. In particular, radial compressive stress can provide sup-

TABLE 1—*Interfacial shear strength.*

Fiber	$\frac{L_c}{R}$	τ MPa	τ ksi
AS-4	57 ± 7	86	12
Kevlar 49	75 ± 5	23	3.3

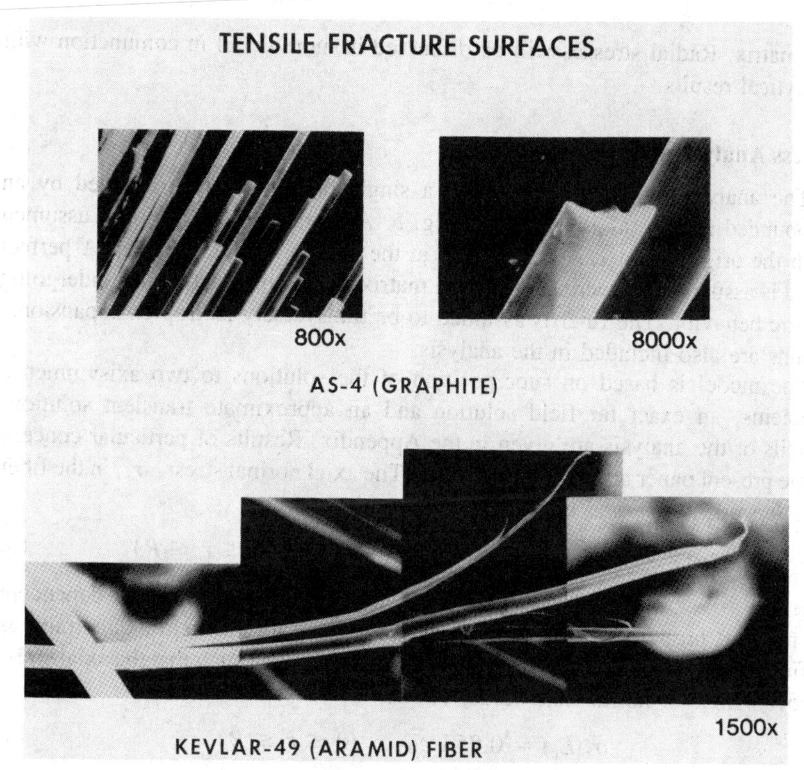

FIG. 3—*Comparison of fiber failure modes.*

port to the oriented fibrils of the fiber substructure. The longitudinal Poisson's ratio of Kevlar 49 fibers are, however, very close to the value of the matrix. In the case of AS-4 fibers, the longitudinal Poisson ratio is considerably less than

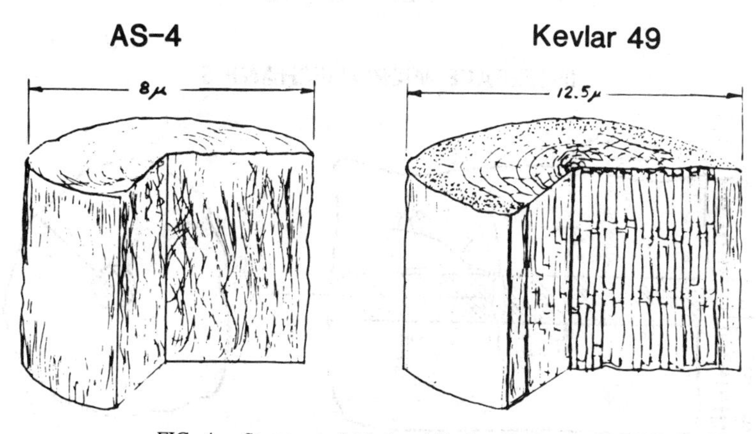

FIG. 4—*Structure of AS-4 and Kevlar 49 fibers.*

the matrix. Radial stresses will be discussed in more detail in conjunction with analytical results.

Stress Analysis Model

The analytical model is based on a single broken fiber surrounded by an unbounded matrix, as illustrated in Fig. 5. Axisymmetric behavior is assumed with the origin of an x, r axis system at the broken end of the fiber. A perfect bond is assumed between the fiber and matrix with both constituents undergoing elastic behavior. The fiber is assumed to be transversely isotropic. Expansional strains are also included in the analysis.

The model is based on superposition of the solutions to two axisymmetric problems, an exact far field solution and an approximate transient solution. Details of the analysis are given in the Appendix. Results of particular concern to the present paper are summarized here. The axial normal stress, σ_x, in the fiber is independent of r and is of the form

$$\sigma_x = [1 - (4.75\bar{x} + 1)e^{-4.75\bar{x}}]A_1\varepsilon_0 \qquad (0 \leq r \leq R) \tag{3}$$

where $\bar{x} = x/L_c$, ε_0 is the far field axial strain, and A_1 is a constant dependent on material properties, the expansional strains, and the far field axial strain, as defined in the Appendix. The critical length L_c is defined such that the axial stress recovers 95% of its far field value, that is,

$$\sigma_x(L_c) = 0.95A_1\varepsilon_0 \qquad (0 \leq r \leq R) \tag{4}$$

The interfacial shear stress is given by the relationship

$$\tau_{xr}(\bar{x}, R) = -4.75\mu A_1\varepsilon_0\bar{x}e^{-4.75\bar{x}} \tag{5}$$

where

$$\mu = \sqrt{\frac{G_m}{E_{1f} - 4\nu_{12f}G_m}} \tag{6}$$

INTERFACE MICROMECHANICS

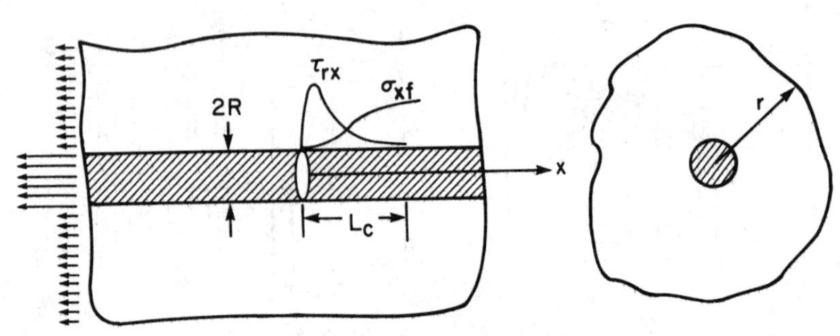

FIG. 5—*Stress analysis model.*

The axial fiber modulus is denoted by E_{1f}, the longitudinal Poisson ratio of the fiber ν_{12f} is determined by measuring the radial contraction under a uniaxial tensile load in the fiber direction, while G_m denotes the matrix shear modulus. Note the negative sign in front of Eq 5. In most "shear lag" type analysis, the interfacial shear stress is positive. The difference is simply the definition of positive shear which is not, in the case of "shear lag" models, consistent with classical theory of elasticity, while Eq 5 is.

The radial stress σ_r at the interface is of the form

$$\sigma_r(\bar{x}, R) = [A_2 + \mu^2 A_1(4.75\bar{x} - 1)e^{-4.75\bar{x}}]\varepsilon_0 \tag{7}$$

where A_2 is a constant dependent on material properties, the expansional strains, and the far field axial strain, as defined in the Appendix. Numerical results are normalized by σ_0, which represents the far field fiber stress in the absence of expansional strains. In particular,

$$\sigma_0 = A_3 \varepsilon_0 \tag{8}$$

with

$$A_3 = E_{1f} + \frac{4K_f \nu_{12f} G_m (\nu_{12f} - \nu_m)}{(K_f + G_m)} \tag{9}$$

where ν_m is the Poisson ratio of the matrix and K_f is the plane strain bulk modulus of the fiber and is related to classical engineering constants through the relationship

$$K_f = \frac{E_m}{2(2 - E_{2f}/2G_{2f} - 2\nu_{2f}E_{2f}/E_{1f})} \tag{10}$$

In Eq 10, E_{2f} denotes the radial modulus of the fiber and G_{2f} denotes the shear modulus relative to the plane of the cross section of the fiber.

Numerical Results

Numerical results are presented for an AS-4 fiber and Kevlar 49 fiber embedded in an Epon 828 epoxy matrix. Assumed material properties are listed in Table 2. The modulus of Epon 828 was measured, while the remaining properties were obtained from Refs *8* and *9*.

TABLE 2—*Material properties.*

Property	Epon 828	AS-4	Kevlar 49
E_1, GPa (Msi)	3.8 (0.55)	241 (35)	124 (18)
E_2, GPa (Msi)	3.8 (0.55)	21 (3)	6.9 (1)
ν_{12}	0.35	0.25	0.33
G_{23}, GPa (Msi)	1.4 (0.20)	8.3 (1.2)	2.6 (0.38)
α_1, 10^{-6}/°C (10^{-6}/°F)	68 (32)	−0.11 (−0.5)	−0.11 (−0.5)
α_2, 10^{-6}/°C (10^{-6}/°F)	68 (32)	8.5 (4)	64 (30)

Equations 3, 5, and 7 normalized by σ_0, are plotted in Figs. 6 through 8 for the graphite/epoxy system. The results in Figs. 6 and 7 are relatively insensitive to thermal strains caused by cool down from the cure temperature. As can be seen in Fig. 8, however, the radial stress is quite sensitive to thermal strains. The specimens used to produce the data in Table 1 were cured at 75°C (167°F) and postcured at 125°C (257°F). The difference between room temperature, 21°C (70°F), and the postcure temperature yields $\Delta T = -104$°C. This value of ΔT represents a worst case as far as residual stresses caused by cure in conjunction with the single-fiber test are concerned. The value of $\Delta T = -75$°C was chosen for the present examples, as it is assumed that some relieving of residual stresses occur during cool down from the postcure temperature.

Similar behavior compared to the results in Figs. 6 through 8 can be obtained in conjunction with Kevlar 49/Epon 828 material.

A comparison between radial stresses for the AS-4/Epon 828 system and the Kevlar 49/Epon 828 system under tensile axial load is made in Fig. 9. The potential fiber support supplied by the radial stress makes this comparison of particular interest.

Note that in the absence of expansional strains, ε_0 is normalized out of the numerical results. For the results containing thermal stresses, the ratio $\Delta T/\varepsilon_0$ must be input into the normalized results. Thus, for those cases, a precise value of ε_0 must be chosen. Since the results in Fig. 9 are not normalized, it is obvious that ε_0 must be prescribed. The value of $\varepsilon_0 = 1\%$ utilized in the numerical results represents the onset of observed fiber breakage in the single-fiber test.

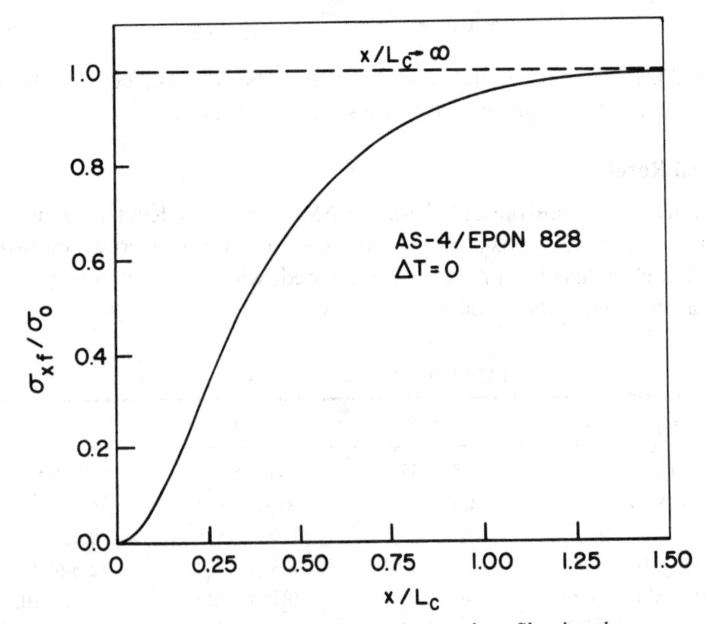

FIG. 6—*Fiber axial stress distribution along fiber length.*

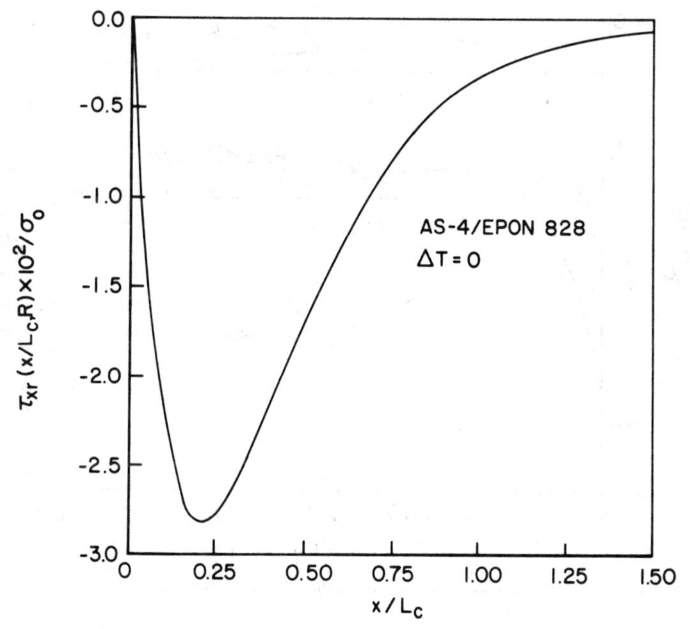

FIG. 7—*Interfacial shear stress distribution along fiber length.*

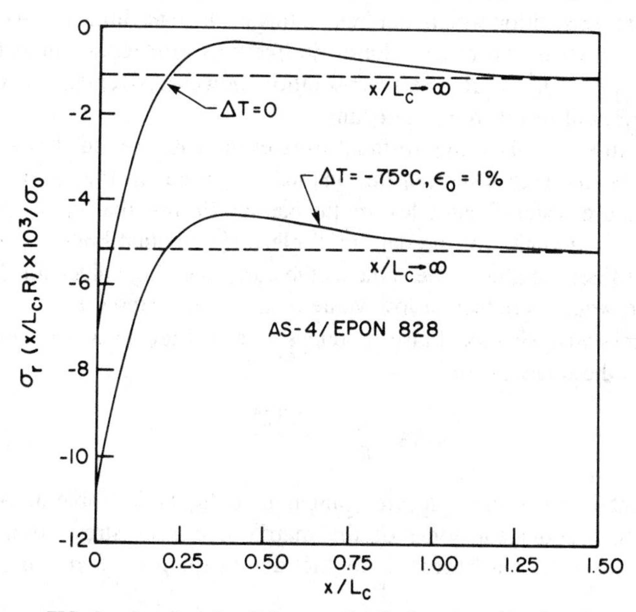

FIG. 8—*Interfacial radial stress distribution along fiber length.*

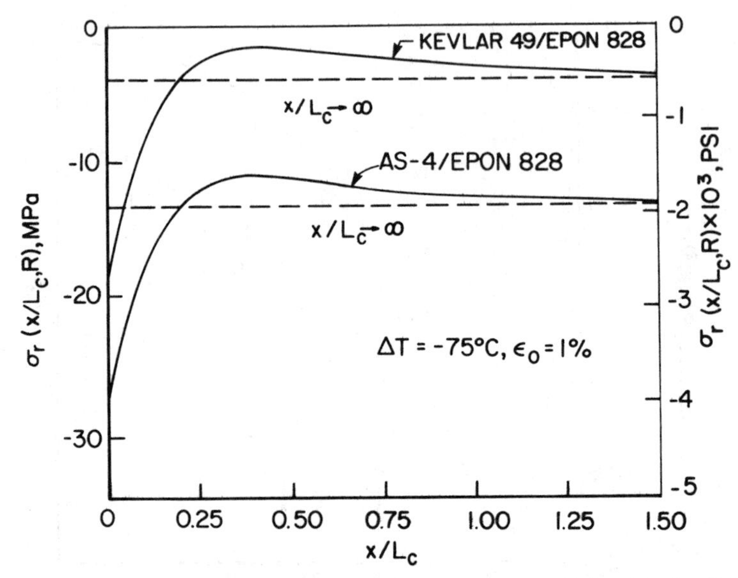

FIG. 9—*Comparison of interfacial radial stress distribution between AS-4 and Kevlar 49 fibers.*

Discussion

The maximum interfacial shear stress in Fig. 7 does not occur at the end of the fiber as in classical "shear lag" type analysis. This is due to the fact that the free end boundary condition which requires τ_{xr} to vanish on the broken end of the fiber is exactly satisfied. An exact solution is likely to produce a singularity at the coordinates $x = 0, r = R$. The exact solution, however, over the interval $x = 0$, $0 \leq r \leq R$ will result in τ_{xr} vanishing.

Results in Fig. 8 show that residual stresses increase the radial pressure at the fiber matrix interface when tensile loading is applied. In Fig. 9, however, the radial pressure is significantly less on the Kevlar 49 fiber than on the AS-4 fiber. A cursory examination of the data in Table 2 reveals that both ν_{12f} and α_2 for Kevlar 49 fiber are almost the same as the corresponding values for the matrix. Thus, one would anticipate a low value of interfacial pressure.

The stress analysis model also produces a critical length which can be calculated from the relationship

$$\frac{L_c}{R} = \frac{2.375}{\mu} \tag{11}$$

Experimental values of L_c/R are compared to Eq 11 in Table 3. An attempt is made to account for some of the matrix nonlinear stress-strain behavior by applying a secant modulus to the determination of μ. In particular,

TABLE 3—*Critical lengths.*

	AS-4	Kevlar 49
L_c/R (experiment)	57	75
L_c/R (theory)	31	22
L'_c/R (theory)	42	30

Eq 11 becomes

$$\frac{L'_c}{R} = \frac{2.375}{\mu'} \tag{12}$$

where

$$\mu' = \frac{G'_m}{E_{1f} - 4\nu_{12f}Gm'} \tag{13}$$

and G'_m is the secant modulus at $\varepsilon_0 = 3\%$. The value of G'_m was determined by using a measured value of the tensile secant modulus in conjunction with $\nu_m = 0.35$, that is,

$$G'_m = \frac{E'_m}{2(1 + \nu_m)} \tag{14}$$

where $E'_m = 2.1$ GPa (0.3 Msi) yields a value $G'_m = 0.77$ GPa (0.11 Msi).

As discussed previously, Eqs 11 and 13 yield a lower bound to L_c/R. Results in Table 3 reveal that the measured value of critical length for AS-4 is close to the theoretical value, while the theoretical value for Kevlar 49 is considerably below the experimental value. Again, the substructure of the Kevlar 49 fiber in conjunction with the low radial pressure at the fiber matrix interface makes this aramid fiber susceptible to longitudinal splitting which results in relatively low "apparent" interfacial shear strength and a high value of critical length.

Summary

An approximate solution based on classical theory of elasticity has been developed for determining the axisymmetric state of stress in a system consisting of a single broken fiber surrounded by an unbounded matrix. The analytical model is applicable to the stress analysis of a single-fiber interfacial shear strength test specimen. This model provides an improvement over existing "shear lag" models in that it provides an estimate of all stress components, is applicable to orthotropic fibers of the transversely isotropic class, includes the effects of expansional strains as a result of moisture and temperature, and provides a shear stress distribution that satisfies the free edge boundary conditions at the broken end of the fiber.

Numerical results indicate both the importance of radial stress and the impact of temperature caused by cure cycle cool down on the radial stress. Experimental results also indicate the importance of radial stress as a mechanism for providing support to the fiber. As discussed, little support is provided for Kevlar fibers through radial pressure as a result of the similarity between the fiber and matrix longitudinal Poisson's ratio and transverse coefficient of thermal expansion.

APPENDIX

Stress Analysis

Consider the model illustrated in Fig. 5 with the origin of an x, r coordinate system located at the end of the broken fiber. The fiber is assumed to be transversely isotropic and the matrix isotropic. A perfect fiber/matrix interfacial bond is also assumed.

The analysis is based on the superposition of an exact far field solution and an approximate local transient solution.

Far Field Solution

The far field solution represents the stress distribution away from the end of the broken fiber. Consider an infinite continuous, circular fiber surrounded by an unbounded matrix. The fiber and matrix are assumed to be subjected to a uniform axial strain ε_0.

The displacement field is assumed to be of the form

$$u = x\varepsilon_0, \qquad v = F(r), \qquad w = 0 \tag{15}$$

where u, v, and w denote displacements in the axial, radial, and hoop directions, respectively. Equation 15 leads to the following strains

$$\varepsilon_x = x\varepsilon_\theta, \qquad \varepsilon_r = \frac{dF}{dr}, \qquad \varepsilon_\theta = \frac{F}{r} \tag{16}$$

$$\gamma_{xr} = \gamma_{x\theta} = \gamma_{r\theta} = 0 \tag{17}$$

where ε_x, ε_r, and ε_θ denote normal strains in the axial, radial, and hoop directions, respectively, while γ_{xr}, $\gamma_{x\theta}$, and $\gamma_{r\theta}$ denote shear strains in $x - r$, $x - \theta$, and $r - \theta$ planes, respectively. Taking Eq 17 into account, the stress-strain relations for a transversely isotropic material in cylindrical coordinates are of the form

$$\sigma_x = (E_1 + 4K\nu_{12}^2)(\varepsilon_x - \bar{\varepsilon}_x) + 2K\nu_{12}(\varepsilon_r + \varepsilon_\theta - \bar{\varepsilon}_r - \bar{\varepsilon}_\theta) \tag{18}$$

$$\sigma_r = 2K\nu_{12}(\varepsilon_x - \bar{\varepsilon}_x) + (K + G_{23})(\varepsilon_r - \bar{\varepsilon}_r) + (K - G_{23})(\varepsilon_\theta - \bar{\varepsilon}_\theta) \tag{19}$$

$$\sigma_\theta = 2K\nu_{12}(\varepsilon_x - \bar{\varepsilon}_x) + (K - G_{23})(\varepsilon_r - \bar{\varepsilon}_r) + (K + G_{23})(\varepsilon_\theta - \bar{\varepsilon}_\theta) \tag{20}$$

$$\tau_{r\theta} = \tau_{x\theta} = \tau_{xr} = 0 \tag{21}$$

where σ_x, σ_r, and σ_θ are the normal stresses in the axial, radial, and hoop directions, respectively. The shear stresses in the $r - \theta$, $x - \theta$, and $x - r$ planes are denoted by $\tau_{r\theta}$, $\tau_{x\theta}$, and τ_{xr}, respectively. Overbars denote expansional strains.

The equilibrium equations of classical theory of elasticity are reduced to the single differential equation

$$\frac{d\sigma_r}{dr} + \frac{\sigma_r - \sigma_\theta}{r} = 0 \tag{22}$$

Combining Eqs 16, 18, 19, 20, and 21, we obtain the second order differential equation

$$\frac{d^2F}{dr^2} + \frac{1}{r}\frac{dF}{dr} - \frac{F}{r^2} = 0 \tag{23}$$

which yields a solution of the form

$$F = a_0 r + \frac{a_1}{r} \tag{24}$$

where a_0 and a_1 are constants. To avoid a singularity at $r = 0$,

$$a_{1f} = 0 \tag{25}$$

where the subscript f refers to the fiber, that is, $0 \leq r \leq R$. In addition, the radial and hoop stresses must dissipate in the matrix. Thus, a cursory examination of Eqs 16, 19, 20, and 24 reveals that

$$a_{0m} = (1 + \nu_m)\bar{\varepsilon}_m - \nu_m \varepsilon_0 \tag{26}$$

where the subscript m denotes the matrix, that is, $R \leq r$, and $\bar{\varepsilon}_m$ is the matrix expansional strain. The coefficients a_{0f} and a_{1m} can be determined from the continuity conditions

$$v_f(R) = v_m(R) \tag{27}$$

$$\sigma_{rf}(R) = \sigma_{rm}(R) \tag{28}$$

Combining Eqs 15, 16, 19, and 24 to 26 with Eqs 27 and 28, we obtain two algebraic equations from which a_{0f} and a_{1m} can be determined. These solutions lead to the following stresses and displacements

$$\sigma_{xf} = A_1 \varepsilon_0 \tag{29}$$

$$\sigma_{rf} = \sigma_{\theta f} = A_2 \varepsilon_0 \tag{30}$$

$$\sigma_{xm} = E_m(\varepsilon_0 - \bar{\varepsilon}_m) \tag{31}$$

$$\sigma_{rm} = -\sigma_{\theta m} = A_2\left(\frac{R}{r}\right)^2 \varepsilon_0 \tag{32}$$

$$v_f = A_4 r \varepsilon_0, \qquad v_m = \left(A_5 r + \frac{A_6}{r}\right)\varepsilon_0 \tag{33}$$

where

$$A_1 = E_{1f}(1 - \bar{\varepsilon}_{1f}/\varepsilon_0) + \frac{4K_f G_m \nu_{12f}}{(K_f + G_m)}\{\nu_{12f} - \nu_m + [(1 + \nu_m)\bar{\varepsilon}_m - \bar{\varepsilon}_{2f} - \nu_{12f}\bar{\varepsilon}_{1f}]/\varepsilon_0\}$$

$$A_2 = 2\frac{K_f G_m}{(K_f + G_m)}\{\nu_{12f} - \nu_m + [(1 + \nu_m)\bar{\varepsilon}_m - \bar{\varepsilon}_{2f} - \nu_{12f}\bar{\varepsilon}_{1f}]/\varepsilon_0\}$$

$$A_4 = \frac{-\{K_f \nu_{12f} + \nu_m G_m - [K_f(\bar{\varepsilon}_{2f} + \nu_{12f}\bar{\varepsilon}_{1f}) + G_m(1 + \nu_m)\bar{\varepsilon}_m]/\varepsilon_0\}}{(K_f + G_m)}$$

$$A_5 = -\nu_m + (1 + \nu_m)\bar{\varepsilon}_m/\varepsilon_0$$

$$A_6 = \frac{R^2 K_f}{(K_f + G_m)}\{\nu_m - \nu_{12f} + [\bar{\varepsilon}_{2f} + \nu_{12f}\bar{\varepsilon}_{1f} - (1 + \nu_m)]/\varepsilon_0\}$$

The fiber axial and radial expansional strains are denoted by $\bar{\varepsilon}_{1f}$ and $\bar{\varepsilon}_{2f}$, respectively. For the case of thermal expansion

$$\bar{\varepsilon}_m = \alpha_m \Delta T, \qquad \bar{\varepsilon}_{1f} = \alpha_{1f}\Delta T, \qquad \bar{\varepsilon}_{2f} = \alpha_{2f}\Delta T \tag{34}$$

where α_m, α_{1f}, and α_{2f} denote the linear coefficient of thermal expansion for the matrix, fiber axial direction, and fiber radial direction, respectively, and ΔT is the temperature change.

Transient Solution—For axisymmetric loading with x dependence, the equilibrium equations are of the form

$$\frac{\partial \sigma_x}{\partial x} + \frac{\partial \tau_{rx}}{\partial r} + \frac{\tau_{rx}}{r} = 0 \tag{35}$$

$$\frac{\partial \tau_{rx}}{\partial x} + \frac{\partial \sigma_r}{\partial r} + \frac{\sigma_r - \sigma_\theta}{r} = 0 \tag{36}$$

$$\frac{\partial \tau_{\theta x}}{\partial x} + \frac{\partial \tau_{r\theta}}{\partial r} + \frac{2\tau_{r\theta}}{r} = 0 \tag{37}$$

Equation 37 is exactly satisfied by the assumption

$$\tau_{\theta x} = \tau_{r\theta} = 0 \tag{38}$$

while Eqs 35 and 36 can be satisfied by utilizing a stress function ϕ such that

$$\sigma_x = \frac{\partial^2 \phi}{\partial r^2} + \frac{1}{r}\frac{\partial \phi}{\partial r} \tag{39}$$

$$\sigma_r = \sigma_\theta = \frac{\partial^2 \phi}{\partial x^2} \tag{40}$$

$$\tau_{xr} = -\frac{\partial^2 \phi}{\partial x \partial r} \tag{41}$$

Based on the results of the classical "shear lag" type analysis, we know that the character of σ_x and τ_{xr} at the fiber/matrix interface must be similar to the distributions illustrated in Fig. 5. Thus, we assume

$$\phi_f = B_0 r^2(1 + \lambda x)e^{-\lambda x} + F(x) \tag{42}$$

$$\phi_m = \frac{B_1}{r^2}(1 + \lambda x)e^{-\lambda x} \tag{43}$$

where B_0 and λ are constants. Since the broken end of the fiber must be free

$$\sigma_{xf}(0, r) = -A_1\varepsilon_0 \tag{44}$$

$$\tau_{xrf}(0, r) = 0 \tag{45}$$

Equation 44 assures that the total solution produces zero axial stress at the broken end of the fiber. Substituting Eq 42 into Eqs 39 and 41, taking into account Eq 44, we find that

$$B_0 = -\frac{A_1\varepsilon_0}{4} \tag{46}$$

$$\sigma_{xf} = -A_1\varepsilon_0(1 + \lambda x)e^{-\lambda x} \tag{47}$$

$$\tau_{xrf} = -\frac{A_1\varepsilon_0\lambda^2}{2} rxe^{-\lambda x} \tag{48}$$

The conditions in Eqs 44 and 45 are exactly satisfied by Eqs 47 and 48. Continuity conditions at the fiber/matrix interface require that

$$\tau_{xrf}(x, R) = \tau_{xrm}(x, R) \tag{49}$$

$$\sigma_{rf}(x, R) = \sigma_{rm}(x, R) \tag{50}$$

Combining Eqs 41 and 43 with Eq 49, taking into account Eq 48, we find

$$B_1 = \frac{A_1 R^4 \varepsilon_0}{4} \tag{51}$$

Substituting Eqs 42 and 43 into Eq 40, taking into account Eqs 46 and 51, we obtain the radial stresses

$$\sigma_{rf} = \frac{A_1\varepsilon_0\lambda^2}{4} r^2(1 - \lambda x)e^{-\lambda x} + f(x) \tag{52}$$

$$\sigma_{rm} = -\frac{A_1 R^4 \lambda^2}{4r^2}(1 - \lambda x)e^{-\lambda x} \tag{53}$$

where

$$f(x) = \frac{d^2F}{dx^2} \tag{54}$$

Substituting Eqs 52 and 53 into Eq 50, we obtain the result

$$f(x) = \frac{A_1 \varepsilon_0 R^2}{2}\lambda^2(1 - \lambda x)e^{-\lambda x} \tag{55}$$

The displacements are assumed to be of the form

$$u_f = \left[c_1\left(1 - \frac{r^2}{R^2}\right) + c_2 x\right]e^{-\lambda x} \tag{56}$$

$$u_m = c_2\left(\frac{R}{r}\right)^2 x e^{-\lambda x} \tag{57}$$

$$v_f = v_m = w_f = w_m = 0 \tag{58}$$

Equation 58 implies that

$$\gamma_{x\theta} = \gamma_{r\theta} = 0 \tag{59}$$

Continuity of u at the fiber/matrix interface is assured by Eqs 56 and 57. The constants c_1 and c_2 are chosen such that the fiber axial strain-stress relations are satisfied at the center of the fiber. In particular,

$$\varepsilon_{xf}(x, 0) = \frac{\partial u_f(x, 0)}{\partial x} = \frac{1}{E_{1f}}[\sigma_{xf}(x, 0) - \nu_{12f}\sigma_{rf}(x, 0)] \tag{60}$$

Substituting Eq 56 and the appropriate stresses into Eq 60, we can solve the resulting relationship for the undetermined constants, with the result

$$c_1 = \frac{2A_1 \varepsilon_0}{\lambda E_{1f}} \tag{61}$$

$$c_2 = \frac{A_1 \varepsilon_0}{E_{1f}}(1 + \nu_{12f}\lambda^2 R^2) \tag{62}$$

Note two important features of the displacement functions chosen in Eqs 56 and 57. In particular, the fiber displacement is parabolic at $x = 0$, which appears to be physically logical, and the matrix displacements damp out as r increases, which is required in this transient solution.

To obtain a relationship for λ, we consider the shear strain-stress relations in the matrix. Thus,

$$\gamma_{xrm} = \frac{\partial u_m}{\partial r} = \frac{\tau_{xrm}}{G_m} \tag{63}$$

Combining Eqs 41, 43, and 51, substituting the resulting expression for τ_{xrm} into Eq 63, along with Eqs 57 and 62, we obtain the algebraic equation

$$(E_{1f} - 4\nu_{12f}G_m)R^2\lambda^2 - 4G_m = 0 \tag{64}$$

from which we find

$$\lambda = \frac{2\mu}{R} \tag{65}$$

where

$$\mu = \sqrt{\frac{G_m}{E_{1f} - 4\nu_{12f}G_m}}$$

Determination of Critical Length —The critical length is defined as the length along the fiber required for the axial stress in the fiber to recover 95% of its far field value. Combining results from the far field solution and the transient solution, we find

$$\sigma_{xf} = [1 - (1 + \lambda x)e^{-\lambda x}]A_1\varepsilon_0 \tag{66}$$

The critical length is defined such that

$$\sigma_{xf}(L_c) = 0.95A_1\varepsilon_0 \tag{67}$$

Combining Eqs 66 and 67, we arrive at the relationship

$$(1 + \lambda L_c)e^{-\lambda L_c} = 0.05 \tag{68}$$

Iteration yields the result

$$\lambda L_c = 4.75 \tag{69}$$

or by utilizing Eq 65

$$L_c = \frac{4.75}{\lambda} = \frac{2.375R}{\mu} \tag{70}$$

Total Solution —Combining the far field and the transient solution leads to the results

$$\sigma_{xf} = [1 - (1 + 4.75\bar{x})e^{-4.75\bar{x}}]A_1\varepsilon_0 \tag{71}$$

$$\sigma_{rf} = \sigma_{\theta f} = \left[A_2 - A_1\mu^2\left(2 - \frac{r^2}{R^2}\right)(1 - 4.75\bar{x})e^{-4.75\bar{x}}\right]\varepsilon_0 \tag{72}$$

$$\tau_{xrf} = -4.75\mu A_1\varepsilon_0\left(\frac{r}{R}\right)\bar{x}e^{-4.75\bar{x}} \tag{73}$$

$$u_f = \left\{2.375E_{1f}\bar{x} + A_1\left[1 - \left(\frac{r}{R}\right)^2 + 2.375 \times (1 + 4\nu_{12f}\mu^2)\bar{x}\right]e^{-4.75\bar{x}}\right\}\frac{R\varepsilon_0}{\mu E_{1f}} \tag{74}$$

$$\sigma_{xm} = \left[E_m(1 - \bar{\varepsilon}_m/\varepsilon_0) - A_1\left(\frac{R}{r}\right)(1 - 4.75\bar{x})e^{-4.75\bar{x}}\right]\varepsilon_0 \tag{75}$$

$$\sigma_{rm} = [A_2 - A_1\mu^2(1 - 4.75\bar{x})e^{-4.75\bar{x}}]\left(\frac{R}{r}\right)^2\varepsilon_0 \tag{76}$$

$$\sigma_{\theta m} = -[A_2 + A_1\mu^2(1 - 4.75\bar{x})e^{-4.75\bar{x}}]\left(\frac{R}{r}\right)^2\varepsilon_0 \tag{77}$$

$$\tau_{xrm} = -4.75\mu\left(\frac{R}{r}\right)^3\bar{x}A_1\varepsilon_0 e^{-4.75\bar{x}} \tag{78}$$

$$u_m = 2.375\frac{R\varepsilon_0}{\mu E_{1f}}\left[E_{1f}\bar{x} + A_1(1 + 4\nu_{12f}\mu^2)\left(\frac{R}{r}\right)^2\bar{x}e^{-4.75\bar{x}}\right] \tag{79}$$

Both ν_f and ν_m are given by Eq 33.

References

[1] Cox, H. L., *British Journal of Applied Physics,* Vol. 3, No. 1, 1952, pp. 72–79.

[2] Rosen, B. W., *Fiber Composite Materials,* Chap. 3, American Society of Metals, pp. 37–75.

[3] Kelly, A. and Tyson, W. R., *Journal of Mechanics and Physics of Solids,* Vol. 10, No. 2, 1963, pp. 199–208.

[4] Amirbayat, J. and Hearle, W. S., *Fiber Science and Technology,* Vol. 2, No. 2, 1969, pp. 123–141.

[5] Drzal, L. T., Rich, M. J., Camping, J. D., and Park, W. J., in *Proceedings of the 35th Annual Technical Conference of the Reinforced Plastics/Plastics Institute,* The Society of the Plastics Industry, Inc., Chicago, 1980, Section 20-C, pp. 1–7.

[6] Diefendorf, R. J. and Torkarsky, E. W., *The Relationship of Structure to Properties in Graphite Fibers,* U.S. Air Force Report AFML-TR-72-133, 1972.

[7] Pruneda, C. O., Steele, W. J., Kershaw, R. P., and Morgan, R. J., *Polymer Reprints,* American Chemical Society, New York, Vol. 22, 1981, pp. 216–228.

[8] Ashton, J. E., Halpin, J. C., and Petit, P. H., *Primer on Composite Materials Analysis,* Technomic Publishing Co., Inc., Stamford, CT, 1969.

[9] *Fiber Composite Analysis and Design, Vol. 1, Composite Materials and Laminates,* prepared by Materials Science Corporation, U.S. Department of Transportation, Federal Aviation Administration, 1984.

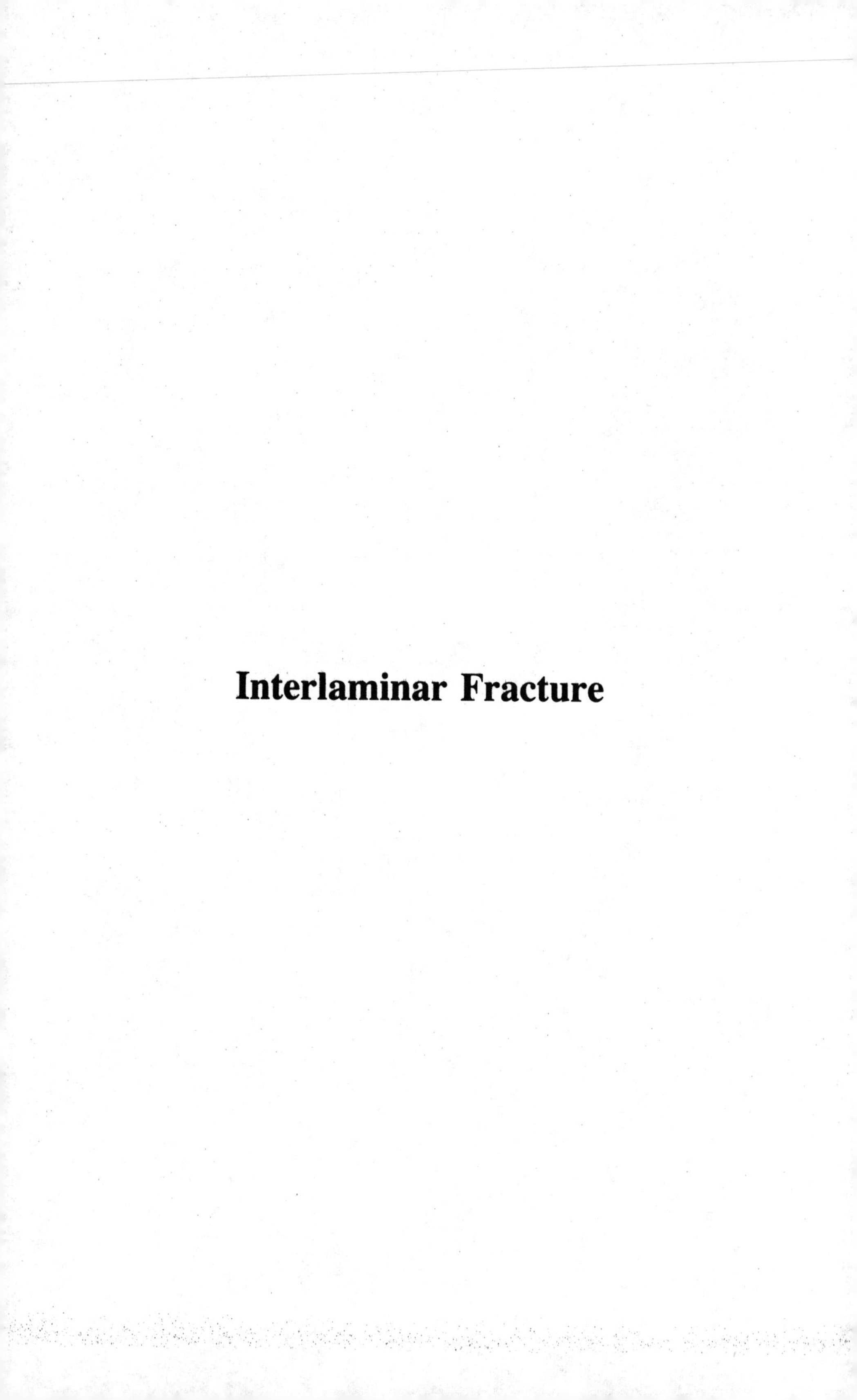

Interlaminar Fracture

T. Kevin O'Brien,[1] Norman J. Johnston,[2] Ivatury S. Raju,[3]
Don H. Morris,[4] and Robert A. Simonds[4]

Comparisons of Various Configurations of the Edge Delamination Test for Interlaminar Fracture Toughness

REFERENCE: O'Brien, T. K., Johnston, N. J., Raju, I. S., Morris, D. H., and Simonds, R. A., **"Comparisons of Various Configurations of the Edge Delamination Test for Interlaminar Fracture Toughness,"** *Toughened Composites, ASTM STP 937,* Norman J. Johnston, Ed., American Society for Testing and Materials, Philadelphia, 1987, pp. 199-221.

ABSTRACT: Various configurations of edge delamination tension (EDT) test specimens were manufactured and tested to assess the usefulness of each configuration for measuring interlaminar fracture toughness. Tests were performed on both brittle (T300/5208) and toughened-matrix (T300/BP907) graphite reinforced composite laminates. The mixed-mode interlaminar fracture toughness G_c was measured during tension tests of $(30/-30_2/30/90_n)_s$, n = 1 or 2; $(35/-35/0/90)_s$; and $(35/0/-35/90)_s$ layups designed to delaminate at low tensile strains. Laminates were made without inserts so that delaminations would form naturally between the central 90° plies and the adjacent angle plies. Laminates were also made with Teflon® inserts implanted between the 90° plies and the adjacent angle (θ) plies at the straight edge to obtain a planar fracture surface. In addition, Mode I interlaminar tension fracture toughness G_{Ic} was measured from laminates with the same layups but with inserts in the midplane, between the central 90° plies, at the straight edge. All of the EDT configurations were useful for ranking the delamination resistance of composites with different matrix resins. Furthermore, the variety of layups and configurations available yield interlaminar fracture toughness measurements, both pure Mode I and mixed mode, needed to generate delamination failure criteria.

The influence of insert thickness and location, and coupon size on G_c values were evaluated. For toughened-matrix composites, laminates with 1.5-mil (38.1-μm) thick inserts yielded interlaminar fracture toughness numbers consistent with data generated from laminates without inserts. Coupons of various sizes yielded similar G_c values. The influence of residual thermal and moisture stresses on calculated strain energy release rate for edge

[1]Senior research scientist, Aerostructures Directorate, U.S. Army Research and Technology Activity (AVSCOM), NASA Langley Research Center, MS 188E, Hampton, VA 23665.

[2]Senior research scientist, NASA Langley Research Center, MS 226, Hampton, VA 23665.

[3]Senior scientist, Analytical Services and Materials, NASA Langley Research Center, 28 Research Dr., Hampton, VA 23665.

[4]Professor and instructor, respectively, Virginia Polytechnic Institute, Blacksburg, VA 24061.

delamination was also reviewed. Edge delamination data may be used to quantify the relative influence of residual thermal and moisture stresses on interlaminar fracture for different composite materials.

KEY WORDS: composite materials, delamination, fracture, toughness, graphite/epoxy

Nomenclature

Δa	Finite element size at delamination front
E	Axial modulus
E_{LAM}	Laminate modulus
E^*	Modulus of laminate completely delaminated along one or more interfaces
E_{11}, E_{22}	Lamina moduli
G_{12}	Lamina shear moduli
G	Strain energy release rate
G_I, G_{II}, G_{III}	Strain energy release rate components as a result of opening, sliding shear, and tearing shear fracture modes
G^M, G^{M+T}, G^{M+T+H}	Strain energy release rate as a result of mechanical, mechanical plus thermal, and mechanical plus thermal plus hygroscopic loads
G_C	Critical strain energy release rate for delamination onset
G_{Ic}	Critical Mode I strain energy release rate for delamination onset
ΔH	Percentage moisture weight gain
h	Ply thickness
N	Number of plies
N_x, N_y, N_{xy}	In-plane stress resultants
M_x, M_y, M_{xy}	Moment resultants
$(\overline{Q})$	Transformed reduced stiffness matrix
ΔT	Temperature difference between stress-free temperature and test temperature
t	Laminate thickness
t_A	Thickness of adhesive bond
u	Strain energy density
$\varepsilon_x, \varepsilon_y, \gamma_{xy}$	In-plane strains
$\kappa_x, \kappa_y, \kappa_{xy}$	Out-of-plane curvatures
ν_{12}	Lamina Poisson's ratio
σ	Stress
θ	Fiber orientation angle

Introduction

A simple tension test was proposed for measuring the mixed-mode interlaminar fracture toughness of composites [1–5]. In this test, laminates are loaded in tension to develop high interlaminar tensile and shear stresses at the straight edge causing delamination. For these laminates, a noticeable change in the linear load-deflection curve occurs at the onset of edge delamination. The strain at delamination onset is substituted into a closed form equation for strain energy release rate G to obtain the critical value G_c for edge delamination. This G_c value is a measure of the interlaminar fracture toughness of the composite material. Finite element analyses are performed to obtain the contribution of the crack-opening G_I, sliding-shear G_{II}, and tearing-shear G_{III} fracture modes to the total strain energy release rate.

The edge delamination tension (EDT) test has been used to rank the relative delamination resistance of composites with brittle and toughened-resin matrices, and determine their fracture mode dependence. Toughness measurements from other interlaminar fracture toughness test configurations agree well with EDT measurements [*1–5*]; however, the accuracy of interlaminar fracture toughness measurements generated from such tests has been questioned [*6*]. Self-similarity of delamination growth, accuracy of the finite element analysis of mixed-mode ratios, and the influence of residual thermal and moisture stresses on critical strain energy release rates G_c are some of the concerns that have been raised. Recently, a pure Mode I version of the EDT test, with Teflon® inserts embedded in the midplane at the straight edge, was proposed to overcome some of these concerns [*7*].

This paper will examine interpretation of data for three configurations of the EDT test, one without inserts, one with midplane inserts, and one with inserts at the $\theta/90$ interfaces. Four different layups were tested: $(30/-30_2/30/90_n)_s$, $n = 1, 2$; $(35/-35/0/90)_s$; and $(35/0/-35/90)_s$. These layups were designed to yield the lowest delamination onset strain to measure a given G_c [*4*]. Laminates were tested in the three configurations: (1) the pure Mode I configuration with midplane inserts, (2) a mixed-mode configuration with inserts at the interface between the central 90° plies and the adjacent angle plies, and (3) the original mixed-mode configuration where delaminations form naturally at the $\theta/90$ interfaces. Data generated from EDT tests with different coupon sizes and insert thicknesses were compared for composites with graphite fibers (Thornel T300)[5] in both brittle (Narmco 5208) and tough (Cycom BP907) matrix resins. The accuracy of finite element analysis of mixed-mode ratios and the significance of residual thermal and moisture stresses to strain energy release rates were also addressed.

Materials

Composite panels of two graphite/epoxy materials, Thornel 300 (T300) fibers in Narmco 5208 matrix and T300 fibers in American Cyanamid BP907 matrix, were fabricated. Table 1 lists the basic lamina properties ($E_{11}, E_{22}, G_{12}, \nu_{12}$) measured for these two materials using the procedure outlined in Ref *2*. Panels were

[5]Use of manufacturer's tradename does not constitute endorsement, either expressed or implied, by NASA or AVSCOM.

TABLE 1—*Lamina material properties.*

	T300/5208	T300/BP907
E_{11}, Msi[a]	18.2	15.0
E_{22}, Msi	1.23	1.23
G_{12}, Msi	0.832	0.700
ν_{12}	0.292	0.314

[a] MSI = 6.89 GPa.

made with the following layups: $(30/-30_2/30/90_n)_s$, where n = 1 or 2; $(35/-35/0/90)_s$; and $(35/0/-35/90)_s$. Thin strips of Teflon were inserted at selected locations in each panel using a template. As shown in Fig. 1, panels were constructed so that coupons with and without inserts were cut from the same panel. The Teflon strips were either 1.5 or 3.5 mil (38.1 or 88.9 μm) thick, and were placed either at a single $\theta/90$ interface or at the midplane between the two central 90° plies. Coupons were cut from the panels with inserts extending either throughout the width, for determining laminate modulus with the interface completely delaminated, or with inserts extending partially through the width from both edges, for measuring interlaminar fracture toughness. Table 2 lists the five coupon sizes that were tested. Unless otherwise specified, 5-in. (127-mm) long by 1-in. (25 mm) wide Size E coupons were tested.

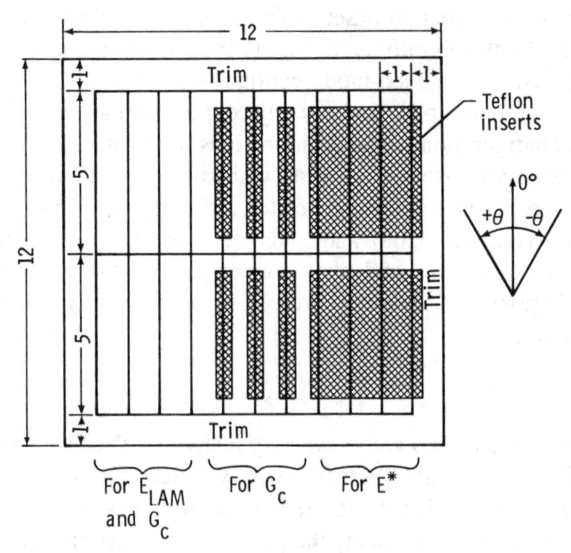

FIG. 1—*EDT panel showing coupon location and inserts (dimensions are in inches, 1 in. = 2.54 cm).*

TABLE 2—*Specimen dimensions.*[a]

Coupon Size	Length, in.	Width, in.	Grip Distance, in.	Gage Length, in.
A	10	1.5	7	4
B	10	1.0	7	4
C	10	0.5	7	4
D	5	0.5	3	1
E	5	1.0	3	1

[a] 1 in. = 2.54 cm.

Test Procedure

Coupons were loaded in tension, through friction grips, in either a screwdriven or hydraulic machine at a relatively slow crosshead speed. Tests were conducted under ambient laboratory conditions, that is, at a nominal room temperature of 70°F (21°C) and a relative humidity of 60%. In most cases, a minimum of five replicate tests were performed for each laminate orientation. Longitudinal strain was measured using extensometers, either a single clip-gage or a pair of direct current differential transducers (DCDTs) mounted on the centers of the front and back faces of the coupon. Table 2 lists the extensometer gage lengths for the various specimen sizes tested. Load and strain were continuously monitored and recorded on an *x-y* recorder. Coupons without inserts were loaded until delaminations formed on the edge and the corresponding abrupt jump in the load deflection curve was observed [*1–5*]. Coupons with inserts extending partially through the width from either edge were loaded until a noticeable change in slope or nonlinearity was observed in the load-deflection curve. A zinc-iodide solution was injected in the delaminated interface, and an X-ray radiograph was taken to confirm that a delamination had extended from the Teflon insert. Coupons with inserts extending throughout the laminate width were loaded until a load deflection curve was obtained for measuring laminate modulus with the interface delaminated throughout.

Analysis

Laminated Plate Theory

The interlaminar fracture toughness G_c of a composite laminate is the critical value of the strain energy release rate G required to grow a delamination. A closed-form equation was derived for the mixed-mode strain energy release rate for edge delamination growth in a composite laminate [*1*]. This equation

$$G = \frac{\varepsilon^2 t}{2}(E_{LAM} - E^*) \tag{1}$$

where

ε = nominal tensile strain,
t = laminate thickness,
E_{LAM} = laminate modulus, and
E^* = modulus of a laminate completely delaminated along one or more interfaces

is independent of delamination size. The strain energy release rate depends on the laminate layup and the location of the delaminated interface, which determines $(E_{LAM} - E^*)$. If the lamina properties are known, then E_{LAM} and E^* can be calculated from laminated plate theory and the rule of mixtures [1–5]. The $(30/-30_2/30/90_n)_s$ layups delaminate at the 30/90 interfaces.

As outlined in Ref 1, after delamination these layups are modeled as three sublaminates, two $(30/-30)_{2S}$ and one $(90)_{2n}$ laminate, loaded in parallel to account for the loss in transverse contraction as delaminations grow under an applied strain. Thus,

$$E^* = \frac{8E_{(30/-30)} + 2nE_{(90)}}{8 + 2n} \tag{2}$$

where $E_{(90)}$ is equal to E_{22}, and $E_{(30/-30)}$ can be calculated either from laminated plate theory or measured from a tension test of a $(30/-30)_s$ laminate [1–5]. The $(35/-35/0/90)_s$ and $(35/0/-35/90)_s$ layups delaminate between the 0/90 and −35/90 interfaces, respectively. After delamination, these laminates are modeled as two $(35/0/-35)_s$ sublaminates and one $(90)_2$ laminate, yielding

$$E^* = \frac{6E_{(35/0/-35)} + 2E_{(90)}}{8} \tag{3}$$

where $E_{(35/0/-35)}$ may be calculated either from laminated plate theory or measured from a tension test of a $(35/0/-35)_s$ laminate. However, assuming the sublaminates to be symmetric yields a slightly different axial modulus than if they are modeled as $(35/-35/0)_T$ and $(35/0/-35)_T$ asymmetric laminates as a result of the bending-extension coupling and twist-extension coupling present in these two asymmetric layups, respectively. The axial modulus of an asymmetric layup may be calculated from laminated plate theory by assuming N_y, N_{xy}, κ_x, M_y, and κ_{xy} are all zero for a constant ε_x [8–10]. This technique allows for a nonzero κ_y and yields a slightly different axial modulus for the asymmetric configuration than for the symmetric configuration.

Table 3 compares the axial modulus calculated from laminated plate theory for the (35/−35/0) and (35/0/−35) sublaminates for both the symmetric and asymmetric configurations using lamina properties from Table 1. A small difference in modulus was obtained for the $(35/-35/0)_T$ layup compared to the $(35/0/-35)_s$ layup, but no significant difference was observed for the $(35/0/-35)_T$ layup.

TABLE 3—*Influence of asymmetry on sublaminate moduli.*

Layup	E, msi	
	T300/5208	T300/BP907
$(35/0/-35)_s$	9.699	8.051
$(35/0/-35)_T$	9.698	8.053
$(35/-35/0)_T$	9.562	7.927
$(35/-35/0/90_2)_T$	6.468	5.436
$(35/0/-35/90_2)_T$	6.664	5.604
$(30/-30)_s$	7.030	5.899
$(30/-30_2/30/90_2)_T$	5.640	4.770
$(30/-30_2/30/90_4)_T$	4.885	4.150

Because the delaminations that formed naturally (that is, without artificially implanted inserts) at $\theta/90$ interfaces in all the layups tested wandered from one $\theta/90$ interface to its symmetric counterpart (Fig. 2*a*), these laminates were all modeled as a set of three symmetric sublaminates after delamination (Fig. 2*b*), [*1–5*]. However, for the laminates that contained Teflon inserts in one $\theta/90$ interface (Fig. 2*c*), the laminates were modeled as two asymmetric sublaminates after delamination. Hence, the equations for the delaminated modulus E^* for the $(30/-30_2/30/90_n)_s$, $(35/-35/0/90)_s$, and $(35/0/-35/90)_s$ layups become

$$E^* = \frac{4E_{(30/-30)_s} + (4 + 2n)E_{(30/-30_2/30/90_{2n})_T}}{8 + 2n} \quad (4)$$

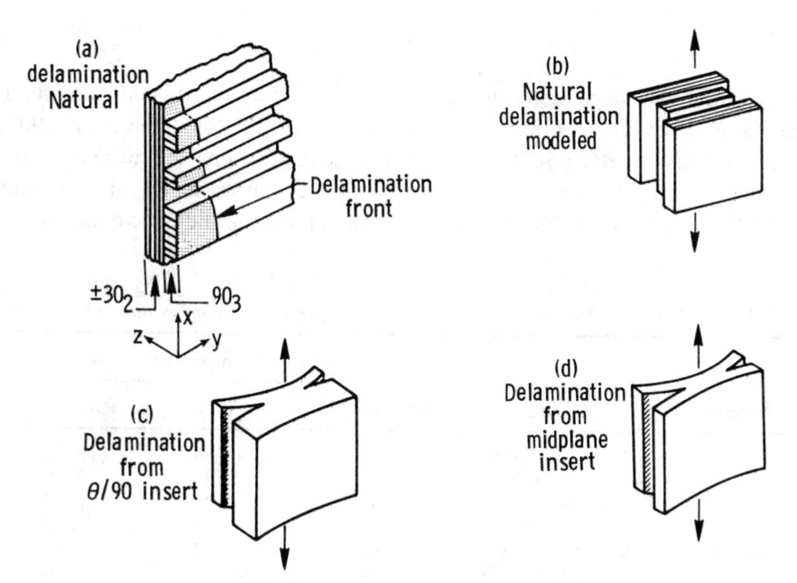

FIG. 2—*EDT test configurations.*

$$E^* = \frac{3E_{(35/-35/0)_T} + 5E_{(35/-35/0/90_2)_T}}{8} \quad (5)$$

$$E^* = \frac{3E_{(35/0/-35)_T} + 5E_{(35/0/-35/90_2)_T}}{8} \quad (6)$$

respectively. The asymmetric sublaminate moduli in Eqs 4 to 6 were calculated using lamina properties from Table 1 and are listed in Table 3. Table 4 compares the delaminated modulus E^* calculated for the natural delamination to E^* values calculated for the single artificially delaminated $\theta/90$ interface. The differences among E^* values, and hence the corresponding differences among G values from Eq 1, illustrate that for the natural delamination case the delamination is driven only by a mismatch in transverse (Poisson) contraction between the sublaminates, but for the artificially delaminated case, the delamination is driven by a combination of Poisson mismatch and the curvature assumed by the asymmetric sublaminates before the delamination grows from the insert.

If the insert is placed at the midplane between the two central 90° plies (Fig. 2*d*), as was proposed in Ref 7, then no Poisson mismatch results, and the delamination is driven entirely by the curvature assumed by the asymmetric sublaminates before the delamination grows from the insert. For this midplane delamination case, the delaminated moduli of the $(30/-30_2/30/90_n)_s$, $(35/-35/0/90)_s$, and $(35/0/-35/90)_S$ layups become

$$E^* = E_{(30/-30_2/30/90_n)_T} \quad (7)$$

$$E^* = E_{(35/-35/0/90)_T} \quad (8)$$

$$E^* = E_{(35/0/-35/90)_T} \quad (9)$$

respectively. The asymmetric moduli in Eqs 7 to 9 were calculated using lamina properties from Table 1 and listed in Table 4 as E^* for a midplane (90/90 interface) insert. Because these midplane delaminations are driven entirely by asymmetric sublaminate curvature with no Poisson mismatch, the delamination is purely an opening Mode I fracture. Therefore, for midplane delamination,

TABLE 4—*Delaminated modulus* E* *for different EDT configurations.*

		Delaminated Modulus E^*, msi		
Material	Layup	Natural $\theta/90$	$\theta/90$ Insert	90/90 Insert
T300/5208	$(30/-30_2/30/90)_s$	5.870	6.196	6.420
	$(30/-30_2/30/90_2)_s$	5.097	5.600	5.640
	$(35/-35/0/90)_s$	7.582	7.628	7.550
	$(35/0/-35/90)_s$	7.582	7.802	7.855
T300/BP907	$(30/-30_2/30/90)_s$	4.965	5.222	5.404
	$(30/-30_2/30/90_2)_s$	4.343	4.733	4.770
	$(35/-35/0/90)_s$	6.346	6.370	6.310
	$(35/0/-35/90)_s$	6.346	6.522	6.570

Eq 1 becomes

$$G_I = \frac{\varepsilon^2 t}{2}(E_{LAM} - E^*) \tag{10}$$

where E^* is calculated from one of Eqs 7 to 9, for the particular layup tested.

Recently, an analysis was developed that incorporates the influence of residual thermal and moisture stresses to the strain energy release rate for edge delamination [*10*]. This analysis yielded the following equation for the total strain energy release rate

$$G = t_{LAM}u_{LAM} - t_{SUB}u_{SUB} - t_{90}u_{90} \tag{11}$$

where t is the thickness and u is the strain energy density of the original laminate (LAM), the sublaminates (SUB), and the 90° plies (90). The strain energy density is defined as

$$u = \frac{1}{2N}\sum_{k=1}^{N}\{\varepsilon\}_k^t\{\sigma\}_k \tag{12}$$

where N is the number of plies, and $\{\varepsilon\}_k^t$ is the transpose of the total strain vector for the kth ply, which includes contributions from mechanical loading, thermal gradients (ΔT), and hygroscopic (moisture) percentage weight gain (ΔH). The stress vector for the kth ply in Eq 12 is given by

$$\{\sigma\}_k = [\overline{Q}]_k\{\varepsilon\}_k \tag{13}$$

where $[\overline{Q}]_k$ is the transformed reduced stiffness matrix of the kth ply as defined in laminated plate theory. Therefore, Eq 11 requires a ply-by-ply evaluation of the strain energy density in the laminated and delaminated regions to account for the biaxial thermal and moisture stresses present in the laminate.

Figure 3 shows the influence of residual thermal and moisture stresses on G for edge delamination in the $-30/90$ interfaces of the eleven-ply (30/−30/

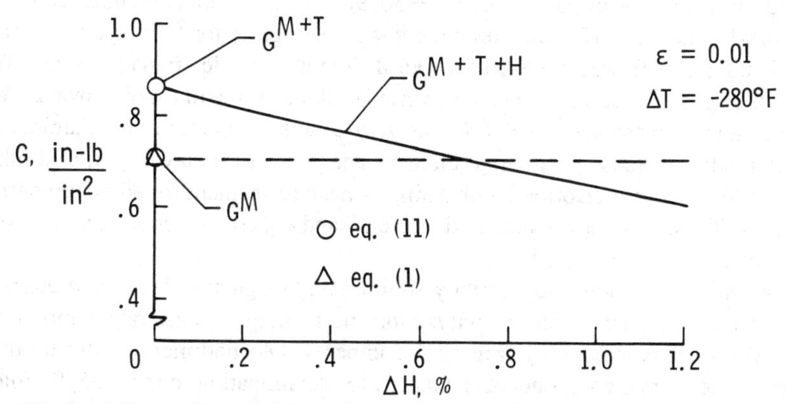

FIG. 3—*Influence of residual thermal and moisture stresses on strain energy release rate.*

$30/-30/90/\overline{90})_s$ laminate with an applied mechanical strain of 0.01 and $\Delta T = -280°F$ ($-173.3°C$). As shown on the ordinate, the strain energy release rate resulting from mechanical loading only G^M calculated from Eq 11 is identical to G calculated from Eq 1. However, if the residual thermal strain is included, the strain energy release rate G^{M+T} is higher than G^M for the same applied mechanical strain. If the laminate also absorbs moisture, the residual thermal stresses are relaxed and the strain energy release rate G^{M+T+H} decreases depending on the percentage of moisture weight gain ΔH. For the case shown in Fig. 3, the residual thermal stresses are completely relaxed after a moisture weight gain of approximately 0.7% where G^{M+T+H} is equal to G^M. Epoxy matrix composites may absorb nearly this much water from the ambient laboratory air in a matter of weeks [*10*]. Therefore, the influence of residual thermal stresses may be relatively small at ambient conditions, but may become more significant under dry or water-saturated conditions. Furthermore, composites that are manufactured at higher temperatures but absorb very little moisture may require that thermal and moisture effects be included in the G analysis for edge delamination. However, the relative contribution of residual thermal and moisture stresses to G is smaller for toughened-matrix composites that delaminate at high strains because a large mechanical strain at delamination onset has a much greater contribution to the strain energy released than ΔT or ΔH.

The tests in this study were conducted on graphite/epoxy materials in the ambient laboratory environment described earlier. Therefore, the influence of residual thermal and moisture stresses were not included in the data reduction for these tests.

Finite Element Analysis

A quasi three-dimensional finite element analysis was performed with the virtual crack extension technique to determine the G_I, G_{II}, and G_{III} components of the total strain energy release rate for several configurations of the edge delamination test [*1–5*]. In Ref *1*, the G_I, G_{II}, and G_{III} components were calculated for an eleven-ply $(30/-30/30/-30/90/\overline{90})_s$ layup that delaminated in the $-30/90$ interfaces. The G_{III} component was negligible for this layup. The delamination growth was modeled for four different initial delamination sizes. The results indicated that the G_I/G_{II} ratio varied with delamination size; however, the finite element mesh used in Ref *1* was very coarse for the longest delamination sizes modeled. Subsequent finite element analyses of this layup [*2*], and other layups [*4,5*], were performed with a single-mesh refinement for all delamination lengths. These analyses indicated that the G_I and G_{II} components were independent of delamination length.

Recently, an anisotropic elasticity solution and singular hybrid finite element formulation were employed to analyze the strain energy release rate components for edge delamination [*11*]. Figure 4 compares the nondimensionalized strain energy release rate components calculated for delamination in the $-35/90$ interfaces of a $(0/35/-35/90)_s$ laminate using both the displacement-based, eight-

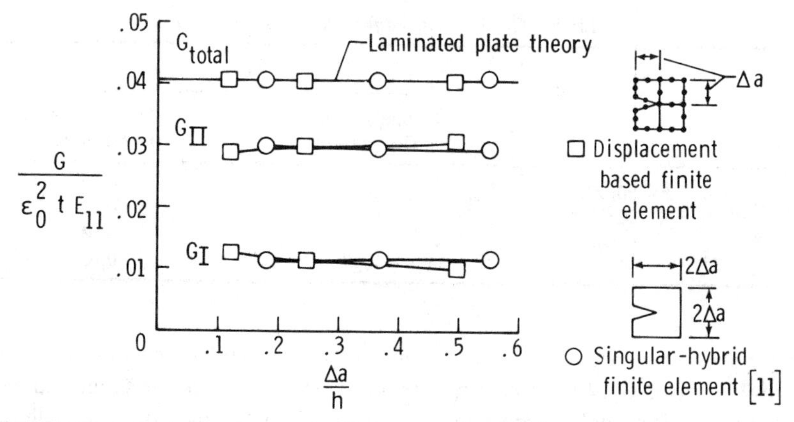

FIG. 4—*Convergence comparison for displacement-based and singular-hybrid finite element analyses of* $[0/\pm35/90]_s$ *laminates.*

noded square, parabolic finite elements and the singular hybrid element at the delamination front. Both analyses were performed with several different mesh refinements, and the results have been plotted as a function of element size at the delamination front Δa normalized by ply thickness h. Between $0.18 < \Delta a/h < 0.55$, the singular hybrid element yields constant G_I and G_{II} values. However, in Ref *11,* the singular hybrid analysis yielded variable G_I and G_{II} values for singularity element sizes $\Delta a/h < 0.18$, and for $\Delta a/h > 0.55$. The reasons for these variations are the following. First, for $\Delta a/h < 0.18$, the neighboring regular eight-noded elements are also subjected to the singular stress field. Thus, the crack tip element is too small. Second, for $\Delta a/h > 0.55$, the crack tip singular elements are required to capture both the singular and far-field components, which the singular element is unable to handle. Thus, the crack tip element is too large. However, when the size of the singular element is $0.18 < \Delta a/h < 0.55$, the element is not subjected to these extreme requirements and is able to delineate the stress field accurately and yield accurate G_I and G_{II} values. Therefore, the singular hybrid analysis with mesh refinements in this range may be used as a bench mark solution to compare to other solutions.

In contrast to the results for the singular hybrid element, the G_I and G_{II} values calculated with the eight-noded displacement-based element at the delamination front vary continuously with $\Delta a/h$. Therefore, a converged solution is never obtained for this element using the virtual crack extension technique. However, if an element size of $\Delta a/h = 0.25$ is used, the G components calculated with the eight-noded element agree fairly well with the singular-hybrid element results. Hence, four square elements through the ply thickness, with dimensions $\Delta a/h = 0.25$, appear to be a good choice for the displacement-based finite element mesh at the delamination front. Table 5 lists the ratio of G_I to the total G calculated for the four layups tested in this study with either natural delamination, where both $\theta/90$ interface delaminations are modeled, or for a single

TABLE 5—G_I/G *calculated from finite element analysis.*

Layup	G_I/G	
	Single $\theta/90$ Delamination	Double $\theta/90$ Delamination
$(30/-30_2/30/90)_s$	0.68	0.64
$(30/-30_2/30/90_2)_s$	0.66	0.64
$(35/-35/0/90)_s$	0.76	0.94
$(35/0/-35/90)_s$	0.49	0.63

$\theta/90$ delamination growing from an insert. These G_I/G ratios were calculated using the displacement-based finite element analysis with the suggested mesh refinement. The total G consisted of G_I and G_{II} only since the calculated G_{III} component was negligible for each case.

Results

Test data were compared for laminates with various insert thicknesses, insert locations, and coupon sizes to identify if these differences in configuration influenced interlaminar fracture toughness measurement. Because previous studies using the edge delamination test on graphite/epoxy composites indicated that the G_I component alone may control the onset of delamination, the G_I components of the measured G_c for different layups were compared first [*4,5*]. In addition, G_c measurements were plotted as a function of the G_I and G_{II} components assuming a linear failure criterion.

Variation in G_c *with Insert Thickness*

Because the interlaminar fracture toughness is measured at the onset of delamination from the insert embedded at the straight edge, the thickness of the insert will determine the relative sharpness of the delamination front. If the insert is too thick, the delamination may behave as if the crack tip was blunted and had a finite notch root radius. This blunted crack would yield higher apparent toughness values than a sharp crack. Therefore, EDT coupons were made with two different insert thicknesses, and data were compared to adhesive bond toughness data with comparable bond thicknesses to determine if interlaminar fracture toughness values could be obtained from coupons with inserts.

Figure 5 compares interlaminar fracture toughness measurements for $(30/-30_2/30/90_2)_s$ laminates made of T300/5208 and T300/BP907. Tests were conducted on laminates with 3.5- and 1.5-mil (88.9- or 38.1-μm) inserts at the midplane, and on laminates without inserts. All three configurations showed the improved toughness of the T300/BP907 compared to the T300/5208 material. For both materials, the laminates with the thicker inserts yielded higher apparent toughness values than the laminates with the thinner inserts.

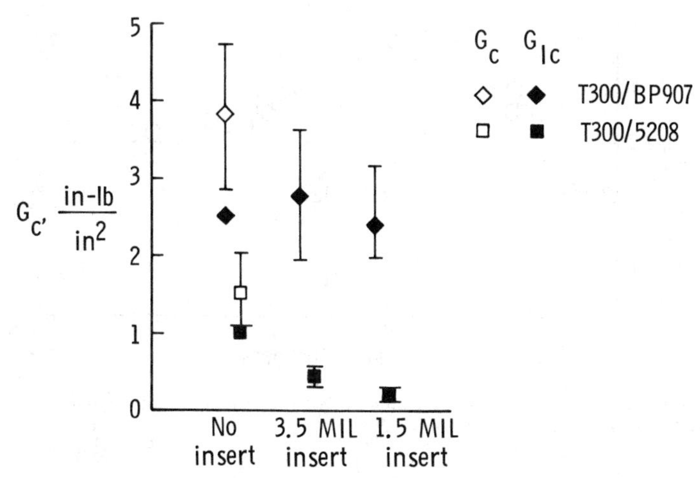

FIG. 5—*Interlaminar fracture toughness from* $[+30/-30_2/+30/90_2]_s$ *EDT tests with and without midplane inserts.*

As illustrated in Fig. 6, these results may be compared to fracture toughness measurements of adhesive bonds assuming that the resin pocket that forms at the end of the insert is analogous to an adhesive bond with a thickness t_A equal to the insert thickness. Previous work on adhesive bond fracture indicated that the bond thickness must be below a certain value to achieve a realistic fracture toughness measurement [*12*]. Figure 7 shows fracture toughness measurements determined from double cantilever beam (DCB) adhesive bond tests, with BP907 as the adhesive, as a function of bond thickness. The data indicate that fracture toughness is constant for bond thicknesses below 2.5 mil (63.5 μm). For bond thicknesses greater than 2.5 mil (63.5 μm), fracture toughness measurements are unrealistically high as a result of the relaxed constraint on the resin allowing greater localized plastic deformation near the crack tip. Using the adhesive bond analogy, the G_{Ic} results shown in Fig. 5 for T300/BP907 EDT tests may be artificially elevated for the laminates with 3.5-mil (88.9-μm) inserts, but G_{Ic}

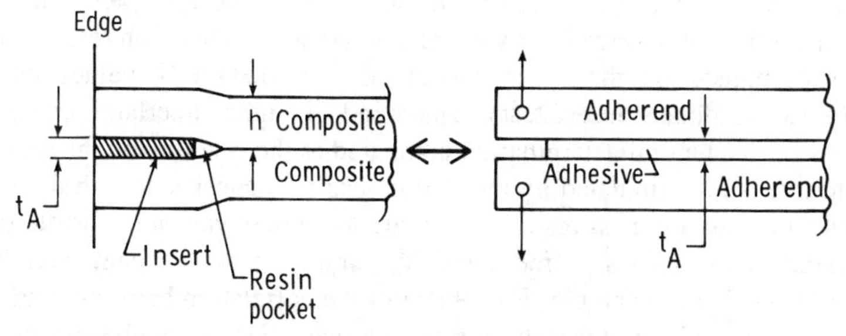

FIG. 6—*Adhesive bond analogy for delamination growing from insert.*

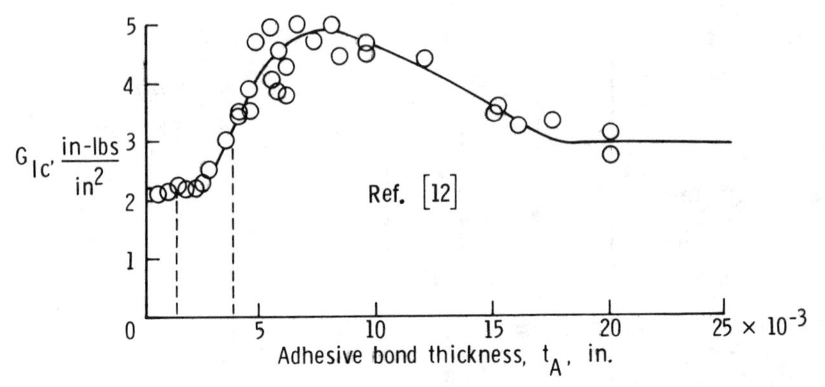

FIG. 7—*Fracture toughness as a function of bond thickness for BP907 adhesive in double cantilever beam with aluminum adherends.*

values for laminates with 1.5-mil (38.1-μm) inserts should be representative of G_{Ic} for delamination growth between plies.

Figure 5 also shows G_c results for laminates without inserts (open symbols) and their G_I components calculated from finite element analysis (Table 5). For the T300/BP907, the G_I component of the natural delamination mixed-mode test agrees well with the G_{Ic} measurement from the laminate with the 1.5-mil (38.1-μm) insert and G_{Ic} measurements from DCB tests on thin adhesive bonds (Fig. 7) [*12*].

For the T300/5208 laminates, the G_I component of the natural delamination mixed-mode test was higher than the G_{Ic} measurements from laminates with both the 1.5- and 3.5-mil (38.1- and 88.9-μm) midplane inserts. However, these natural delamination G_c values were higher than the G_c values measured previously on eleven-ply layups [*1*], and they had considerably more scatter than the G_{Ic} measurements, which may indicate that extensive matrix cracking may have been present in the four central 90° plies before delamination occurred [*4*]. Therefore, these experiments were repeated on ten-ply $(30/-30_2/30/90)_s$ laminates that were less likely to experience extensive matrix cracking before delamination because of the reduced number of 90° plies.

Figure 8 shows results obtained from the ten-ply T300/5208 laminates. The total G_c measurements were slightly lower and had less scatter than results for the twelve-ply laminate, but the G_I component still exceeded the G_{Ic} values obtained from the two midplane insert tests. The trend of higher interlaminar fracture toughness for the natural delamination compared to the fracture toughness of the thin adhesive bonds simulated by the Teflon inserts is consistent with the trends noted when comparing neat resin G_{Ic} fracture toughness values for brittle resins to interlaminar G_{Ic} values as measured by composite double cantilever beam (DCB) tests [*13*]. For example, Fig. 9 shows the correlation between neat resin G_{Ic} and composite G_{Ic} for a variety of resin matrices. For the tougher resins, neat resin G_{Ic} exceeds composite G_{Ic} because of the large plastic zones that form in

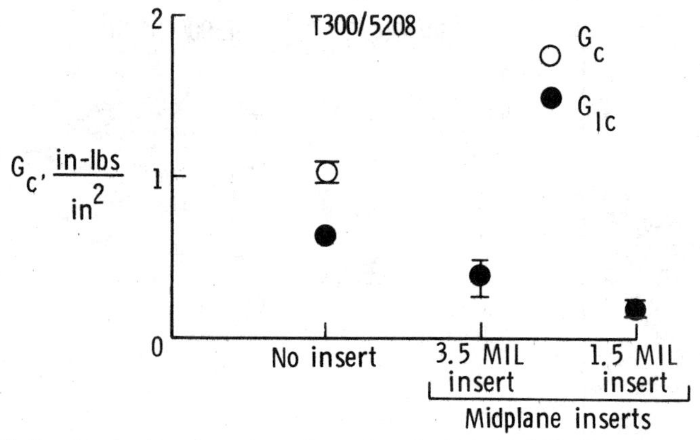

FIG. 8—*Interlaminar fracture toughness from* $[+30/-30_2/+30/90]_s$ *EDT tests.*

neat resin fracture tests. However, for the brittle resin matrices, neat resin G_{Ic} is less than G_{Ic} for the composite. Apparently, the close proximity of the fibers in the composite, which is analogous to a very thin bond line, does not significantly lower toughness by increasing constraint for the brittle resin, but may actually increase the toughness as a result of the interaction of the crack front with the fibers creating more plastic flow locally at the fibers than was observed in neat resin fracture tests [*14*].

All subsequent test data reported was generated with the 1.5-mil (38.1-μm) inserts and compared to data generated from coupons without inserts.

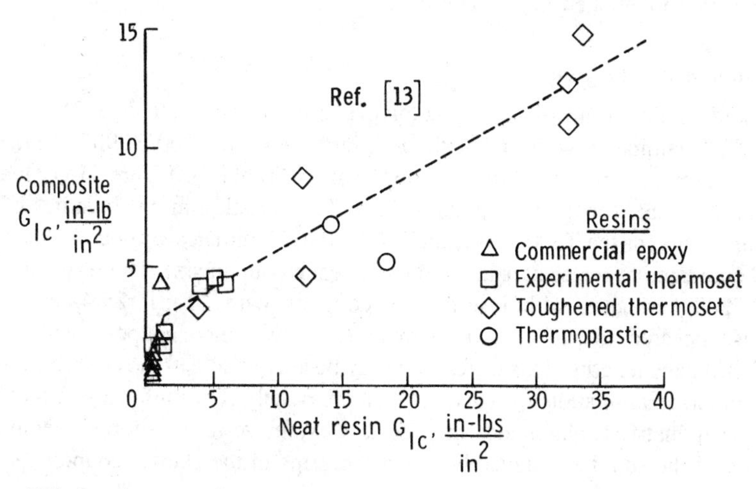

FIG. 9—*Composite* G_{Ic} *versus neat resin* G_{Ic} *for various resin matrix materials.*

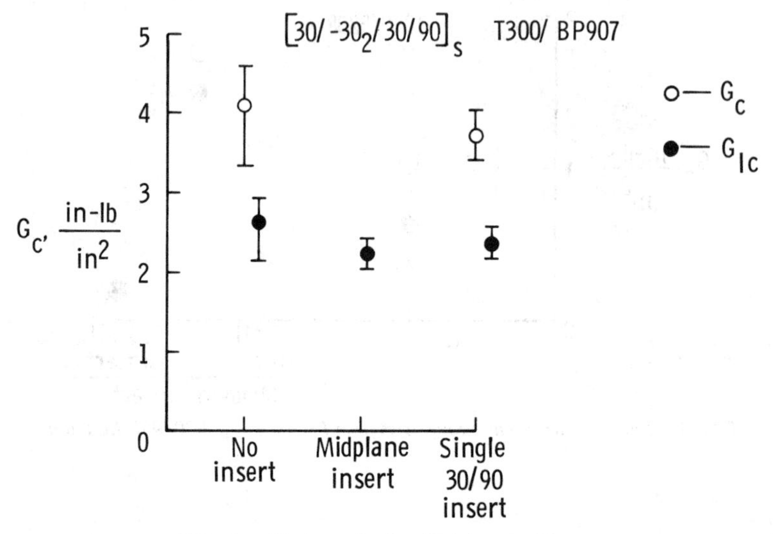

FIG. 10—*Variation in* G$_c$ *with insert location.*

Variation in G$_c$ *with Insert Location*

Figure 10 compares the G_{Ic} values for midplane delamination of the T300/BP907 ten-ply $(30/-30_2/30/90)_s$ layup with the G_I components of G_c for the natural mixed-mode delamination, and for mixed-mode delamination from inserts in a single 30/90 interface. These G_I values are in excellent agreement. Therefore, all three configurations of this 30/90 layup yield similar results for the T300/BP907 toughened-matrix composite.

Variation in G$_c$ *with Coupon Size*

Mixed-mode delamination tests were conducted on $(35/-35/0/90)_s$ T300/5208 laminates with and without inserts, and on T300/BP907 laminates without inserts, using five different coupon sizes (Table 2). Figure 11 compares G_c measurements for the five coupon sizes. The variation in mean values of G_c measurements for the T300/5208 and T300/BP907 laminates without inserts was small compared to the scatter in the data for each coupon size. However, for the T300/5208 laminates with inserts, the coupons with 10-in. (25.4-cm) gage lengths appeared to yield slightly lower G_c values than coupons with 5-in. (12.7-cm) gage lengths. This difference may be attributable to the contribution of curvature to delamination growth discussed previously. The uniform κ_y curvature in the asymmetric sublaminates may be less extensive in the shorter specimens because of the smaller distance between the grips in the shorter coupons.

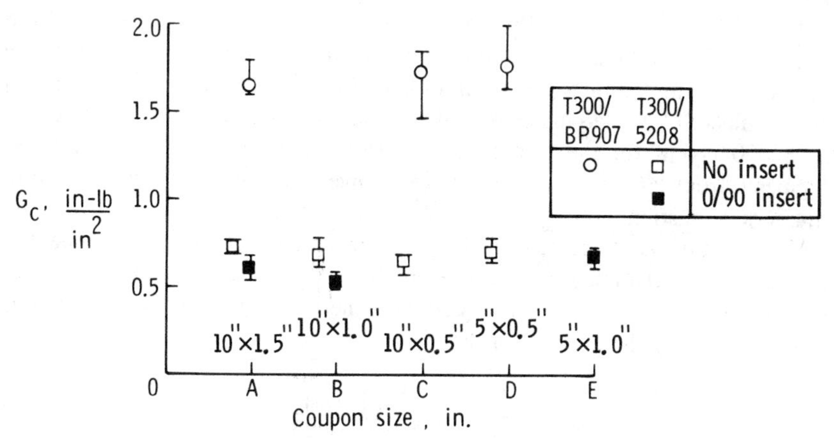

FIG. 11—*Variation in* G_c *with coupon size for* $[\pm 35/0/90]_s$ *EDT tests.*

Variation in G_c *with Layup*

Figure 12 compares G_c and G_{Ic} data for $(35/-35/0/90)_s$ and $(35/0/-35/90)_s$ T300/5208 laminates with no inserts, with midplane inserts, and with inserts at a single $\theta/90$ interface. For the mixed-mode configurations, the G_c values for the two layups do not agree. Table 5 shows that the G_I/G ratios for these two layups are different. Although the two layups have different Mode I percentages, Fig. 12 indicates that the G_I components for delamination onset from the insert

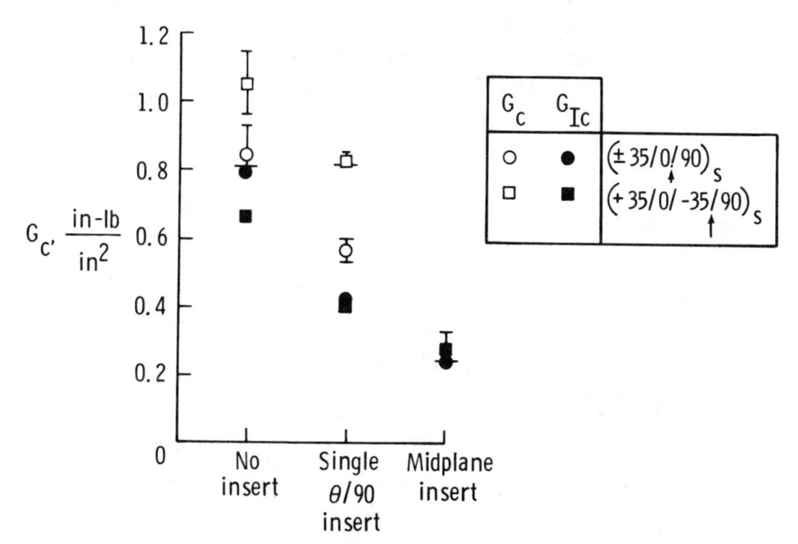

FIG. 12—*Variation in* G_c *with EDT layup and configuration for T300/5208.*

in the $\theta/90$ interface are nearly identical for both layups. The G_{Ic} values from coupons of the two layups containing midplane inserts also agree. However, the G_{Ic} values from laminates with midplane inserts were lower than the G_I components of G_c for laminates with $\theta/90$ interface inserts. As noted earlier for the 30/90 layup, for the brittle 5208 matrix composite the toughness measurements from laminates with inserts are lower than the measurements from natural delamination.

Although the data generated in this study indicates that the G_I component is responsible for delamination growth even under mixed-mode loading, the criterion for mixed-mode delamination may be generally expressed as a failure envelope defined by the polynomial

$$\left(\frac{G_I}{G_{Ic}}\right)^m + \left(\frac{G_{II}}{G_{IIc}}\right)^n = 1 \tag{14}$$

In Ref *15* interlaminar fracture data in the literature was plotted and indicated that a linear failure criterion, where $m = 1$ and $n = 1$, provided the best fit to the data. Figures 13 and 14 show similar plots for T300/5208 and for T300/BP907 using the data generated in this study along with edge delamination data for T300/BP907 from Ref *16,* and G_{IIc} data from end-notched flexure tests [*17*]. These plots also indicate that a linear failure criterion may be appropriate, however data from tests with other G_I/G_{II} ratios are needed to determine accurately the shape of the failure envelope. Because the G_{IIc} values are nearly an order of magnitude larger than the G_{Ic} values for these two materials, the failure envelope is almost horizontal over the range of G_I/G_{II} ratios tested. Therefore, even if delamination failure is governed by a linear failure criterion as depicted in Figs. 13 and 14, the failure appears to be controlled by the G_I component alone when the data are plotted as shown in Figs. 5, 10, and 12.

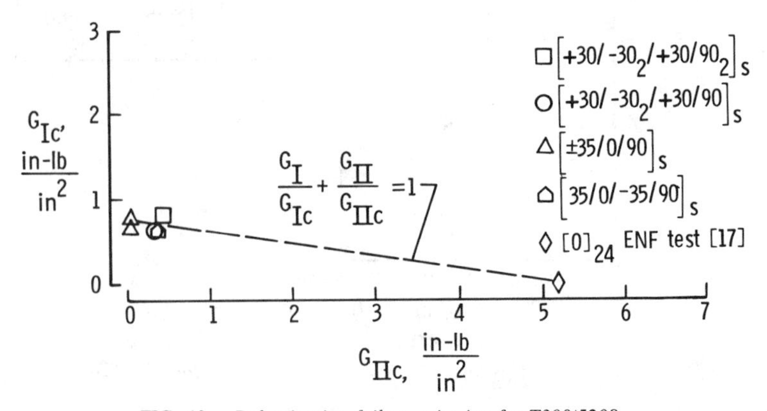

FIG. 13—*Delamination failure criterion for T300/5208.*

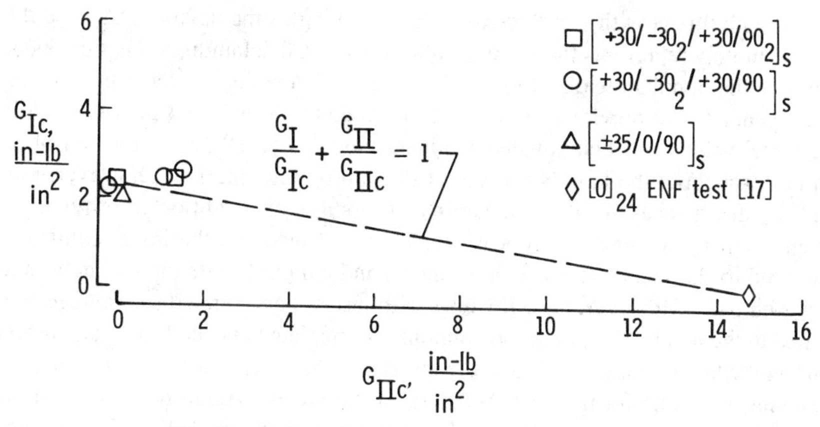

FIG. 14—*Delamination failure criterion for T300/BP907.*

Discussion

This discussion will summarize some of the advantages and disadvantages of the edge delamination tension (EDT) test configurations with and without inserts. Some advantages and disadvantages are common to both configurations. The EDT test involves a simple loading, does not require a measurement of delamination size, may be conducted on a variety of layups to provide a range of mixed-mode ratios, yields data consistent with other interlaminar fracture tests, and provides a ranking of the relative interlaminar fracture toughness of different composite materials. However, for EDT layups with 0° plies, G_c measurement is limited by the failure strain of the fibers, whereas for layups without 0° plies, toughened-matrix composites may exhibit nonlinearity in the load-displacement curve before delamination onset [*2,4*]. Alternate layup designs such as $(35/-35_2/35/0_2/90)_s$, where the increased laminate thickness reduces the strain required to measure a given G_c, may overcome these limitations. No closed-form elasticity solution exists for edge delamination that yields G_I, G_{II}, and G_{III} for arbitrary layups, however, a singular hybrid finite element analysis yields a bench mark solution for the various G components. The G_c measurements from the EDT test may be influenced by residual thermal and moisture stresses, which can be included in the data reduction but would require measurement of stress free temperature, moisture content, and moisture and thermal coefficients of expansion.

One motivation for including inserts at the edge was to remove the uncertainties in assuming the delaminations that naturally wander from one $\theta/90$ interface to another (Fig. 2*a*) can be modeled as three sublaminates loaded in parallel (Fig. 2*b*). Although the formation of the pattern shown in Fig. 2*a* along the edge is random, once the pattern is formed it remains unchanged as the delamination grows through the laminate width. Therefore, the delamination growth through the width is self-similar, and the strain energy release rate associated with this growth is reflected in Eq 1, as long as the delaminated modulus E^* accurately represents the modulus after the natural delamination has extended through the laminate width. Plots of modulus as a function of delamination size were generated in previous studies and indicated that Eqs 2 to 4 provide a fairly accurate estimate of delaminated laminate modulus [*1,18,19*]. Inclusion of an insert throughout the laminate width at the appropriate interface, however, provides a direct measure of the delaminated modulus, in addition to providing a single planar delamination front for EDT tests. Therefore, the insert eliminates the need for lamina property measurements and laminate plate theory analysis to determine E^*. However, the EDT tests with inserts have some disadvantages not found in the natural delamination coupons. A template is needed to locate inserts during the layup of the panel, and the insert material may deform during the cure resulting in nonuniform insert thickness in the panel. Nonuniformity of insert thickness may cause uncertainty in the determination of E^* and G_c. In addition, the deviation from the linear load-displacement curve is not as abrupt for laminates with inserts. Because the delamination grows from an embedded insert, delamination onset cannot be visually verified. Hence, the delamination onset strain is more difficult to determine in laminates with inserts. Furthermore, because implanted delaminations at one asymmetrically located interface or at the midplane result in a bending-extension coupling contribution to E^*, G_c measurements may vary slightly with specimen size.

Table 6 summarizes the advantages and disadvantages of the edge delamination test. Most of the concerns about accurate G_c measurement with the EDT test may be overcome by choosing appropriate layups, thicknesses, and coupon sizes, or by implanting inserts at selected interfaces. However, for all the configurations of the edge delamination test, residual thermal and moisture stresses will contribute to the strain energy release rate for edge delamination.

The significance of residual thermal and moisture stresses to strain energy release rates ultimately depends on how these measurements are used. If toughness measurements are used to compare materials for improved delamination resistance, then these thermal and moisture effects become of secondary importance. This is especially true if tests are conducted at room temperature ambient conditions, and the difference in toughness measurements for different materials is large [*3,5*]. For example, the 7% error in G_c calculated in Ref *10* as a result of neglecting thermal and moisture effects for T300/5208 EDT tests is insignificant compared to the tenfold increase in G_c measured for composites with toughened matrices [*3,5*]. If, however, these interlaminar toughness measurements are used as delamination failure criteria to predict delamination growth in composite structures of the same material, but with different geometries and loadings, then these thermal and moisture effects may become more significant. Other factors may need to be addressed to accurately calculate G. For example, assuming a constant ΔT from the cure temperature over which there exists a constant coefficient of thermal expansion may be physically unrealistic. In addition, assuming that the average moisture content of the laminate is representative

TABLE 6—*Advantages and disadvantages of the EDT test.*

Natural Delamination No Inserts	Artificial Delamination with Inserts	Both Configurations
	ADVANTAGES	
• Easy to manufacture • Distinct jump in load-displacement curve and delamination visible at onset • No size effect on G_c • Delamination typical of those in structure	• well-defined delamination plane on edge • E^* measured directly	• simple loading • G independent of delamination size • several layups for range of mixed-mode conditions • data consistent with other toughness tests • provides ranking of interlaminar fracture toughness of composites
	DISADVANTAGES	
• Irregular delamination forms on edge • E^* must be calculated	• requires template to make panel • E^* measurement affected by insert uniformity • delamination onset hard to detect • Some size effect on G_c	• G_c measurement limited by fiber failure • Nonlinear behavior may occur before delamination onset • no closed-form solution for G components • residual thermal and moisture stresses may influence G_c

of the moisture content at the delamination front may also be in error. Some knowledge of the moisture distribution through the laminate may be needed. The detailed information required for carefully conducted laboratory tests may not be available to analyze the strain energy release rate for the delamination growing in the structure. Nevertheless, conducting edge delamination tests where these effects can be quantified, and compared to data from other interlaminar fracture toughness tests where these effects are not present, would help document the relative influence of residual thermal and moisture stresses on the interlaminar fracture of composite materials.

Conclusions

Edge delamination tension (EDT) tests were performed on both brittle (T300/5208) and toughened-matrix (T300/BP907) graphite reinforced composite laminates designed to delaminate at the straight edge. The mixed-mode interlaminar fracture toughness G_c was calculated from straight edge delamination data measured during tension tests of $(30/-30_2/30/90_n)_s$, n = 1 or 2; $(35/-35/0/90)_s$; and $(35/0/-35/90)_s$ laminates without inserts, and laminates with inserts at the $\theta/90$ interface. In addition, Mode I interlaminar tension fracture toughness G_{Ic} was measured from laminates with the same layups but with inserts in the midplane at the straight edge. The influence of insert thickness and location, coupon size, and layup on G_c measurement was evaluated. Based on the results of this study, the following conclusions were reached:

1. All configurations of the EDT test were useful for ranking the delamination resistance of composites with different matrix resins.
2. Strain energy release rate components may be accurately calculated with displacement-based elements, using the virtual crack extension technique, if eight-noded square parabolic elements are used at the delamination front with side dimensions equal to one quarter of the ply thickness.
3. For toughened-matrix composites, laminates with 1.5-mil (38.1-μm) thick inserts yielded interlaminar fracture toughness numbers consistent with data generated from laminates without inserts.
4. Coupons of various sizes yielded similar results.
5. Delamination appeared to be governed by a linear failure criterion relating G_I and G_{II}.

References

[1] O'Brien, T. K., "Characterization of Delamination Onset and Growth in a Composite Laminate," in *Damage in Composite Materials, ASTM STP 775,* K. L. Reifsnider, Ed., American Society for Testing and Materials, Philadelphia, 1982, pp. 140–167.

[2] O'Brien, T. K., Johnston, N. J., Morris, D. H., and Simonds, R. A., "A Simple Test for the Interlaminar Fracture Toughness of Composites," *SAMPE Journal,* Vol. 18, No. 4, July/Aug. 1982, p. 8.

[3] Johnston, N. J., O'Brien, T. K., Morris, D. H., and Simonds, R. A., "Interlaminar Fracture Toughness of Composites II—Refinement of the Edge Delamination Test and Application to Thermoplastics," in *Proceedings of the 28th National SAMPE Symposium and Exhibition,* Society for the Advancement of Material and Process Engineering, Corina, CA, 1983, p. 502.

[4] O'Brien, T. K., "Mixed-Mode Strain-Energy-Release Rate Effects on Edge Delamination of Composites," in *Effects of Defects in Composite Materials, ASTM STP 836,* American Society for Testing and Materials, Philadelphia, 1984, pp. 125–142.

[5] O'Brien, T. K., Johnston, N. J., Morris, D. H., and Simonds, R. A., "Determination of Interlaminar Fracture Toughness and Fracture Mode Dependence of Composites using the Edge Delamination Test," in *Proceedings of the International Conference on Testing, Evaluation, and Quality Assurance of Composites,* T. Feest, Ed., Butterworth Scientific, Ltd., Kent, England, 1983, p. 223.

[6] Whitney, J. M., Browning, C. E., and Hoogsteden, W., "A Double Cantilever Beam Test for Characterizing Mode I Delamination of Composite Materials," *Journal of Reinforced Plastics and Composites,* Vol. 1, No. 4, 1982, p. 297.

[7] Whitney, J. M. and Knight, M., "A Modified Free-Edge Delamination Specimen," in *Delamination and Debonding of Materials, ASTM STP 876,* W. S. Johnson, Ed., American Society for Testing and Materials, Philadelphia, 1985, pp. 298–314.

[8] Ho, T. and Schapery, R. A., "The Effect of Environment on the Mechanical Behavior of AS/3501-6 Graphite Epoxy Material—Phase IV," Naval Air Systems Command (ATC) Report R-92000/3CR-9, Feb. 1983.

[9] Whitcomb, J. D. and Raju, I. S., "Analysis of Interlaminar Stresses in Thick Composite Laminates With and Without Edge Delamination," in *Delamination and Debonding of Materials, ASTM STP 876,* W. S. Johnson, Ed., American Society for Testing and Materials, Philadelphia, 1985, pp. 69–94.

[10] O'Brien, T. K., Raju, I. S., and Garber, D. P., "Residual Thermal and Moisture Influences on the Strain Energy Release Rate Analysis of Edge Delamination," *Journal of Composites Technology and Research,* Vol. 8, No. 2, Summer 1986, pp. 37–47.

[11] Wang, S. S., "The Mechanics of Delamination in Fiber-Reinforced Composite Materials—Part III," NASA CR, 1985.

[12] Chai, H., "Bond Thickness Effect in Adhesive Joints and Its Significance for Mode I Interlaminar Fracture of Composites," in *Composite Materials: Testing and Design (Seventh Conference), ASTM STP 893,* J. M. Whitney, Ed., American Society for Testing and Materials, Philadelphia, 1986, pp. 209–231.

[13] Hunston, D. L., "Composite Interlaminar Fracture: Effect of Matrix Fracture Energy," *Composites Technology Review,* Vol. 6, No. 4, Winter 1984, pp. 176–186.

[14] Bascom, W. D., Boll, D. J., Fuller, B., and Phillips, P., "Fractography of the Interlaminar Fracture of Carbon Fiber Epoxy Composites," in this volume, pp. 131–149.

[15] Johnson, W. S., and Mangalgari, P. D., "Influence of Resin on Interlaminar Fracture," in this volume, pp. 295–315.

[16] Adams, D. F., Zimmerman, R. S., and Odem, E. M., "Determining Frequency and Load Ratio Effect on the Edge Delamination Test in Graphite/Epoxy Composites," in this volume, pp. 242–259.

[17] Murri, G. B. and O'Brien, T. K., "Interlaminar G_{IIc} Evaluation of Toughened-Resin Matrix Composites Using the End-Notched Flexure Test," in *Proceedings of the 26th AIAA/ASME/ASCE/AHS Structures, Structural Dynamics, and Materials Conference,* American Institute of Aeronautics and Astronautics, New York, April 1985, pp. 197-202.

[18] O'Brien, T. K., "The Effect of Delamination on the Tensile Strength of Unnotched, Quasi-Isotropic, Graphite/Epoxy Laminates," in *Proceedings of the SESA/JSME 1982 Joint Conference on Experimental Mechanics, Hawaii, Part I,* Society for Experimental Stress Analysis, Brookfield, CT, May 1982, p. 236.

[19] O'Brien, T. K., Ryder, J. T., and Crossman, F. W., "Stiffness, Strength, and Fatigue Life Relationships for Composite Laminates," in *Proceedings of the Seventh Annual Mechanics of Composites Review,* U.S. Airforce Wright Aeronautical Laboratory (AFWAL), Dayton, OH, Oct., 1981.

Anoush Poursartip [1]

The Characterization of Edge Delamination Growth in Laminates Under Fatigue Loading

REFERENCE: Poursartip, A., **"The Characterization of Edge Delamination Growth in Laminates Under Fatigue Loading,"** *Toughened Composites, ASTM STP 937,* Norman J. Johnston, Ed., American Society for Testing and Materials, Philadelphia, 1987, pp. 222-241.

ABSTRACT: It is generally accepted that the onset and stable growth of edge delaminations under quasi-static loading can be characterized using the strain energy release rate G and delamination resistance curves (R curves). This study addresses the problem of edge delamination growth under fatigue loading. To characterize the fatigue growth successfully, it is shown that it is not sufficient to consider only the maximum energy (G_{max}) released as the delamination grows. Instead, the energy released must be compared to the increasing resistance to further growth. This increasing resistance to delamination growth G_R can be explained best in terms of the associated matrix cracking, which is observed under both static and fatigue loading. Using results from a $[\pm30/\pm30/90/\overline{90}]_s$ T300-5208 and a $[45/0/-45/90]_s$ XAS-914 graphite-epoxy laminate, separate power law correlations between the edge delamination fatigue growth rates and the term (G_{max}/G_R) are obtained. The measured exponents are in the range 4.1 to 6.

KEY WORDS: composite materials, graphite-epoxy, delamination, matrix cracking, stiffness loss, fracture mechanics, strain energy release rate, R curve, fatigue, growth law

Nomenclature

a	Strip delamination size
a_{max}	Maximum delamination size
A, β, n	Empirical coefficients
b	Half-width of laminate
da/dN	Fatigue growth rate of delamination
E	Axial stiffness of partially delaminated laminate
E_{LAM}	Axial stiffness of undamaged laminate
E^*	Axial stiffness of completely delaminated laminate

[1]Assistant professor, Department of Metallurgical Engineering, The University of British Columbia, 309–6350 Stores Rd., Vancouver, B.C., V6T 1W5 Canada.

F	Work done by external forces on system
K	Stress intensity factor
G	Strain energy release rate
G_I, G_{II}, G_{III}	Opening, in-plane shear, and antiplane shear strain energy release rate components, respectively
G_c	Critical strain energy release rate for the onset of delamination
G_{max}	Maximum strain energy release in fatigue loading
G_R	Delamination resistance
N	Number of fatigue cycles
R	Fatigue load ratio $= \sigma_{min}/\sigma_{max}$ or $\varepsilon_{min}/\varepsilon_{max}$
U	Elastic energy stored in system
W	Energy required for crack formation

Introduction

The onset and growth of delaminations under fatigue loading has recently received much attention, for they have been singled out as an important failure mechanism [1]. The presence of delaminations leads to a redistribution of stresses within the laminate, and consequent losses in stiffness and strength. Furthermore, the growth of other damage mechanisms, such as matrix cracking and fiber breakage, may be enhanced.

Delaminations may initiate at free edges such as holes and cutouts, at internal interfacial flaw sites, or as a result of impact loading. Though they are difficult to detect in service, they can be grown and monitored fairly easily in the laboratory. Furthermore, the analysis of their behavior can draw heavily on the more established field of fracture mechanics, as they grow in a self-similar manner. The best results have been achieved using the strain energy release rate G, rather than the stress intensity factor K. The analytical difficulties in determining the value of the singularity at the crack tip [1] make the use of G most attractive. The strain energy release rate has been determined using both compliance solutions and numerical virtual crack extension techniques [2-8].

A significant feature of edge delamination behavior is that edge delaminations grow in a stable manner under quasi-static loading. Linear elastic fracture mechanics [9] applied to edge delaminations [4] shows that as a delamination grows, G is either constant or increasing, depending on the loading conditions. This should fuel further growth and lead to fracture. If a delamination is observed to grow stably, then the resistance to growth, once initiated, must be increasing. One way to characterize this increasing resistance is to use a delamination resistance curve (R curve) [4,9].

Edge delaminations also grow under fatigue loading. Unfortunately, there are problems in using the strain energy release rate to characterize their growth. For example, in tests under strain control, the maximum G is constant [4]. Therefore, based on our experience with other materials, we would expect a constant growth rate [9]. However, the observed delamination growth rate actually decreases with cycling [4,8]. This work proposes to explain this effect, while still using strain energy release rate concepts.

In a metal, the increasing resistance to crack growth under static loading is a result of the increasing plasticity at the crack tip because of overloading. Under fatigue loading, this effect is not present. On the other hand, it will be shown that the increasing resistance to delamination growth in a laminate is due to the associated matrix cracking, which is present under both static and fatigue loading. In both materials, the energy released, and thus available to propagate the crack, is increasing, but in the case of the laminate, the amount of work that must be done to propagate the crack is increasing at an even faster rate. Thus, the laminate delamination growth rate decreases, while the metal crack growth rate increases.

Data for two different laminates are presented. The data from the $[\pm30/\pm30/90/\overline{90}]_s$ T300-5208 laminates were first presented in Ref *4*. The other results are from $[45/0/-45/90]_s$ XAS-914 laminates. In both cases, the fatigue growth rates are shown to be a function of (G_{max}/G_R), where G_R is the current resistance to crack growth.

Experimental Procedure

$[\pm30/\pm30/90/\overline{90}]_s$ Laminates

Full details of the experimental method are presented in Ref *4*. The specimens were 254 mm long by 38 mm wide, with a ply thickness of 0.14 mm. Static tension and tension-tension fatigue tests were conducted under strain control, with strain measured using linear variable displacement transducers. Delaminations were monitored using in situ dye-penetrant radiography.

$[45/0/-45/90]_s$ Laminates

Unnotched $[45/0/-45/90]_s$ XAS-914 graphite-epoxy laminates were tested in static tension and load-controlled tension-tension fatigue tests, $R = 0.1$. Specimens were 203 mm long by 19 mm wide, with a total thickness of 1 mm. Strains were monitored using an 80-mm gauge length clip gauge mounted on anvils clamped to the specimen.

Delaminations and matrix cracks were monitored using zinc iodide enhanced radiography. The zinc iodide solution was injected along the free edges of the specimen and allowed to infiltrate the cracks while still under load. The specimen was then removed from the loading grips, placed in an *X*-ray machine, and exposed to a 30-kV, 10-mA beam for 45 s. Polaroid Type 55 positive-negative film was used. Delamination sizes were then measured on enlarged prints using a Leitz® Tas Plus Image Analyzer. Specimen edges were polished and inspected to identify the delaminating interfaces.

Stiffness was monitored during all tests and correlated with both delamination size and matrix crack densities. Stiffness was determined from the line of best fit to the stress-strain curve between $\sigma = 20$ and 100 MPa. In static tests, this involved loading and unloading the specimen, while in fatigue tests, the cycling was automatically interrupted at frequent intervals and the load ramped to 100 MPa quasi-statically. All tests were carried out on a computer-controlled MTS 810 servohydraulic fatigue machine.

Damage Development

$[\pm30/\pm30/90/\overline{90}]_s$ Laminates

It is reported in Ref *4* that similar damage developed under both quasi-static and fatigue loading. A few cracks formed in the 90° plies, followed by delaminations along the edge. The number of 90° ply cracks increased significantly in the delaminated zone. Delaminations grew much more rapidly along the length of the specimen than through the width. The delaminations grew in the −30/90 interfaces, shifting from one interface to the other through 90° ply cracks.

$[45/0/-45/90]_s$ Laminates

Damage development in this system was also similar under both quasi-static (Fig. 1) and fatigue loading (Fig. 2). Cracks first formed in the 90° layer. The initial density, before the onset of delamination, was about 0.9 cracks/mm. Next, delaminations initiated at either the mid-plane 90/90 interface or the −45/90 interfaces. The delaminations grew rapidly along the length of the specimen, and more slowly across the width. The delaminations jogged back and forth between the interfaces through the 90° layer cracks. As the delaminations grew, the density of the 90° cracks increased considerably in the delaminated length, with the highest measured crack density being about 1.8 cracks/mm (Fig. 3). Cracks in the −45° plies also appeared, also initiating at the free edge. Cracks in the +45° plies appeared later, and were fewer in number (Fig. 4). Most of the 90° cracks extended beyond the delamination front, and across most of the specimen, while the −45 and +45° cracks were densest in the delaminated zone. For example, no +45° cracks were observed down the centerline of any specimen. A likely explanation for the difference in crack densities in the two-angle ply orientations is that only the −45° plies are adjacent to the delaminated interfaces.

An approximate technique [*10*] was used to predict the interlaminar normal stresses at the free edge (see Table 1). The two interfaces that delaminate experience the highest interlaminar normal stresses. As a result of symmetry conditions, the midplane 90/90 interface will not be subject to any interlaminar shear stresses, but this is not true for the −45/90 interfaces.

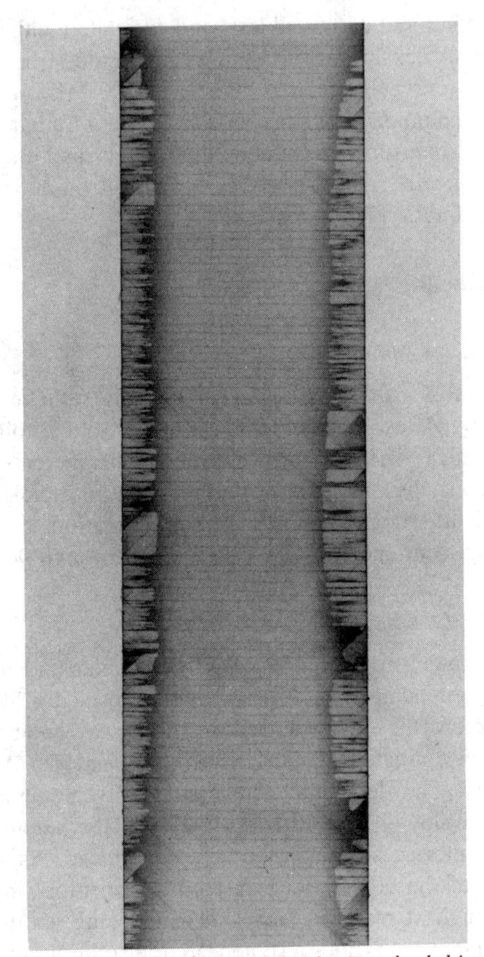

FIG. 1—*Radiograph of [45/0/−45/90]$_s$ XAS-914 laminate loaded in static tension.*

Stiffness Loss

It is well documented that matrix cracking and delamination lead to a loss in stiffness [*4,11,12,13*]. O'Brien [*4*] proposed a rule-of-mixtures analysis to relate the stiffness reduction to delamination.

His analysis assumes that the stiffness will vary linearly between the original undelaminated value E_{LAM}, and a stiffness corresponding to a set of sublaminates, created by the total delamination of the relevant interfaces. He showed that an equation of the form

$$E = E_{LAM} - (E_{LAM} - E^*)\frac{a}{b} \tag{1}$$

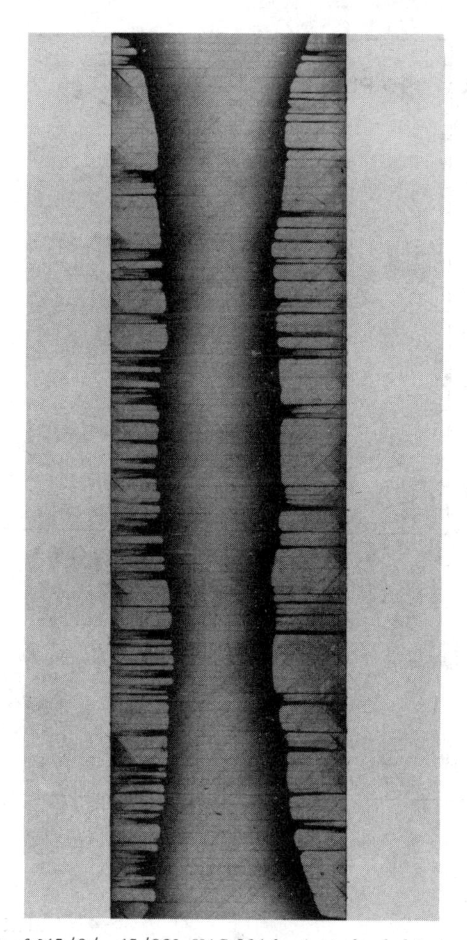

FIG. 2—*Radiograph of* $[45/0/-45/90]_s$ *XAS-914 laminate loaded in tension-tension fatigue.*

where E^* is the stiffness of the sublaminates under the same axial strain, is in very good agreement with experimental data for the $[\pm30/\pm30/90/\overline{90}]_s$ laminate. For that laminate, matrix cracking occurred only in the 90° plies, and the contribution to the stiffness loss by this mechanism was small.

In the case of the $[45/0/-45/90]_s$ laminate, the system is more complicated, as in addition to delamination, extensive cracking in the off-axis plies is observed. We consider first the experimental results plotted in Fig. 5 of the normalized stiffness as a function of delamination size (a/b). The initial reduction of $0.04E_{LAM}$ is due to cracking in the 90° plies before any delamination. Thereafter, we observe a linear reduction in stiffness with delamination size, where

$$E = 0.960E_{LAM} - 0.253E_{LAM} \times \left(\frac{a}{b}\right) \quad (2)$$

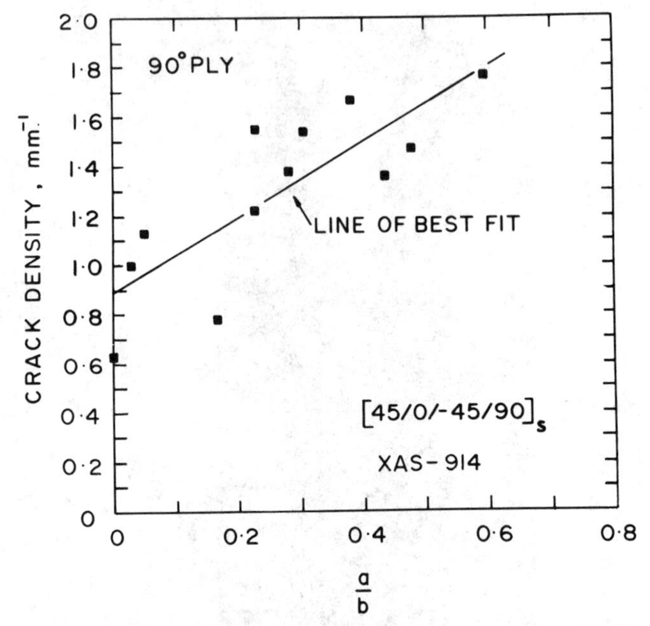

FIG. 3—*Variation of 90° ply crack density with delamination size* (a/b).

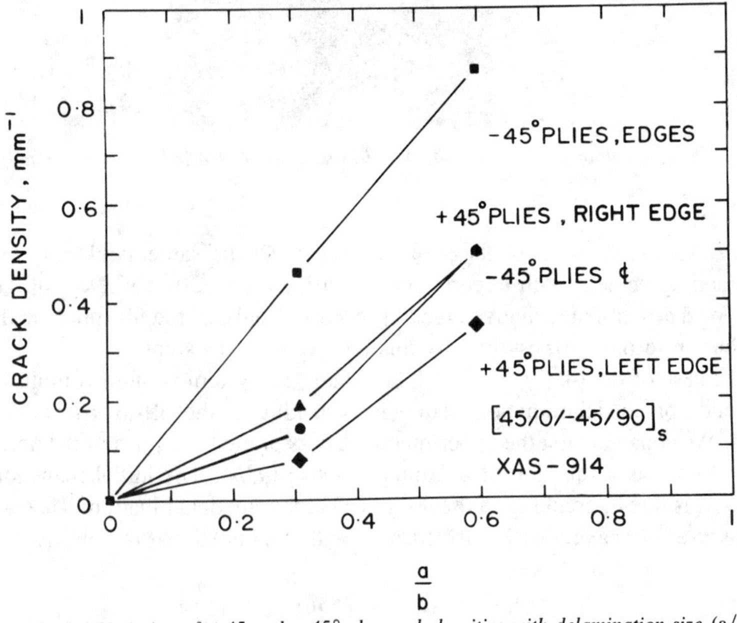

FIG. 4—*Variation of +45 and −45° ply crack densities with delamination size* (a/b).

TABLE 1—*Interlaminar normal stresses* σ_Z.[a,b]

Interface	90/90	90/−45	−45/0	0/45
σ_z/σ_0[c]	0.317[d]	0.238[d]	0.14	0.037

[a]Method of Ref *10*.
[b]Unidirectional moduli are E_L = 145 GPa, E_T = 9.5 GPa, G_{LT} = 5.6 GPA, ν_{LT} = 0.31, $\alpha_L = 0.5 \times 10^{-6}$/°C, $\alpha_T = 35 \times 10^{-6}$/°C.
[c]Assuming a −100°C temperature change on cooling, and an applied axial load σ_0 = 600 MPa.
[d]Interfaces which were observed to delaminate.

Appendix I shows that the stiffness reduction is most likely a result of the associated matrix cracking (Figs. 3 and 4). Equation 2 can therefore be seen as a rule-of-mixtures equation where the loss in stiffness is due to matrix cracking, rather than loss of coupling as a result of delamination. Similar behavior has been observed with a $[45/90/-45/0]_s$ laminate [*12*] made of the same material. The data for that laminate, which delaminated at the 90/−45 interfaces, are shown in Fig. 6. The stiffness also decreases linearly with growing delamination size, but to a value of $0.69E_{LAM}$, rather than $0.707E_{LAM}$. The extra stiffness reduction is due to more complete cracking of the off-axis plies.

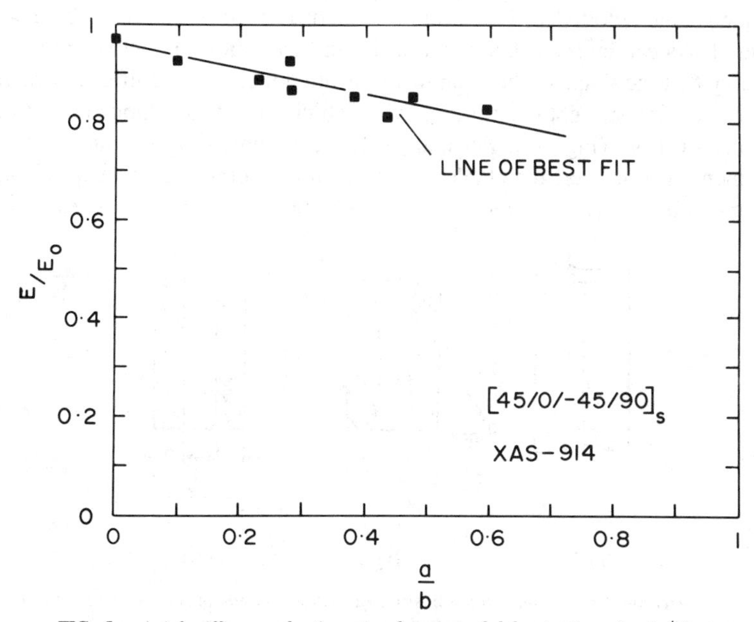

FIG. 5—*Axial stiffness reduction as a function of delamination size* (a/b).

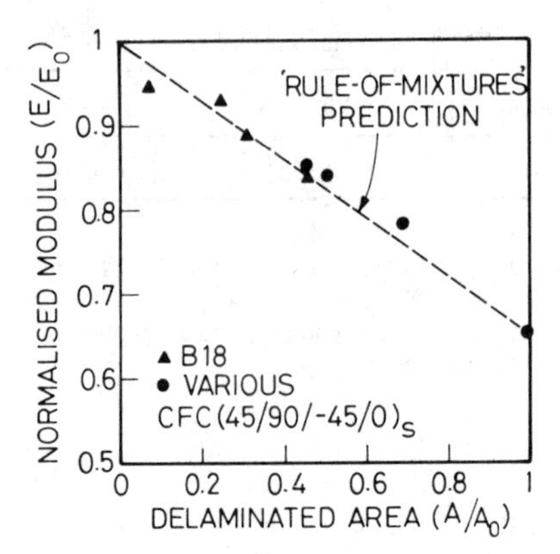

FIG. 6—*Axial stiffness reduction as a function of delamination size* (a/b). *Reproduced from Ref* 12.

Delamination Growth

[±30/±30/90/$\overline{90}$]$_s$ Laminates

The data presented in this work are in terms of the maximum delamination length a_{max} (see Fig. 7*b*), whereas the same data in Ref *4* are presented in terms of a, where a is an equivalent delamination size, calculated to give a delamination area equal to that observed (Fig. 7*c*). We have had to use a_{max}, as only a_{max} was measured continuously throughout the fatigue tests. However, a_{max} and a are interchangeable. Plotted in Fig. 8 are a_{max} and a, from tests where both were measured. As can be seen, there is a fairly linear relationship between the two, indicating that the shape of the delaminations did not change significantly as they grew. Plotted is a straight line for $a_{max} = a$, which would correspond to a straight delamination front (Fig. 7*c*). For convenience, we now drop the subscript.

As mentioned earlier, the stable growth of a delamination under quasi-static loading can be shown on a delamination resistance curve (R curve). At a given

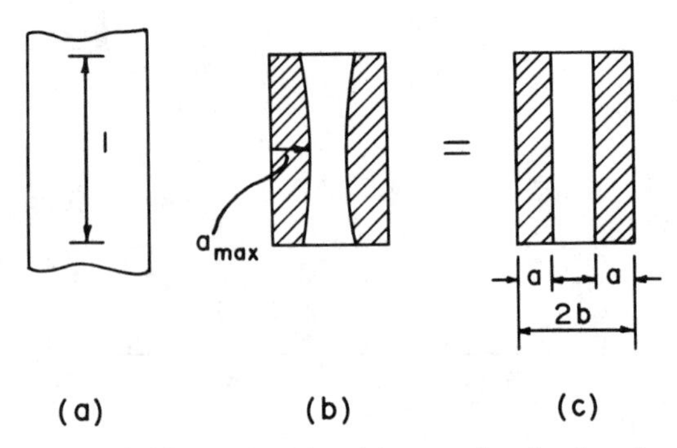

FIG. 7—*Determination of delamination size:* (a) *gauge length of strain measuring device,* (b) *actual delamination shape, and* (c) *equivalent strip delamination.*

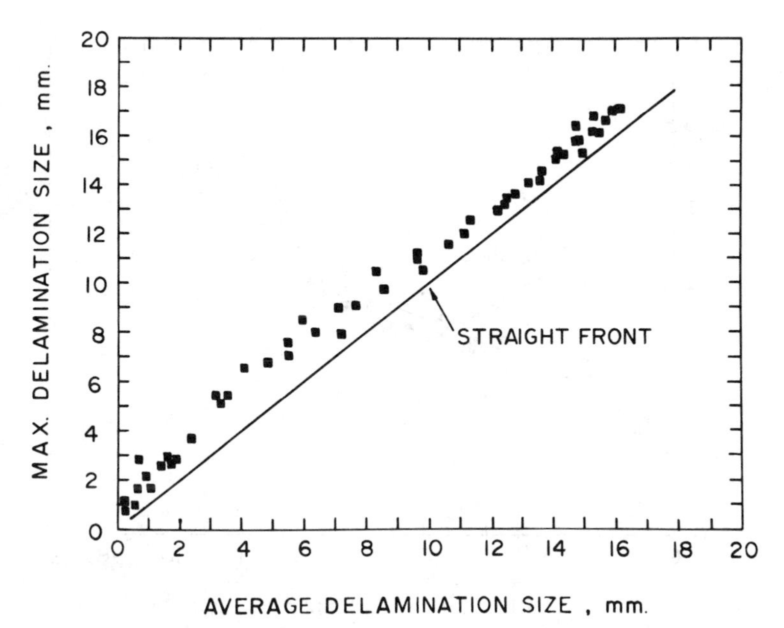

FIG. 8—*Correlation between maximum delamination size and average delamination size. Data from Ref* 4.

load and stable delamination size, the strain energy release rate G (calculated using Eq 4 below) must be just equal to the delamination resistance G_R. Otherwise, the delamination would grow even further. By increasing the applied load, and monitoring the growing delamination, a plot of G_R as a function of a can be generated. The first point on the R curve is G_c, the critical strain energy release rate for the onset of delamination.

Figure 9 shows the results for the $[\pm30/\pm30/90/\overline{90}]_s$ laminate. G_R initially increases roughly linearly with delamination size, and then rises sharply as the delamination approaches the specimen centerline. Also shown is a fitted curve used to characterize the increasing resistance numerically. This curve was used in the subsequent fatigue analysis.

Under constant maximum strain fatigue loading, the delamination grows at a decreasing rate. An example is shown in Fig. 10. Data for three maximum strain levels of 3000, 3250, and 3500 $\mu\varepsilon$ were used.

[45/0/−45/90]$_s$ Laminates

The behavior of the $[45/0/-45/90]_s$ laminate is similar to that of the $[\pm30/\pm30/90/\overline{90}]_s$ laminate. Plotted in Fig. 11 is the R curve for quasi-static loading. In contrast to the T300-5208 data, every point on this figure corresponds to a different specimen. The specimens had to be removed from the grips to be X-rayed, and problems in realigning the specimen and clip gauge precluded their reuse. The resistance to crack growth increases even more rapidly than in the previous case, and it is notable that the absolute values are higher.

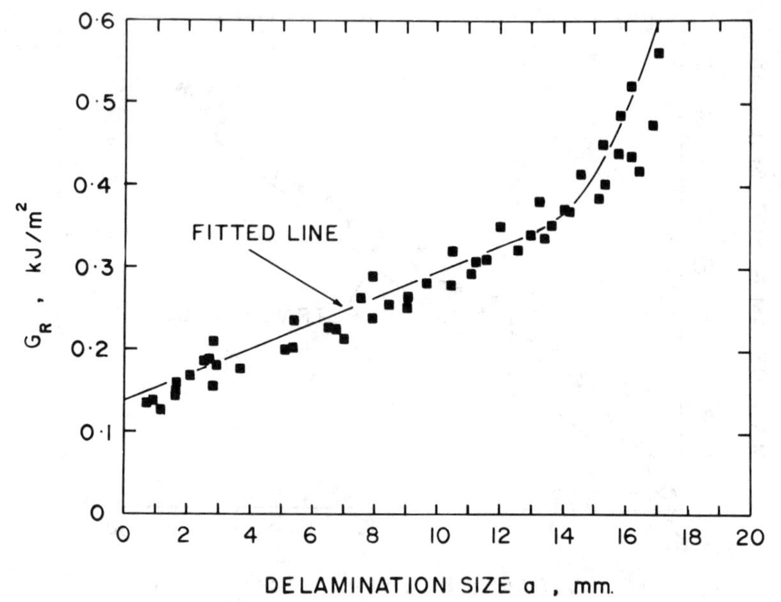

FIG. 9—*Quasi-static delamination resistance curve (*R*-curve) for* $[\pm30/\pm30/90/\overline{90}]_s$ *T300-5208 laminate. Data from Ref* 4.

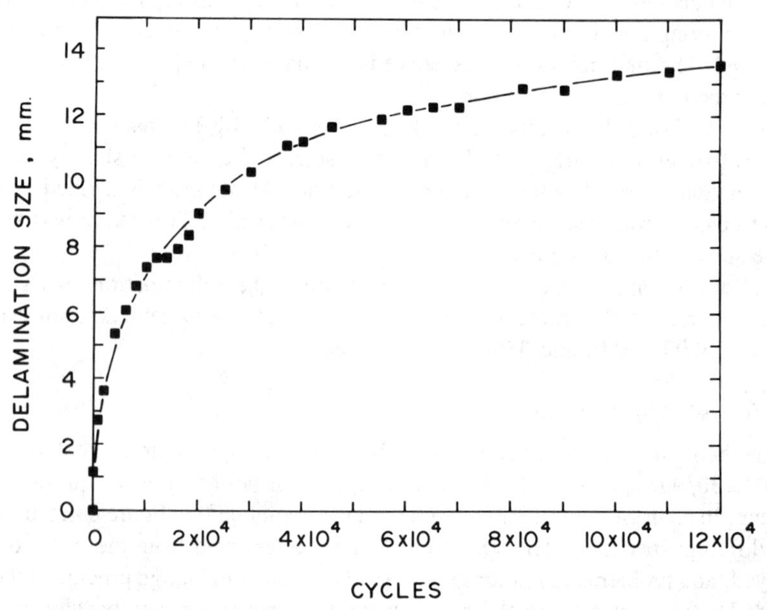

FIG. 10—*Delamination size as function of fatigue cycles for* $[\pm30/\pm30/90/\overline{90}]_s$ *T300-5208 laminate under strain-controlled fatigue loading. Data from Ref* 4.

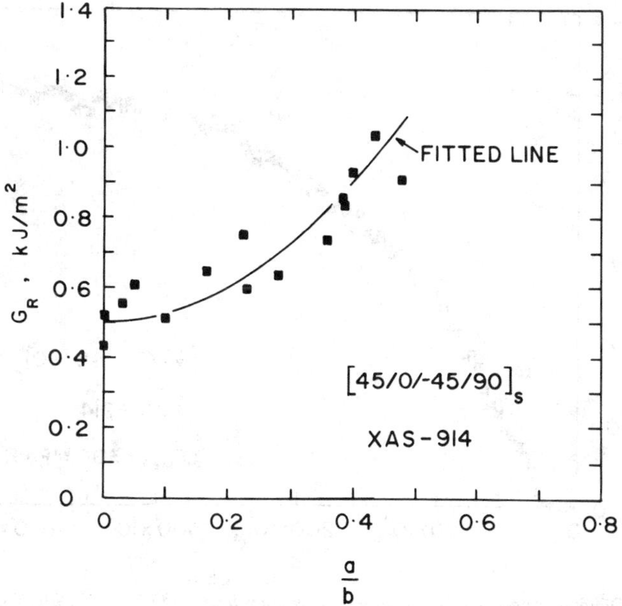

FIG. 11—*Quasi-static delamination resistance curve (*R*-curve) for* $[45/0/-45/90]_s$ *XAS-914 laminate.*

Generally, the T300-5208 system is considered to be relatively brittle [*8*], and the XAS-914 system relatively tough [*14*]. However, the different lay-ups of the present two laminates leads to differing ratios of opening mode (Mode I) and shear mode (Modes II and III) strain energy release rate components [*8*]. Therefore, it is not possible to compare the behavior of the two laminates in terms of material properties alone.

The fatigue behavior is similar (Fig. 12) to the other laminate. The delamination size is an inferred size from the stiffness reduction, using Eq 2. As in situ radiography was not possible, the only means of monitoring the delaminations without terminating the test was to measure the stiffness. However, delamination sizes measured when fatigue tests were terminated before failure correlated well with Eq 2. Tests were conducted for maximum load levels of 250, 300, 350, and 400 MPa, respectively, at a load ratio of $R = 0.1$.

Analysis and Discussion

In linear elastic fracture mechanics, the condition for crack growth is [*9*]

$$d(U - F + W)/da = 0 \tag{3}$$

where U is the elastic energy stored in the system, F the work done by the external forces, and W is the energy for crack formation. The term $G = d(F - U)/da$ is the strain energy release rate, or crack extension force, and the

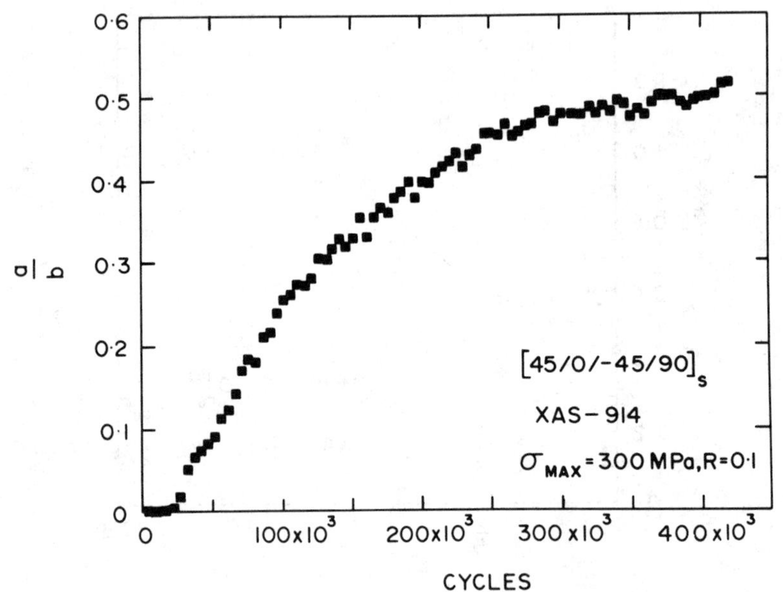

FIG. 12—*Delamination size as a function of fatigue cycles for [45/0/−45/90]$_s$ XAS-914 laminate under load-controlled fatigue loading.*

term $G_R = dW/da$ is the crack resistance (force). G can be expressed in terms of the compliance (or stiffness) of the system, and the geometry, such as that for the case of an edge delamination [*4*],

$$G = \varepsilon^2 \times (E_{\text{LAM}} - E^*) \times \frac{t}{2} \tag{4}$$

where ε is the strain, and t the laminate thickness.

In a brittle material, the energy required for crack growth is the surface energy to form the new surfaces. If the only surface being created were the delaminated surface of the interfaces, then the resistance would be constant, and once that value had been exceeded (that is, $G > G_c$) catastrophic growth would occur. However, as the R curves in Figs. 9 and 11 show, the energy required to propagate the delamination increases significantly. As the surface energy of the delaminating interfaces is unlikely to be changing, the explanation may lie in the associated matrix cracking. Two mechanisms are likely. First, the total matrix crack surface area (which is comparable in size to the delamination interfacial surface area) may be increasing at an incrementally faster rate than the delamination itself. Secondly, the work done in creating matrix cracks will increase as the weaker sites and paths are exhausted early in the process. Such a mechanism has recently been analyzed for the case of just matrix cracking under static loading [*15,16*].

The growth of cracks in metals under fatigue loading is a function of the stress intensity factor range ΔK [9],

$$\frac{da}{dN} = f(\Delta K) \tag{5}$$

with the most widely used form being the empirical power law function,

$$\frac{da}{dN} = A(\Delta K)^{\beta} \tag{6}$$

Some work has been done in composites using this equation [17,13], but a preferred form is the analogous power law equation in terms of the strain energy release rate [7]. Considering the behavior at constant load ratio R, then

$$\frac{da}{dN} = A(G_{max})^{n} \tag{7}$$

As described earlier in Ref *4*, this equation could only model the last 2 mm of delamination growth from a total of 12 mm. However, Eqs 5 to 7 consider only the driving forces for crack growth, and implicitly assume a constant resistance. Though this is quite valid for most materials, they should be modified to reflect the changing resistance in laminates, such that Eq 5 becomes, in terms of the maximum strain energy release rate, and constant load ratio R,

$$\frac{da}{dN} = f(G_{max}, G_{R}) \tag{8}$$

The equivalent power law relation to Eq 7 would then be

$$\frac{da}{dN} = A\left(\frac{G_{max}}{G_{R}}\right)^{n} \tag{9}$$

This has the right form, in that when G_R is constant, Eq 9 reduces to Eq 7, with G_R taken into the constant A. Note that Eq 9 is dimensionally correct, while Eq 7 implicitly contains the dimensions in the constant.

Analysis of Results

Data for both laminates were analyzed in the following manner: G_{max} was calculated from Eq 4 for a given load and crack length. G_R was calculated by evaluating the fitted equation to the R curve data for the same crack length. The slope of the crack-length-versus-cycles curve at that crack length was then determined, and plotted against the value of G_{max}/G_R.

In the case of the data of O'Brien [4], tests were run under strain control, and therefore G_{max} was constant during a test. With the present results, tests were run under load control, and as a test progressed, the value of G_{max} increased. However, in both cases G_R increased rapidly, and overall G_{max}/G_R decreased.

Results for the $[\pm30/\pm30/90/\overline{90}]_s$ laminate are shown in Fig. 13 and for the $[45/0/-45/90]_s$ laminate in Fig. 14. In both cases, the power law equation describes the results well. Thus, for the $[\pm30/\pm30/90/\overline{90}]_s$ laminate

$$\frac{da}{dN} = 3.864 \times 10^{-2} \left(\frac{G_{max}}{G_R}\right)^6 \text{ mm/cycle} \tag{10}$$

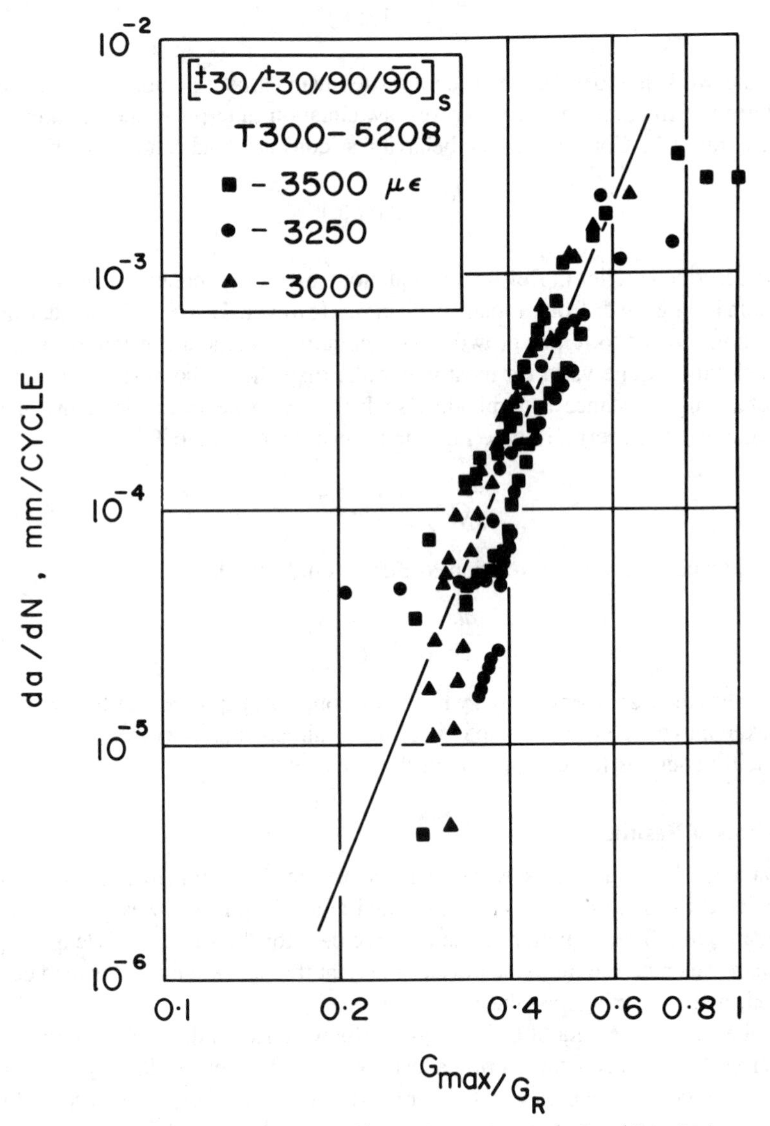

FIG. 13—*Delamination growth rate* da/dN *as a function of* (G_{max}/G_R). *Both axes are logarithmic.*

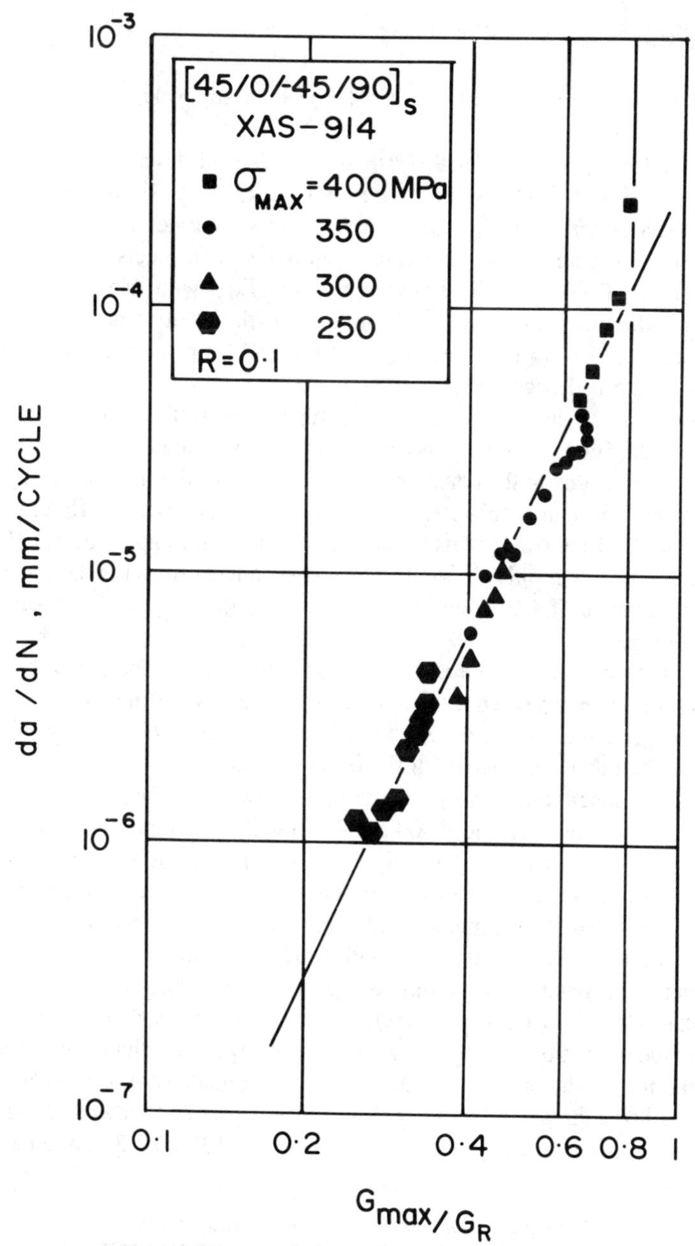

FIG. 14—*Delamination growth rate* da/dN *as a function of* (G_{max}/G_R). *Both axes are logarithmic.*

and for the $[45/0/-45/90]_s$ laminate

$$\frac{da}{dN} = 2.547 \times 10^{-4} \left(\frac{G_{max}}{G_R}\right)^{4.1} \text{mm/cycle} \quad (11)$$

The data for the $[\pm 30/\pm 30/90/\overline{90}]$ laminate is well described by Eq 10 over the range $0.35 < G_{max}/G_R < 0.75$. Below $G_{max}/G_R \approx 0.35$, the rates drop off rapidly. Data gathered for $G_{max}/G_R \gtrsim 0.75$ are less accurate, as insufficient crack length readings were available to calculate da/dN with precision.

The original T300-5208 data [*4*] were fitted by Eq 7 with an exponent of 3.11. As described earlier, only the last 2 mm of growth out of some 12 mm of total growth were used from each specimen. The first 10 mm were discounted, until the growth rate had become roughly constant.

The discrepancy in exponents (3.11 compared to 6.0) can be explained by noting that the final crack lengths are longer at higher applied G_{max}. Therefore, G_R cannot correspond to the constant in Eq 7. Hence, in the final stage of growth, G_R is roughly constant within a test, but varies between tests at different applied G_{max} (Table 2). To allow for this variation is equivalent to applying Eq 10 to the final 2 mm of growth only. If we do so, a least squares fit to the data in Table 2 yields an exponent of 4.9, which is much closer to the value of 6 measured over the whole range.

All the preliminary results for the XAS-914 laminate are fitted well by Eq 11. The tests were conducted under load control, and as the stiffness decreased with cycling, G_{max} increased (Eq 4). As a result, the change in G_{max}/G_R during a test is smaller than that observed with the other laminate.

Both the exponent and constant in the power law fit for the XAS-914 laminate are lower than those for the T300-5208 laminate. This might be seen as an indication of higher resistance to fatigue growth. However, as noted earlier, the lay-ups are different, and the observed behavior may be due to differing G_I/G_{II} ratios, rather than differing material resistance. It would be necessary to compare the same lay-up from both systems, with roughly equal G_I/G_{II} ratios, to isolate the influence of material properties. Keeping this in mind, we make the observation that when $G_{max}/G_R = 1$, effectively when under quasi-static loading, the XAS-914 delamination will grow at a much slower rate than the T300-5208 delamination. On the other hand, the XAS-914 laminate shows no indication of having reached a threshhold value, below which the fatigue growth rate drops off significantly, even at $G_{max}/G_R = 0.3$, while the T300-5208 laminate shows

TABLE 2—*Constant crack growth rate measurements.*

G_{max}	G_R[a]	G_{max}/G_R	da/dN,[b] mm/cycle
102	318	0.321	2.46×10^{-5}
119	326	0.365	3.82×10^{-5}
138	360	0.384	6.3×10^{-5}

[a]Calculated value from fitted line on Fig. 9.
[b]Average observed value.

threshhold behavior at about $G_{max}/G_R = 0.35$. Whether these effects are primarily geometry dependent, or material dependent, is not clear.

Much work needs to be done to characterize fully the behavior of the different systems. However, it has been suggested that the use of G, and R curves, allows the prediction of the static behavior of different lay-ups from a small data base [*18*]. If so, it may also prove possible to characterize their fatigue behavior in the same fashion, using the present technique. Such a possibility is to be desired, as otherwise laminates are an infinite set of materials [*1*].

Conclusions

The modified power law equation proposed in this study was successful in characterizing edge delamination growth under fatigue loading for two different carbon-fiber epoxy laminates. The reasoning behind the equation can be explained in terms of the energy available for crack growth and the energy required, using fracture mechanics concepts. As such, the method should have more generality than a purely phenomenological model.

Acknowledgments

The author would like to thank Dr. T. K. O'Brien of NASA Langley for generously providing the data on the $[\pm30/\pm30/90/\overline{90}]_s$ T300-5208 laminates.

The author would also like to acknowledge many interesting discussions with Drs. J. S. Nadeau and E. Teghtsoonian. The help of Mr. R. G. Bennett with the experimental work and in preparing the figures, and the help of Mrs. J. Kitchen in preparing the manuscript, are much appreciated.

APPENDIX I

Stiffness Loss of the $[45/0/-45/90]_s$ Laminate

Given the complex damage state observed, the stiffness reduction measured may be due to a variety of mechanisms. The different mechanisms and their respective stiffness reductions are shown in tabular form in Table 3.

TABLE 3—*Stiffness reduction mechanisms.*

Mechanism	E^*/E_{LAM}
90/90 interface delaminated:	
(1) with no coupling	1
(2) with coupling	0.667
−45/90 interface delaminated:	
(1) with no coupling	0.872
(2) with coupling	0.823
90° ply cracked	0.964
±45° plies cracked	0.692
90, ±45° plies cracked	0.646

Delamination

We observe delaminations at both the −45/90 and 90/90 interfaces, with the crack shifting back and forth among the interfaces. If we assume complete delamination of the 90/90 interface, there will be no calculated stiffness loss unless we allow for the bending/extensional coupling effect [*19*] of the now unbalanced half laminates. Then the stiffness will reduce to $E^* = 0.667E_{LAM}$. If we assume that delaminations occur at the −45/90 interfaces, using the rule of mixtures method [*4*], the stiffness will reduce to either $0.872E_{LAM}$ with no coupling and $0.823E_{LAM}$ with coupling. As argued in Ref *4*, the shifting of the delamination from one interface to another suggests that the true bending/ extensional coupling contribution to the stiffness reduction is smaller than that calculated by assuming that the interfaces are cleanly delaminated.

Matrix Cracking

The effect of matrix cracking on the stiffness can be estimated using the ply discount scheme [*19*]. This technique consists of recalculating the laminate stiffness, and assigning zero axial and shear stiffness to the cracked laminae. This result is the maximum stiffness reduction possible. In practice, the cracked layers are still load carrying, and will contribute to the laminate stiffness.

The observed initial drop of $0.04E_{LAM}$ was measured before the initiation of any delaminations, and during the period when the only damage was the cracking of the 90° plies. It is exactly equal to the expected loss of axial stiffness contribution from the cracked 90 plies. The effect of cracking all the off-axis plies is to reduce the laminate stiffness to $0.646E_{LAM}$. This predicted reduction of $0.354E_{LAM}$ is higher than the observed final value of $(0.04 + 0.253)E_{LAM} = 0.293E_{LAM}$. However, we observe (Figs. 2 to 4) that though the 90 and −45° plies are fairly heavily cracked, the +45° plies are lightly cracked, and should therefore retain a measurable part of their load-bearing ability. The $[45/90/-45/0]_s$ laminate (Fig. 6) shows a greater stiffness reduction of $0.31E_{LAM}$, though still shy of $0.354E_{LAM}$. As in this lay-up, both ±45 and −45° plies are adjacent to the delaminated interfaces, they are more susceptible to cracking, and will lose more of their load-bearing capacity.

References

[*1*] Hashin, Z., "Analysis of Composite Materials—A Survey," *Journal of Applied Mechanics,* Vol. 50, Sept. 1983, p. 481.

[*2*] *Fatigue of Fibrous Composite Materials, ASTM STP 723,* American Society for Testing and Materials, Philadelphia, 1981.

[*3*] Chai, H., *Composites,* Vol. 15, No. 4, Oct. 1984, p. 277.

[*4*] O'Brien, T. K., in *Damage in Composite Materials, ASTM STP 775,* K. L. Reifsnider, Ed., American Society for Testing and Materials, Philadelphia, 1982, pp. 140–167.

[*5*] Rybicki, E. F., Schmueser, D. W., and Fox, J., *Journal of Composite Materials,* Vol. 11, Oct. 1977, p. 470.

[*6*] Wang, A. S. D. and Crossman, F. W., "Initiation and Growth of Transverse Cracks and Edge Delaminations in Composite Laminates, Parts 1 and 2," *Journal of Composite Materials,* Supplement, Vol. 14, 1980, pp. 71–106.

[7] Wilkins, D. J., Eisenmann, J. R., Camin, R. A., Margolis, W. S., and Benson, R. A., in *Damage in Composite Materials, ASTM STP 775,* K. L. Reifsnider, Ed., American Society for Testing and Materials, Philadelphia, 1982, pp. 168–183.
[8] O'Brien, T. K., "Interlaminar Fracture of Composites," NASA Technical Memorandum 85768, USAAVSCOM Technical Report TR-84-B-2, June 1984.
[9] Broek, D., *Elementary Engineering Fracture Mechanics,* third revised ed., Martimus Nijhoff, The Hague, 1982, Chaps. 5 and 10.
[10] Pagano, N. J. and Pipes, R. B., *International Journal of Mechanical Sciences,* Vol. 15, 1973, p. 679.
[11] Highsmith, A. L. and Reifsnider, K. L., in *Damage in Composite Materials, ASTM STP 775,* K. L. Reifsnider, Ed., American Society for Testing and Materials, Philadelphia, 1982, p. 103.
[12] Poursartip, A., Ashby, M. F., and Beaumont, P. W. R., *Composites Science and Technology,* Vol. 25, No. 3, 1986, pp. 193–218.
[13] Ogin, S. L., Smith, P. A., and Beaumont, P. W. R., *Composites Science and Technology,* Vol. 22, No. 1, 1985, p. 23.
[14] Bader, M. G. and Boniface, L., "The Assessment of Fatigue Damage in CFRP Laminates," in *Proceedings of the International Conference on Testing, Evaluation, and Quality Control of Composites,* T. Feest, Ed., Butterworths Press, 1983, Sevenoaks, Kent, U.K., p. 66.
[15] Wang, A. S. D., Chou, P. C., and Lei, S. C., *Journal of Composite Materials,* Vol. 18, May 1984, p. 239.
[16] Fukumaga, H., Chou, T. W., Peters, P. W. M., and Schulte, K., *Journal of Composite Materials,* Vol. 18, July 1984, p. 339.
[17] Wang, S. S. and Wang, H. T., *Transactions, American Society of Mechanical Engineers,* Vol. 101, 1979, p. 34.
[18] O'Brien, T. K., "The Effect of Delamination on the Tensile Strength of Unnotched, Quasi-Isotropic, Graphite/Epoxy Laminates," in *Proceedings of 1982 SESA/JSME Joint Conference on Experimental Mechanics,* Society for Experimental Stress Analyses (SESA), Brookfield Center, CT, 1982, p. 236.
[19] Jones, R. M., *Mechanics of Composites Materials,* Scripta McGraw-Hill, Washington, DC, 1975.

Donald F. Adams,[1] *Richard S. Zimmerman,*[1] *and Edwin M. Odom*[2]

Frequency and Load Ratio Effects on Critical Strain Energy Release Rate G_c Thresholds of Graphite/Epoxy Composites

REFERENCE: Adams, D. F., Zimmerman, R. S., and Odom, E. M., **"Frequency and Load Ratio Effects on Critical Strain Energy Release Rate G_c Thresholds of Graphite/Epoxy Composites,"** *Toughened Composites, ASTM STP 937,* Norman J. Johnston, Ed., American Society for Testing and Materials, Philadelphia, 1987, pp. 242–259.

ABSTRACT: Graphite/epoxy composite laminates of T300/BP907 and AS6/HST-7 in layup orientations of $[\pm35/0/90]_s$ and $[\pm30_2/90/\overline{90}]_s$ were axial tension fatigue tested. Tests were conducted at 5 and 10 Hz, and at loading ratios of $R = 0.1$ and $R = 0.5$. Edge delamination as a function of number of fatigue cycles was monitored by monitoring stiffness reduction during fatigue testing. Delamination was confirmed and documented using dye-enhanced X-ray and optical photography. Critical strain energy release rates were then calculated. The composites delaminated readily, with loading ratio having a significant influence. Frequency effects were negligible.

KEY WORDS: graphite/epoxy composites, cyclic fatigue, edge delamination test, load ratio effects, frequency effects

The fatigue loading response of composite materials has long been a topic of interest. Various factors affecting fatigue response have been studied to help better understand this composite material behavior. Accumulated damage during fatigue loading initially involves microcracking of the matrix between fibers or between laminae of the composite laminate or both, leading to gross cracking and delamination between laminae. Composite stiffness and strength are typically degraded as a result of this accumulated damage, which also results in an increase in overall composite strain.

[1]Professor of mechanical engineering and staff engineer, respectively, Composite Materials Research Group, University of Wyoming, Mechanical Engineering Department, Box 3295, Laramie, WY 82071.

[2]Composites engineer, General Electric Corp., 336 Woodward Rd. S.E., Albuquerque, NM 87102.

Changes in the strain energy release rate G as a function of the loading ratio and test frequency were the principal parameters studied here. These effects were determined by conducting an ordered sequence of fatigue tests on two different graphite/epoxy composite systems, namely, T300/BP907 (recently renamed CYCOM 907) and AS6/HST-7, each in two different laminate configurations, $[\pm 30_2/90/\overline{90}]_s$ and $[\pm 35/0/90]_s$. The T300/BP907 consists of Union Carbide Thornel T300 graphite fiber [*1*] in an American Cyanamid BP907 (CYCOM 907) epoxy matrix [*2*]. This is a 177°C cure modified epoxy formulated for improved toughness. The AS6/HST-7 consists of Hercules AS6 graphite fiber [*3*] in an American Cyanamid CYCOM HST-7 epoxy matrix [*4*]. The HST-7 achieves its improved toughness via a combination of two discrete layers, namely, a higher modulus resin on the graphite fibers to provide structural properties and environmental resistance, and a separate high strain epoxy resin layer applied to one side of the prepreg to impart toughness to the composite laminate. This system is also cured at 177°C.

The two layup orientations were selected to promote free edge delamination under axial tensile loading. The delaminations should occur at the interface between the 90° plies and the adjacent ply.

Two testing frequencies, 5 and 10 Hz, and two tensile loading ratios, $R = 0.1$ and $R = 0.5$, were used to evaluate frequency or load effects or both on the four laminates. It was initially planned to use strain ratio as the variable in the test matrix, but it was found to be impractical to test the specimens in strain control because of the viscoelastic response of the materials, which caused load to pass into compression during a test. Load control was therefore used for all testing.

All test specimens were instrumented with a strain gage extensometer to monitor strain response and to provide a means of computer control in suspending a fatigue test after a specified stiffness loss caused by damage accumulation. Stiffness monitoring in this manner was the means used to detect delamination in a laminate. The fatigue test was then interrupted, a visual inspection was made to verify the existence of a delamination, and a dye-enhanced radiograph was taken to document the extent of delamination. Stiffness and strain data were later analyzed to determine the effects of cyclic frequency and load ratio on the calculated strain energy release rate.

Test Procedures

All materials used in this program were supplied by NASA-Langley in flat plate form. An edge delamination test specimen geometry was used for all testing. These specimens were 152 mm (6 in.) long, 12.7 mm (0.5 in.) wide, and approximately 1.8 mm (0.070 in.) thick.

All static testing was performed in an Instron Model 1125 electromechanical universal test machine. All unidirectional specimens were instrumented with a longitudinal extensometer to measure strain in the specimen, and to allow recording of the complete stress-strain curve to failure. Additionally, a transverse extensometer was used on the 0° tension specimens of both material systems to allow calculation of Poisson's ratio. Data were recorded on an HP-21MX-E minicomputer for later reduction and plotting.

All tension-tension fatigue edge delamination testing was performed using an MTS Model 810 servohydraulic test system. All fatigue specimens were instrumented with a dynamic-rated Instron extensometer, to measure strain in the specimen and to allow for calculation of a modulus during each cycle of the fatigue test. The extensometer gage length was set to 51 mm (2 in.), to measure any change of stiffness over the largest length reasonable for the 152-mm (6-in.) long edge delamination specimens. The extensometer arms were bonded to the surface of each specimen using a 5-min cure DEVCON two-part epoxy adhesive. This bonding was done to prevent extensometer slippage during the fatigue testing, which would have invalidated the strain data.

Specimens were held in the test machine using MTS Model 647 hydraulically actuated grips. Their use enabled a constant gripping force to be applied throughout a fatigue test, which would not be true for wedge-action grips. Wedge-action grips tend to tighten over the duration of a fatigue test, sometimes crushing the ends of the test coupon.

Data were recorded, and control of each test was performed, using an HP-21MX-E minicomputer. Data acquisition software combined a real time data acquisition and reduction routine and a logarithmic data storage routine. The data acquisition routine is capable of reading 8 channels of data for a fatigue test conducted at 10 cycles/s at a sampling rate of 320 samples/s/channel. Additionally, the data acquisition for each cycle was synchronized to begin sampling at the beginning of the cycle, that is, at $\overline{\sigma}_{min}$, the minimum value of the cyclic applied stress. This allowed a very efficient data storage format for reduction purposes. After the stress and strain data were acquired for a complete cycle, the dynamic stiffness was calculated using a linear regression curve-fit beginning at the start of each cycle. A decision based upon this value was then made to continue or interrupt the load cycling. If the instantaneous dynamic stiffness indicated a 5% drop from the previously stored cycle value, then the test was interrupted to allow for specimen inspection and a dye-enhanced radiograph to be taken. Large delaminations were typically detected and the test terminated.

Since stiffness reductions preceded specimen failure, it would be quite possible for the specimen to start to fail before a scheduled data storage cycle. To prevent the above condition, the dynamic modulus was checked for decay ten times between each storage cycle and all values temporarily held in memory. If the specimen had not yet failed at the next designated storage cycle, the ten data set values were deleted from temporary memory and the data set values for the data storage cycle were stored. If the specimen failed between designated storage cycles, then all values stored in temporary memory since the last storage cycle were transferred to permanent storage on disk in the data file. This method ensured the maximum amount of useful data being collected without the data files becoming too large and cumbersome.

Dye-enhanced radiography was used to verify the extent of edge delamination in many of the test specimens. A PANTEK Model HF75 Industrial X-ray unit was used to take all radiographs. This particular unit is designed specifically for use with composite materials and was relatively easy to set up and use based on the low energy requirements and digital controls employed. Some study was done of previous X-ray work performed by a number of researchers to identify the best dye penetrant to enhance the radiograph contrast [*5,6*]. Zinc iodide in solution with methanol, water, and a wetting agent was found to be the least toxic and easiest to use [*5*]. Previous work by Sendeckyj [*7*] provided a procedure for mixing and using the zinc iodide dye solution. Temperature and relative humidity were also recorded during the entire testing phase. Full details of the test procedures are included in Ref *8*.

Experimental Results

Static Tests

Unidirectional composites and the two laminate orientations were statically tested to obtain engineering constants for use during the edge delamination fatigue testing, and for later data reduction and plotting of strain energy release rate for the laminates. Five specimens of each test type were tested. Average static test values for the T300/BP907 are presented in Table 1. Average static test results for the AS6/HST-7 composite are given in Table 2.

The unidirectional material constants were used as input to a laminated plate theory computer program, Program AC-3 [*9*], to predict the laminate and sublaminate stiffness properties needed to calculate the strain energy release rate G to be discussed later. All property predictions for the full laminates were higher than measured values. The predicted sublaminate stiffness properties were therefore adjusted by the ratio of measured to predicted full laminate values. This method was found to provide the best estimate of sublaminate properties. These stiffness values were then used as the properties of the sublaminates, that is, the sublaminates [$\pm 30_2$], [$\pm 35/0$], and [± 35]. The sublaminates being used in the edge delamination test were not supplied as actual composites and thus could not be tested to confirm the predicted values for these two material systems. Such testing should be a part of future work since (linearly elastic) laminated plate theory was found to not be adequate in predicting the full laminate stiffness properties and, therefore, probably does not predict sublaminate properties with sufficient accuracy either. The adjustment of sublaminate properties based on full laminate test data as performed here presumably improved the accuracy of the predictions of the sublaminate properties, but this should be verified by actual sublaminate testing.

Transverse coefficient of thermal expansion (CTE) testing was performed, using a quartz-glass tube dilatometer apparatus, on both unidirectional com-

TABLE 1—*Average static material properties for T300/BP907 graphite-epoxy in the room temperature, dry condition.*

	UNIDIRECTIONAL MATERIAL CONSTANTS		
E_{11}	136 GPa		(19.7 Msi)
E_{22}	8.55 GPa		(1.24 Msi)
G_{12}	4.16 GPa		(0.60 Msi)
ν_{12}		0.281	
σ_{11}	1430 MPa		(208 ksi)
σ_{22}	71.7 MPa		(10.4 ksi)
τ_{12}	124 MPa		(18.0 ksi)

	Measured Values	Predicted Stiffness [9]	Adjustment Ratio
	LAMINATE MATERIAL CONSTANTS		
Laminate A: $[\pm 30_2/90/\overline{90}]_s$			
E_x	40.3 GPa (6.19 Msi)	54.2 GPa (7.86 Msi)	0.788
$\sigma_{x\text{ult}}$	360 MPa (52.1 ksi)		
$\sigma_{x\text{Delam}}$	305 MPa (44.2 ksi)		
$\varepsilon_{x\text{ult}}$	0.0150		
$\varepsilon_{x\text{Delam}}$	0.0072		
Laminate B: $[\pm 35/0/90]_s$			
E_x	51.0 GPa (7.39 Msi)	64.6 GPa (9.37 Msi)	0.789
$\sigma_{x\text{ult}}$	443 MPa (64.2 ksi)		
$\sigma_{x\text{Delam}}$	428 MPa (62.0 ksi)		
$\varepsilon_{x\text{ult}}$	0.0085		
$\varepsilon_{x\text{Delam}}$	0.0080		

	Predicted Value [9]	Adjusted Value
	SUBLAMINATE MATERIAL CONSTANTS	
Sublaminate A: $[\pm 30_2]_s$		
E_x	43.7 GPa (6.34 Msi)	34.4 GPa (4.99 Msi)
Sublaminate B: $[\pm 35/0]_s$		
E_x	67.4 GPa (9.77 Msi)	53.1 GPa (7.70 Msi)

posites, over a temperature range from −40 to 100°C. The T300/BP907 composite averaged 33.4×10^{-6}/°C while the AS6/HST-7 averaged 40.3×10^{-6}/°C. The AS6/HST-7 composite probably exhibited a higher transverse CTE than the T300/BP907 as a result of the high strain epoxy present between each ply. The additional epoxy material effectively reduced the fiber volume and raised the CTE accordingly.

An attempt was made, using a laser interferometer apparatus, to also measure the axial CTE of both laminates. Adequate thermal control could not be achieved, however, to measure the very low values sufficiently accurately, and work was eventually suspended. More work will be required before this capability becomes operational within the Composite Materials Research Group (CMRG) at the

TABLE 2—*Average static material properties for AS6/HST-7 graphite-epoxy in the room temperature, dry condition.*

	UNIDIRECTIONAL MATERIAL CONSTANTS		
E_{11}	105 GPa		(15.2 Msi)
E_{22}	6.62 GPa		(0.96 Msi)
G_{12}	3.17 GPa		(0.46 Msi)
ν_{12}		0.300	
σ_{11}	1560 MPa		(226 ksi)
σ_{22}	61.4 MPa		(8.9 ksi)
τ_{12}	112 MPa		(16.3 ksi)

	Measured Values	Predicted Stiffness [9]	Adjustment Ratio
	LAMINATE MATERIAL CONSTANTS		
Laminate A: $[\pm 30_2/90/\overline{90}]_s$			
E_x	38.8 GPa (5.63 Msi)	41.8 GPa (6.06 Msi)	0.929
σ_{xult}	295 MPa (42.8 ksi)		
σ_{xDelam}	276 MPa (40.0 ksi)		
ε_{xult}	0.0085		
ε_{xDelam}	0.0076		
Laminate B: $[\pm 35/0/90]_s$			
E_x	45.5 GPa (6.61 Msi)	49.8 GPa (7.22 Msi)	0.914
σ_{ult}	632 MPa (91.6 ksi)		
σ_{xDelam}	520 MPa (75.4 ksi)		
ε_{xult}	0.0150		
ε_{xDelam}	0.0111		
Laminate C: $[\pm 35/90/0]_s$			
E_x	45.5 GPa (6.60 Msi)		
σ_{ult}	781 MPa (113 ksi)		
σ_{xDelam}	781 MPa (113 ksi) (no delamination before failure)		
ε_{xult}	0.0170		
ε_{xDelam}	0.0170 (no delamination before failure)		

	Predicted Value [9]	Adjusted Value
	SUBLAMINATE MATERIAL CONSTANTS	
Sublaminate A: $[\pm 30_2]_s$		
E_x	33.7 GPa (4.88 Msi)	31.2 GPa (4.53 Msi)
Sublaminate B: $[\pm 35/0]_s$		
E_x	51.9 GPa (7.52 Msi)	47.4 GPa (6.87 Msi)
Sublaminate C: $[\pm 35]_s$		
E_x	22.6 GPa (3.28 Msi)	20.7 GPa (3.00 Msi)

University of Wyoming. The quartz-glass tube dilatometer equipment is capable of measuring coefficient of thermal expansion values down to about 2.0 ×

10^{-6}/°C. The laser interferometer theoretically should be able to measure values down to 5.0×10^{-8}/°C. This sensitivity is necessary when measuring axial expansion of unidirectional graphite/epoxy composites.

Thermal expansion data are discussed here for general information since the laminate cooldown stresses induced by ply expansion mismatches undoubtedly also influence the onset of edge delamination.

Fatigue Tests

Tension-tension fatigue testing was performed on both graphite fiber-reinforced composite materials, using two different layups for the T300/BP907 material and three layups for the AS6/HST-7 material. Only two different layups were scheduled to be tested for both materials, but an error in plate layup orientation of one of the AS6/HST-7 plates received resulted in a third layup inadvertently being tested. The two $[\pm35/0/90]_s$ plates of AS6/HST-7 as supplied appeared to be identical, but were later discovered to have been laid up in a different stacking sequence. Both plates had been marked with a 0° reference direction, and were cut up using that reference. Well into the program, after all testing was completed and while a microscopic study was being conducted to identify the delamination locations, the anomaly was discovered. The orientation of one of the supposedly $[\pm35/0/90]_s$ plates was found to actually be $[\pm35/90/0]_s$. This was detected by observing the location of the delamination and then recognizing that the 0 and 90° plies were reversed for some of the test specimens. Both plates had been laid out and cut up very carefully, and each individual specimen had been prominently marked as to which plate it originated from. One of the plates was thus determined to have a $[\pm35/0/90]_s$ orientation, as desired, while the other was found to be $[\pm35/90/0]_s$. If the error had been discovered earlier, a replacement plate could have been tested. But at such a late date in the study it was decided to proceed, and to determine the effect on results of the improper layup. An analysis was conducted to calculate E^*, the stiffness of a laminate completely delaminated along one or more interfaces, and G_c, the critical strain energy release rate, for the anomalous laminate also. The axial stiffness of the laminate after complete delamination along the interface between the anomalous −35 and 90° plies was calculated according to the rule-of-mixtures formula [*8–11*]

$$E^* = \frac{4E_{[\pm35]_s} + 2E_{22} + 2E_{11}}{8} \tag{1}$$

to be 38.3 GPa (5.54 Msi) for the anomalous $[\pm35/90/0]_s$ laminate versus 37.2 GPa (5.40 Msi) for the desired $[\pm35/0/90]_s$ laminate. This represents only a 2.6% difference in E^* values. These E^* values were calculated based on physical observations of failed specimens to determine the sublaminates which were formed by the $[\pm35/90/0]_s$ laminates. The anomalous laminate delaminated primarily into $[\pm35]$ and $[90/0]_s$ sublaminates, although the crack jumped to the 90/0 interface in regions along the specimen length.

The E^* calculated for the anomalous laminate was then used to calculate the G_c for specimens cut from that laminate. The E^* and G_c values for the appropriate layup were used to plot all the test data.

All of the T300/BP907 laminates, and the AS6/HST-7 $[\pm 30_2/90/\overline{90}]_s$ laminates, were found to be correctly laid up.

In summary, the T300/BP907 graphite/epoxy was supplied by NASA-Langley in two layups, and the AS6/HST-7 was inadvertently supplied in three layups, for a total of five layups. Two different frequencies and two load ratios were used in testing these five laminates. One fatigue curve was generated for the T300/BP907 graphite/epoxy $[\pm 30_2/90/\overline{90}]_s$ layup at a testing frequency of 2 Hz, $R = 0.1$ to evaluate fully the effect of frequency on the tension fatigue testing. There were no frequency dependent effects seen when the 2-, 5-, and 10-Hz fatigue curves were compared for the T300/BP907. All further testing for this program was therefore conducted at only 5 and 10 Hz to expedite the testing. Because of lack of sufficient material of the AS6/HST-7, however (because of variations in fabricated plate thickness, as discussed in Ref *8*), only fourteen fatigue curves were actually generated. Fortuitously, frequency effects were found to be nil, and thus, the lack of material did not prevent the attainment of the goals of this study. One of the frequency iterations was simply omitted from the testing for each of the AS6/HST-7 laminates. The two curves not generated were the 5-Hz, $R = 0.5$ curves for the two AS6/HST-7 laminates. Results from the 10-Hz, $R = 0.5$ curves should be very close to what the 5-Hz curves would have been, since frequency appeared to have minimal effect on these materials in these laminate orientations.

Strain values at the onset of delamination were taken from the computer data files for the plots. Critical strain energy release rates were then calculated using the static stiffness values and the strain values from the fatigue data files. Equations 2 or 3 were used to calculate G_c [*10–13*].

For the $[\pm 30_2/90/\overline{90}]_s$ and $[\pm 35/0/90]_s$ laminates:

$$G_c = \frac{\varepsilon_c^2 t}{2}(E_{\mathrm{LAM}} - E^*) \tag{2}$$

For the $[\pm 35/90/0]_s$ laminate:

$$G_c = \frac{\varepsilon_c^2 t}{4}(E_{\mathrm{LAM}} - E^*) \tag{3}$$

where

ε_c = axial strain at delamination onset;
t = laminate thickness;
E_{LAM} = laminate axial stiffness, as determined from static tests or from laminated plate theory [*9,14*]; and
E^* = axial stiffness of a laminate completely delaminated along one or more interfaces.

For the $[\pm 30]_s$ sublaminate, E^* was calculated using

$$E^* = \frac{8E_{[\pm 30]_s} + 3E_{22}}{11} \tag{4}$$

For the $[\pm 35/0]_s$ sublaminate, E^* was calculated using

$$E^* = \frac{6E_{[\pm 35/0]_s} + 2E_{22}}{8} \tag{5}$$

Equations 2 and 3 are derived and discussed in detail in Ref *11*. Briefly, elastic strain energy per unit volume, in terms of strain, is defined as $u = E\varepsilon^2/2$. Thus, the total strain energy U is obtained by multiplying u by the specimen volume, that is, $V = 2blt$, where $2b$ is the specimen width, t its thickness, and l the gage length over which axial displacements are to be measured. By definition, $G = -dU/dA$, where A is the delaminated area, namely, $A = 2(la)$, a being the depth of the delamination in from each edge. Thus,

$$G = -\frac{(2blt)\varepsilon^2}{2(2l)}\frac{dE}{da} \tag{6}$$

But E, the axial stiffness of a partially delaminated laminate, can be expressed as $(E^* - E_{LAM})\frac{a}{b} + E_{LAM}$, using a rule of mixtures assumption (Eq 4 of Ref *11*). Thus, Eq 2 is obtained. The factor of 4 instead of 2 in the denominator of Eq 3 reflects the fact that twice the delamination area is created for this laminate.

Figures 1 through 4 show the load ratio effect on both of the T300/BP907 laminates. In all cases, the load ratio $R = 0.5$ results in a higher critical strain level to runout than the $R = 0.1$ ratio. Critical strain energy release rates are also plotted on these curves. Strain levels for the $[\pm 35/0/90]_s$ laminates are somewhat higher than for the $[\pm 30_2/90/\overline{90}]_s$ laminates.

Figures 5 and 6 show the load ratio effect observed in the AS6/HST-7 laminates. There are no 5-Hz comparison curves because of the previously mentioned lack of material to complete these curves. The $R = 0.5$ curves show a higher critical strain level to runout, as might be expected since the range of applied cyclic stress is less than for $R = 0.1$. Also, the $[\pm 35/0/90]_s$ laminate has a higher strain level to runout than does the $[\pm 30_2/90/\overline{90}]_s$ laminate. A comparison of Fig. 1 with Figs. 2 or 3 with Fig. 4 indicates the lack of frequency effect.

For the T300/BP907 composite, critical strains were consistently higher for the $[\pm 35/0/90]_s$ laminate than for the $[\pm 30_2/90/\overline{90}]_s$ laminate. However, the $[\pm 30_2/90/\overline{90}]_s$ laminate exhibited a consistently higher G_c than the $[\pm 35/0/90]_s$ laminate, for both frequencies and both load ratios. This is reversed from the trend of the critical axial strain curves, being a result of the definition of G_c, as reflected by Eqs 2 and 3.

For the AS6/HST-7 composite, the $[\pm 35/0/90]_s$ laminate exhibited much higher critical strain levels than the $[\pm 30_2/90/\overline{90}]_s$ laminate at the two load ratios. The critical strain energy release rate curves indicate a more consistent

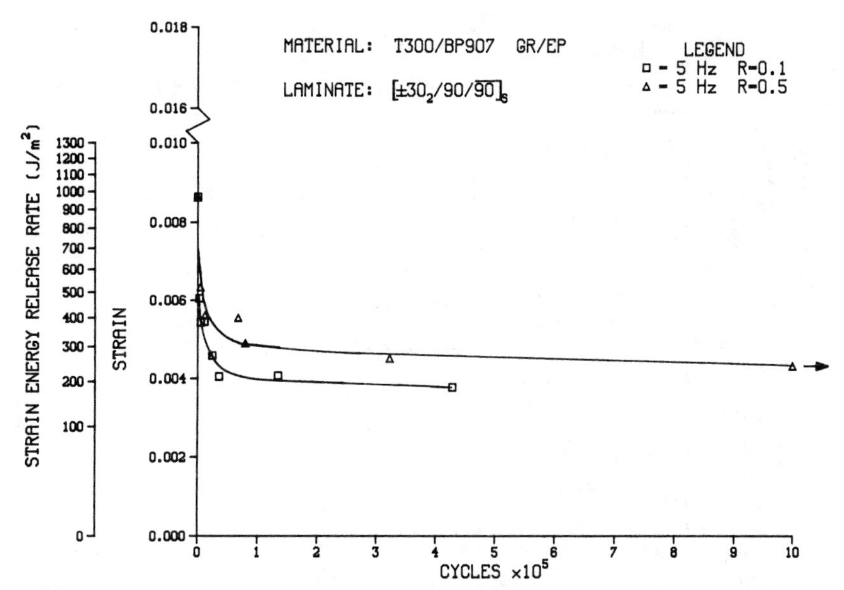

FIG. 1—*Critical strain and critical strain energy release rate versus number of fatigue cycles for T300/BP907 graphite/epoxy* $[\pm30_2/90/\overline{90}]_s$ *laminate at 5 Hz and load ratios of* R = *0.1 and* R = *0.5.*

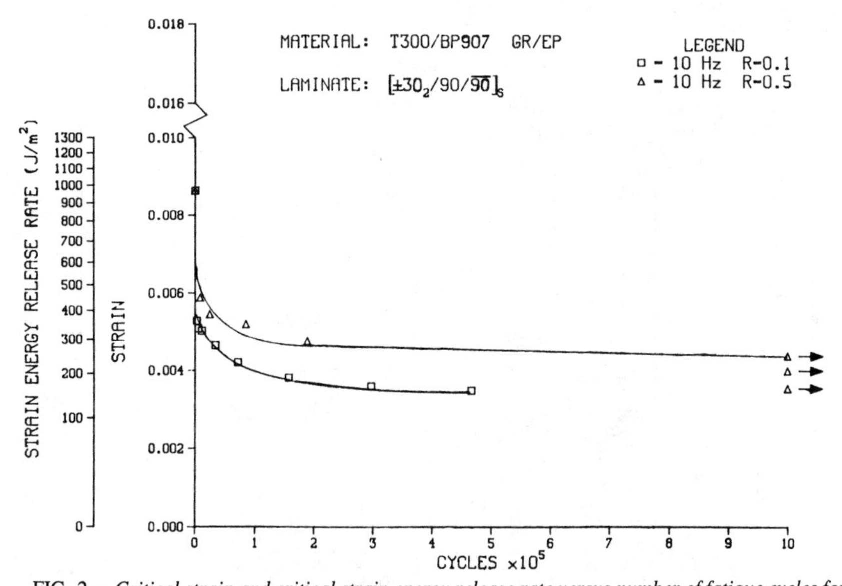

FIG. 2—*Critical strain and critical strain energy release rate versus number of fatigue cycles for T300/BP907 graphite/epoxy* $[\pm30_2/90/\overline{90}]_s$ *laminate at 10 HZ and load ratios of* R = *0.1 and* R = *0.5.*

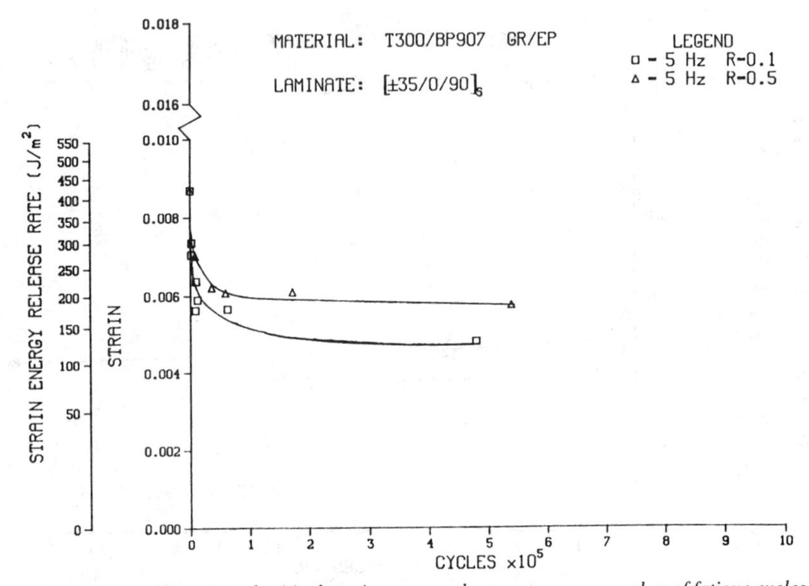

FIG. 3—*Critical strain and critical strain energy release rate versus number of fatigue cycles for T300/BP907 graphite/epoxy* $[\pm 35/0/90]_s$ *laminate at 5 Hz and load ratios of* R = *0.1 and* R = *0.5.*

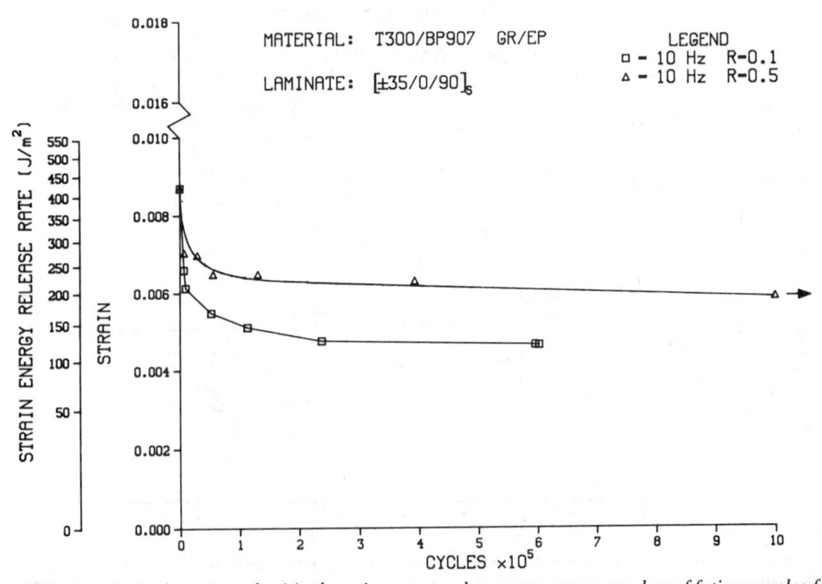

FIG. 4—*Critical strain and critical strain energy release rate versus number of fatigue cycles for T300/BP907 graphite/epoxy* $[\pm 35/0/90]_s$ *laminate at 10 Hz and load ratios of* R = *0.1 and* R = *0.5.*

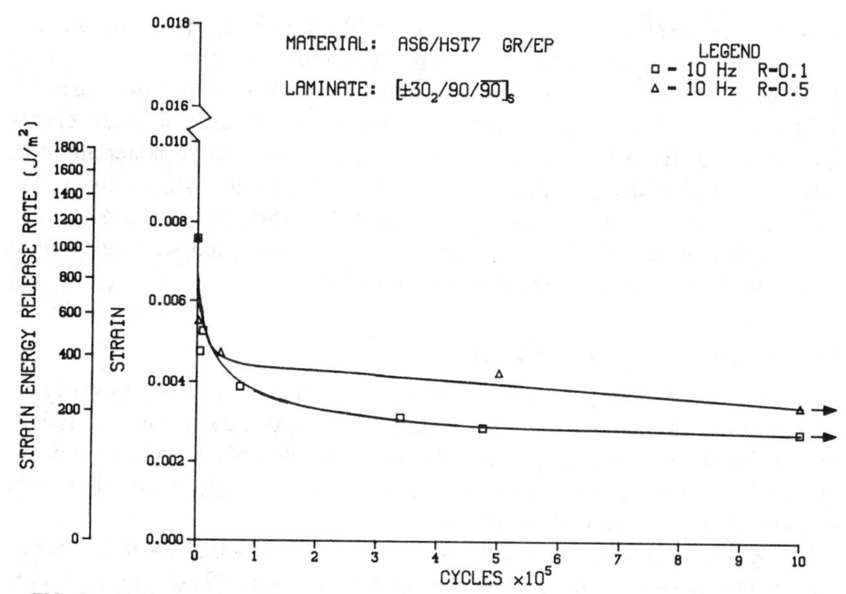

FIG. 5—*Critical strain and critical strain energy release rate versus number of fatigue cycles for AS6/HST-7 graphite/epoxy* $[\pm 30_2/90/\overline{90}]_s$ *laminate at 10 Hz and load ratios of* R = *0.1 and* R = *0.5.*

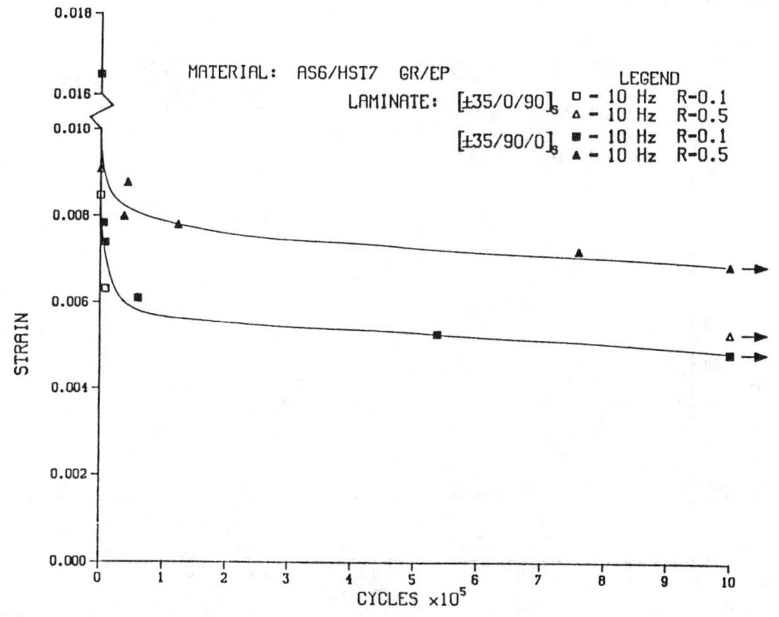

FIG. 6—*Critical strain versus number of fatigue cycles for AS6/HST-7 graphite/epoxy* $[\pm 35/0/90]_s$ *and* $[\pm 35/90/0]_s$ *laminates at 10 Hz and load ratios of* R = *0.1 and* R = *0.5.*

trend between the two AS6/HST-7 laminates. There was very little difference in G_c between the two laminates at the two load ratios.

Comparison plotting of the data was also performed for the two materials at analogous conditions. For example, the plots of the critical strain energy release rate G_c for both materials in Figs. 7 and 8 show these materials to be relatively equal in delamination resistance in the $[\pm30_2/90/\overline{90}]_s$ laminate. The T300/BP907 material is slightly superior for the $[\pm35/0/90]_s$ laminates.

A number of additional comparison plots, of delamination strain as well as strain energy release rate, are presented in Ref *8*.

Radiograph and Visual Observations

Many X-ray photographs were taken over the course of this study to verify the delamination sensed by the stiffness monitoring computer controlling the fatigue tests and seen visually during a test. As previously described, an X-ray opaque zinc iodide dye solution was used to enhance the delaminations, and thus show them much more vividly in the radiographs.

The X-rays were taken at 29 kV and 9 mA for 11 s at a film focal distance of 46 cm (18 in.) for the T300/BP907 composites. The AS6/HST-7 laminates were slightly thicker, requiring a longer exposure time and higher energy setting on the X-ray unit. X-rays of the AS6/HST-7 were taken at 30 kV and 10 mA for 15 s at a film focal distance of 46 cm (18 in.). Another visual inspection was then made and an optical photograph of the specimen edge was taken to identify the ply where delamination occurred.

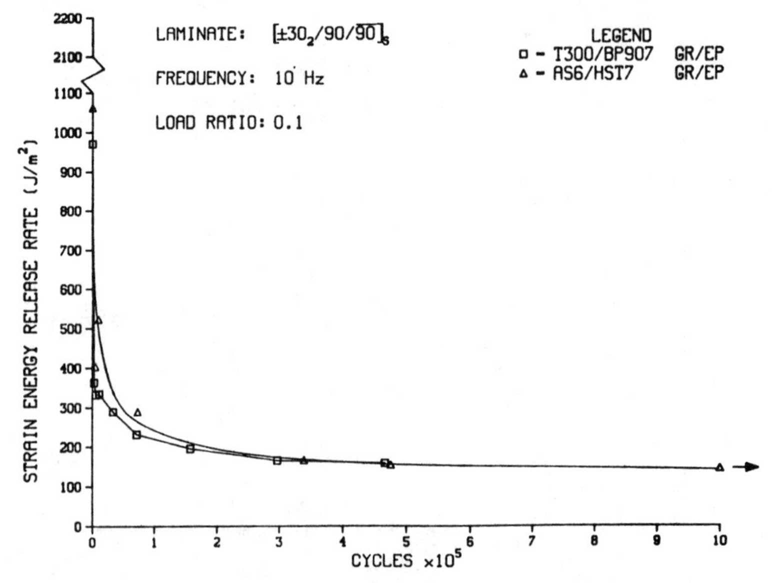

FIG. 7—*Critical strain energy release rate versus number of fatigue cycles for T300/BP907 and AS6/HST-7 graphite/epoxy* $[\pm30_2/90/\overline{90}]_s$ *laminates at 10 Hz and load ratio of* R = *0.1.*

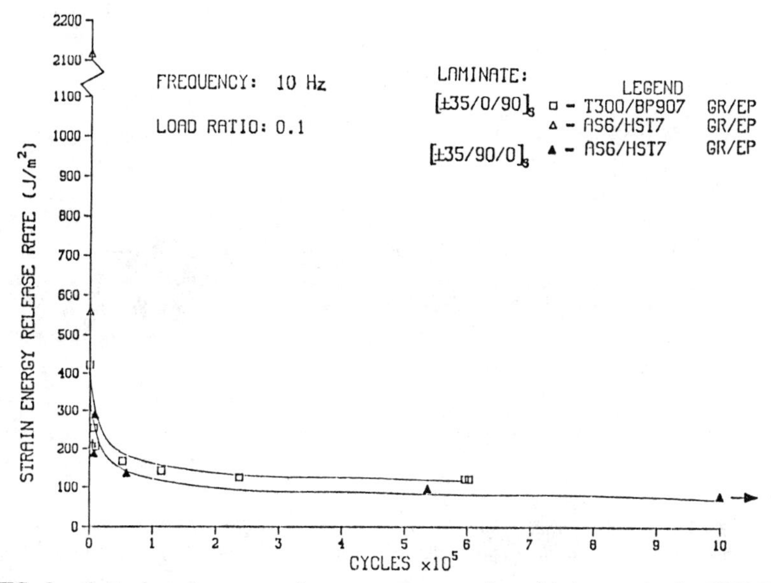

FIG. 8—*Critical strain energy release rate versus number of fatigue cycles for T300/BP907 [±35/0/90]$_s$ and AS6/HST-7 [±35/0/90]$_s$ and [±35/90/0]$_s$ graphite/epoxy laminates at 10 Hz and load ratio of* R = *0.1.*

Figures 9 and 10 are typical dye-enhanced radiographs of failed specimens, showing the extent of delamination normally seen in the five different laminates tested. No significant differences were observed between the five laminates tested regarding their delamination patterns. As can be seen in the radiographs, only a

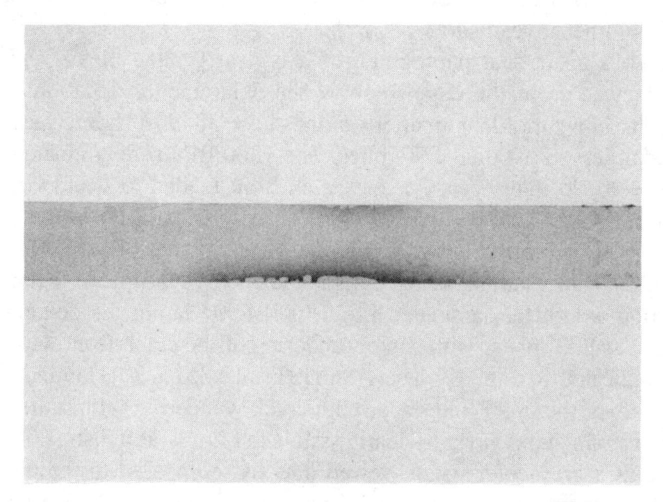

FIG. 9—*Dye-enhanced radiograph of a typical T300/BP907 [$\pm30_2/90/\overline{90}$]$_s$ laminate showing typical delamination zones at the edges of the specimen.*

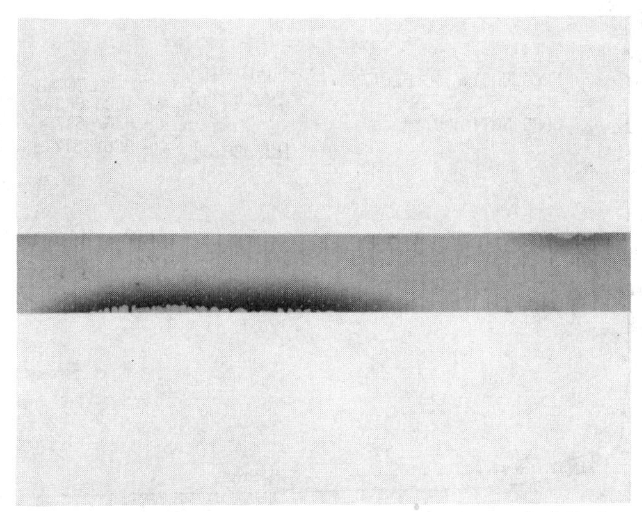

FIG. 10—*Dye-enhanced radiograph of a typical AS6/HST-7 $[\pm 35/0/90]_s$ laminate showing typical delamination zones at the edges of the specimen. Cracks in the 90° plies can also be seen.*

slight delamination occurred before suspending any test as a result of the 95% residual stiffness limit detected by the computer program. Additional radiographs are included in Ref *8*, and more fully illustrate the extent of delamination for the five laminates tested.

Visual examination of the tested laminates was performed to identify the ply where the delaminations occurred. The $[\pm 30/90/\overline{90}]_s$ laminates consistently delaminated between the −30 and the 90° ply. The delamination path wandered back and forth through the three 90° plies in a pattern similar to that previously observed by other investigators [*11,15*].

Figure 11 is an optical photograph of a typical $[\pm 30_2/90/\overline{90}]_s$ AS6/HST-7 laminate viewed from the edge to show the delamination location within the laminate. It shows the delaminations along the −30/90 ply interfaces and the wandering observed within the 90° plies. The T300/BP907 delaminated similarly along the −30/90 interfaces, but the crack front tended to wander more. The constraint in the AS6/HST-7 $[\pm 30_2/90/\overline{90}]_s$ was probably a result of the high strain epoxy layer present between plies.

The $[\pm 35/0/90]_s$ laminates were also studied to identify the location of the delamination within the laminate. The T300/BP907 laminates delaminated between the 0 and 90° plies, with some wandering of the crack front within the 90° plies, as indicated in Fig. 12. The AS6/HST-7 $[\pm 35/0/90]_s$ laminate also delaminated along the 0/90° interface, with crack wandering within the 90° plies.

The high strain epoxy layer is clearly visible in Fig. 11 as a distinct layer which stands out as a grey color when viewed directly, compared to the black of the graphite fiber-reinforced epoxy plies. There tended to be a high degree of thickness variability in the high strain epoxy layers, and a pattern of voids present.

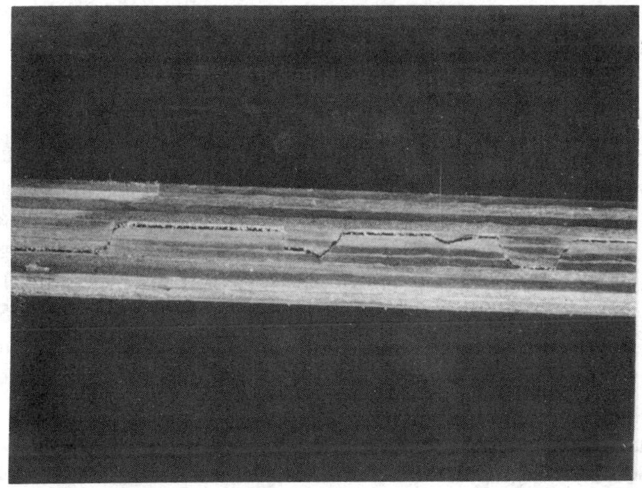

FIG. 11—*Edge view of a typical* $[\pm 30/90/\overline{90}]_s$ *AS6/HST-7 laminate, showing the delamination crack at the* −30/90 *interface and in the* 90° *plies.*

Delaminations tended to propagate from these voids to some extent when they occurred at the −30/90 or 0/90 interfaces. During the fatigue testing, some flaking of the high strain epoxy layer at the specimen edges was also noticed; these flakes accumulated on the testing machine base, below the test specimen. The distinct grey color was quite visible during and subsequent to each fatigue

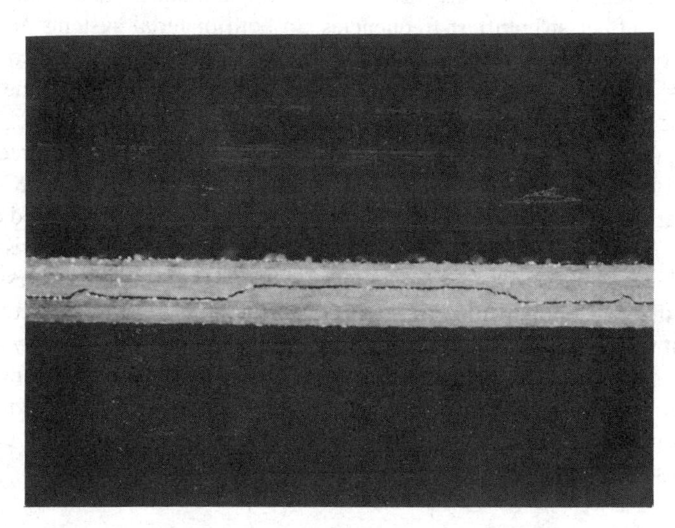

FIG. 12—*Edge view of a typical* $[\pm 35/0/90]_s$ *T300/BP907 laminate, showing the delamination crack at the* 0/90 *interface and in the* 90° *plies.*

test performed on the AS6/HST-7 material. It is not known to what extent this flaking affected test results.

Discussion

All laminates tested in fatigue exhibited a stiffness reduction because of delamination, that is, the release of strain coupling between sets of plies. This permitted correlations between cycles to failure, axial strain at delamination onset, and critical strain energy release rate G_c. The critical strain energy release rate of the AS6/HST-7 graphite/epoxy was greater than or equal to that of the T300/BP907 graphite/epoxy at both frequencies and load ratios. The AS6/HST-7 composite is an epoxy prepreg interleaved with a high strain epoxy film between each ply [*1*]. The high strain epoxy is intended to improve the fracture toughness of the composite, and did appear to provide a slight improvement over the conventional epoxy matrix.

The T300/BP907 graphite/epoxy in the $[\pm30_2/90/\overline{90}]_s$ laminate orientation exhibited a higher critical strain energy release rate than in the $[\pm35/0/90]_s$ laminate orientation, at both test frequencies and load ratios, demonstrating the influence of interlaminar stresses at the free edges of laminates. For the AS6/HST-7, the $[\pm35/0/90]_s$ laminate exhibited a critical strain energy release rate which was equal to or greater than that of the $[\pm30_2/90/\overline{90}]_s$ laminate, at both test frequencies and load ratios. The (anomalous) $[\pm35/90/0]_s$ laminate exhibited a critical strain energy release rate which was nearly half that of the $[\pm30_2/90/\overline{90}]_s$ and $[\pm35/0/90]_s$ laminates because of the twofold crack area generated in the anomalous laminate.

Test frequency had negligible effect on the value of critical strain energy release rate for both material systems and both laminates. Load ratio had a pronounced effect at both test frequencies, in both material systems and laminates. A load ratio of $R = 0.5$ resulted in a higher G_c in all cases when compared to an $R = 0.1$. This effect was not unexpected, however, because of the reduction in the alternating load component in the higher load ratio case.

Classical linearly elastic laminated plate theory [*14*] was used to predict the laminated composite stiffness properties, using Computer Program AC-3 [*9*]. This linear theory predicted values somewhat higher than were measured experimentally. The BP907 and HST-7 are unique matrix material systems, possessing a somewhat higher fracture toughness and a more highly nonlinear stress-strain response than most polymer matrix systems. This nonlinear stress-strain response is thought to account for at least a portion of the variance from (linearly elastic) laminated plate theory predictions. The very high Poisson's ratios exhibited by the sublaminates (for example, values greater than 1.10) also influenced the overall laminate response, by affecting the shear transfer between plies during a tension test.

Acknowledgment

This work was sponsored by NASA-Langley Research Center, Hampton, Virginia, under Grant NAG-1-351, with Dr. T. Kevin O'Brien as the NASA Technical Officer. The advice and assistance of Dr. O'Brien is greatly appreciated.

References

[*1*] "Typical Properties of 'Thornel' High Modulus Carbon Yarns and Their Composites," Bulletin 465-252, Union Carbide Corp., Danbury, CT, 1982.

[*2*] "CYCOM 907—A 350°F Cure Modified Epoxy Laminating Material," Data Sheet BPT207A, American Cyanamid Co., Havre de Grace, MD, 1983.

[*3*] "Magnamite Graphite Fibers," Product Data Sheet 860, Hercules Inc., Magna, UT, 1983.

[*4*] "CYCOM HST-7 Tough Advanced Composites," Data Sheet BPT 762, American Cyanamid Co., Havre de Grace, MD, 1983.

[*5*] Crossman, F. W., Warren, W. J., Wang, A. S. D., and Law, G. E., Jr., "Initiation and Growth of Transverse Cracks and Edge Delamination in Composite Laminates, Part 2: Experimental Correlation," *Journal of Composite Materials-Supplement,* Vol. 14, 1980, pp. 88–100.

[*6*] Ulman, D. A., Henneke, E. G., et al., "Nondestructive Evaluation of Damage in Metal Matrix Composites," ARPA Order 3388, Contract No. N00014-78-C-0153, June 1980.

[*7*] Sendeckyj, G. P., "Characterization, Analysis, and Significance of Defects in Composite Materials," in *AGARD Conference Proceedings, No. 355, Meeting of the Structures and Materials Panel (56th),* 1983, pp. 2–1 to 2–22.

[*8*] Zimmerman, R. S., Adams, D. F., and Odom, E. M., "Load Ratio and Frequency Effects on Strain Energy Release Rate During Tensile Fatigue Testing Utilizing the Edge Delamination Test," Report UWME-DR-401-109-1, Department of Mechanical Engineering, University of Wyoming, Laramie, WY, Dec. 1984.

[*9*] "Advanced Composites Design Guide, Vol. II-Analysis," Air Force Flight Dynamics Laboratory, Dayton, OH, Jan. 1973, p. 2. B.1.1.

[*10*] "Standard Tests for Toughened Resin Composites," NASA Reference Publication 1092, revised edition, compiled by ACEE Composites Project Office, NASA-Langley Research Center, Hampton, VA, 1983, p. 8.

[*11*] O'Brien, T. K., "Characterization of Delamination Onset and Growth in a Composite Laminate," in *Damage in Composite Materials, ASTM STP 775,* K. L. Reifsnider, Ed., American Society for Testing and Materials, Philadelphia, 1982, pp. 140–167.

[*12*] O'Brien, T. K., "Mixed-Mode Strain-Energy-Release Rate Effects on Edge Delamination of Composites," in *Effects of Defects in Composite Materials, ASTM STP 836,* American Society for Testing and Materials, Philadelphia, 1984, pp. 125–142.

[*13*] O'Brien, T. K., Johnston, N. J., Morris, D. H., and Simonds, K. A., "A Simple Test for the Interlaminar Fracture Toughness of Composites," *SAMPE Journal,* Vol. 18, No. 4, July/Aug. 1982, pp. 8–15.

[*14*] Jones, R. M., *Mechanics of Composite Materials,* McGraw-Hill, New York, 1975.

[*15*] O'Brien, T. K., "An Approximate Stress Analysis for Delamination Growth in Unnotched Composite Laminates," presented at the 6th Annual Mechanics of Composites Review, Air Force Materials Laboratory, Dayton, OH, Oct. 1980.

Isaac M. Daniel, [1] *Igbal Shareef,* [2] *and Abdu A. Aliyu* [3]

Rate Effects on Delamination Fracture Toughness of a Toughened Graphite/Epoxy

REFERENCE: Daniel, I. M., Shareef, I., and Aliyu, A. A., **"Rate Effects on Delamination Fracture Toughness of a Toughened Graphite/Epoxy,"** *Toughened Composites, ASTM STP 937,* Norman J. Johnston, Ed., American Society for Testing and Materials, Philadelphia, 1987, pp. 260–274.

ABSTRACT: The objective of this study was to determine the effects of loading rate on interlaminar fracture toughness of T300/F-185 graphite/epoxy composite, having an elastomer-modified epoxy resin matrix. Mode I interlaminar fracture was investigated by means of uniform width and width-tapered double cantilever beam (DCB) specimens. Hinged tabs were used to insure unrestrained rotation at the free ends. Specimens were loaded at quasi-static deflection rates of up to 8.5 mm/s (20 in./min) corresponding to crack extension rates of up to 21 mm/s (49.6 in./min). Crack extension was monitored by means of strain gages mounted on the surface of the specimen. Continuous records were obtained of load, deflection, and crack extension for determination of the strain energy release rate. The latter was calculated by means of the area method and beam analysis method, and expressed as a power law of the crack extension velocity. Results indicate that the strain energy release rate decreases with crack velocity by over 20% over three decades of crack velocity.

KEY WORDS: graphite/epoxy, toughened composites, fracture toughness, rate effects, strain energy release rate, double cantilever beam, crack propagation, test methods

Delamination or interlaminar cracking is considered one of the predominant types of damage in composite materials. The initiation and growth of delamination may result in progressive stiffness degradation and eventual failure of composite structures. Interlaminar crack propagation can occur under opening, shearing, tearing, or a combination thereof; therefore, delamination fracture toughness can be characterized by stress intensity factors or strain energy release rates in Modes I, II, or III. Several test methods for Mode I and mixed Modes

[1]Professor, Theoretical and Applied Mechanics, Department of Civil Engineering, Northwestern University, Evanston, IL 60201.
[2]Assistant professor, Bradley University, Peoria, IL 61606.
[3]Graduate student, Ahmadu Bello University, Zaria, Nigeria.

I and II have been developed. The most commonly used specimens for Mode I characterization are the double cantilever beam (DCB) specimen, discussed by Whitney et al. [*1*], and the width-tapered double cantilever beam (WTDCB) specimen introduced by Mostovoy et al. [*2*]. These specimens have been used by a number of investigators for determination of Mode I delamination fracture toughness [*1-6*]. Mixed mode delamination (Modes I and II) has been studied by Wilkins et al. [*6*] using a coupon with a step change in thickness, and by O'Brien et al. [*7*] using an edge delamination specimen.

The importance of delamination fracture toughness has stimulated a number of investigations of the effects of various parameters on this property. The effects of fiber orientation and layup were studied by Wilkins et al. [*6*] and Hunston and Bascom [*8*]. The latter also investigated effects of temperature and loading rate. Devitt et al. [*5*] studied the effects of viscoelastic behavior. Rate effects on delamination fracture toughness were also investigated by Miller et al. [*9*] and the authors [*10*]. Some investigators have not found any noticeable rate effects in the composite, even when the matrix resin is rate sensitive [*8,9*]. Hunston and Bascom [*8*] found a pronounced rate dependence in an elastomer-modified epoxy, with the fracture energy decreasing with increasing loading rate. In the case of the composite, however, they found a slight increase in fracture energy with loading rate. Their results may be influenced by the fact that they used glass cloth instead of unidirectional plies. The authors [*10*] found that the strain energy release rate in Mode I delamination for a unidirectional AS-4/3501-6 graphite/epoxy increases with strain rate. The matrix material in this case is a brittle one with a relatively low fracture toughness.

To improve the damage tolerance of composites, research has been directed in recent years towards tougher matrix resins. Elastomer-modified epoxies have been proposed as toughened matrices. The neat resin in such cases has a much higher fracture toughness than the unmodified epoxy. The relative increase in fracture toughness for the composite is not as high for the bulk resin [*7,8*]. These toughened resins exhibit very high rate effects in bulk form, but it is not well known yet how they behave within the composite.

The objective of this study was to determine the effects of loading rate on Mode I interlaminar fracture toughness of F-185/T300 graphite/epoxy having an elastomer-modified matrix epoxy. Uniform (UDCB) and width-tapered double cantilever beam specimens of unidirectional material were tested at various loading rates covering three decades of crack propagation velocity.

Analysis

A review of various analysis methods of the double cantilever beam test was given by Whitney et al. [*1*]. Methods described were the area method, beam analysis method, and a generalized empirical method. Further discussion of analysis methods was given by the authors [*10*]. A general expression for the strain energy release rate was obtained accounting for the contributions of the transverse shear modulus G_{13} and the kinetic energy expressed in terms of the crack propagation velocity $\dot{a}$.

In the beam analysis method, the specimen is assumed to consist of two identical cantilever beams with built-in ends and length equal to the length of the crack (Fig. 1). Neglecting the contributions of the transverse shear modulus and kinetic energy, the strain energy release rate is given by

$$G_I = \frac{12P^2a^2}{E_{11}b^2h^3} \tag{1}$$

where

P = applied load,
a = crack length,
b = specimen width,
h = cantilever beam thickness, and
E_{11} = longitudinal modulus (in fiber direction).

To account for the energy stored in the uncracked portion of the beam, an empirical formula based on an analysis of a beam on elastic foundation has been proposed:

$$G_I = \frac{12P^2a^2}{E_{11}b^2h^3}\left(1 + k_1\frac{h}{a}\right) \tag{2}$$

where k_1 is a constant determined experimentally.

The uniform DCB specimens were analyzed by the so-called area method. The energy increase per unit crack extension is simply calculated as

$$G_I = \frac{1}{2b\Delta a}(P_1\delta_2 - P_2\delta_1) \tag{3}$$

where load P_1 corresponding to opening deflection δ_1 drops to level P_2 corresponding to deflection δ_2 after an increment Δa in crack length.

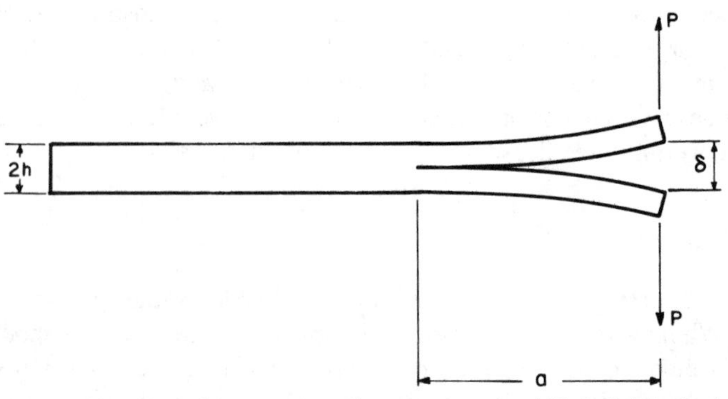

FIG. 1—*Double cantilever beam specimen for measurement of Mode I delamination fracture toughness in composites.*

One of the most difficult tasks in high rate testing is the measurement of crack length. To eliminate the dependence of the results on crack length, the width-tapered double cantilever beam (WTDCB) specimen was used (Fig. 2). If the beam is loaded at the apex of the taper, the specimen compliance C varies with the square of the crack length as follows:

$$C = \frac{\delta}{P} = \frac{12a^2k}{E_{11}h^3} \tag{4}$$

where

$k = \frac{a}{b}$, and

b = beam width at crack length a.

In this case, the strain energy release rate is independent of crack length, varying only with the square of the applied load

$$G_{\mathrm{I}} = \frac{P^2}{2b}\frac{dC}{da} = \frac{12P^2k^2}{E_{11}h^3} \tag{5}$$

It has been suggested that the initial rotation and displacement at the crack tip violate the ideal conditions of a built-in end [2]. To correct for this possible deviation, a small increment βh is added to the crack length such that

$$\frac{dC}{da} = \frac{24(a + \beta h)k}{E_{11}h^3} \tag{6}$$

from which we obtain

$$G_{\mathrm{I}} = \frac{12P^2k^2}{E_{11}h^3}\left(1 + \beta\frac{h}{a}\right) \tag{7}$$

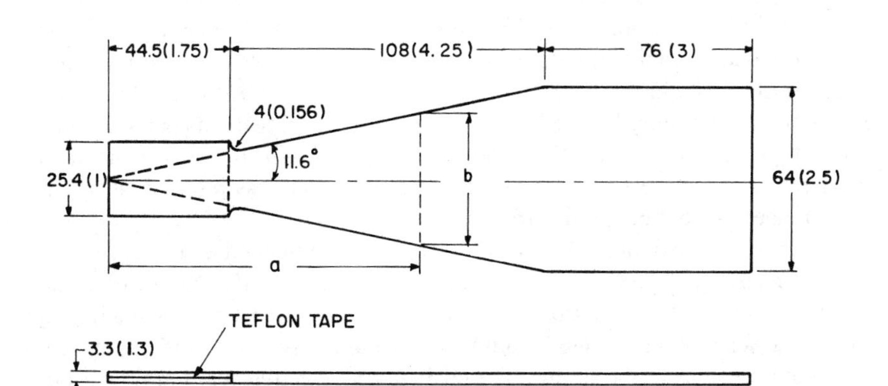

FIG. 2—*Geometry of the width-tapered double cantilever beam (WTDCB) specimen.*

The value of β is obtained experimentally by measuring the crack length and compliance. In the case of the specimens tested, β was found to be nearly zero.

Besides the beam theory, the area method was also used for determination of the strain energy release rate as follows:

$$G_I = \frac{P_1\delta_2 - P_2\delta_1}{(a_2^2 - a_1^2)} k \tag{8}$$

where a_1 and a_2 are crack lengths corresponding to loads P_1, P_2, respectively.

Experimental Procedure

The material used in this investigation was T300/F-185 consisting of an elastomer-modified epoxy resin matrix (F-185, Hexcel Corp.) and Thronel T300 fibers (Union Carbide). The prepreg material and neat resin were procured from the Hexcel Corp. Twenty-four ply unidirectional plates were fabricated for the DCB specimens. Standard layup and bagging procedures were used and the recommended curing schedule was followed.

A 3.8.-cm (1.5-in.) wide and 0.025-mm (0.001-in.) thick Teflon® film was inserted at the mid-surface of the laminate along one edge to initiate the crack. The uniform DCB specimens were 22.9 cm (9 in.) long, 2.54 cm (1 in.) wide, and 3.56 mm (0.140 in.) thick with an initial artificial crack of 3.81-cm (1.5-in.) length at one end. Metallic hinges were bonded to the cracked end of the specimen to allow for unrestrained rotation at that end during load introduction (Fig. 3). The WTDCB specimens were 22.9-cm (9-in.) long with a 2.54-cm (1-in.) wide uniform segment of 3.81-cm (1.5-in.) length, a tapered segment 10.80 cm (4.25 in.) long, and another rectangular segment at the end of dimensions 7.6 by 6.4 cm (3 by 2.5 in.). These specimens were also loaded through bonded metallic hinges like the uniform DCB specimens.

All specimens were tested in an Instron testing machine at crosshead rates ranging from 0.0085 to 8.5 mm/s (0.02 to 20 in./min). (At least five specimens were tested at each loading rate.) Continuous records of the load and opening deflection, and intermittent records of crack extension, were obtained for determination of the strain energy release rate. At the very low loading rates, crack extension was monitored visually. The edge of the specimen was painted white with a typewriter correction fluid and marked at 1.27-cm (0.50-in.) intervals. Whenever the crack reached one of these marks, a trace was inserted electronically on the load-deflection record.

For the higher loading rates, two strain gages were mounted on one surface of the specimens along the centerline at distances of 1.27 and 2.54 cm (0.5 and 1 in.) from the initial crack tip. The two gages were connected to two adjacent arms of a Wheatstone bridge resulting in a single channel record giving the algebraic difference of the strains. The positive and negative peaks of this record correspond to the arrival of the crack front at the corresponding gage locations. Under continuous loading, load-versus-time and strain-versus-time records were

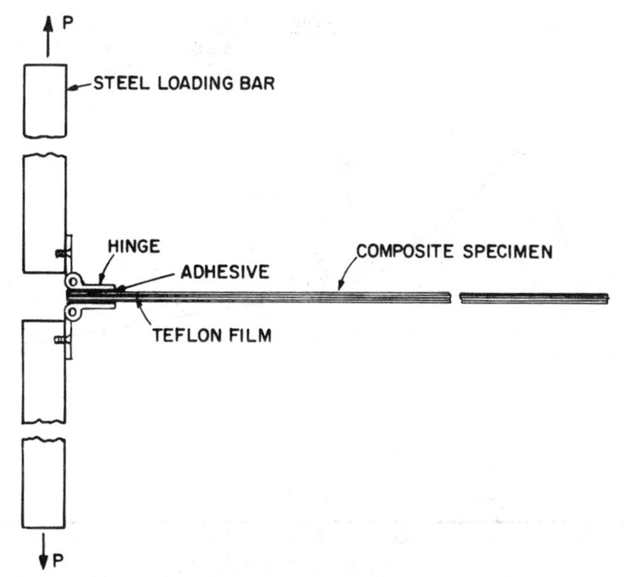

FIG. 3—*Double cantilever beam specimen with load introduction attachments.*

obtained with an electronic two-channel recorder (Bascom-Turner) operating in the sweep mode. From these records, the loads and deflections corresponding to two crack lengths were determined, and the strain energy release rate computed by the area and the beam analysis methods.

Results and Discussion

Visual monitoring of crack extension was possible for crosshead rates up to 0.85 mm/s (2 in./min). Typical load-deflection curves with marks indicating crack extension at 1.27-cm (0.5-in.) intervals are shown in Figs. 4 to 6 for the UDCB specimens and Figs. 7 and 8 for the WTDCB specimens. For higher crosshead rates, the load-deflection records were supplemented with strain records from two strain gages mounted on one of the specimen surfaces. The difference of the strains from the two strain gages was obtained as a function of time, converted to a strain versus deflection record, and superimposed on the load-deflection record. Typical load-deflection and strain-deflection records are shown in Figs. 9 to 12.

The critical strain energy release rate was computed using the area method and the beam analysis method. Five to ten specimens were tested at each loading rate. Several computations were made for each specimen at the lower loading rates.

Average results for all the specimens tested, including values of standard deviation for G_{Ic}, are tabulated in Table 1. There is a definite trend for the critical strain energy release rate to decrease with increasing loading rate or crack propagation velocity. G_{Ic} decreases by approximately 20% over three decades of crack

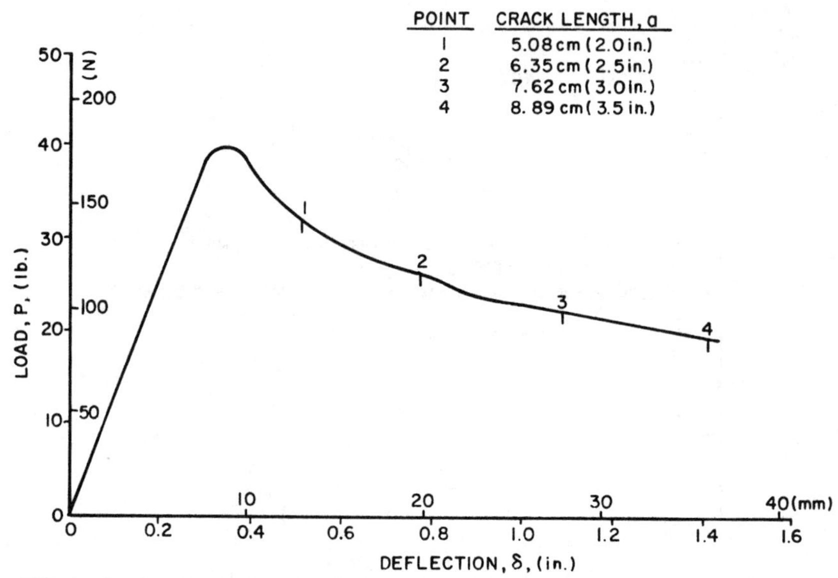

FIG. 4—*Load versus crack opening deflection for UDCB specimen loaded at a crosshead rate of 0.0085 mm/s (0.02 in./min).*

velocity. The average values of G_{Ic} were plotted versus crack velocity $\dot{a}$ on a semi-log and on a log-log scale in Figs. 13 and 14. In both cases, a straight line was fitted through the data points yielding two possible forms of the relationship

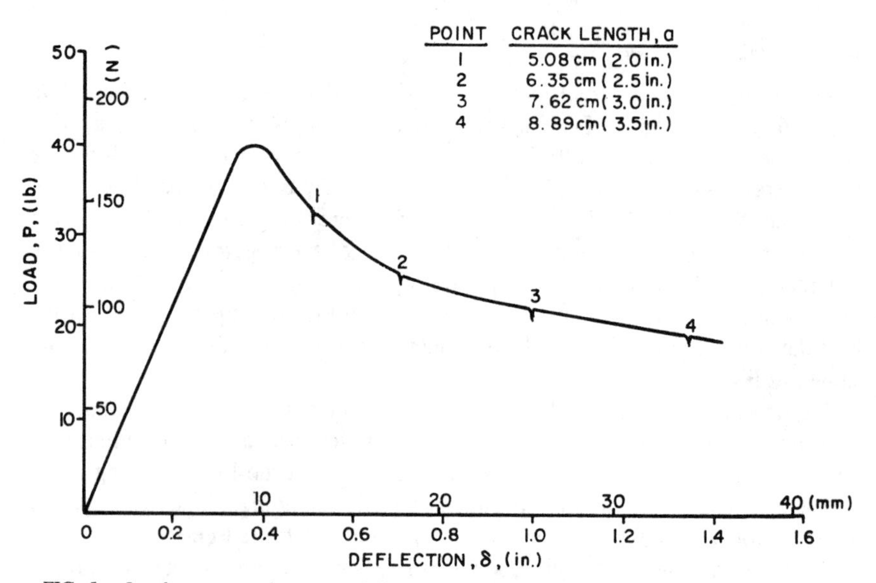

FIG. 5—*Load versus crack opening deflection for UDCB specimen loaded at a crosshead rate of 0.085 mm/s (0.2 in./min).*

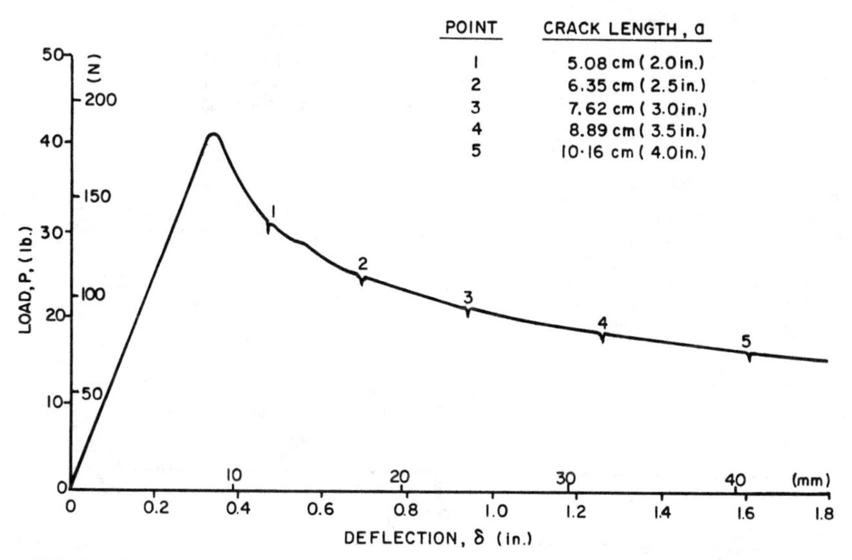

FIG. 6—*Load versus crack opening deflection for UDCB specimen loaded at a crosshead rate of 0.85 mm/s (2 in./min).*

between G_{Ic} and $\dot{a}$:

$$G_{Ic} = A \log \dot{a} + B$$

or

$$G_{Ic} = k\dot{a}^n$$

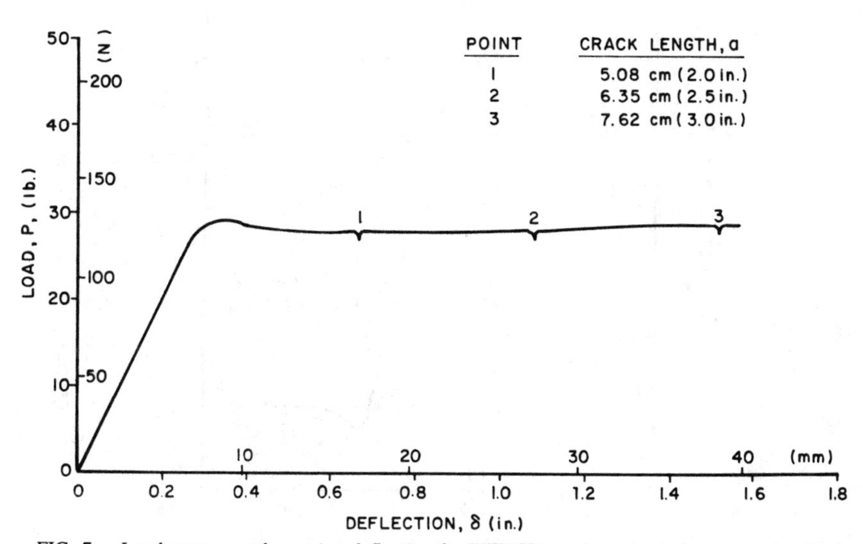

FIG. 7—*Load versus crack opening deflection for WTDCB specimen loaded at a crosshead rate of 0.0085 mm/s (0.02 in./min).*

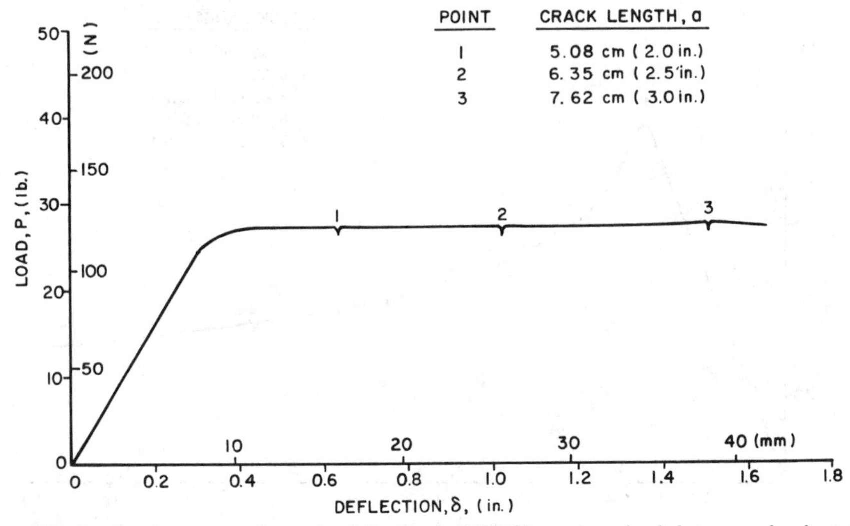

FIG. 8—*Load versus crack opening deflection for WTDCB specimen loaded at a crosshead rate of 0.85 mm/s (2 in./min).*

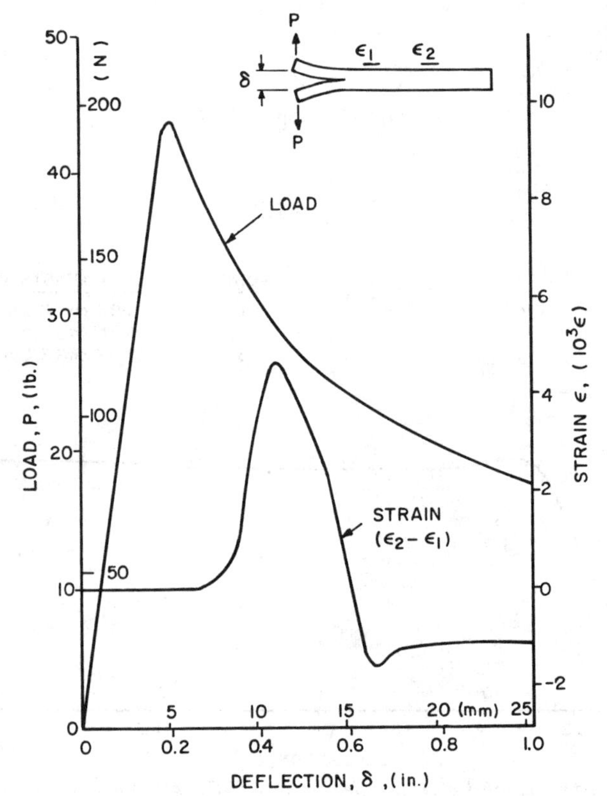

FIG. 9—*Load-deflection and corresponding strain record for a UDCB specimen loaded at a crosshead rate of 0.85 mm/s (2 in./min).*

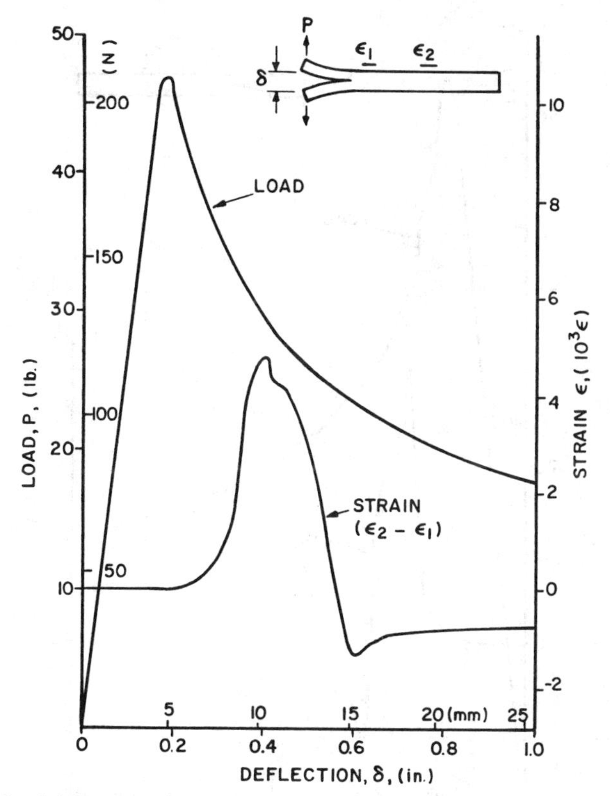

FIG. 10—*Load-deflection and corresponding strain record for a UDCB specimen loaded at a crosshead rate of 4.23 mm/s (10 in./min).*

where

$A = -0.098$,
$B = 1.625$,
$k = 1.625$, and
$n = -0.0271$.

if $\dot{a}$ is in mm/s and G_{Ic} in kJ/m^2.

At this point the relations above are curve-fitting schemes and no physical significance is attached to them.

Results are in qualitative agreement with those described by Hunston and Bascom for the neat resin [*8*]. They showed that the fracture energy of an elastomer-modified epoxy increases sharply with temperature and decreases with loading rate. In the case of the composite, as predicted by Hunston and Bascom, the dependence on rate of loading is not as dramatic.

The influence of the elastomer modifiers was studied by Bascom et al. [*11*] who showed that the value G_{Ic} increases from 0.20 kJ/m^2 for the unmodified resin (F-205) to 3.8 kJ/m^2 for the elastomer-modified resin (F-185), a nearly

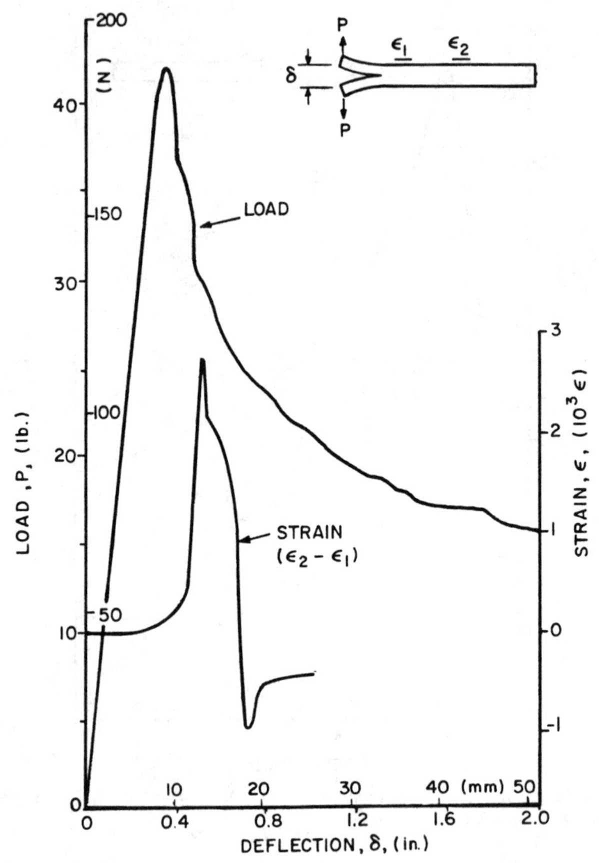

FIG. 11—*Load-deflection and corresponding strain record for a UDCB specimen loaded at a crosshead rate of 8.5 mm/s (20 in./min).*

twentyfold increase. This increase is attributed to the increased energy absorption through the high deformation or debonding or both of the rubber particles. In the case of the composite, the increase in G_{Ic} is approximately tenfold. The reason for the nonproportional increase of G_{Ic} in the composite may be the fact that a large portion of the fracture toughness of the composite is due to fiber bridging and not to the matrix itself.

Subsequent work since the completion of this paper has shown that G_{Ic} continues to decrease with increasing crack velocity up to at least $\dot{a} = 1$ m/s following a power law relation [*12*]. This may be attributed to the fact that the deformation to failure, hence absorbed energy, of the rubber-filled epoxy decreases with increasing strain rate. These results are in contrast to those for AS 4/3501-6 graphite/epoxy with an unmodified brittle matrix [*10,13*]. Results for this material system showed that G_{Ic} increases with crack propagation velocity up to a value of approximately 1 m/s, and thereafter it decreases.

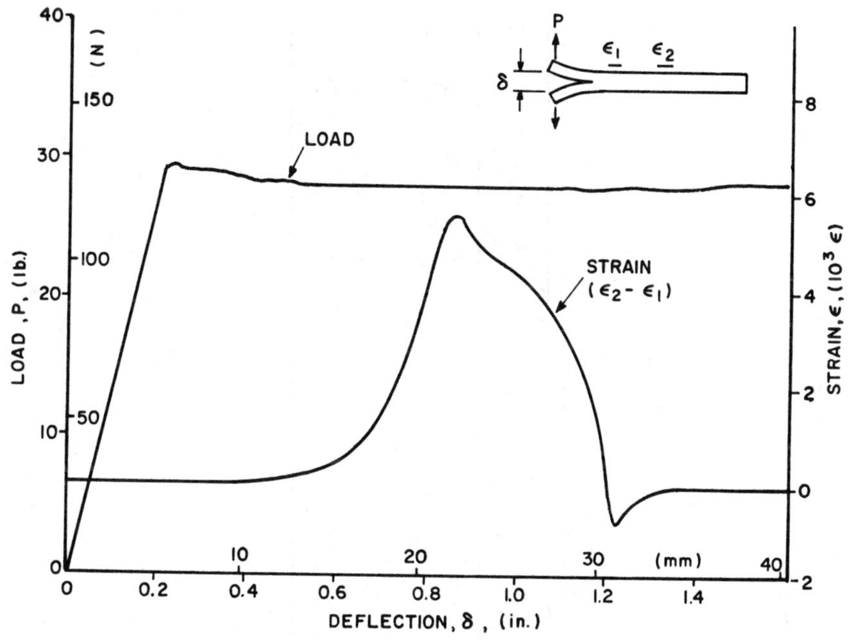

FIG. 12—*Load-deflection and corresponding strain record for a WTDCB specimen loaded at a crosshead rate of 8.5 mm/s (20 in./min).*

The results above can be explained in terms of rate sensitivity behavior of polymers, since the composite fracture toughness measured is a matrix-dominated property. It has been observed that polymers can exhibit three types of behavior under varying loading rates: rate insensitivity, positive rate sensitivity, and negative rate sensitivity. It is possible for a given polymer to exhibit more than one type of rate sensitivity depending on the temperature, molecular structure, and range of crack propagation velocity. Examples of this type of behavior for polymers have been discussed by Williams [*14*].

Conclusions

The effects of loading rate were determined on interlaminar fracture toughness of a toughened graphite/epoxy, F-185/T300. Uniform and width-tapered double cantilever beam specimens were tested at opening deflection rates between 0.0085 and 8.5 mm/s (0.02 and 20 in./min) corresponding to crack extension rates of up to 21 mm/s (0.83 in./s). At the lower rates, crack extension was monitored visually. At higher rates, crack extension was monitored by means of strain gages mounted on the surface of the specimen.

Results covering three decades of loading rate show that the critical strain energy release rate decreases by up to 20% for the range of rates considered here. The strain energy release rate was expressed in two forms, a linear function

TABLE 1—*Critical strain energy release rate for delamination fracture of T300/F-185 graphite/epoxy.*

Number of Specimens	Type of Specimen	Deflection Rate, mm/s (in./min)	Crack Velocity $\dot{a}$, mm/s (in./min)	G_{Ic}, kJ/m^2 (in. · lb/in.^{2})
2	WTDCB	0.0085 (0.02)	0.010 (0.023)	1.88 ± 0.04 (10.71 ± 0.22)
8	UDCB	0.0085 (0.02)	0.015 (0.036)	1.81 ± 0.12 (10.31 ± 0.69)
5	UDCB	0.085 (0.20)	0.179 (0.423)	1.69 ± 0.06 (9.64 ± 0.34)
2	WTDCB	0.85 (2.00)	0.795 (1.88)	1.71 ± 0.03 (9.76 ± 0.16)
6	UDCB	0.85 (2.00)	1.67 (3.94)	1.62 ± 0.02 (9.23 ± 0.14)
2	WTDCB	8.50 (20.00)	11.02 (26.05)	1.52 ± 0.01 (8.68 ± 0.08)
5	UDCB	8.50 (20.00)	21.0 (49.6)	1.50 ± 0.06 (8.58 ± 0.36)

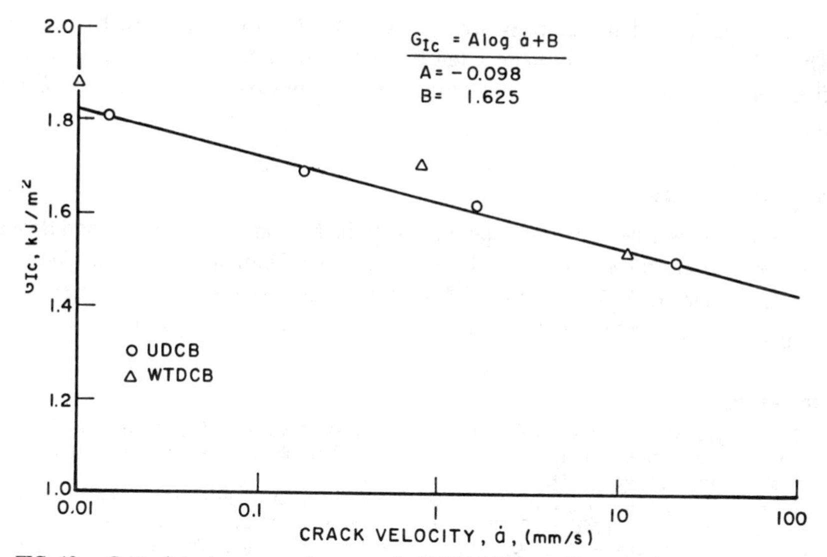

FIG. 13—*Critical strain energy release rate for F-185/T300 graphite/epoxy as a function of crack velocity.*

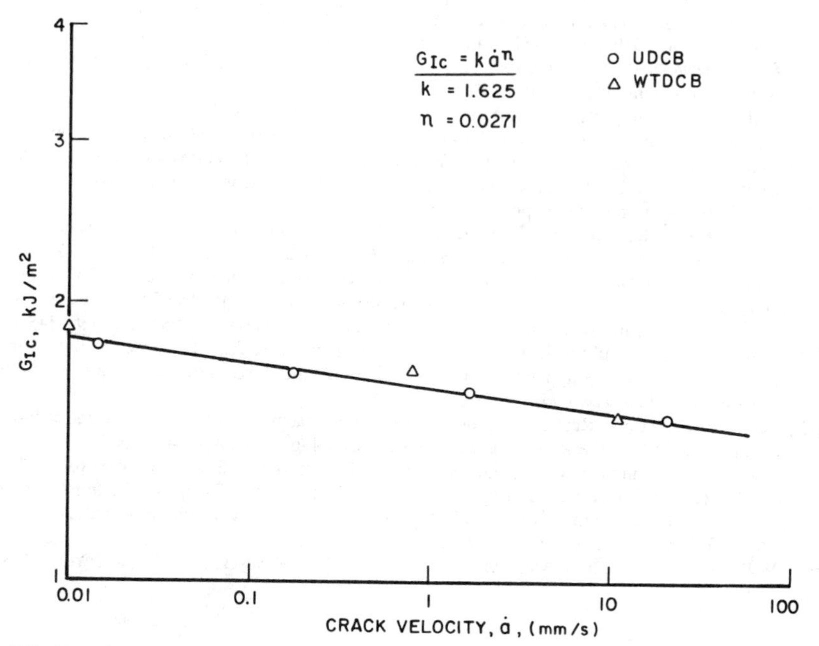

FIG. 14—*Critical strain energy release rate for F-185 /T300 graphite/epoxy as a function of crack velocity.*

of the logarithm of the crack velocity and a power law of the crack velocity. The value of G_{Ic} is approximately ten times that for a composite with unmodified epoxy matrix. The composite toughness is approximately half that of the matrix resin.

Acknowledgments

The work described here was sponsored by the National Aeronautics and Space Administration (NASA)-Langley Research Center, Hampton, Virginia. We are grateful to Dr. J. D. Whitcomb of NASA-Langley for his encouragement and cooperation, and to Dr. S. Mostovoy of IIT for helpful discussions.

References

[1] Whitney, J. M., Browning, C. E., and Hoogsteden, W., "A Double Cantilever Beam Test for Characterizing Mode I Delamination of Composite Materials," *Journal of Reinforced Plastics and Composites,* Vol. 1, Oct. 1982, pp. 297–313.

[2] Brussat, T. R., Chin, S. T., and Mostovoy, S., "Fracture Mechanics for Structural Adhesive Bonds," Phase II, AFML-TR-77-163, Wright-Patterson Air Force Base, OH, 1978.

[3] Bascom, W. D., Bitner, R. J., Moulton, R. J., and Siebert, A. R., "The Interlaminar Fracture of Organic-Matrix Woven Reinforced Composites," *Composites,* Vol. 11, Jan. 1980, pp. 9–18.

[4] deCharentenay, F. X. and Benzeggagh, M., "Fracture Mechanics of Mode I Delamination in Composite Materials," in *Advances in Composite Materials, Proceedings of ICCM 3,* Vol. 1, Pergamon Press, New York, 1980, pp. 186–197.

[5] Devitt, D. F., Schapery, R. A., and Bradley, W. L., "A Method for Determining Mode I Delamination Fracture Toughness of Elastic and Viscoelastic Composite Materials," *Journal of Composite Materials,* Vol. 14, 1980, pp. 270–285.

[6] Wilkins, D. J., Eisenmann, J. R., Camin, R. A., Margolis, W. S., and Benson, R. A., "Characterizing Delamination Growth in Graphite-Epoxy," in *Damage in Composite Materials: Basic Mechanisms, Accumulation, Tolerance, Characterization, ASTM STP 775,* American Society for Testing and Materials, Philadelphia, 1982, pp. 168–183.

[7] O'Brien, T. K., Johnston, N. J., Morris, D. H., and Simonds, R. A., "A Simple Test for the Interlaminar Fracture Toughness of Composites," *SAMPE Journal,* July/Aug. 1982, pp. 8–14.

[8] Hunston, D. L., and Bascom, W. D., "Effects of Layup, Temperature, and Loading Rate in Double Cantilever Beam Tests of Interlaminar Crack Growth," *Composites Technology Review,* Vol. 5, No. 4, Winter 1983, pp. 118–119.

[9] Miller, A. G., Hertzberg, P. E., and Rantala, V. W., "Toughness Testing of Composite Materials," *Proceedings of the 12th National SAMPE Technical Conference,* 1980, pp. 279–293.

[10] Aliyu, A. A. and Daniel, I. M., "Effects of Strain Rate on Delamination Fracture Toughness of Graphite/Epoxy," in *Delamination and Debonding of Materials, ASTM STP 876,* W. S. Johnson, Ed., American Society for Testing and Materials, Philadelphia, 1985, pp. 336–348.

[11] Bascom, W. D., Moulton, R. J., Rowe, E. H., and Siebert, A. R., "The Fracture Behavior of an Epoxy Polymer Having a Bimodal Distribution of Elastomeric Inclusions," in *Proceedings of Meeting of American Chemical Society,* 1978.

[12] Auser, J. W., "Load Rate Effects on Delamination Fracture Toughness of Graphite/Epoxy Composites," M. S. thesis, Illinois Institute of Technology, Chicago, May 1985.

[13] Yaniv, G. and Daniel, I. M., "Height-Tapered Double Cantilever Beam Specimen for Study of Rate Effects on Fracture Toughness of Composites," presented at Composite Materials Testing and Design: Eighth Symposium, Charleston, SC, April 1986, sponsored by American Society for Testing and Materials, Philadelphia.

[14] Williams, J. G., *Fracture Mechanics of Polymers,* John Wiley and Sons, New York, 1984, p. 179.

Alan J. Russell[1] *and Ken N. Street*[1]

The Effect of Matrix Toughness on Delamination: Static and Fatigue Fracture Under Mode II Shear Loading of Graphite Fiber Composites

REFERENCE: Russell, A. J. and Street, K. N., "**The Effect of Matrix Toughness on Delamination: Static and Fatigue Fracture Under Mode II Shear Loading of Graphite Fiber Composites,**" *Toughened Composites, ASTM STP 937,* Norman J. Johnston, Ed., American Society for Testing and Materials, Philadelphia, 1987, pp. 275–294.

ABSTRACT: Both the static and fatigue behavior of delaminations subjected to pure Mode II shear loading were investigated using end-delaminated flexure specimens. By carrying out the constant amplitude sine-wave fatigue tests at fatigue load ratios (R) of $R = 0$ and $R = -1$, the importance of shear reversal was evaluated. The effect of improved matrix toughness on the Mode II properties was determined by testing four different graphite fiber composite systems having widely different Mode I fracture energies. The results indicated that the benefits of improved matrix toughness on the composite properties were less under Mode II loading than under Mode I and were reduced further still, or eliminated entirely, under Mode II fatigue conditions. It was found also that full shear reversal had an accelerating effect on the Mode II fatigue crack growth rates, especially at low values of ΔG_{II}. These findings are discussed in terms of the failure mechanisms taking place. The significance of the results to damage tolerance and long-term durability design considerations are also addressed.

KEY WORDS: delamination, shear, fracture, fatigue, growth rate, toughness, graphite fiber

Sound experimental evidence has now established that the criticality of delaminations in laminated composite structures depends not only on the opening tensile stresses present but rather on the entire out-of-plane tensile/shear (Modes I/II and III) fracture envelope. For both brittle systems [*1–3*] and toughened systems [*4,5*], it has been shown that the presence of Mode II reduces the value of the

[1]Research scientist and composite group leader, respectively, Defence Research Establishment Pacific, F. M. O., Victoria, B. C. VOS 1BO, Canada.

Mode I strain energy release rate (G_I) necessary for mixed-mode interlaminar fracture. Furthermore, in the absence of significant peel stresses, Mode II alone can result in fracture, albeit at higher values of strain energy release rate than those for pure Mode I. Indeed, for first generation graphite/epoxy, G_{IIc} is often four to five times greater than G_{Ic} [*1*]. However, there is also evidence that as G_{Ic} increases as a result of improved matrix toughness, the corresponding mixed-mode interlaminar fracture energy increase is proportionately *less*, and furthermore, it decreases as the percentage of Mode II increases [*4,5*]. Despite this unexplained and significant inequality, G_{IIc} is not a parameter that is currently considered when evaluating new matrix-toughened composite materials.

Considerably less attention has been given to delamination growth under cyclic loading, perhaps because the available data indicate that subcritical crack growth requires a minimum threshold strain energy release rate ΔG_{TH} which is seldom less than 20% of the fracture energy G_c [*2,3,6,7*]. This suggests that laminated composite structures could be designed for no interlaminar fatigue crack growth using the fatigue threshold ΔG_{TH} with only a small penalty relative to a design based on the static value G_c. However, ΔG_{TH} has been found also to be a smaller fraction of G_c for mixed-mode fracture than for pure Mode I [*2,7*]. In addition, almost all measurements of interlaminar fatigue crack growth have been made at fatigue load ratios R greater than zero (zero-tension, zero-shear). Yet, numerous fatigue studies, including those employing high shear stresses [*8*], have shown that testing with $R = -1$ can greatly reduce the number of cycles to failure. These facts suggest that greatly enhanced rates of interlaminar fatigue crack growth may occur under pure Mode II loading at an R ratio of -1.

In this paper, a method is described for measuring the Mode II fatigue crack growth rates of delaminations subjected to both full shear reversal ($R = -1$) and positive shear only ($R = 0$). The relative effects of improved matrix toughness on G_{Ic}, G_{IIc}, and the Mode II fatigue behavior are evaluated by investigating four different graphite fiber composite systems having widely different values of G_{Ic}.

Mode II Fatigue Test

A fully automated testing procedure was developed to control and monitor the Mode II fatigue crack growth. End notched cantilever beam specimens (ENCB) were chosen because of the simplicity of both specimen geometry and loading and their ability to produce long delamination extensions (Fig. 1). Moreover, a large change in specimen compliance with crack growth made possible the accurate monitoring of delamination length by periodically measuring the specimen compliance. This was accomplished during a relatively slow ramp cycle which followed each block of constant amplitude sine-wave cycles. The two loading pins mounted on top of the load cell were adjusted to allow a gap of 0.3 mm more than the thickness of the specimen so that free rotation of the delaminated end of the specimen could occur. As a consequence, the load cell output was not truly sinusoidal, but contained a small plateau in the region of zero load (Fig. 2). Use was made of this plateau to correct for thermal drift in both the

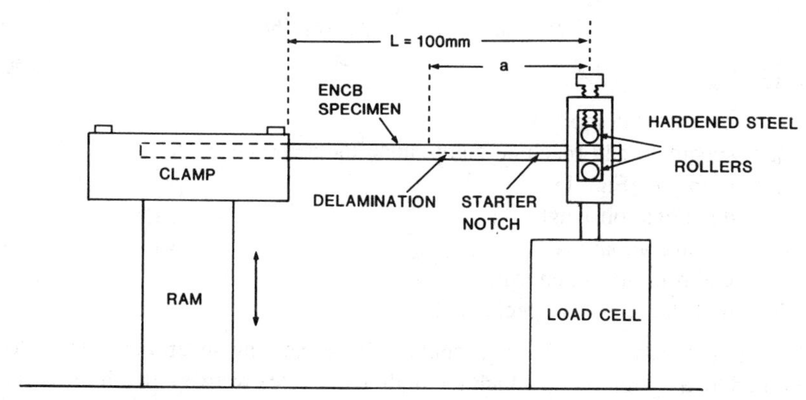

FIG. 1—*Loading arrangement of the end notched cantilever beam Mode II fatigue specimen.*

load cell and ram linear variable differential transformer (LVDT) outputs throughout the duration of each test.

The following relation between the ENCB specimen compliance and delamination length can be derived from beam theory as shown in Appendix A.

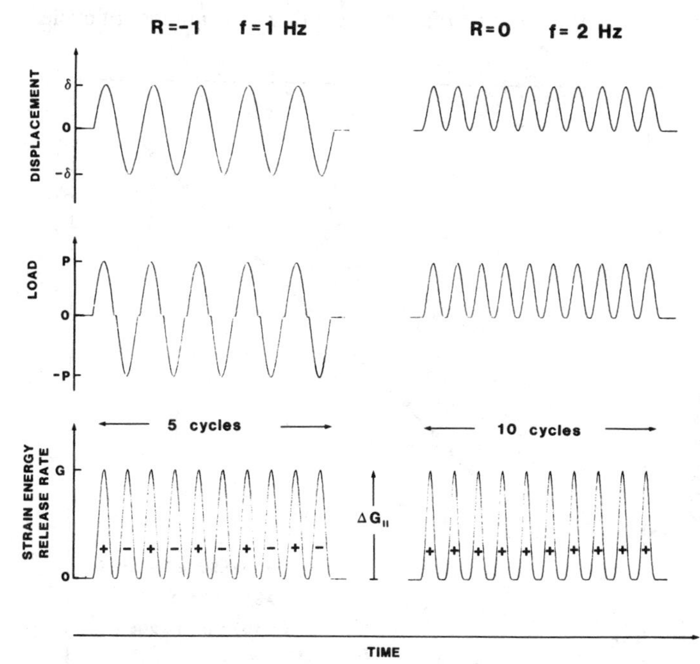

FIG. 2—*Relationship between displacement, load, and strain energy release rate for the* R = -1 *and* R = 0 *fatigue tests. The + and − signs refer to the shear direction. Note that for the same displacement, both tests have the same* ΔG_{II}*, but that the* R = -1 *tests have twice as many* ΔG_{II} *peaks per cycle.*

$$C = C_0 + (L^3 + 3a^3)/2Ebh^3 \quad (1)$$

where

C = measured compliance,
C_0 = specimen and machine compliance independent of a,
L = span (see Fig. 1),
a = delamination length,
E = flexural modulus,
b = width of specimen, and
h = half thickness of specimen.

Equation 1 was verified experimentally by subjecting several specimens to a modified loading regime in which ten high load cycles were applied immediately after each slow ramp cycle. This produced a series of distinct bands on the fracture surface whose position could be readily measured and plotted against the corresponding compliance, as shown in Fig. 3. The straight line fit to the data validates Eq 1 and the intercept on the y axis yields the value of C_0. The final unknown term in Eq 1, Ebh^3, was evaluated for *each specimen* using the initial crack length which was measured on the fracture surface after each test was completed and the initial compliance. After calculating the value of a for each periodic compliance measurement, the crack growth rates da/dN were obtained by curve fitting and differentiating the crack length versus number of cycles N data.

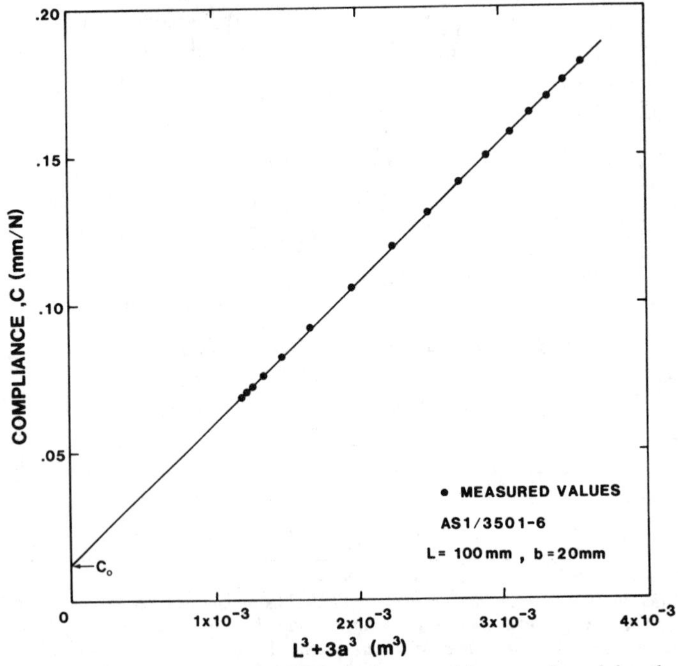

FIG. 3—*Relationship between ENCB specimen compliance and crack length.*

During the last ten cycles in each block, peak and valley measurements of both load P and displacement δ were made. The strain energy release rate G corresponding to each load and displacement was calculated by differentiating Eq 1 with respect to a and substituting into the standard compliance equation for G,

$$G = \frac{P^2}{2b}\frac{dC}{da}$$

to give

$$\Delta G_{II} = 9a^2(\Delta P)^2/(4Eb^2h^3) \tag{2}$$

or, in terms of δ and C,

$$\Delta G_{II} = 9a^2(\Delta\delta)^2(1 - C_0/C)/2b(L^3 + 3a^3)C \tag{3}$$

Since ΔG_{II} is proportional to the square of ΔP, it is positive for both positive and negative shear loadings and hence, for $R = -1$, two maxima in ΔG_{II} of equal value occur during each cycle (see Fig. 2). On the other hand for $R = 0$, there is only one maximum per cycle. Because of this difference and the need to minimize any strain rate effects, the $R = -1$ tests were run at 1 Hz while the $R = 0$ tests were carried out at a frequency of 2 Hz.

From Eq 3 it can be seen that for constant amplitude cycling, ΔG_{II} increases initially as a result of the a^2 term, but that at longer crack lengths the $L^3 + 3a^3$ term becomes dominant resulting in a reduction in ΔG_{II} after going through a maximum at $a = 0.58L$. For a span L of 100 mm, tests were initiated from the end of the insert with $a = 25$ mm and terminated at a crack length of 95 mm. Thus, for most values of ΔG_{II}, two measurements of da/dN were obtained from each specimen, before and after the above maximum.

Specimen Fabrication and Static Tests

The four composite systems studied in this investigation were, in order of increasing toughness:

1. a first generation high temperature graphite/epoxy—Hercules AS1/3501-6,
2. a second generation high temperature graphite/epoxy—Hercules AS4/2220-3,
3. an intermediate temperature, rubber toughened graphite/epoxy—Hexcel C6000/F155, and
4. a high temperature graphite/thermoplastic—ICI VICTREX APC2 that is, AS4 fiber/poly(etheretherketone) (PEEK) matrix.

Unidirectional laminates nominally 3 mm thick were layed-up and processed according to the prepreg manufacturers' recommended laminating procedures. In addition, the AS1/3501-6 was postcured for 8 h at 177°C. Nonadhesive inserts, 0.025 mm thick, were located on the midplane at the end of the laminates to act as delamination starter notches. Specimens, 150 mm long and 20 mm wide were cut from the laminates and stored in a dessicator until immediately before testing at ambient temperature and humidity. Both the static and fatigue tests were carried out on specimens from the same laminate, thus minimizing material variability effects.

For the Mode I tests, tabs were bonded onto both sides of the specimen at the end adjacent to the starter notch to form double cantilever beam specimens (DCB) with an initial crack length of 45 mm. The delamination was advanced from the end of the insert in increments of a few millimetres at a crack opening displacement rate of 2 mm/min. The load-displacement data were reduced to the form of a crack growth resistance curve as described previously [*1*]. G_{IIc} measurements were made using end notched flexure specimens (ENF) having a half span length of 50 mm and an initial delamination length of 25 mm which included a 1 to 2-mm long wedge-opened precrack. The ram displacement rate was 2 mm/min and testing and data reduction were carried out as described in Ref *1*. For each composite system, a minimum of three DCB specimens and five ENF specimens were tested.

Results and Discussion

Static Tests

The results of the static tests are given in Table 1. The Mode I delamination growth was quite stable except for an initial pop-in extension in the two tougher (F155 and PEEK) systems. Typical load-displacement curves were essentially linear to maximum load as shown in Fig. 4. Fiber bridging was observed in all but the rubber toughened epoxy system (possibly because of its lower fiber volume fraction) and resulted in an increase in crack growth resistance (Fig. 5). The G_{Ic} values listed in Table 1 refer to the *initial* or onset of delamination fracture energies, whereas the G_{Is} values refer to the stabilized or plateau fracture energies measured after the full development of fiber bridging. These values of G_{Ic} and G_{Is} are in good agreement with other published values of the Mode I fracture energies of these materials [*5,9*], except for the AS4/APC2 values which are somewhat lower than those reported by Leach and Moore [*10*].[2] The Mode

TABLE 1—*Static Test Data.*

Material	Fiber Volume Fraction	G_{Ic}, J/m^2 [a]	G_{Is}, J/m^2 [a]	G_{IIc}, J/m^2 [a]	G_{IIc}/G_{Ic}
AS1/3501-6	0.62	110 ± 5	190 ± 10	605 ± 30	5.5
AS4/2220-3	0.61	160 ± 10	255 ± 15	750 ± 25	4.7
C6000/F155	0.51	495 ± 25	510 ± 20	900 ± 50	1.8
AS4/APC2	0.66	1330 ± 85	1540 ± 60	1765 ± 235	1.2

[a] Values given are mean ± standard deviation.

[2] Recent tests on APC2 laminates fabricated by ICI have yielded higher G_{Ic} values in agreement with those in Ref *10*.

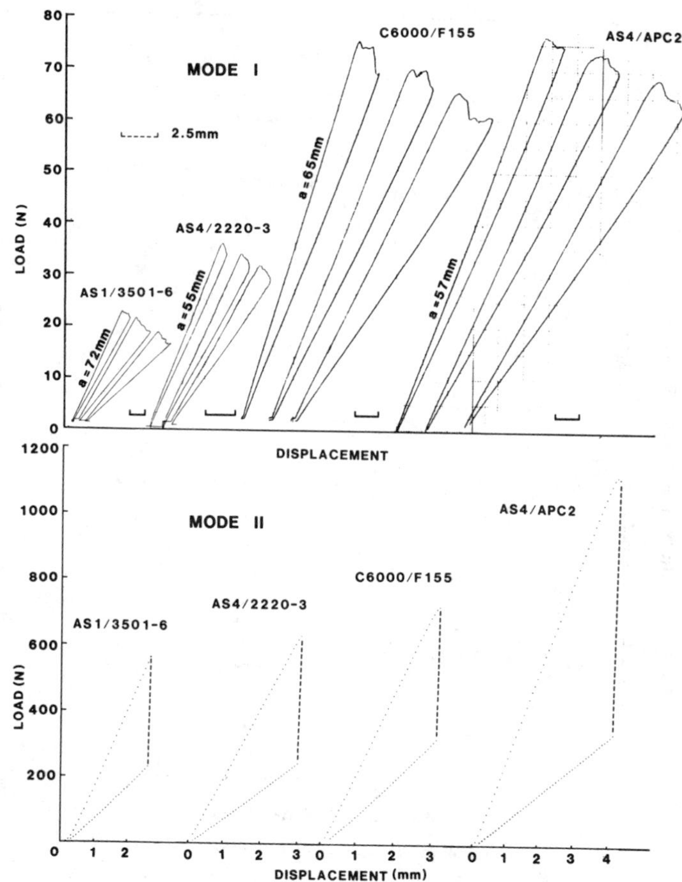

FIG. 4—*Representative load-displacement curves for the static tests. Analog plots are given for Mode I, whereas the Mode II curves represent the digital data stored during testing.*

II results follow the same trend as the Mode I data except that the effect of enhanced matrix toughness on composite G_{IIc} is significantly less than on G_{Ic}. Whereas G_{Ic} for the composite increases by more than tenfold by replacing the 3501-6 matrix with a PEEK matrix, G_{IIc} increases by less than a factor of three.

Representative fractographs of the static test specimens are shown in Fig. 6. Little difference was observed between the 3501-6 and 2220-3 Mode I fracture surfaces. However, the presence of the large loosely bonded rubber particles in the F155 epoxy resulted in a breakup of the fracture plane, presumably by their acting as multiple initiation sites for localized fracture of the epoxy. The F155 system also showed evidence of fiber/matrix debonding, whereas the PEEK system displayed no interfacial failure and featured dimples resulting from plastic deformation and necking of the matrix.

In Mode II, both the 3501-6 and 2220-3 fracture surfaces contained extensive hackling, whereas hackle formation in the F155 was less well developed and in

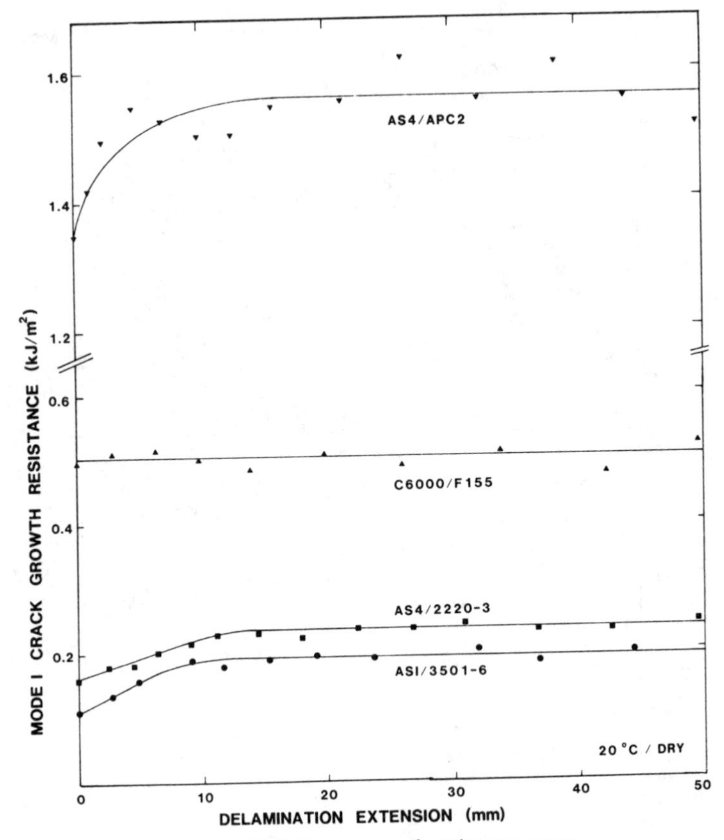

FIG. 5—*Mode I crack growth resistance curves.*

the PEEK was completely absent having been replaced by plastic flow and necking in the shear direction.

Thus, the high values of G_{IIc}/G_{Ic} for the two brittle systems can be regarded as being the result of the high energy hackle dominated Mode II fracture surface. In the tougher systems, hackle formation is reduced or eliminated entirely, resulting in a lower energy fracture closer to G_{Ic}. The relative effectiveness of multiple matrix microcracking versus matrix plastic deformation in providing enhanced composite toughness is clearly indicated by these results.

Mode II Fatigue

The Mode II fatigue data are shown in Figs. 7 to 10. In general, the fatigue crack growth data obeyed a relationship of the form

$$da/dN = B(\Delta G_{II})^n \tag{4}$$

where B and n are constants dependant on the material, test environment, and the

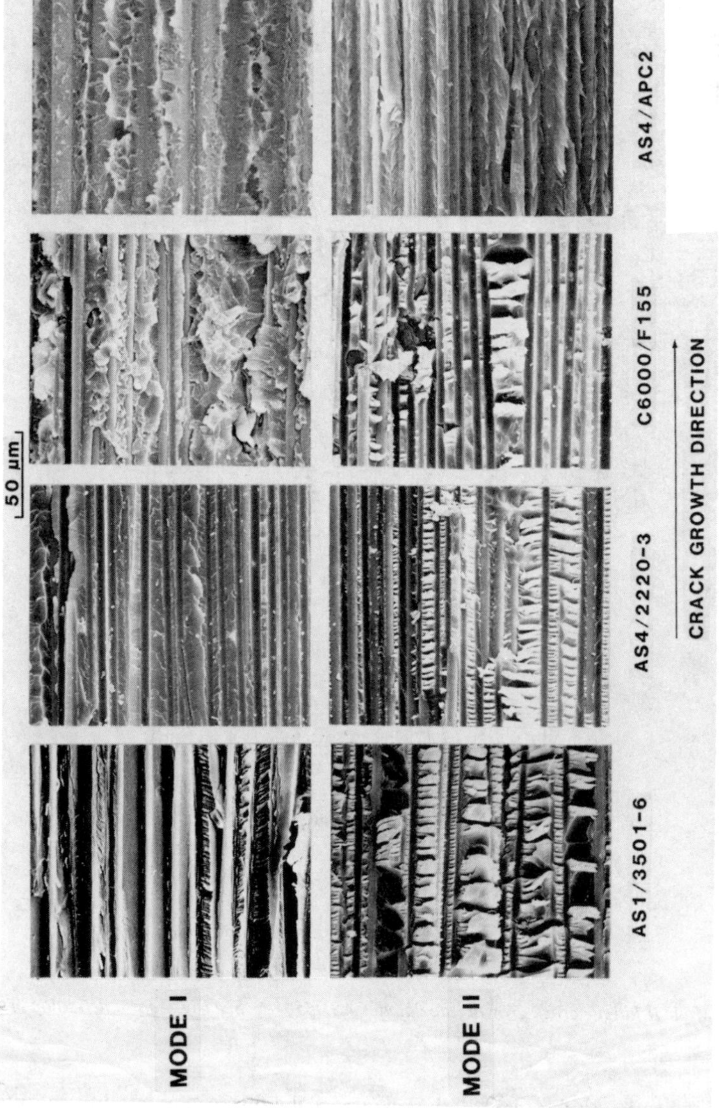

FIG. 6—*Representative fractographs of the failed static specimens. The Mode II surfaces have been sheared from left to right with respect to the opposite surface.*

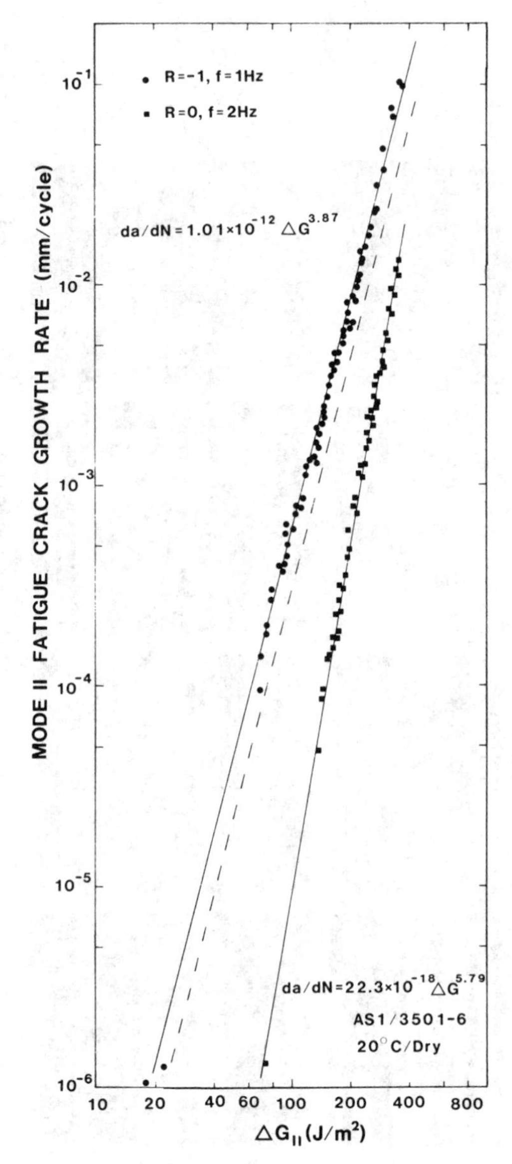

FIG. 7—*Mode II fatigue crack growth rate data for AS1/3501-6. See Fig. 2 for a definition of* ΔG_{II}.

fatigue load ratio. In the two tougher systems, the curves deviated from linearity at low values of G_{II}, suggesting a possible threshold value G_{TH}. The values of B and n given in Figs. 7 to 10 were obtained by least squares fitting the linear portions of the fatigue crack growth rate curves. In all four systems, the effect of fatigue load ratio was the same; namely, a reduction in the value of n for $R = -1$

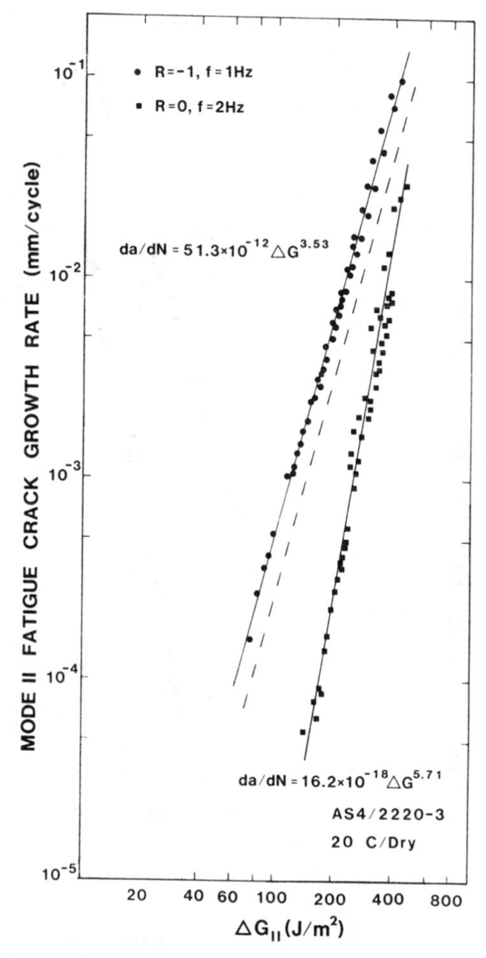

FIG. 8—*Mode II fatigue crack growth rate data for AS4/2220-3. See Fig. 2 for a definition of* ΔG_{II}.

compared to $R = 0$. In fact, if the ratio of the two slopes is calculated, a value close to 0.63 is obtained in all cases except for the PEEK system which had a somewhat lower value. Two factors contributed to the greater da/dN values of the $R = -1$ data. First, a factor of two in da/dN can be attributed to the fact that each $R = -1$ amplitude cycle contains two maxima in ΔG, whereas each $R = 0$ cycle contains only one (see Fig. 2), and this is indicated by the dashed curves in Figs. 7 to 10. Second, the differences between the dashed curves and the $R = 0$ curves represent the direct effect of shear reversal in the former and positive only shear in the latter case. Clearly, the micromechanical damage mechanisms that control crack extension are more degrading when $R = -1$. It should be pointed out that if the data in Figs. 7 to 10 were replotted against the

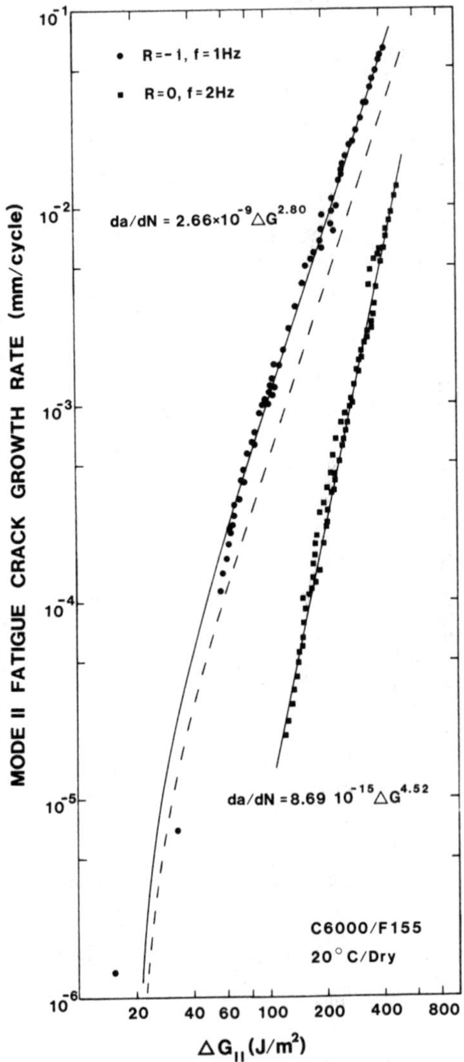

FIG. 9—*Mode II fatigue crack growth rate data for C6000/F155. See Fig. 2 for a definition of* ΔG_{II}.

stress intensity factor ΔK_{II}, then the effect of fatigue load ratio would be less since ΔK_{II}, unlike ΔG_{II}, is twice as great for $R = -1$ as it is for $R = 0$ (for the same maximum load). However, the ratio of the slopes would be unaltered.

For both $R = 0$ and $R = -1$, the principal effect of increasing the matrix toughness was a reduction in the slope n as can be seen clearly in Fig. 11. This effect is so pronounced that were it not for the downward curvatures of the $R = -1$ plots of the tougher systems the fatigue threshold would appear to decrease significantly with increasing toughness. However, the results to date

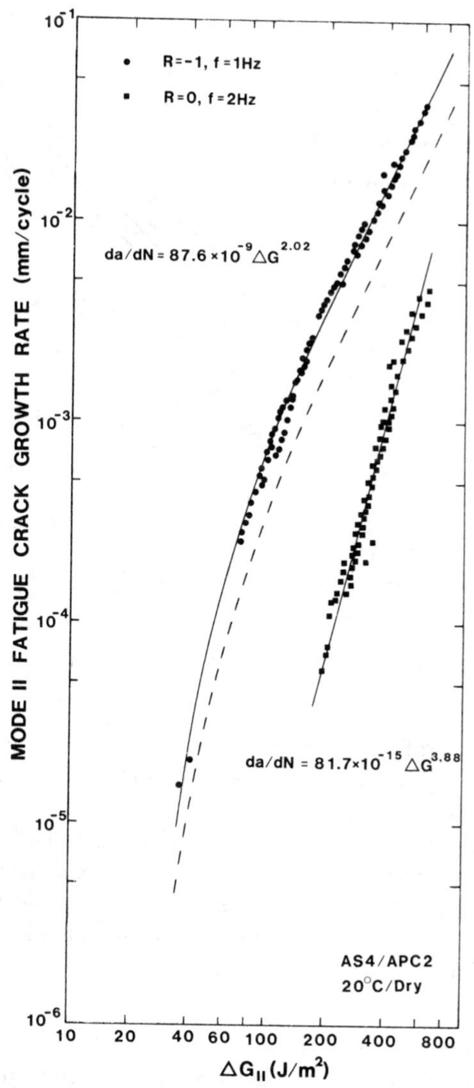

FIG. 10—*Mode II fatigue crack growth rate data for AS4/APC2. See Fig. 2 for a definition of* ΔG_{II}.

indicate that the effective threshold values (defined as ΔG_{II} corresponding to 10^{-6} mm/cycle) increase slightly with increasing G_{IIc}.

Owing to the scarcity of published data on Mode II interlaminar fatigue crack growth rates, only a limited comparison can be made with the current results. Data obtained on crack lap shear specimens [7] (80 to 85% Mode II, $R = 0.05$) for first generation 177°C curing graphite/epoxy, gave a value for n of 6.07 and an effective threshold of 68 J/m^2. These are quite comparable with the present pure Mode II values for AS1/3501-6 at $R = 0$ of $n = 5.8$ and $\Delta G_{\text{TH}} = 66$ J/m^2.

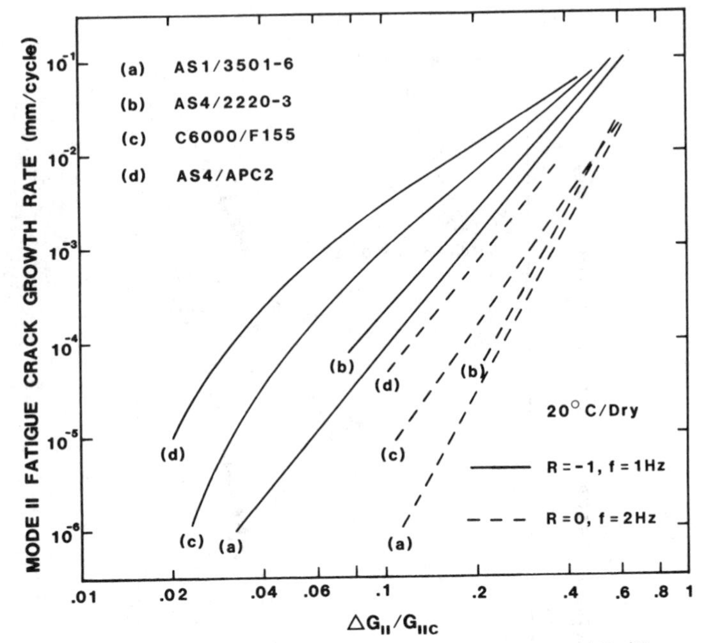

FIG. 11—*Fatigue data from Figs. 7 to 10 replotted against* $\Delta G_{II}/G_{IIc}$.

Fractography of the fatigue specimens gave few clues as to the mechanisms controlling the propagation behaviors because of the fretting of the fracture surfaces which both obscured many of the original features and probably introduced new ones. Nevertheless, one major difference between the Mode II static and fatigue surfaces was the greater amount of fiber/matrix debonding in the latter, particularly in the case of the PEEK system (see Fig. 12). As a consequence, it may be necessary to increase the strength of the fiber/matrix interface to optimize composite interlaminar fatigue properties.

General

The relative effects of matrix toughness on the composite properties measured in this investigation are summarized in Fig. 13 along with some additional matrix data. The matrix values for fracture energy and tensile modulus are from published reports [*5,9–11*], whereas the Rockwell E hardness measurements were made on extracted matrix samples according to ASTM Test for Rockwell Hardness of Plastics and Electrical Insulating Materials (D 785). It should be pointed out that the PEEK matrix data were measured on a grade of PEEK different from that in the composite (APC2). Furthermore, the crystal structure in a solid matrix sample will likely differ from that in the composite as a result of the absence in the former case of solidification nucleating sites on the fiber surface. The poor transfer of Mode I toughness from the matrix to the composite has been well documented

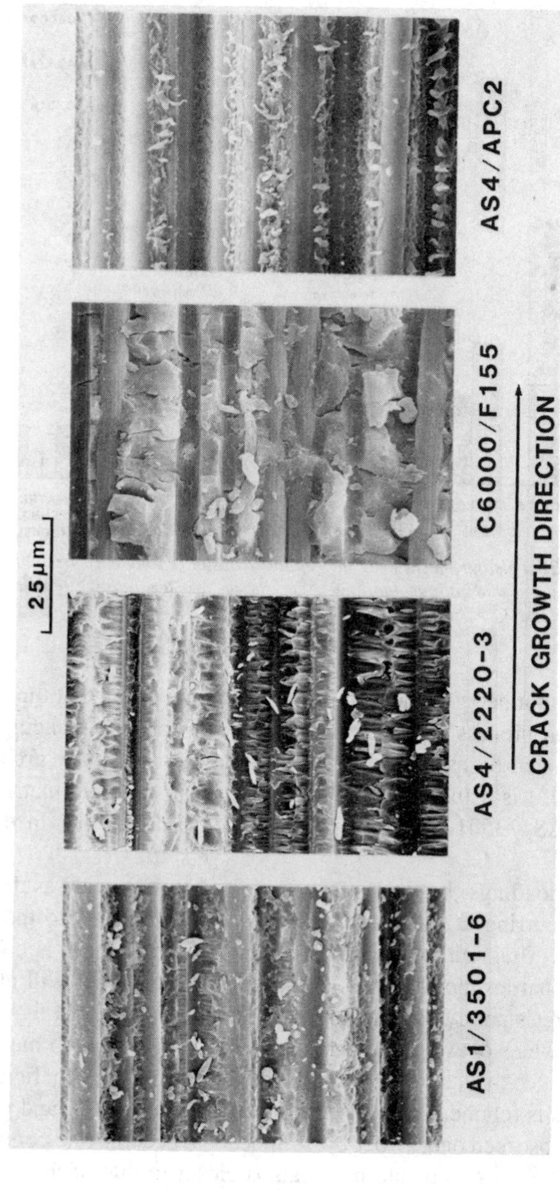

FIG. 12—*Representative fractographs of* R = −1 *Mode II fatigue specimens.*

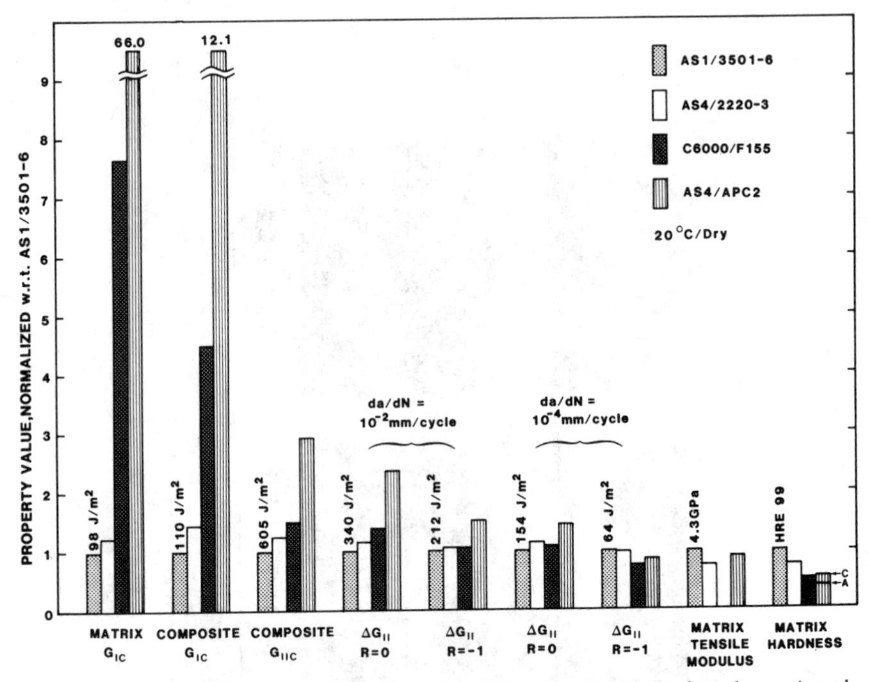

FIG. 13—*Normalized histogram showing the relative transfer of toughness from the matrix to the composite for both static and fatigue loading. Also shown are the relative tensile modulus and hardness of the matrices* (C = *crystalline,* A = *amorphous).*

[*12,13*], and is explained generally by the constraining effect that the fibers have on the size of the process or plastic zone at the tip of an advancing crack. A similar argument can be used for G_{IIc} except that the apparently greater loss in toughness (Fig. 13) is actually due to the presence of multiple matrix cracking in the baseline AS1/3501-6 system rather than to a lower G_{IIc} in the tougher materials.

Under fatigue loadings, both the resistance to fracture as well as the extent of microdamage occurring at the crack tip per cycle contributed to the measured values of da/dN. Thus, although the softer more ductile PEEK is considerably tougher than the harder more brittle 3501-6 epoxy, the former will also experience more hysteresis per cycle, and hence, more microdamage at the crack tip is likely. At large values of ΔG_{II}, the damage zone is large for both materials, and hence, the resistance to static fracture predominances and the beneficial effect of matrix toughness is retained. However, at lower values of ΔG_{II}, and particularly at $R = -1$, the absorbed energy per cycle appears to become rate controlling and the ranking of the four composite materials is closer to that of the matrix hardness values.

In terms of structural integrity, there is little doubt that the increases in G_{Ic} and G_{IIc} resulting from improved matrix toughness will not only enhance the resistance of the composite to impact but will also provide a greater capability to tolerate larger delaminations without catastrophic failure. However, as far as the long-term performance of the structure is concerned, the use of tougher matrix materials, such as the ones examined here, will do little to reduce the growth rate of delaminations and, under certain fatigue loading conditions, as shown in Fig. 13, may in fact be detrimental to the life of the structure. These findings could have a profound effect on design methods aimed at incorporating damage tolerance and long-term durability into high performance composite materials structures.

Conclusions

The results of the static and fatigue tests support the following conclusions concerning the effect of matrix toughness on delamination behavior.

1. Improvements in interlaminar fracture energy as a result of increasing the matrix toughness are significantly less under Mode II shear loading than under Mode I tensile loading. This is due to a change in the Mode II failure mechanism from the energy absorbing multiple matrix microcracking of the brittle matrix systems to the constrained plastic deformation of the toughened thermoplastic matrix system.
2. For both brittle and toughened matrix materials, shear reversal ($R = -1$) results in a large increase in the Mode II interlaminar fatigue crack growth rates, particularly at low values of ΔG_{II} compared to $R = 0$ behavior.
3. Increasing the matrix toughness results in a reduction in the slope of the da/dN versus ΔG_{II} curves for both $R = 0$ and $R = -1$. As a consequence, under certain loading conditions, the tougher matrix composite systems are more sensitive to interlaminar fatigue crack growth than the brittle matrix systems. At low values of ΔG_{II}, matrix hardness measurements appear to provide a better ranking of the fatigue resistance.
4. The early indications of significantly low values of fatigue threshold fracture energy ΔG_{TH} presented in this paper suggest that structural components made from laminated composites must allow for possible finite lifetimes based on shear dominated interlaminar fatigue crack growth rates. This is equally valid for both brittle and toughened matrix composites.
5. Regarding future work, the significance of moisture, temperature, and frequency on the behavior of toughened matrix composites should be investigated to ascertain whether next generation composite structures may necessitate fatigue critical design methods.

APPENDIX A

Compliance Determination of the End Notched Cantilever Beam Specimens

With reference to Fig. A1 and using beam theory, the flexural deflection along the undelaminated segment of the specimen is

$$\Delta(x) = x^2(3L - x)P/4Ebh^3 \quad \text{(A1)}$$

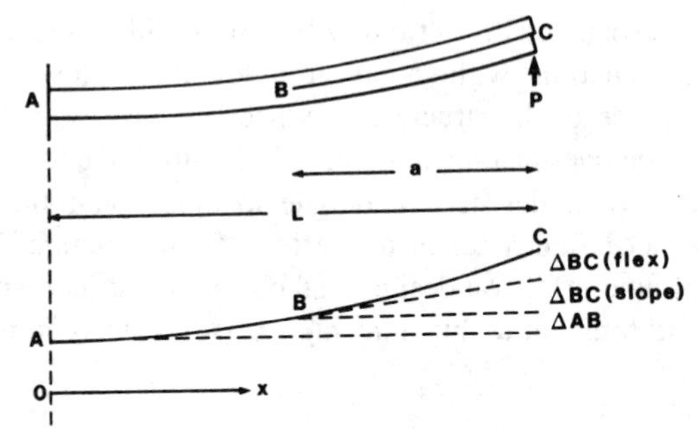

FIG. A1—*Specimen deflections.*

at $x = L - a$

$$\Delta AB = (2L^3 - 3aL^2 + a^3)P/4Ebh^3 \quad \text{(A2)}$$

A portion of the deflection of the delaminated region of the specimen results from the slope at B which is obtained by differentiating Eq A1.

$$d\Delta/dx = (6Lx - 3x^2)P/4Ebh^3 \quad \text{(A3)}$$

which at $x = L - a$,

$$= 3(L^2 - a^2)P/4Ebh^3 \quad \text{(A4)}$$

resulting in a deflection as a result of the slope at B of

$$\Delta BC(\text{slope}) = 3a(L^2 - a^2)P/4Ebh^3 \quad \text{(A5)}$$

The flexural deflection of the delaminated part of the specimen, assuming both halves carry equal loads, is given simply by

$$\Delta BC(\text{flex}) = 8a^3P/4Ebh^3 \quad \text{(A6)}$$

Provided that the total deflection is small, these three terms can be summed arithmetically to obtain the total deflection

$$\Delta AC = P(L^3 + 3a^3)/2Ebh^3$$

and hence, the flexural compliance of the specimen is given by

$$C = \frac{\Delta AC}{P} = (L^3 + 3a^3)/2Ebh^3$$

Acknowledgments

The authors wish to thank G. D. Dorey and co-workers at the Royal Aircraft Establishment, Farnborough, England for supplying the AS4/APC2 laminate at short notice. The diligent technical expertise of E. Jensen and the manuscript preparation of J. Stringer are also gratefully acknowledged.

References

[1] Russell, A. J. and Street, K. N., in *Delamination and Debonding, ASTM STP 876,* W. S. Johnson, Ed., American Society for Testing and Materials, Philadelphia, 1986, pp. 349–370.
[2] Wilkins, D. J., Eisenmann, J. R., Camin, R. A., Margolis, W. S. and Benson, R. A., in *Damage in Composite Materials, ASTM STP 775,* K. L. Reifsnider, Ed., American Society for Testing and Materials, Philadelphia, 1982, pp. 168–183.
[3] O'Brien, T. K., in *Effects of Defects, ASTM STP 836,* American Society for Testing and Materials, Philadelphia, 1984, pp. 125–142.
[4] O'Brien, T. K., "Interlaminar Fracture of Composites," W. S. Johnson, Ed., NASA Technical Memorandum 85768, June 1984.
[5] Bradley, W. L. and Cohen, R. N., in *Delamination and Debonding, ASTM STP 876,* W. S. Johnson, Ed., American Society for Testing and Materials, Philadelphia, 1986, pp. 389–410.
[6] Bathias, C. and Laksimi, A., in *Delamination and Debonding, ASTM STP 876,* W. S. Johnson, Ed., American Society for Testing and Materials, Philadelphia, 1986, pp. 217–237.
[7] Ramkumar, R. L. and Whitcomb, J. D., in *Delamination and Debonding, ASTM STP 876,* W. S. Johnson, Ed., American Society for Testing and Materials, Philadelphia, 1986, pp. 315–335.
[8] Bevan, L. G., *Composites,* Vol. 8, No. 4, Oct. 1977, pp. 227–232.
[9] Bascom, W. D., Bullman, G. W., Hunston, D. L., and Jensen, R. M., "The Width-Tapered Double Cantilever Beam for Interlaminar Fracture Testing," in *Proceedings of the 29th National SAMPE Symposium,* Society for the Advancement of Material and Process Engineering, Covina, CA, April 1984.
[10] Leach, D. C., Curtis, D. C., and Tamblin, D. R., "Delamination Behavior of Aromatic Polymer Composite ASP-2," in this volume, pp. 358–380.
[11] Johnston, N. J., *ACEE Composite Structures Technology,* NASA Conference Publication 2321, Aug. 1984.
[12] Bascom, W. D., Bitner, J. L., Moulton, R. J., and Siebert, A. R., *Composites,* Vol. 11, No. 1, Jan. 1980, pp. 9–18.
[13] Scott, J. M. and Phillips, D. C., *Journal of Material Science,* Vol. 10, No. 4, April 1975, pp. 551–562.

DISCUSSION

G. M. Newaz[1] *(written discussion)* —I am quite disappointed to see throughout the conference, many researchers, including yourself, either discuss or attempt to relate static toughness of the resin to the long-term performance of the composite, for example, fatigue. This is very misleading because one is essentially dealing with two completely different loading conditions. I do not see a physical basis for relating static toughness of the resin to delamination growth in fatigue. To assume this is to say that the polymer thermomechanical behavior in static and fatigue loading are the same. Would it not be more appropriate to establish fatigue crack growth rate of the neat resin, and correlate this to the delamination growth rate of the composite in fatigue loading?

[1]Owens-Corning Fiberglas Technical Center, Granville, OH.

A. J. Russell and K. N. Street (authors' closure)—We certainly agree that there is no reason to assume that static and fatigue properties should be directly related mechanistically. However, we consider that a comparison of the two properties is essential if a damage tolerant composite structure is required. A large improvement in composite toughness may be of little benefit to a heavily fatigue loaded component if it is accompanied by an increase in interlaminar fatigue crack growth rate. It was the intention of this paper to show not only that static toughness does *not* correlate with interlaminar fatigue resistance, but also that a change in static loading mode alone (from tension to shear) can greatly *reduce* the apparent benefits of a tougher resin. We also agree that a comparison of neat resin fatigue crack growth rate data with those of the composites would help to clarify the fatigue mechanisms responsible for the observed differences in interlaminar fatigue crack growth rates.

W. S. Johnson[1] *and P. D. Mangalgiri*[1]

Influence of the Resin on Interlaminar Mixed-Mode Fracture

REFERENCE: Johnson, W. S. and Mangalgiri, P. D., "**Influence of the Resin on Interlaminar Mixed-Mode Fracture,**" *Toughened Composites, ASTM STP 937,* Norman J. Johnston, Ed., American Society for Testing and Materials, Philadelphia, 1987, pp. 295–315.

ABSTRACT: Both literature review data and new data on toughness behavior of seven matrix and adhesive systems in four types of tests were studied to assess the influence of the resin on interlaminar fracture. Mixed-mode (that is, various combinations of opening Mode I [G_I] and shearing Mode II [G_{II}]) fracture toughness data showed that the mixed-mode relationship for failure appears to be linear in terms of G_I and G_{II}. The study further indicates that fracture of brittle resins is controlled by the G_I component, and that fracture of many tough resins is controlled by total strain energy release rate G_T. Regarding the relation of polymer structure and the mixed-mode fracture: high Mode I toughness requires resin dilatation; dilatation is low in unmodified epoxies at room temperature/dry conditions; dilatation is higher in plasticized epoxies, heated epoxies, and in modified epoxies; modification improves Mode II toughness only slightly compared with Mode I improvements. Analytical aspects of the cracked lap shear test specimen were explored. Geometric nonlinearity must be addressed in calculating the G_I/G_{II} ratio. The ratio varies with matrix modulus, which in turn varies with moisture and temperature.

KEY WORDS: interlaminar fracture toughness, adhesives, double cantilever beam specimen, edge delamination tension specimen, cracked lap shear specimen, end notched flexure specimen, mixed-mode fracture, strain energy release rate, bonded joints, composites

Fracture mechanics technology is being applied extensively to composite materials and adhesively bonded joints, primarily to determine the matrix and adhesive material toughness so as to aid in material development, screening, selection, and design. Efforts to determine toughness make use of the fact that the delamination failure mode in polymer matrix composites is very similar to the debonding of an adhesive joint: both the delamination and debond are usually "captured" between two boundary plies in the case of delamination, or adherends in the case of debonding. This physical restraint, in the presence of a mixture of

[1]Senior research engineer and NRC resident research associate, respectively, NASA Langley Research Center, MS 188E, Hampton, VA 23665.

external loads, can result in a variety of loading modes at the delamination or debond tip. These modes may range from pure Mode I (opening or peel) through various combinations of Mode I and Mode II (sliding or shear) to pure Mode II loading. Mode III (tearing) may even be present. Considerable effort has been made to identify proper test specimens and testing techniques for measuring the in situ toughness of composite matrix materials and adhesives. In particular, the relative influence of the Mode I and Mode II components on fracture has been of interest.

The purpose of this paper is to compare in situ toughness behavior of seven matrix and adhesive systems (as listed in Table 1) to assess the influence of resin on interlaminar fracture. By interlaminar fracture, the authors are referring to both delamination toughness of a composite laminate and debond toughness of an adhesive joint. Both brittle and tough systems will be addressed. Particular emphasis will be placed on the mixed-mode influence on fracture. Although most of the data have been taken from the literature, some new data will be presented. The observed behavior of the different systems will be related to their chemical structures. Some analytical considerations for assessing mixed-mode ratios will be presented.

Experiments

There are currently four tests widely used to measure interlaminar fracture toughness in terms of strain energy release rate G. Each test's specimen type is shown in Fig. 1. Shown first is the double cantilever beam (DCB) specimen, used to determine pure Mode I toughness. Second is the edge delamination tension (EDT) test specimen, used to examine a variety of mixed-mode conditions, ranging from $5.7 > G_I/G_{II} > 0.4$, depending on specimen lay-up and geometry. Third is the cracked lap shear (CLS) specimen, which is also used to test a variety of mixed-mode conditions, ranging from $0.6 > G_I/G_{II} > 0.2$, again depending on the specimen geometry. Pure Mode II toughness is found by using the fourth specimen type in the end-notched flexure (ENF) test. All of these tests, except

TABLE 1—*Resin systems.*

Resins[a]	Polymer Structure	Supplier[a]
5208	epoxy—unmodified	Narmco
3501-6	epoxy—unmodified	Hercules
Hx205	epoxy—chain extended	Hexcel
F-185[b]	epoxy—chain extended/ rubber modified	Hexcel
EC 3445[b]	epoxy—rubber modified	3-M
FM-300[b]	epoxy—rubber modified	American Cyanamid
PEEK	semicrystalline	Imperial Chemical Industries

[a] Use of trade or manufacturer names does not constitute an official endorsement, expressed or implied, by the National Aeronautics and Space Administration.

[b] Commercial adhesive.

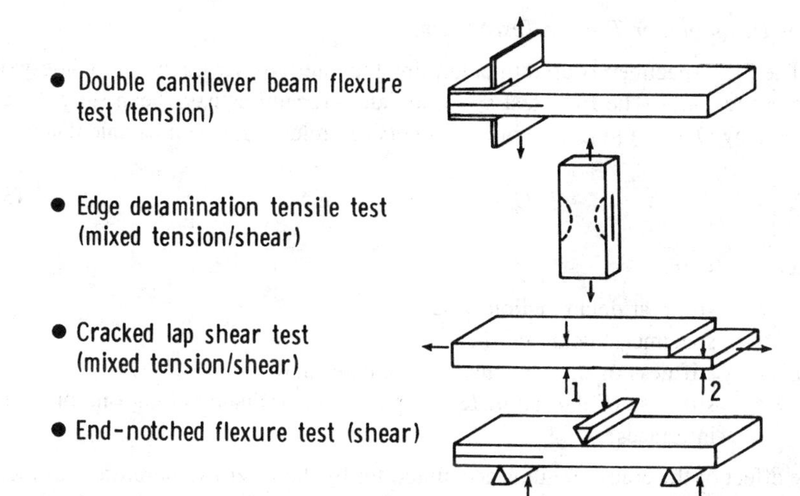

FIG. 1—*Four specimen types used for determination of interlaminar fracture toughness.*

the EDT, are also used to characterize adhesives in bonded joints. All of the specimens, except the EDT, are calibrated by means of the general relationship between strain energy release rate G and specimen compliance C

$$G = \frac{P^2}{2b}\frac{dC}{da} \tag{1}$$

where P is the load, b the width, and a the crack length [*1*].

Although data acquired from tests involving all four of the above specimen types will be presented and discussed, only the CLS specimens were used to generate new data for this paper.

Double Cantilever Beam Specimen

The pure Mode I DCB specimen has been well documented for testing adhesives [*2–4*] and for testing interlaminar toughness of composites [*5*]. Linear beam theory can be used to determine G if the crack length is sufficiently long compared to the specimen adherend thickness h. The compliance is commonly expressed as $C = 8a^3/bEh^3$ for plane stress conditions, where E is the longitudinal modulus of the adherends. This expression is valid as long as modulus is taken as apparent modulus. Ashizawa [*6*] presents correction factors for the flexural modulus. By expressing the derivative dC/da in terms of C and substituting into Eq *1*, an expression for strain energy release rate becomes

$$G = \frac{3P^2C}{2ba} \tag{2}$$

The dC/da term can be calculated directly from the load-displacement plots of the experimental data. This technique is explained in Refs *5* and *7*.

Edge Delamination Tension Test Specimen

The EDT specimen is used to determine the interlaminar fracture toughness of composites only. The EDT test specimen and procedures have been fully documented by O'Brien [8]. The total strain energy release rate can be calculated by

$$G = \frac{\varepsilon^2 t}{2}(E_{\mathrm{LAM}} - E^*) \tag{3}$$

where

ε = strain at delamination onset,
t = specimen thickness,
E_{LAM} = stiffness of the undamaged laminate, and
E^* = stiffness of the laminate completely delaminated along one or more interfaces.

The effect of the crack length is accounted for by the E terms; therefore the crack length does not appear in Eq 3. The $G_{\mathrm{I}}/G_{\mathrm{II}}$ ratio must be calculated by finite element analysis of the specimen lay-up of interest [8].

Cracked Lap Shear Specimen

The CLS specimen was first used by Brussat et al. [9] for testing adhesively bonded metallic joints. Mall et al. [7,10] used the CLS specimen to study adhesively bonded composite joints. Wilkins [5] was the first successfully to use the CLS specimen to measure mixed-mode interlaminar fracture toughness in composites. An estimate for the total strain energy release rate for the CLS specimen is given by

$$G = \frac{P^2}{2b^2}\left[\frac{1}{(Eh)_2} - \frac{1}{(Eh)_1}\right] \tag{4}$$

where the subscripts 1 and 2 refer to the sections indicated in Fig. 1. The total G can also be found from the experimental compliance measurements and Eq 1.

To get an accurate measure of the G_{I} and G_{II} components, a geometric nonlinear finite element model must be used [11]. This is because the lack of symmetry in the CLS specimen causes out-of-plane displacements and rotations. (This will be further discussed under Finite Element Analysis of CLS Specimens.)

End-Notched Flexure Specimen

The ENF specimen is used to determine Mode II toughness and was introduced by Russell [12]. The specimen is essentially the same as the DCB specimen, except that it is loaded in three-point bending. This loading condition results in pure shear loading at the crack tip. The Teflon® starter strip supposedly allows no significant friction to occur between crack flanks. Under the bending, the crack does not open; thus Mode I loading does not occur. Simple beam theory was used [12] to obtain the expression

$$G = \frac{9P^2a^2C}{2b(2L^3 + 3a^3)} \tag{5}$$

where L is the distance between the center and outer loading pins.

Finite Element Analysis of CLS Specimens

The cracked lap shear specimens (used to generate new data for this paper) were analyzed with the finite element program GAMNAS [*11*] to determine the strain energy release rate for given geometry, debond length, and applied load. This two-dimensional analysis accounts for the geometric nonlinearity associated with the large rotations in the unsymmetrical cracked lap shear specimen.

The finite element mesh consisted of about 1700 isoparametric 4-node elements and had about 3700 degrees of freedom. The strip of matrix material was modeled with four layers of elements. A multipoint constraint was applied to the loaded end of the model to prevent rotation (that is, all of the axial displacements along the ends are equal to simulate actual grip loading of the specimen). Plane-strain conditions were assumed in the finite-element analysis. The strain energy release rate was computed using a virtual crack closure technique. The details of this procedure are given in Ref *13*.

Influence of the Polymer Structure

For a polymer to have a high Mode I toughness, the material must be able to dilatate (increase volume) under plane-strain tensile loading. This volume increase can be from elastic, nonlinear elastic, or inelastic expansion. Mechanisms such as chain extension, crazing, void formation followed by shear band formation, and plasticity can contribute to the volume expansion. Those polymeric materials that are highly rigid as a result of a high degree of cross-linking (such as unmodified epoxies) are not normally able to exhibit much ability to dilatate. Therefore, they will have a rather low Mode I toughness. Figure 2 offers a schematic of resin stress-strain behavior and the stress concentrations at the crack tip in two resin systems, one tough and one brittle. Assuming that the resin at the crack tip is in a state of plane strain, the nonlinear stress-strain curve shown for the tough material can occur only if the material's volume increases. The plane strain state implies that the material cannot neck-down to maintain a constant volume. Notice that the shaded area under the two stress-strain curves is representative of the energy required for failure (that is, the toughness). As can be seen in the fatigue, the plastic deformation (which must be accompanied by a volume expansion) requires considerable energy. Further, one should notice that the stress concentration at the crack tip is lower for those materials that exhibit plastic deformation, thus spreading the load over a larger area.

On the other hand, the polymer need not dilatate under the shearing deformation of Mode II loading. Therefore, a highly cross-linked epoxy may have a much higher Mode II toughness than Mode I toughness. Specimens with a high percentage of Mode II (shear) present have hackles on their fracture surfaces

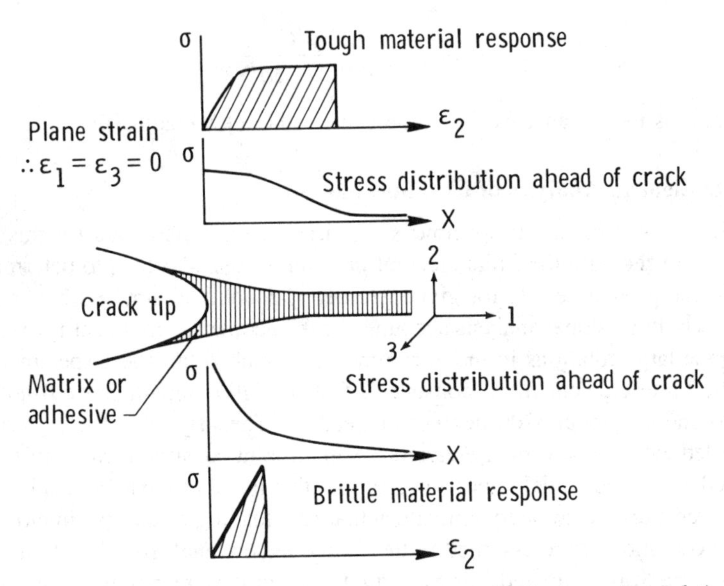

FIG. 2—*Schematic of resin stress-strain behavior and stress distribution at the crack tip.*

[*14*]. Pure shear stress can be resolved into a combination of tensile and compressive stresses. A simple Mohr's circles analysis will show that for pure shear stress the maximum tensile stress acts on a 45° plane. As a peel stress is added to the shear stress, the angle decreases. The maximum normal stress is expressed as

$$\sigma_{\max} = \frac{\sigma_1 + \sigma_2}{2} + \sqrt{\left(\frac{\sigma_1 - \sigma_2}{2}\right)^2 + \tau_{12}^2} \tag{6}$$

The maximum principal stress will act on a plane at $\theta°$ to the adherend surface where

$$\tan 2\theta = \frac{\tau_{12}}{(\sigma_1 - \sigma_2)/2} \tag{7}$$

Therefore, when only shear stresses are present $\theta = 45°$. When only peel stresses (σ_1) are present $\theta = 0°$. Pure Mode II loading will result in the sharpest hackles. As the percent of Mode I (peel) component increases, the hackles become less obvious, reflecting the fact that the angle of maximum tensile stress is decreasing toward 0°. The matrix material would be expected to fracture in a plane nearly perpendicular to the maximum tensile stress.

Both moisture and heat may increase the ability of polymers to dilatate, thus increasing the Mode I toughness, but probably causing little improvement in the Mode II toughness. Moisture may infuse into the polymer, separating the polymer molecules, thus reducing their secondary bond attraction, thereby making it easier for the molecules to move past each other. Moisture may also increase the critical strain levels. This process is a form of plasticization. Heat creates a similar effect. The higher temperature, especially approaching the glass transition temperature (T_g), causes the material to expand, thus creating more free volume. This increase in free volume, for a given resin system, allows the molecules to move past each other more easily and will probably result in a higher Mode I toughness.

Rubber toughened epoxies have a propensity to dilatate by the stretching of the rubber particles and the formation of internal voids. Thus rubber toughened epoxies such as FM-300 and EC 3445 could have critical Mode I strain energy release rate (G_{Ic}) values as high as the G_{IIc} values. Bascom et al. [*4*] suggest that the rubber modified epoxies exhibit large increases in toughness with relatively little loss in high tensile strength, modulus, or thermal mechanical resistance of the epoxy matrix resin. They further suggest that the rubber particles allow a much larger volume for plastic deformation at the crack tip than allowed by the unmodified epoxy resins. This deformation involves a more or less symmetrical dilatation of the rubber particles accompanied by plastic flow of the epoxy. The problem with the highly rubber modified epoxies and the linear systems is that they become so ductile in the presence of heat and moisture that they usually cannot function as a structural material.

Dilatation can be seen in polysulfones, linear systems that are essentially masses of polymer chains that are intertwined but not physically linked together as epoxies are. These linear systems are more ductile and can dilatate more than the unmodified epoxies. (Polysulfone data are not presented in this paper.)

Other modifications can also contribute to improving the material toughness; one example is chain-extended epoxies such as the Hx205 resin. Although the material is still cross-linked, the extended chains can behave somewhat like linear systems, thus improving the G_I toughness. At the same time, they are expected to display more viscoelastic behavior and to be more sensitive to temperature and moisture conditions.

Crystalline resins such as polyetheretherketone (PEEK) have a toughening mechanism consisting of chains unfolding. They also have a strengthening mechanism consisting of the crystals slipping to achieve a structure oriented in the direction of the stress.

Results and Discussion

Toughness Data

This section will present mixed-mode interlaminar fracture toughness results for brittle systems (T300/5208 and AS1/3501-6 composites) and tougher systems (FM-300 and EC 3445 adhesives, and Hx205, F-185, and PEEK matrix composites). Figures displaying toughness data will be labeled as to specimen type, G_I/G_{II} ratio, and source reference. Where no reference is given, the data were generated by the authors.

The authors recognize that the difference in resin thickness between the composite delamination and adhesive joints could result in fracture toughness values that are resin thickness dependent. However, the emphasis of this paper is the mixed-mode interaction behavior and not the absolute value of toughness.

Brittle Systems

T300/5208 —The interlaminar fracture envelope for T300/5208 composites is shown in Fig. 3. The G_I component at fracture is plotted against the G_{II} component at fracture. The sum of the G_I and G_{II} components is the total strain energy release rate G_T. The value of the critical Mode I strain energy release rate (G_{Ic}) is approximately one tenth of the critical Mode II strain energy release rate (G_{IIc}). The mixed-mode fractures appear to be controlled by the G_I component. This is a good example of an unmodified epoxy with limited dilatational ability having a much lower G_{Ic} than G_{IIc}. The scatter bands represent the maximum, minimum, and average values of the data. Notice that the source is given for each specimen type. The CLS data were generated for this paper and are presented in Table 2.

AS1/3501-6 —Another rather brittle unmodified epoxy system is 3501-6. The AS1/3501-6 system has been extensively studied by Russell [*12*], Russell and Street [*17*], Wilkins [*5*], Law and Wilkins [*18*], and Jurf and Pipes [*19*] to determine interlaminar fracture toughness. Figure 4 shows some of Russell and Street's [*17*] results using the DCB, CLS, and ENF specimens. The data spread is plus or minus one standard deviation from the mean of their results. Their data indicate that G_{IIc} is about three and a half times higher than G_{Ic}. This G_{Ic} value is about 50% higher than the G_{Ic} for the 5208 matrix.

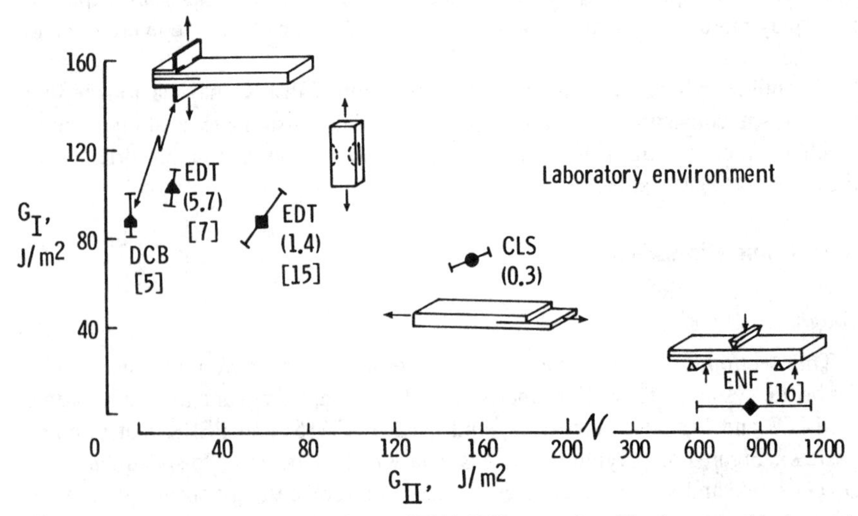

FIG. 3—*Interlaminar fracture toughness of T300/5208 composites. The data spread is the maximum and minimum values.*

TABLE 2—*Cracked lap shear fracture toughness tests data for T300/5208 (three plies to three plies).*

Specimen	Delamination Length a, mm	Critical Load P_c, kN	Strain Energy Release Rate		
			Mode I G_I, J/m^2	Mode II G_{II}, J/m^2	Total G_c, J/m^2
1	36.8	5.40	52.6	126.4	178
	80.3	5.51	53.6	131.4	185
	101.6	5.27	49.0	120.0	169
	119.1	5.68	53.1	129.9	183
2	35.6	5.84	76.2	181.8	256
	53.3	5.36	62.1	151.9	214
	88.9	5.40	63.8	156.2	220
	110.5	5.45	64.7	158.3	223
	130.8	5.34	62.1	151.9	214
	143.5	5.49	65.8	161.2	227
3	35.6	5.74	71.9	176.1	248
	62.2	5.51	66.1	161.9	228
	68.6	5.57	67.9	166.1	234
	78.1	5.67	70.2	171.8	242
	116.8	5.41	64.1	156.9	221
	128.9	5.54	67.0	164.0	231
	166.4	5.31	61.5	150.5	212
4	34.3	5.79	72.2	176.8	249
	85.6	5.52	65.8	161.2	227
	104.6	5.47	64.7	158.3	223
	128.0	5.56	66.7	163.3	230

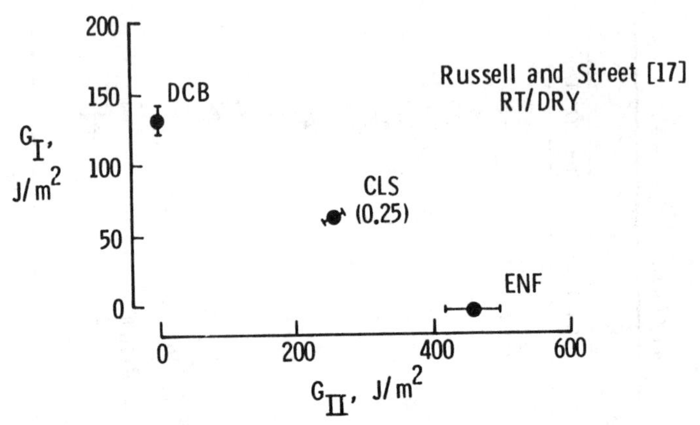

FIG. 4—*Interlaminar fracture toughness of AS1/3501-6 composites. The data spread is one standard deviation from the mean.*

Law and Wilkins [*18*] tested DCB specimens and three geometries of the CLS specimen to assess the effects of heat and moisture on mixed-mode fracture. The DCB data were 100% Mode I. The CLS specimens had G_I/G_{II} ratios of 0.20, 0.42, and 0.59. The specimens with 0.20 and 0.42 ratios were tested wet and dry at 93 and 22°C, as were the DCB specimens. These data are plotted in Fig. 5. The 22°C/dry data are almost horizontal, indicating a G_I-dominated fracture mode. (The 22°C/dry mixed mode data is noticeably different than the Russell and

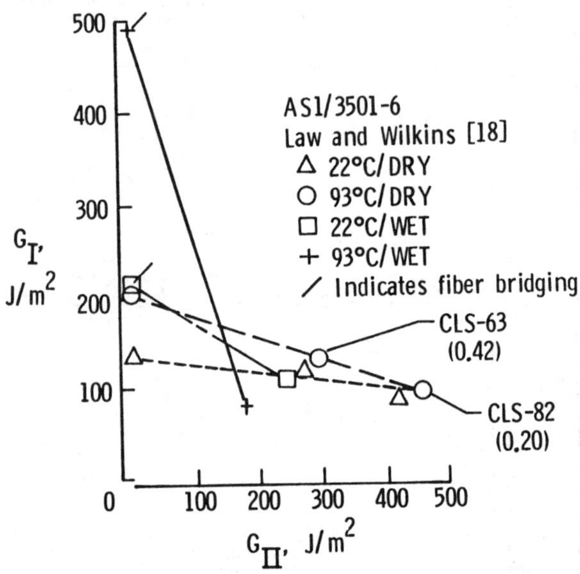

FIG. 5—*Interlaminar fracture toughness of AS1/3501-6 composites under various environmental conditions.*

Street data in Fig. 4. This may be due to data scatter, material variability, or differences in data reduction and analysis.) The G_I is very low compared to the G_{II} components. However, as heat is applied (93°C/dry), the G_I increases while the G_{II} components remain about the same. When moisture is introduced (22°C/wet), the G_I increases and the G_{II} components appear to decrease. When both heat and moisture are introduced (93°C/wet), G_I increases markedly, while the G_{II} appears to be substantially reduced.

These results by Law and Wilkins [*18*] are somewhat clouded by the fact that the DCB specimens showed considerable fiber bridging, particularly at the higher temperature and wet conditions. Therefore, the higher temperature and wet condition DCB values are probably artificially higher than those of the matrix material. The CLS specimens did not have this problem. In fact, Russell and Street [*17*] indicated that the DCB specimen toughness did not increase with temperature when only the initial crack extension was used to calculate G_I, thus avoiding the fiber bridging problem encountered at longer crack lengths. Hunston and Bascom also report little increase in toughness with increasing temperature for unmodified epoxies [*20*]. This is perplexing because Russell and Street also showed that the neat specimens of 3501-6 showed improvements in toughness with increasing temperature. Perhaps the fact that these "initial" crack extensions were from the Teflon starter caused the toughness to be insensitive to the environmental effects. The Teflon starter may be rather blunt compared to a natural delamination, thereby causing the toughness to be high.

Tough Systems

FM-300 —FM-300 is a nitride rubber modified, epoxy-structured adhesive widely used in the aerospace industry. It is a mat reinforced film adhesive cured at 176°C; the resulting bondline is 0.25 mm thick. Mall and Johnson [*7*] determined the Mode I and mixed-mode fracture toughness using DCB and two different geometries of the CLS specimen. These CLS specimens had G_I/G_{II} ratios of 0.33 and 0.38. The G_{Ic} toughness, shown in Fig. 6, is seven to ten times greater than previously found for the brittle systems, (that is, 5208 and 3501-6) at room temperature and dry. The apparent G_{IIc} toughness (based on linear extrapolation from the CLS and DCB data) is a little over twice that of the 3501-6 (Fig. 4) and about equal to that of the 5208 (Fig. 3). The rubber toughening greatly improves the G_{Ic}, but is less effective on the shearing mode G_{IIc}. Even though the DCB adherends were made of unidirectional T300/5208 composites, the data reported in Fig. 6 all show cohesive failures in the adhesive; there was, of course, no fiber bridging.

EC 3445 —EC 3445 is also a rubber modified, epoxy-structured adhesive that is currently used in the aerospace industry. It has 121°C cure temperature. The resin is a one-part paste. Glass beads are used to maintain a bond line thickness of 0.10 mm.

Mall and Johnson [*7*] determined the Mode I and mixed-mode fracture toughness for the EC 3445 as they did for FM-300. Figure 6 shows results for both.

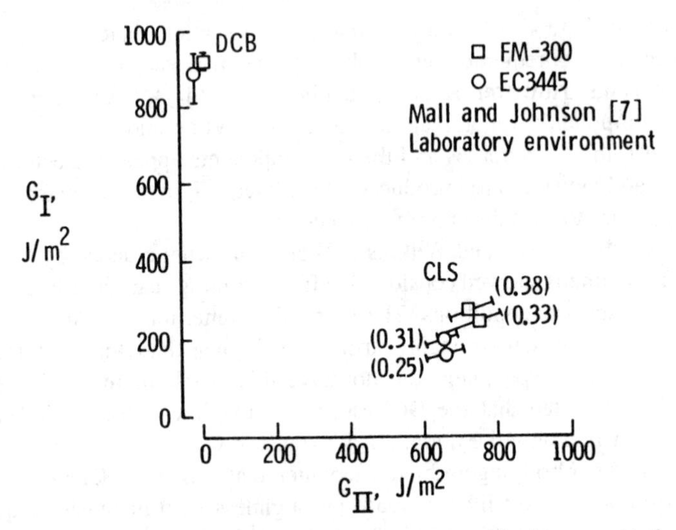

FIG. 6—*Fracture toughness of FM-300 and EC 3445 adhesives. The data spread is the maximum and minimum values.*

As previously discussed for the FM-300, the G_{Ic} and apparent G_{IIc} are practically the same. Two geometries of the CLS specimen were also tested, resulting in G_I/G_{II} ratios of 0.25 and 0.31.

Hx205 —The Hx205 base resin is a standard bisphenol A diglycidyle ether modified with an epoxy novolac and chain extended with additional bisphenols. The Hx205 was chosen for study because it has high toughness. As such, it is an appropriate material to use in evaluating methods for testing interlaminar toughness and mixed-mode fracture. However, it is not used for structural composite matrices because of its low T_g (100°C) and poor hot-wet mechanical properties. Figure 7 presents the fracture envelope for Hx205 matrix composites tested in a laboratory environment. The data spread in Fig. 7 represents the maximum, minimum, and average values of the data for each specimen type. The CLS data for the Hx205 are given in Table 3.

F-185 —F-185 is a commercially available adhesive. It is essentially a rubber modified form of Hx205. Figure 8 shows the DCB data developed by Hunston [*21*] and the EDT data of O'Brien et al. [*22*]. The G_{Ic} is much higher than the G_{Ic} values of the other resin systems presented. The G_{Ic} value is quite close to the apparent G_{IIc} value (extrapolating the data to the G_{II} axis). Apparently, both rubber toughening and extending molecular chains individually contribute to the overall toughness of a material. The effect of extending molecular chains can be seen by the substantial increase in the chain extended Hx205 system over the unmodified epoxies (3501-6 and 5208). F-185 shows a substantial improvement in toughness over Hx205 as a result of rubber toughening.

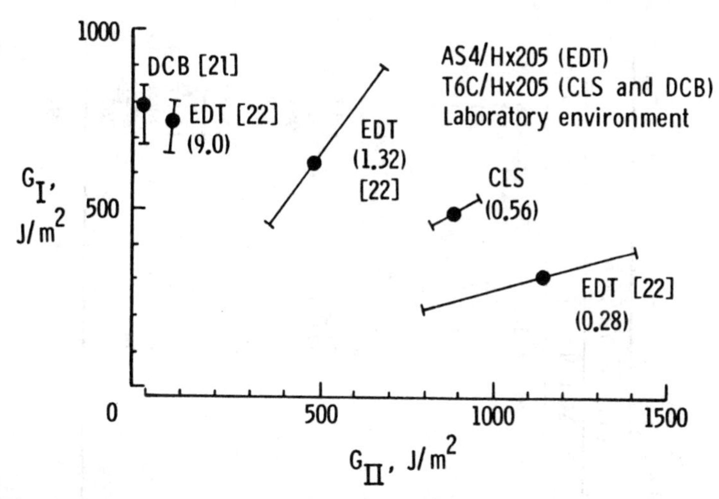

FIG. 7—*Interlaminar fracture toughness of Hx205 matrix composites. The data spread is the maximum and minimum values.*

PEEK—The PEEK matrix material is a polyetheretherketone, a high temperature thermoplastic which has a semicrystalline structure. (This resin is also designated as ICI VICTREX APC2.) Russell and Street [*23*] determined the interlaminar toughness of AS4/PEEK using the DCB and ENF specimens. Their values for the G_{Ic} are 1330 ± 85 J/m^2 (mean ± standard deviation). Using the ENF specimen they determined G_{IIc} to be 1765 ± 235 J/m^2. The value of G_{IIc} is approximately 1.3 times G_{Ic}.

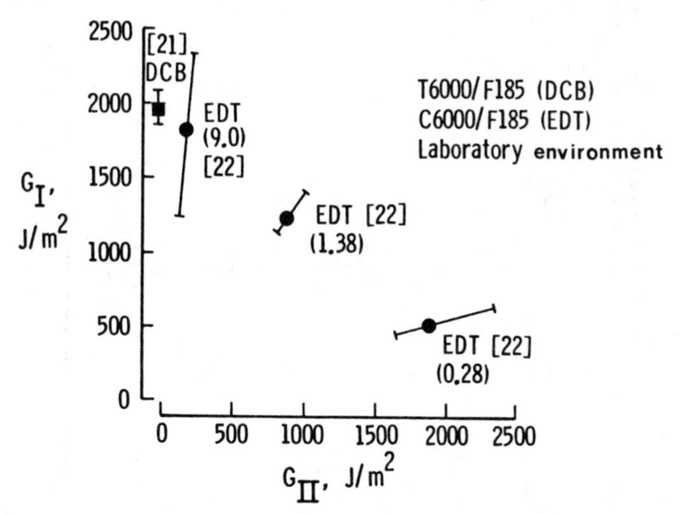

FIG. 8—*Interlaminar fracture toughness of F-185 matrix composites. The data spread is the maximum and minimum values.*

TABLE 3—*Cracked lap shear fracture toughness tests data for T6C/Hx205 (six plies to six plies).*

Specimen	Delamination Length a, mm	Critical Load P_c, kN	Strain Energy Release Rate		
			Mode I G_I, J/m^2	Mode II G_{II}, J/m^2	Total G_c, J/m^2
1	29.7	17.61	376	672	1048
	38.4	19.26	450	803	1253
	68.8	19.02	439	782	1221
	96.0	19.17	446	795	1241
	114.0	19.53	463	825	1288
2	28.7	20.37	490	873	1363
	57.2	19.75	460	821	1281
	82.3	19.79	462	826	1286
	101.1	19.99	472	841	1313
3	27.2	20.14	335	599	934
	39.1	20.46	346	617	963
	51.8	20.18	337	600	937
	70.9	20.08	333	595	928
4	26.9	19.70	462	825	1287
	35.8	20.08	480	857	1337
	53.6	20.17	485	864	1349
	114.8	19.77	466	830	1296
	144.5	19.37	447	797	1244
	166.1	19.55	455	812	1267

Data Trends

Data for a variety of matrix and adhesive systems have been presented and discussed. Mixed-mode fracture toughness for each system is shown in Fig. 9, which is in large part a plot of the mean data points from Figs. 3, 4, 6, 7, and 8. In general, the higher the G_{Ic} value, the closer G_{Ic} is to G_{IIc}. The brittle materials are much more sensitive to the G_I components than are the tougher materials. The tougher materials are almost equally sensitive to G_I and G_{II}.

Static Failure Criteria

A failure criterion based on G has been expressed by several investigators [17,24] in the form of

$$\left[\frac{G_I}{G_{Ic}}\right]^m + \left[\frac{G_{II}}{G_{IIc}}\right]^n = 1 \tag{6}$$

As can be seen from the sampling of resin toughness data shown in Fig. 9, a straight line relationship does a good job of fitting the given data for each material. This simplifies Eq 6 to

$$\frac{G_I}{G_{Ic}} + \frac{G_{II}}{G_{IIc}} = 1 \tag{7}$$

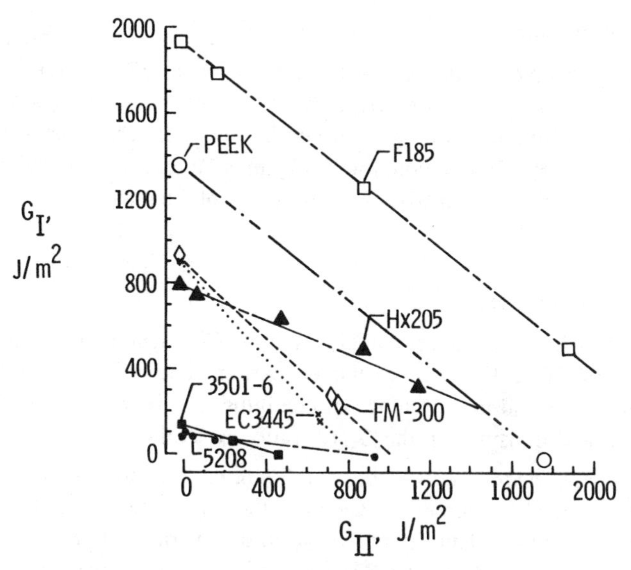

FIG. 9—*Mixed-mode fracture toughness.*

For the tougher resin materials where G_{Ic} is very nearly equal to G_{IIc}, we can assume

$$G_{Ic} = G_{IIc} = G_c \tag{8}$$

Substituting Eq 8 into Eq 7 results in

$$G_I + G_{II} = G_T = G_c \tag{9}$$

where G_c is the critical total strain energy release rate. The total strain energy release rate G_T is much simpler to calculate than individual components G_I and G_{II} [*7,10*]. Simple strength of materials approaches often give adequate estimates of G_T whereas they do poorly predicting G_I and G_{II} components. Therefore, even if G_{Ic} does not exactly equal G_{IIc}, the G_T concept for design purposes may be economically justifiable. Thus, Eq 9 is a reasonable static failure criterion for many tough resin systems while Eq 7 may be appropriate for brittle materials.

Related Data

Jordan and Bradley [*14*] have developed interlaminar toughness data on AS4/3502 and T3T145/F155 composites using symmetrically and asymmetrically loaded split laminates. They found for AS4/3502 material G_{Ic} = 190 J/m^2 and G_{IIc} = 570 J/m^2, and for the T3T145/F155 material G_{Ic} = 431 J/m^2 and G_{IIc} = 1800 J/m^2. As in other cases of somewhat low G_{Ic} values, the ratio G_{Ic}/G_{IIc} is quite low.

Effect of Environment

Further, Fig. 5 gives evidence that for a given resin (3501-6 in this case) the G_{Ic} of the system may increase while the G_{IIc} may remain the same or even decrease under different environmental conditions. Whereas a resin may be G_I sensitive under the cold/dry condition, G_{Ic} may approach G_{IIc} and make the system appear to be G_T sensitive in conditions of increased temperature and moisture.

Analytical Considerations

This section will present several analyses of the CLS specimens using the finite element program GAMNAS as explained earlier. The influence of geometric nonlinear behavior and the influence of changing matrix material modulus as a result of heat and moisture on the G_I/G_{II} ratio are evaluated.

Geometric Nonlinear Effects—Law and Wilkins [*18*] used a geometric nonlinear finite elements program to calculate the G_I/G_{II} ratios of the CLS data shown in Fig. 5. Russell and Street [*17*] used the average value of a strength of materials approach suggested by Brussat et al. [*9*] and a linear finite element program to calculate the G_I/G_{II} ratio for the CLS data shown in Fig. 4.

Figure 10 shows our results using the geometric nonlinear analysis. The results are for a three-ply to three-ply AS1/3501-6 CLS specimen. The G_I/G_{II} ratio is

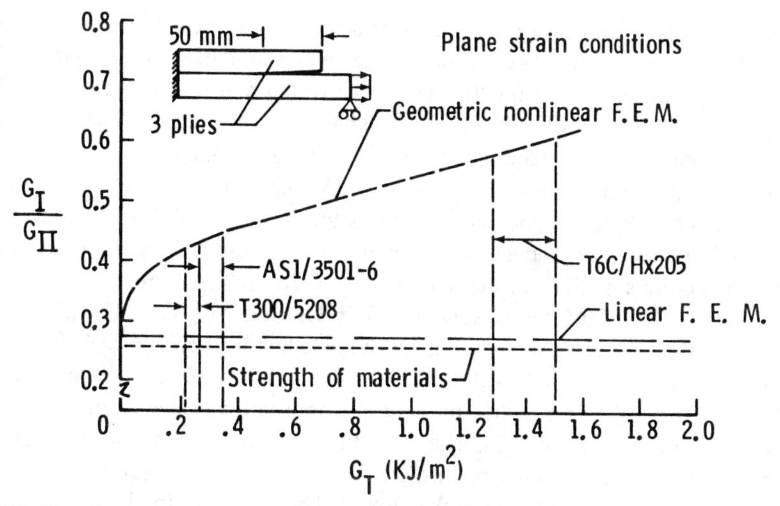

FIG. 10—*Comparison of geometric nonlinear analysis with geometric linear analysis and with the strength of materials approach for calculating* G_I/G_{II} *ratio as a function of total strain energy release rate of the delamination front.*

plotted versus the total strain energy release rate G_T. The G_I/G_{II} ratios from linear finite element analysis and the strength of materials formulation are shown for comparison. Only the geometric nonlinear finite element analysis indicated an increase in G_I/G_{II} ratio with G_T. It is clear that nonlinear analysis should be used for CLS specimens, as explained by Wilkins [*5*] and Dattaguru et al. [*11*]. The tougher matrix requires higher load for fracture, thus resulting in more nonlinear response. Therefore, the tougher the matrix, the greater the error in using a linear finite element program or the strength of materials approach. Russell and Street [*17*] used a G_I/G_{II} ratio of 0.25 for their CLS specimen; according to Fig. 10 the value is closer to 0.42, the value used by Law and Wilkins [*18*]. It should be added that the strength of materials solution in this case is close to the linear finite element analysis only because the adherends are of equal thickness. If the adherends were of different thickness there would not be such a close agreement [*11*].

It should also be clear that if the toughness is influenced by heat and moisture, the G_I/G_{II} ratio is also influenced. For example, if a hot/wet condition causes the matrix toughness to increase, then more load will be required to achieve the higher G_T. From Fig. 10 it is quite clear that the G_I/G_{II} ratio is sensitive to variations in G_T in the proximity of the 3501-6 toughness.

Matrix Modulus Effects—Augl [*25*] has shown that the shear and Young's modulus of 3501-6 can decrease almost an order of magnitude from room temperature/dry to 121°C/wet. The shear modulus decreases by 32% at 60°C/wet. We decided to investigate what effect matrix modulus changes would have on calculated G_I/G_{II} ratios. Using DCB specimens with BP907 adhesive, Chai [*26*] has determined that an adhesive joint bond thickness of 0.064 mm or less would give the same toughness as an interlaminar fracture toughness test in a composite. Therefore, we modeled a thin strip of AS1/3501-6 matrix materials using a finite element method as one would model an adhesive joint [*10*]. The matrix strip was evaluated for thicknesses of 0.025 and 0.0125 mm. Only the matrix material in the adhesive layer was changed to simulate temperature and moisture conditions; the adherend properties were not changed.

Figure 11 shows the results using the CLS-63 (that is, three plies to three plies) specimen geometry and adherend properties that Law and Wilkins [*16*] tested. The G_I/G_{II} ratios were calculated for several load levels and for three different values of Young's moduli for the matrix: 4.28 GPa (21°C/0% relative humidity [RH]); 2.39 GPa (100°C/80% RH); and 0.56 GPa (150°C/80% RH) [*25*].

The calculated results shown in Fig. 11 indicate that little change in G_I/G_{II} ratio occurs at temperatures below 100°C. The decrease of the room temperature [RT]/dry (21°C/0% RH) matrix modulus by approximately one half decreased the G_I/G_{II} ratio by at most 11%. The G_T value remains constant over this range. As the matrix modulus decreases further, the G_I/G_{II} ratio changes become more pronounced, particularly at the higher loads. The G_I/G_{II} ratio changes are primarily attributed to the geometric nonlinear effects, as previously discussed. This may be readily seen by the fact that the G_I/G_{II} ratio changed very little at the lower load levels. The authors have no explanation for the crossover point in the G_I/G_{II} behavior except that perhaps the shear modulus has become so low that shear stresses and deformations dominate the specimen's bondline.

This section showed that decreases in matrix modulus as a result of moderate heat (below 100°C) and moisture do not affect G_I/G_{II} ratio calculations to a significant degree. However, this study did not cover changing adherend properties, nor did it consider material nonlinear effects. Furthermore, 3501-6 is an

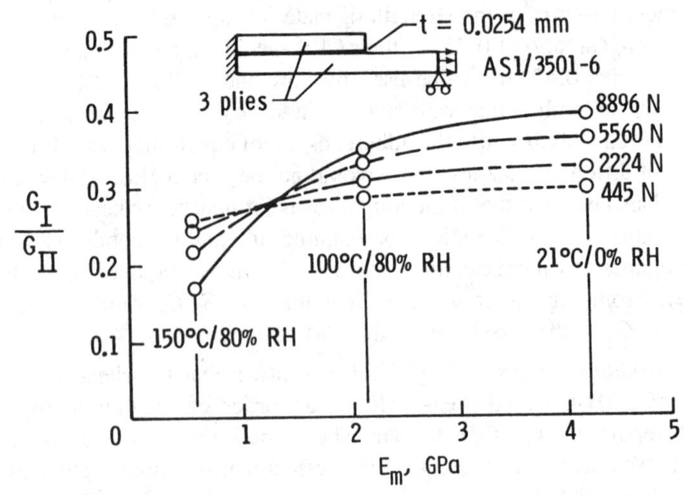

FIG. 11—G_I/G_{II} *ratios as a function of applied load and matrix modules. The adherend properties were held constant.*

unmodified, highly cross-linked epoxy system. One would expect its modulus to be less affected by heat and moisture than those of many modified or rubber toughened epoxies, such as Hx205 or FM-300. In these tougher materials the decrease in modulus may be very significant and should be accounted for in the determination of G_I/G_{II}.

Conclusion

The literature has been reviewed and new data developed for assessing the influence of resin on interlaminar fracture. Data from four specimen types have been studied: double cantilever beam for pure Mode I, the edge delamination tension test for mixed-Mode I and II, the cracked lap shear for mixed Mode I and II, and the end-notched flexure for pure Mode II. Mixed-mode fracture data for seven resins were examined: 5208, 3501-6, FM-300, EC 3445, Hx205, F-185, and PEEK. From the presented mixed-mode fracture data the following was evident:

- The mixed-mode relationship for failure appears to be linear in terms of G_I and G_{II}.
- Fracture of brittle resins (low G_{Ic}) is controlled by the G_I component.
- Fracture of many tough resins (high G_{Ic} where $G_{Ic} \approx G_{IIc}$) is controlled by the total strain energy release rate G_T. Because of the simplicity of calculation, G_T is a good failure criterion for design.

A short explanation of how the polymer structure directly relates to the mixed-mode was given. The following were hypothesized and supported by the data:

- Volume expansion (dilatation) must occur if high opening mode toughness is to be achieved.
- Unmodified epoxies have limited ability to dilatate at room temperature/dry conditions, resulting in low values of G_{Ic}.
- Plasticized epoxies and those at higher temperatures have more free volume; therefore their ability to dilatate increases, resulting in higher values of G_{Ic}.
- Modified epoxies (rubber toughening/extended chains) have increased ability to dilatate; therefore they have higher G_{Ic} values.
- Shear deformation does not require volume dilatation; therefore there is only a limited amount of G_{IIc} improvement by modification.

Several analytical aspects of the cracked lap shear specimen were explored:

- Significant errors in the calculation of G_I/G_{II} ratios could result if geometric nonlinear aspects of the problem were not addressed. The tougher the matrix, the larger the error.
- The G_I/G_{II} ratio was also found to vary with matrix modulus, which in turn varied with moisture and temperature. For 3501-6 resin, this variation was not large below 100°C.

References

[1] Irwin, G. R., "Fracture Mechanics," in *Structural Mechanics*, Goodier and Hoff, Eds., Pergamon Press, New York, 1960, p. 557.

[2] Ripling, E. J., Mostovoy, S., and Patrick, R. L., "Application of Fracture Mechanics to Adhesive Joints," in *Adhesion, ASTM STP 360*, American Society for Testing and Materials, 1964.

[3] Mostovoy, S. and Ripling, E. J., "Fracture Toughness of an Epoxy System," *Journal of Applied Polymer Science*, Vol. 10, 1966, pp. 1351–1371.

[4] Bascom, W. D., Cottington, R. L., and Timmons, C. O., "Fracture Reliability of Structural Adhesives," *Journal of Applied Polymer Science: Applied Polymer Symposium 32*, 1977, pp. 165–188.

[5] Wilkins, D. J., "A Comparison of the Delamination and Environmental Resistance of a Graphite-Epoxy and a Graphite-Bismaleimide," NAV-GD-0037, Naval Air Systems Command, Sept. 1981 (DTIC ADA-112474).

[6] Ashizawa, M., "Improving Damage Tolerance of Laminated Composites Through the Use of New Tough Resins," in *Proceedings of the 6th Conference on Fibrous Composites in Structural Design*, AMMRC MS 83-2, Army Materials and Mechanics Research Center, Watertown, MA, Nov. 1983, pp. IV-21.

[7] Mall, S. and Johnson, W. S., "Characterization of Mode I and Mixed-Mode Failure of Adhesive Bonds Between Composite Adherends," NASA TM 86355, Washington, DC, Feb. 1985 (also in *Composite Materials: Testing and Design (Seventh Conference), ASTM STP 893*, J. M. Whitney, Ed., American Society for Testing and Materials, Philadelphia, 1986, pp. 322–336.

[8] O'Brien, T. K., "Characterization of Delamination Onset and Growth in a Composite Laminate," in *Damage in Composite Materials, ASTM STP 775*, K. L. Reifsnider, Ed., American Society for Testing and Materials, Philadelphia, 1982, pp. 140–167.

[9] Brussat, T. R., Chiu, S. T., and Mostovoy, S., "Fracture Mechanics for Structural Adhesive Bonds," AFML-TR-77-163, Air Force Materials Laboratory, OH, 1977.

[10] Mall, S., Johnson, W. S., and Everett, R. A., Jr., "Cyclic Debonding of Adhesively Bonded Composites," in *Adhesive Joints*, K. L. Mittle, Ed., Plenum Press, New York, 1984, pp. 639–658.

[11] Dattaguru, B., Everett, R. A., Jr., Whitcomb, J. D., and Johnson, W. S., "Geometrically Non-Linear Analysis of Adhesively Bonded Joints," *Journal of Engineering Materials and Technology*, Vol. 106, Jan. 1984, pp. 59–65.

[12] Russell, A. J., "On the Measurement of Mode II Interlaminar Fracture Energies," DREP Materials Report 82-0, Defense Research Establishment Pacific, Victoria, BC, 1982.

[13] Rybicki, E. F. and Kanninen, M. F., "A Finite Element Calculation of Stress Intensity Factors by a Modified Crack Closure Integral," *Engineering Fracture Mechanics*, Vol. 9, No. 4, 1977, pp. 931–939.

[14] Jordan, W. M. and Bradley, W. L., "Micromechanisms of Fracture in Toughened Graphite-Epoxy Laminates," in this volume, pp. 95–114.

[15] O'Brien, T. K., "Mixed Mode Strain-Energy-Release Rate Effects on Edge Delamination of Composites," in *Effects of Defects in Composite Materials, ASTM STP 836*, American Society for Testing and Materials, Philadelphia, 1984, pp. 125–142.

[16] Murri, G. B. and O'Brien, T. K., "Interlaminar G_{IIc} Evaluation of Toughened-Resin Matrix Composites using the End-Notched Flexure Test," *Proceedings of 26th AIAA/ASME/ASCE/AHS Structural Dynamics and Materials Conference, Part I*, pp. 197–202.

[17] Russell, A. J. and Street, K. N., "Moisture and Temperature Effects on the Mixed-Mode Delamination Fracture of Unidirectional Graphite/Epoxy," in *Delamination and Debonding of Materials, ASTM STP 876*, W. S. Johnson, Ed., American Society of Testing and Materials, Philadelphia, 1985, pp. 349–370.

[18] Law, G. E. and Wilkins, D. J., "Delamination Failure Criteria for Composite Structures," NAV-GD-0053, Naval Air Systems Command, Washington, DC, 15 May 1984.

[19] Jurf, R. A. and Pipes, R. B., "Interlaminar Fracture of Composite Materials," *Journal of Composite Materials*, Vol. 16, No. 5, Sept. 1982, pp. 386–394.

[20] Hunston, D. L. and Bascom, W. D., "Effects of Lay-Up, Temperature, and Loading Rate in Double Cantilever Beam Tests of Interlaminar Crack Growth," *Composites Technology Review*, Vol. 5, No. 4, Winter 1983, pp. 118–119.

[21] Hunston, D. L., "Composite Interlaminar Fracture: Effect of Matrix Fracture Energy," *Composites Technology Review,* Vol. 6, No. 4, Winter 1984, pp. 176–180.
[22] O'Brien, T. K., Johnston, N. J., Morris, D. H., and Simonds, R. A., "Determination of Interlaminar Fracture Toughness and Fracture Mode Dependence of Composites using the Edge Delamination Test," in *Proceedings of the International Conference on Testing, Evaluation, and Quality Control of Composites,* T. Feest, Ed., Butterworths, London, 1983, pp. 223–232.
[23] Russell, A. J. and Street, K. N., "The Effect of Matrix Toughness on Delamination: Static and Fatigue Fracture Under Mode II Shear Loading of Graphite Fiber Composites," in this volume, pp. 275–294.
[24] Ramkumar, R. L. and Whitcomb, J. D., "Characterization of Mode I and Mixed-Mode Delamination Growth in T300/5208 Graphite/Epoxy," in *Delamination and Debonding of Materials, ASTM STP 876,* W. S. Johnson, Ed., American Society for Testing and Materials, Philadelphia, 1985, pp. 315–335.
[25] Augl, J. M., "Moisture Effects on the Mechanical Properties of Hercules 3501-6 Epoxy Resin," NSWC TR 79-41, Naval Surface Weapons Center, Silver Springs, MD, 30 March 1979.
[26] Chai, H., "Bond Thickness Effect in Adhesive Joints and Its Significance for Mode I Interlaminar Fracture of Composites" in *Composite Materials: Testing and Design (Seventh Conference), ASTM STP 893,* J. M. Whitney, Ed., American Society for Testing and Materials, Philadelphia, 1986, pp. 209–231.

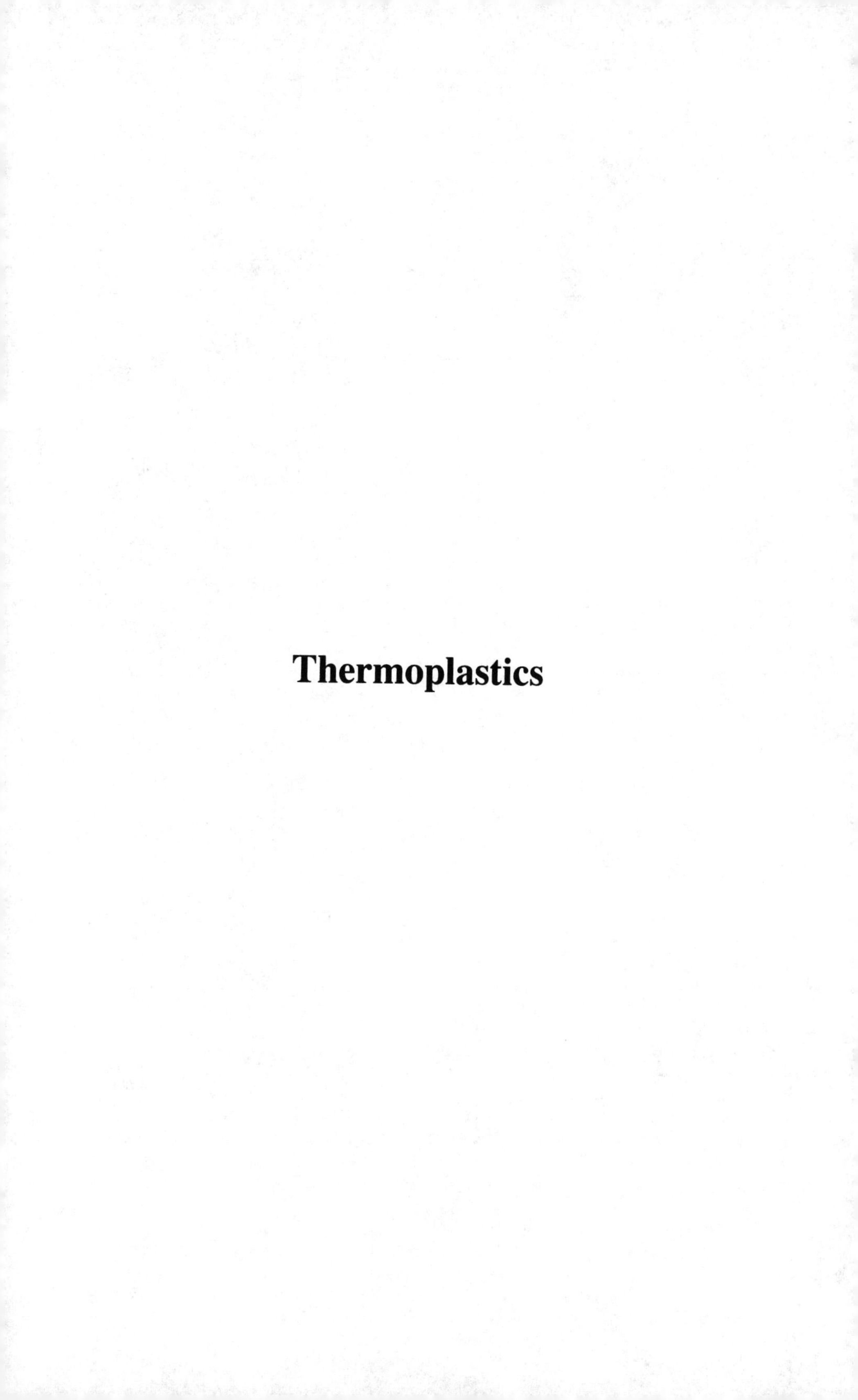

Thermoplastics

William H. Beever,[1] *Charles L. Ryan,*[2] *James E. O'Connor,*[3] *and Alex Y. Lou*[4]

Ryton®-PPS Carbon Fiber Reinforced Composites: The How, When, and Why of Molding

REFERENCE: Beever, W. H., Ryan, C. L., O'Connor, J. E., and Lou, A. Y., **"Ryton®-PPS Carbon Fiber Reinforced Composites: The How, When, and Why of Molding,"** *Toughened Composites, ASTM STP 937,* Norman J. Johnston, Ed., American Society for Testing and Materials, Philadelphia, 1987, pp. 319-327.

ABSTRACT: Ryton®-PPS carbon fiber reinforced composites have excellent temperature stability, chemical resistance, and toughness. As a result, the use of polyphenylene sulfide (PPS) carbon fiber composites is being evaluated throughout the aerospace and related industries. Critical molding parameters such as temperature, pressure, and cooling rate have been investigated both as a fundamental study concerning the polymer matrix and to provide assistance to molders. Initial results indicate that mechanical properties are influenced by the crystallinity of the molded composites. The property variations correlate with the crystallinity data obtained from differential scanning calorimetry (DSC) on the polymer matrix and composites.

KEY WORDS: Ryton®-PPS, composites, properties, crystallinity, molding

Polyphenylene sulfide (PPS) is an engineering polymer with outstanding chemical and thermal resistance, and it is produced by Phillips Petroleum Company employing a proprietary process. Ryton®-PPS is used in many demanding applications where it is combined with short chopped glass and used as an injection molding compound. Recently, PPS has been combined with continuous carbon fiber to afford a thermoplastic prepreg which is being evaluated for use particularly in the aerospace industry.

[1]Senior research chemist, Advanced Composites Division, Phillips 66 Co., 81F, PRC, Bartlesville, OK 74004.

[2]Technical supervisor, Fundamental Polymer Group, Alcoa, Alcoa Labs, Alcoa Center, PA 15069.

[3]Supervisor, Advanced Composites Division, Phillips 66 Co., 105 ARB, PRC, Bartlesville, OK 74004.

[4]Supervisor, Materials Engineering Section, Phillips Petroleum Co., 113, 71C, PRC, Bartlesville, OK 74004.

PPS is a semicrystalline thermoplastic polymer, and, as such, PPS and compounds (composites) produced from it are affected by the degree of crystallinity of the final product [*1–5*]. The extent and type of crystallinity depends primarily on the molding conditions. To assist the molders of PPS-carbon fiber prepreg, this study was designed to determine the optimum molding conditions.

Procedures

Molding

The PPS/carbon fiber prepreg was made by a proprietary process. Hercules AS-4 carbon fibers containing 12 000 filaments were used to make a prepreg tape 4 in. (10 cm) wide and 0.008 in. (0.02 cm) thick. The prepreg contained 68 ± 2% by weight carbon fiber.

Composite laminates were fabricated by first tacking the 4-in. (10-cm) prepreg together with a hot soldering iron and cutting the prepreg to the desired ply size. The prepreg plies were stacked and the corners tacked to give a prepreg laminate. This was then placed in a picture frame mold somewhat smaller than the final thickness of the laminate. The mold and prepreg were covered on top and bottom with nonporous release cloth, placed between two sandblasted stainless steel plates, and put into a preheated platen press at 316°C (600°F). Contact pressure was applied for 4 min followed by 0.7- to 1.0-Mpa (100- to 150-psi) pressure for an additional 3 min. The hot laminate was then transferred directly to a room temperature press and cooled under 0.7- to 1.0-Mpa (100- to 150-psi) pressure to ≤38°C (100°F). This usually took about 1 min. The molded plaques were trimmed and tested in unannealed (amorphous, as-molded) and annealed (crystalline) forms. Annealing was accomplished by subjecting the laminates to a temperature of 200°C (392°F) for a period of 2 h.

Testing

Specimens for testing were cut from the laminates with a water-cooled diamond blade cutoff saw.

Tensile properties were measured according to ASTM Test for Tensile Properties of Fiber-Resin Composites (D 3039). Longitudinal tension specimens were 10 by 0.75 by 0.025 in. (25.4 by 2 by 0.06 cm) and transverse tension specimens were 10 by 1 by 0.125 in. (25.4 by 2.54 by 0.3 cm). Glass reinforced epoxy end tabs were bonded to the grip area of the samples and a strain gage was used to determine the strain. Tensile properties of the PPS sheet were determined by a standard ASTM Test for Tensile Properties of Thin Plastic Sheeting (D 882) procedure using 10-mil (0.254-mm) thick sheet.

Flexural properties were measured by the three-point bending method according to ASTM Tests for Flexural Properties of Unreinforced and Reinforced Plastics and Electrical Insulating Materials (D 790). Specimens were 5 by 0.5 by 0.125 in. (12.7 by 1.3 by 0.3 cm), and a span to depth ratio of 32 to 1 was used.

Interlaminar shear strengths were measured according to ASTM Test for Apparent Interlaminar Shear Strength of Parallel Fiber Composites by Short Beam Method (D 2344). Specimens were 1 by 0.25 by 0.125 in. (2.54 by 0.6 by 0.3 cm) and a span to depth ratio of 4 to 1 was used.

Compressive properties were measured by using the Illinois Institute of Technology Research Institute (IITRI) compressive fixture on an Instron 1125 loading frame following ASTM Test for Compressive Properties of Unidirectional or Crossply Fiber-Resin Composites (D 3410). Specimens were 6 by 1 by 0.125 in. (15 by 2.54 by 0.3 cm) with a gage length of 0.5 in. (1.3 cm). End tabs were applied in these tests to prevent stress concentration in the grip areas.

Fracture toughness (G_{Ic}) values were determined from unidirectional laminates by the double cantilever beam test. Specimens were 10 by 1 by 0.125 in. (25.4 by 2.54 by 0.3 cm) with a 1.5-in. (3.8-cm) starter crack between the central laminae. The G_{Ic} values are an average of several crack growths in each of four specimens prepared from the same laminae.

Thermal analysis was performed using a Perkin-Elmer DSC-2 with the thermal analysis data station (TADS) system. Sample sizes of about 10 mg were heated from 50 to 320°C at a heating rate of 20°C/min. The glass transition temperature T_g was taken as one half the step change at T_g, whereas the crystallization and melting temperatures were obtained from the minimum and maximum of the peaks, respectively.

Results and Discussion

The physical and mechanical properties of a semicrystalline polymer depend on the morphological structure [6]. For a given polymer, lower levels of crystallinity will produce higher elongation and better toughness. Stiffness, thermal stability, and chemical resistance are enhanced by higher levels of crystallinity, but generally at the expense of decreased ductility. Other variables that are important in semicrystalline polymers besides the percent crystallinity are: (a) number and size of spherulites, (b) the crystalline structure, and (c) the crystalline orientation. Orientation can be responsible for anisotropy in the mechanical properties. Crystallite type and size may have an effect on the overall mechanical properties. Larger spherulites are inherently stiffer but less ductile.

Each of these morphological features can be controlled during the molding process. In PPS, a largely amorphous system can be achieved by quick quench from the melt. Smaller crystallites can then be formed by annealing above the T_g (95°C). Larger crystallites are formed by relatively slow cooling the sample from the molten state. Table 1 shows the effects of molding conditions on the properties of 20-mil (0.51-mm) PPS sheet.

Larger crystallites are less ductile and fail at smaller deformation. Hence, tensile strengths and elongation are lower. The amorphous material yields and elongates an order of magnitude more than the annealed material.

Similar to PPS film, the morphological structure of PPS composites is important to the mechanical properties of the materials. A composite with an

TABLE 1—*Effect of thermal history on mechanical properties of unoriented PPS film.*

Molding Condition	Quick Quenched	Quick Quenched	Slow Cooled
Annealed at 200°C	no	yes	no
Density, g/cm^3	1.309	1.346	1.351
% crystallinity[a]	0	32	37
Tensile modulus, MPa	1926	2574	2709
Tensile break, MPa	44.5	80.7	51.3
Elongation at break, %	20.0	4.8	3.4
Tensile yield, MPa	63.6	...	...
Elongation at yield, %	5.0	...	...

[a] From density measurements [7]: ρ_a = 1.31 g/cm^3 = density of 100% amorphous PPS.
ρ_c = 1.43 g/cm^3 = theoretical density of 100% crystalline PPS (from unit cell parameters).

using

$$W_c = \frac{\rho_c}{\rho} \frac{\rho - \rho_a}{\rho_c - \rho_a} = \text{weight fraction crystallinity.}$$

amorphous matrix will be tougher, but at the expense of thermal and chemical resistance. The crystallinity of the matrix can be established from thermal analysis (differential scanning calorimetry [DSC]). A semicrystalline polymer that has been quenched into the amorphous state and can also readily crystallize will have a large step-change thermal transition at the glass transition temperature (T_g), followed by a crystallization exotherm, and finally melting. Figure 1 is a DSC trace of the thermal transitions of a PPS/carbon fiber composite molded in a picture frame mold. The large transition at T_g followed by the crystallization exotherm at 120°C indicates that the matrix is largely amorphous. Following oven annealing for 2 h at 200°C, the matrix is fully crystallized (~35% crystallinity). The DSC trace of the annealed sample shows only a small change at T_g and the absence of any crystallization exotherm (Fig. 2).

In addition to the normal melting peak at about 275°C (548K), there is a peak at about 220°C (490K). This peak is produced when the material is crystallized and held isothermally over a relatively long period (1 + hours). It is often referred to as an annealing peak because the position depends on the annealing temperature. The nature of this peak is not well understood, but it has been seen in other materials such as polyetheretherketone (PEEK) (Fig. 3).

The effects of the two different morphologies on unidirectional carbon fiber composite properties are shown in Table 2. The composite with the amorphous (quenched and unannealed) morphology is tougher as evidenced by the higher G_{Ic} values. In general, though, the differences are not very large except for compressive properties. The softer amorphous matrix allows the fibers to buckle under compression thus giving low compressive strength values. After annealing, the matrix is much stiffer and the compressive strength is almost twice as high. Longitudinal tensile strength and flexural strengths are also somewhat higher for the annealed sample. The value of 0.8 kJ/m^2 (4.4 in. · lb/in.) for G_{Ic} is indicative of the excellent toughness of thermoplastic composites in general.

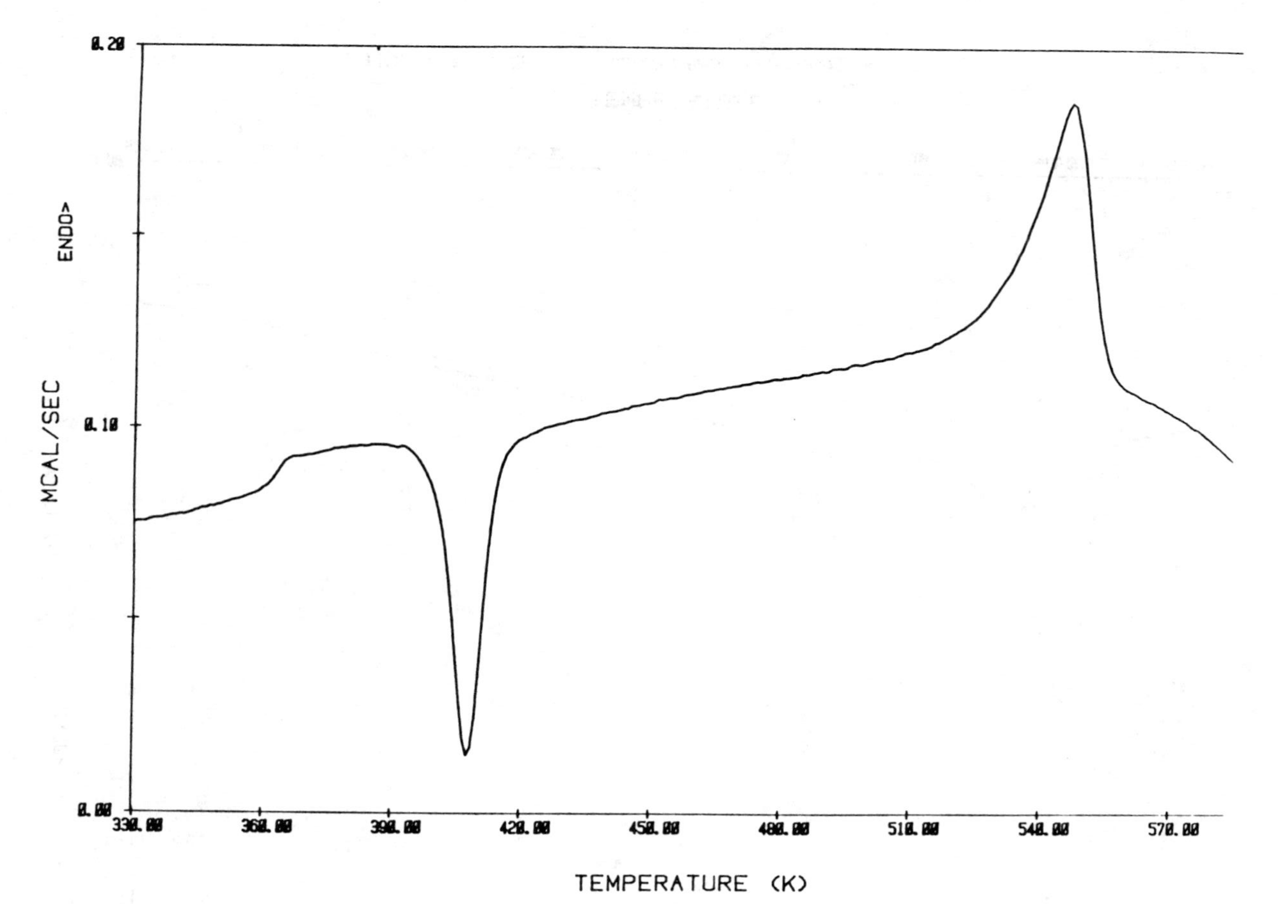

FIG. 1—*DSC analysis of amorphous PPS-carbon fiber composite.*

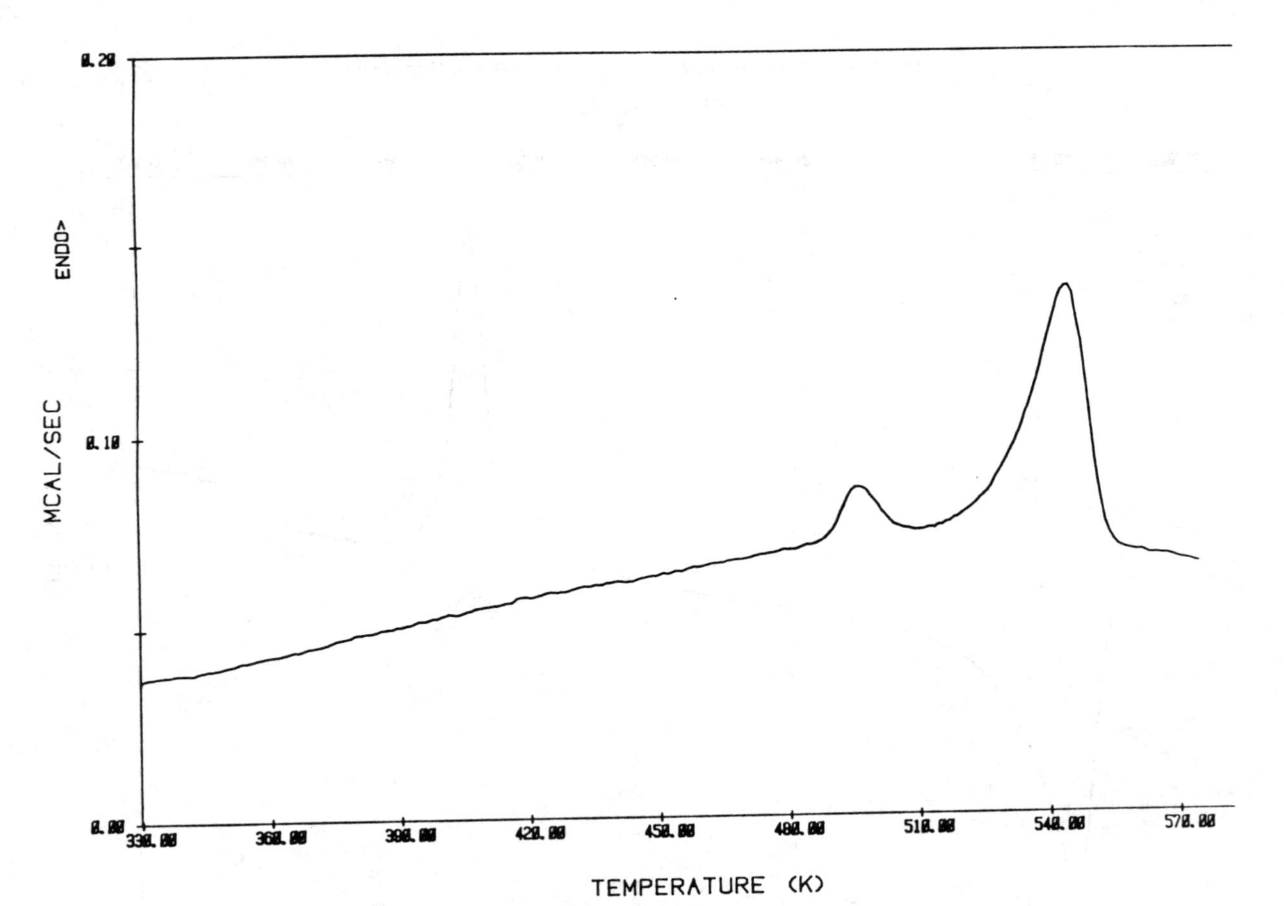

FIG. 2—*DSC analysis of annealed PPS-carbon fiber composite.*

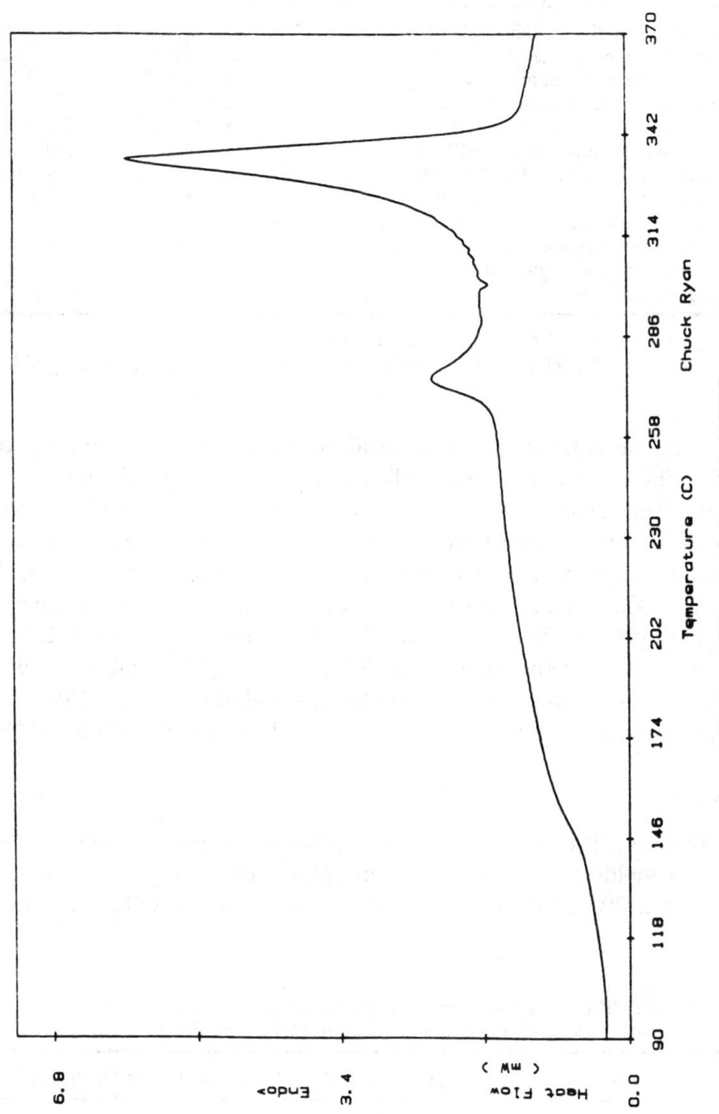

FIG. 3—*DSC analysis of annealed PEEK.*

TABLE 2—*Effect of annealing on mechanical properties of Ryton-PPS/carbon fiber[a] unidirectional laminates.*

Property	Morphology	
	Unannealed	Annealed
Longitudinal tensile modulus, GPa (Msi)	131 (19.0)	135 (19.6)
Longitudinal tensile strength, MPa (ksi)	1490 (216)	1641 (238)
Transverse tensile modulus, GPa (Msi)	9.0 (1.3)	9.0 (1.3)
Transverse tensile strength, MPa (ksi)	36.6 (5.3)	31.7 (4.6)
Longitudinal flexural modulus, GPa (Msi)	118 (17.1)	121 (17.6)
Longitudinal flexural strength, MPa (ksi)	1083 (157)	1290 (187)
Transverse flexural modulus, GPa (Msi)	7.6 (1.1)	9.0 (1.3)
Transverse flexural strength, MPa (ksi)	56.6 (8.2)	53.1 (7.7)
Longitudinal compressive strength, MPa (ksi)	338 (49)	559 (81)
Transverse compressive strength, MPa (ksi)	103 (15)	124 (18)
Short beam shear strength, MPa (ksi)	69 (10)	...
G_{Ic}, kJ/m^2 (in. · lb/in.^{2})	0.8 (4.4)	0.6 (3.4)[b]

[a] Prepreg contains 68 ± 2% by weight carbon fiber.
[b] Values as high as 1.3 kJ/m^2 (7.8 in. · lb/in.^{2}) have been obtained by film stacking [*8*].

Processing variables such as total residence time and maximum temperature reached during molding were also studied. Since these changes should only effect the matrix, their effect on the composite properties should easily be observed by studying a matrix-dominated property such as transverse tensile strength. The results are listed in Table 3. All samples were tested unannealed (quenched and amorphous). Varying the total residence time of the composite in the picture frame mold from 7 to 20 min had no effect on composite properties. Likewise, varying the molding temperature from 291 to 360°C (555 to 680°F) showed no effect. These results show the excellent thermal stability of Ryton-PPS. In general, typical molding temperatures range from 316 to 343°C (600 to 650°F).

Conclusions

Optimum properties for unidirectional PPS/carbon fiber composites are obtained by a molding sequence which includes a quenching step followed by annealing at 204°C (400°F) for 2 h. This gives small crystallites for the PPS

TABLE 3—*Effect of processing variables on transverse tensile properties of Ryton-PPS/carbon fiber composites.[a]*

Total Residence Time, min	Final Temperature, °C (°F)	Transverse Tensile Strength, kPa (ksi)
7	360 (680)	31.0 (4.5)
7	310 (590)	34.5 (5.0)
7	291 (555)	29.7 (4.3)
20	310 (590)	34.5 (5.0)

[a] All samples were unannealed.

morphology as evidenced when PPS film is processed in such a manner. This molding can be carried out in picture frame type molds using a platen press. Because of the excellent thermal stability of PPS, the prepreg can be subjected to an extended mold dwell time and temperatures as high as 360°C (680°F) without a significant loss of composite properties.

Acknowledgments

The authors wish to thank Messrs. C. M. Hall, F. J. Burwell, J. R. Bohannan, and D. L. Straw of the Phillips Research Center for their valuable assistance in materials processing, specimen preparation, and property testing.

References

[1] Brady, D. G., *Journal of Applied Polymer Science,* Vol. 20, 1976, pp. 2541–2551.
[2] Hanmin, Z. and Guoren, H., *Scientia Sinica (B),* Vol. 23, No. 4, 1984, p. 333.
[3] Jog, J. P. and Nadkarni, V. M., *Journal of Applied Polymer Science,* Vol. 30, No. 3, 1985, pp. 997–1009.
[4] Lovinger, A. J., et al., *Polymer,* Vol. 26, No. 11, 1985, pp. 1595–1604.
[5] Murthy, N. S., et al., *Synthetic Metals,* Vol. 9, No. 1, 1984, pp. 91–96.
[6] Billmeyer, F. W., *Textbook of Polymer Science,* 2nd ed., Wiley, New York, 1971.
[7] Tabor, B. J., et al., *European Polymer Journal,* Vol. 7, No. 8, 1971, pp. 1127–1133.
[8] Ma, C. C., O'Connor, J. E., and Lou, A. Y., *National SAMPE Symposium Proceedings,* Vol. 29, 1984, pp. 753–764; also in *SAMPE Quarterly,* Vol. 15, No. 4, 1984, pp. 12–17.

John A. Nairn[1] *and Paul Zoller*[2]

The Development of Residual Thermal Stresses in Amorphous and Semicrystalline Thermoplastic Matrix Composites

REFERENCE: Nairn, J. A. and Zoller P., **"The Development of Residual Thermal Stresses in Amorphous and Semicrystalline Thermoplastic Matrix Composites,"** *Toughened Composites, ASTM STP 937,* Norman J. Johnston, Ed., American Society for Testing and Materials, Philadelphia, 1987, pp. 328–341.

ABSTRACT: The difference between the thermal expansion properties of thermoplastics and high modulus fibers such as graphite and Kevlar® aramid is large. A significant level of residual thermal stresses is expected in thermoplastic matrix composites and these residual stresses will probably play a role in their properties. We first discuss the thermal expansion properties of the two classes of thermoplastics — amorphous thermoplastics and semicrystalline thermoplastics. The relevant thermal expansion data for prediction of the magnitude of the residual stresses in composites is the zero (atmospheric)-pressure thermal expansion data; these data are given for polysulfone and poly(ethylene terephthalate). By following the curvature of cross-ply composite strips, we investigated the build up of residual thermal stresses in thermoplastic matrix composites. In polysulfone composites (that is, amorphous thermoplastic matrix composites), the residual stresses begin to build up at the glass transition temperature. The cross-ply composite strip curvature can be accurately calculated from component properties and the bimetallic strip equation. In poly(ethylene terephthalate) composites (that is, semicrystalline thermoplastic matrix composites), the residual stresses begin to build up at the beginning of crystallization. A calculation of the cross-ply composite strip curvature using the bimetallic strip equation agrees well with the data except for deviations at low temperatures. The low temperature deviations occur when the differential shrinkage is sufficient to cause matrix cracking.

KEY WORDS: residual stresses, thermal stresses, amorphous thermoplastics, semicrystalline thermoplastics, composites, thermoplastic matrices

[1]Formerly, research chemist, Central Research & Development Department, E. I. du Pont de Nemours & Company, Du Pont Experimental Station, Wilmington, DE; now, professor, Materials Science and Engineering, University of Utah, Salt Lake City, UT 84112.

[2]Formerly, Research chemist, Central Research & Development Department, E. I. du Pont de Nemours & Company, Du Pont Experimental Station, Wilmington, DE; now, professor, Mechanical Engineering, University of Colorado, Campus Box 427, Boulder, CO 80309-0427.

The thermal expansion properties of typical matrices and high modulus fibers, such as graphite and Kevlar® aramid, are very different. During composite processing, which always involves some high-temperature step, a considerable amount of differential shrinkage can be expected as the part cools to room temperature. This differential shrinkage can lead to a significant level of residual thermal stresses. The problem of residual thermal stresses is of particular importance when working with thermoplastic matrix composites, because, in general, the differential shrinkage is higher than when working with thermoset epoxy composites [*1*].

The magnitude of the residual thermal stresses in the matrix of thermoplastic and thermoset epoxy composites has been measured by the technique of photoelasticity [*1*]. The use of photoelasticity, however, requires a low volume fraction composite with a transparent matrix that exhibits the photoelastic effect. To extend the study of residual thermal stresses, we have adopted a new technique that is applicable with any matrix. The technique is to construct unbalanced, two-ply, [0/90] composites. Strips of these cross-ply composites will develop curvature that is related to the amount of residual thermal stresses. The use of cross-ply composite strips has already been applied to thermoset polyester and epoxy composites by Bailey and co-workers [*2–4*] to study generation of thermal strains and transverse ply cracking.

In this paper, we apply the technique to study the buildup of residual thermal stresses in amorphous and semicrystalline thermoplastic matrix composites. We will begin by presenting volume-versus-temperature data for representative examples of the two classes of thermoplastics—an amorphous thermoplastic, polysulfone (PSF) and a semicrystalline thermoplastic, poly(ethylene terephthalate) (PET). We next discuss the temperature dependence of the curvature of cross-ply composite strips cut from graphite composites using these two matrices. Finally, a zero-parameter, theoretical analysis using a modified bimetallic strip theory and the component properties of the matrices and the fibers is found to be in good agreement with the data.

Materials and Methods

Thermal Analysis of Thermoplastics

The pressure dilatometer that was used for the thermal expansion experiments was described elsewhere [*5,6*]. In principle, the volume change of a sample together with a known amount of confining fluid (generally mercury) is determined as a function of temperature (to above 400°C) and pressure (to 200 MPa). The volume change of a sample is determined by subtracting out the known volume change of the confining fluid.

The differential scanning calorimetry and the dynamic mechanical analysis were both done on a DuPont 1090 Thermal Analyzer.

Cross-Ply Composite Strip Experiments

The PSF/graphite cross-ply composite strips were made by cutting zero and 90° strips from unidirectional tape, stacking them, and heating in a vacuum bag at 300°C for 5 min. No additional pressure, besides the vacuum pressure, was applied. The resulting specimens were 15 mm wide, 125 mm long, and about 0.6 mm thick. The unidirectional tape was made by covering a steel plate with 0.076-mm Union Carbide P1700 PSF film obtained from Westlake Plastics. The steel plate was wrapped with Union Carbide T300 graphite yarn and covered with another layer of PSF film. The resulting plate was pressed at 300°C and 2 MPa for 1 h.

The PET/graphite cross-ply composite strips were made by filament winding around a steel plate. The steel plate was covered with a 0.11-mm film of Goodyear Cleartuf® PET-type 1002A, and Union Carbide T300 graphite fibers were wrapped on top of the film. A second sheet of PET film covered the first fiber wrap and a second fiber wrap was put on perpendicular to the first fiber wrap. A third sheet of PET film covered the second fiber wrap. The whole plate was dried in a vacuum oven overnight at 130°C. It was then pressed in a vacuum bag at 285°C and about 0.5 MPa for 20 min. The plate was allowed to cool by removing from the press and letting it sit at room temperature. Cutting the resulting composites off the plate resulted in two 200 by 200-mm [0/90] cross-ply composites. The two plaques were cut into cross-ply composite strips about 175 mm long. The curvature of the strips was independent of the strip width over the range we tried—10 to 35 mm. The strips used for the experiments described here were 15 mm wide and about 0.6 mm thick.

The radius of curvature r was determined by measuring the height δ and the chord length l of the arc formed by the composite strip (see Fig. 1) and using the formula

$$r = \frac{(l/2)^2 + \delta^2}{2\delta}$$

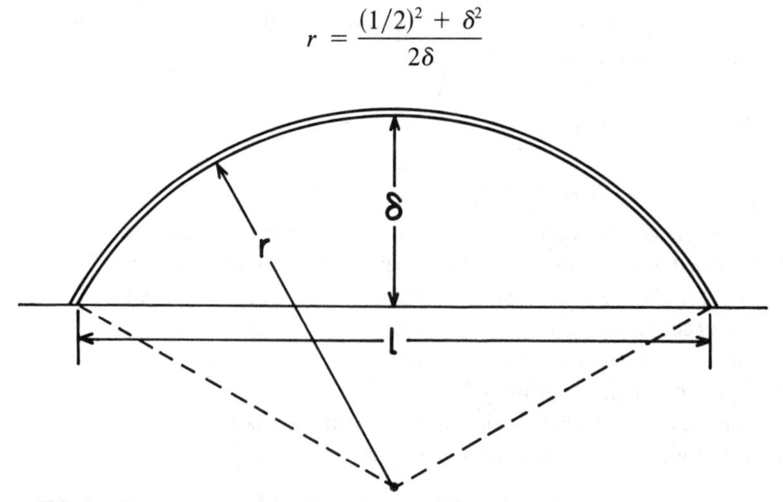

FIG. 1—*Parameters used for determination of the radius of curvature of an arc.*

The height was measured with a traveling microscope and the length was measured with a steel ruler.

Results and Discussion

Pressure-Volume-Temperature Properties of the Matrices

It is well known that the density (or the specific volume) of a polymer depends on the formation conditions (rate of cooling, pressure during cooling, and so forth). For glass-forming polymers, the maximum density differences observed as a result of extremes of formation conditions amount at most to a few tenths of a percent. By contrast, the specific volume of crystallizable polymers can be changed by several percent, most easily by changing the cooling rate from the melt. The differences in density are caused by differences in the level of crystallinity—in fact, it is common practice to determine the level of crystallinity from density measurements.

Figure 2 gives a plot of the zero (atmospheric)-pressure volume of PET [*7*] and polysulfone [*8*] as a function of temperature. In view of the above discussion on the dependence of the specific volume of a solid polymer on formation history, the solid state portion of these curves applies only to a specific sample, the formation history of which is given in the references cited. In the case of PET, the curve applies to highly crystalline polymer, similar to what one might expect to obtain during relatively slow processing, such as molding and laminating of composites. The melt state of a thermoplastic is a true equilibrium state. Its volume is independent of all previous history.

Of course, the volume of a polymer depends also on pressure [*7,8*], but the pressures involved in composites processing (rarely more than 1.5 MPa [200 psi]) are so small that the data taken at zero (atmospheric) pressure is relevant. All fits given below are for zero pressure.

The glassy polymer polysulfone shows a volume-temperature curve typical of all glassy polymers, with a break at the glass transition temperature T_g (about 190°C). The following fits represent the data very accurately:

$$T < 190°\text{C}: \quad V_0 = 0.8051 + 1.756 \times 10^{-4}T \tag{1a}$$

$$T > 190°\text{C}: \quad V_0 = 0.7644 + 3.419 \times 10^{-4}T + 3.126 \times 10^{-7}T^2 \tag{1b}$$

The solid state portion of the PET data can be described by the following three linear fits up to a temperature of 240°C, that is, up to the beginning of melting, which is indicated by the very steep volume change starting at that point.

$$30 < T < 70°\text{C}: \quad V_0 = 0.7083 + 7.60 \times 10^{-5}T \tag{2a}$$

$$70 < T < 160°\text{C}: \quad V_0 = 0.6999 + 1.95 \times 10^{-4}T \tag{2b}$$

$$160 < T < 240°\text{C}: \quad V_0 = 0.6787 + 3.26 \times 10^{-4}T \tag{2c}$$

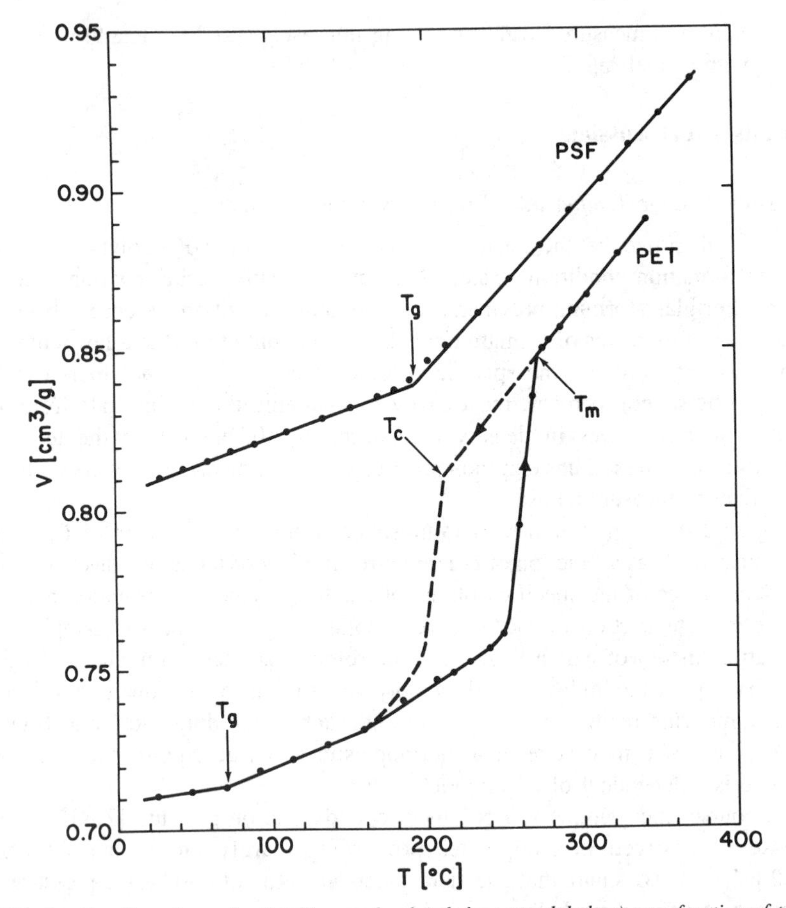

FIG. 2—*Specific volume of polysulfone and poly(ethylene terephthalate) as a function of temperature at zero (atmospheric) pressure.* T_g *is glass transition temperature,* T_m *is the melting temperature, and* T_c *is the crystallization temperature on slow cooling from the melt. The portion of the curve which includes* T_c *is a representative specific volume versus temperature plot during cooling.*

Finally a linear fit also describes the melt portion of the thermal expansion of PET, that is, the data above about 270°C:

$$V_0 = 0.6883 + 5.90 \times 10^{-4}T \tag{3}$$

In all these fits the temperature is in Centigrade and the specific volume in cubic centimetres/gram.

On cooling, the volume versus temperature curve of PET can be expected to follow the melt relationship (Eq 3) to a crystallization temperature T_c below the melting point (dashed line in Fig. 2). All crystallizable polymers show some of this supercooling, but the degree of supercooling can depend on a lot of factors, the most important of which are cooling rate and the presence of nucleating agents. This is discussed further below.

Cross-Ply Composite Strip Curvature

From Timoshenko's analysis of bimetallic strips [*9*], it can be seen that the dimensionless curvature, defined as the strip thickness divided by the radius of curvature (h/r), is directly related to the amount of differential thermal shrinkage which can induce stresses (see Eq 4 below). The dimensionless curvature of a PSF/graphite cross-ply composite strip as a function of temperature while cooling is given in Fig. 3. This data was obtained by heating the cross-ply composite strip until it became flat and then cooling at 1.5°C/min.

The temperature for the onset of the buildup of residual thermal stresses was found by extrapolating the data to zero dimensionless curvature. The result is 190°C. This stress-free temperature is identical to the glass transition temperature of PSF which is 190°C. We conclude that the residual thermal stresses in amorphous thermoplastic matrices begin to form at the glass transition temperature of the matrix. This conclusion is the same as the conclusion reached using the technique of photoelasticity [*1*].

The dimensionless curvature of PSF/graphite cross-ply composite strip does not show any hysteresis after one heating and cooling cycle. The sample used for the data in Fig. 3 had a dimensionless curvature of 0.0057 at 26°C before starting the experiment. After heating and cooling, the dimensionless curvature at 26°C was only slightly changed to 0.0060.

The dimensionless curvature of a PET/graphite cross-ply composite strip during heating is given in Fig. 4. There appears to be a break around 100°C. Above 100°C, the decrease in dimensionless curvature is more rapid and it eventually drops to zero at 257°C. These features correlate with features in the differential scanning calorimeter (DSC) of PET during heating which is given in Fig. 5*a*. The break around 100°C is near the glass transition temperature which is 80 to 90°C. The temperature where the dimensionless curvature drops to zero is nearly identical to the temperature for the end of melting—257 versus 260°C.

The dimensionless curvature on cooling from the melt at 1.5°C/min (see Fig. 4) begins to increase around 240°C. There is a rapid rise until about 180°C. We suspect that this rise is caused by crystallization which, as seen in the DSC during cooling in Fig. 5*b* starts at 227°C and ends at 180 to 200°C. The rapid rise in dimensionless curvature is a result of the large matrix shrinkage that occurs during crystallization. When the matrix shrinkage is high, the differential shrinkage is high and the dimensionless curvature will rapidly increase.

The large shrinkage that occurs during crystallization of PET is shown by the representative cooling curve in Fig. 2. On cooling from the melt, PET will supercool below the equilibrium melting point until crystallization at T_c results in a rapid drop in specific volume. T_c can vary widely with sample conditions. The T_c in Fig. 2 (210°C) and the T_c in Fig. 5*b* (227°C) were both obtained by DSC

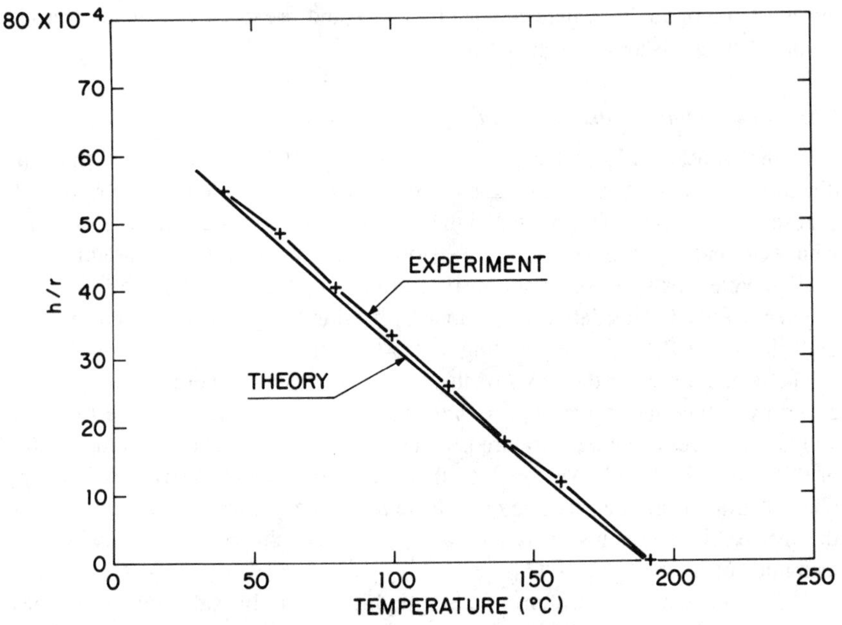

FIG. 3—*Dimensionless curvature as a function of temperature for a polysulfone/graphite cross-ply composite strip. Experimental curve obtained while cooling from 200°C at 1.5°C per minute. Theoretical curve is prediction of dimensionless curvature using the bimetallic strip equation.*

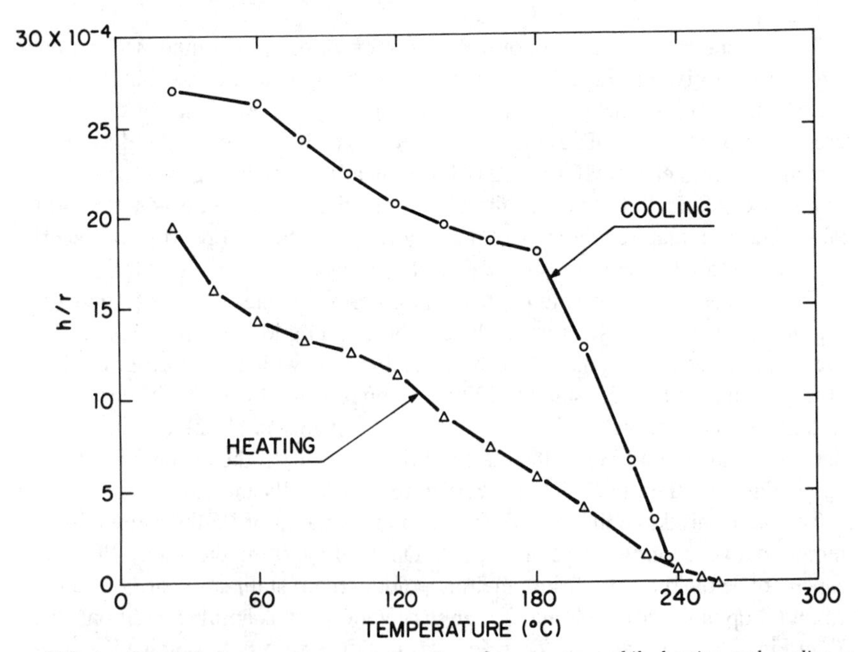

FIG. 4—*Dimensionless curvature as a function of temperature while heating and cooling a poly(ethylene terephthalate)/graphite cross-ply composite strip. Heating rate was 10°C per minute and the cooling rate was 1.5°C per minute.*

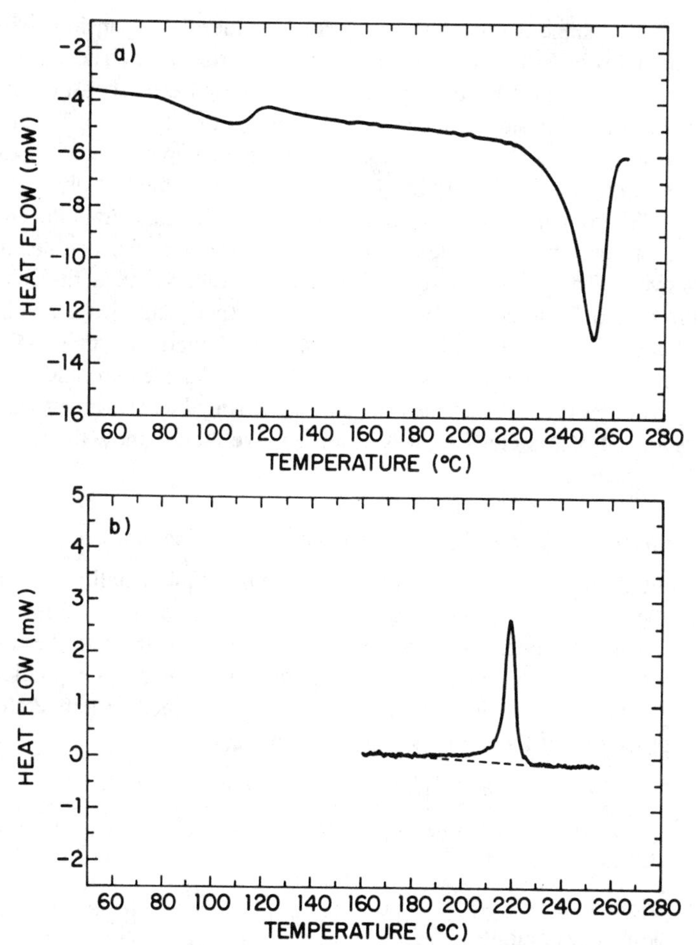

FIG. 5—*Differential scanning calorimetry of poly(ethylene terephthalate).* (a) *DSC while heating at 10°C per minute and* (b) *DSC while cooling at 1.5°C per minute.*

and are both lower than the T_c obtained in the cross-ply composite strip experiment (240°C). A likely explanation for the discrepancy is that the cross-ply composite strips were not heated very far above the melting point before cooling. The ultimate temperature may not have been high enough to eliminate all the nucleation sites; these sites could reduce the amount of supercooling observed. To support this hypothesis, we quote results of DSC experiments on PET/graphite composites. When heated to 265°C, held 5 min, and cooled slowly, the crystallization peak begins at 240°C. When heated to 300°C, held 5 min, and cooled slowly, however, the crystallization peak does not begin until 215°C (data not shown).

Below 180°C, there appears to be two more changes in the slope of the dimensionless curvature plot. The first is at 120°C, where the slope shows a slight increase. This change is related to changes in the thermal expansion coefficient and the modulus of PET as it goes through its glass transition. The second slope change is at 60°C and it is accompanied by the presence of acoustic emissions from the sample. This slope change is probably due to cracking of the matrix as a result of large residual thermal stresses. After the sample has cooled to room temperature, cracks parallel to the fibers can be seen in the 90° ply.

Unlike PSF/graphite cross-ply composite strips, the PET/graphite cross-ply composite strips show a considerable degree of hysteresis after one heating and cooling cycle. The sample used for the data in Fig. 4 had a dimensionless curvature of 0.0020 at 24°C before starting the experiment. After heating and cooling, the dimensionless curvature at 24°C was increased to 0.0027. The cooling during specimen preparation was under 2-MPa pressure and no matrix cracking was observed. In contrast, the cooling during the experiment was under no pressure and matrix cracking was observed. We do not know the reasons for these differences.

Theoretical Analysis of the Cross-Ply Composite Strip Curvature

A two-ply, cross-ply composite strip is an analog of a bimetallic strip. During cooling, the cross-ply composite strip will curve because of a mismatch of the thermal expansion properties of the two plies. The curvature is in the direction of the transverse or 90° ply. The amount of curvature can be predicted by extending Timoshenko's [*9*] bimetallic strip theory to include temperature dependent properties of the composite plies. The result is

$$\frac{h}{r} = \int_{T_0}^{T} \frac{24(\alpha_0 - \alpha_{90})}{(14 + n + 1/n)} dT \tag{4}$$

where

h = strip thickness,
r = radius of curvature,
T_0 = temperature for the onset of stress buildup,
T = temperature, and
n = E_0/E_{90}.

Here α_0 and α_{90} are linear thermal expansion coefficients, E_0 and E_{90} are tensile moduli, and 0 and 90 refer to the fiber directions in the two plies. The α_0 and α_{90} can be estimated using the Schapery equations [*10*]. These equations are exact for composites with isotropic fibers and they have been shown to be adequate for composites with anisotropic fibers [*11*]. The equations are

$$\alpha_0 = \frac{E_L V_f \alpha_L + E_m(T) V_m \alpha_m(T)}{E_L V_f + E_m V_m} \tag{5}$$

$$\alpha_{90} = (1 + \mu_m)\alpha_m(T) V_m + (1 + \mu_f)\alpha_T V_f - \alpha_0(\mu_m V_m + \mu_f V_f)$$

In these equations, the following parameters are assumed to be temperature independent:

V_f = fiber volume fraction,
V_m = matrix volume fraction,
E_L = fiber longitudinal modulus,
μ_f = fiber Poisson's ratio,
μ_m = matrix Poisson's ratio,
α_L = fiber longitudinal thermal expansion coefficient, and
α_T = fiber transverse thermal expansion coefficient.

The remaining parameters, the matrix thermal expansion coefficient $[\alpha_m(T)]$, and the matrix tensile modulus $[E_m(T)]$ are taken to be temperature dependent. $E_m(T)$ will be taken to be the storage modulus found in a dynamic mechanical analyzer (DMA) experiment and $\alpha_m(T)$ will be found by differentiating the zero (atmospheric)-pressure volume versus temperature data $V_m(T)$ for the neat matrix:

$$\alpha_m(T) = \frac{1}{3}\frac{1}{V_m(T)}\frac{dV_m(T)}{dT} \tag{6}$$

For n, we estimate E_0 and E_{90} by using the mechanics of materials approach. The results found in many texts (for example, Jones [*12*]) are:

$$E_0 = E_m(T)V_m + E_L V_f$$

$$\frac{1}{E_{90}} = \frac{V_m}{E_m(T)} + \frac{V_f}{E_T}$$

where E_T is the fiber transverse modulus (assumed temperature independent) and the rest of the parameters are listed above.

We begin with the PSF/graphite cross-ply composite strips. The temperature independent parameters listed in Eq 5 are given in Table 1. The fiber volume fraction was determined by measuring the density of the cross-ply composite strip and the graphite fiber properties come from the work of Adams et al. [*13–15*].

TABLE 1—*Temperature independent parameters of the fibers and the matrices.*

Property	PSF/Graphite Composites	PET/Graphite Composites
V_f	42%	43%
V_m	58%	57%
E_L	220 GPa	220 GPa
E_T	14 GPa	14 GPa
α_L	-0.45×10^{-6}°C^{-1}	-0.45×10^{-6}°C^{-1}
α_T	18×10^{-6}°C^{-1}	18×10^{-6}°C^{-1}
μ_f	0.20	0.20
μ_m	0.33	0.33
T_c	...	239°C
T_{end}	...	184°C

The temperature dependence of the storage modulus of neat PSF is plotted in Fig. 6. Finally, the thermal expansion coefficient of PSF is found by differentiating Eq 1 according to Eq 6. The result of a zero-parameter, modified bimetallic strip calculation using Eq 4 is given in Fig. 3. We find near perfect agreement.

We expect this type of analysis to work for any amorphous thermoplastic matrix composite. In short, a thermoelastic analysis of residual thermal stresses in amorphous thermoplastic matrix composite must include two pieces of information. First, the residual thermal stresses begin to build up at the glass transition temperature. Second, the relevant matrix properties are the zero (atmospheric)-pressure specific volume versus temperature data and the temperature dependence of the mechanical properties. It is unlikely that the range of processing conditions (for example, pressure, cooling rates, annealing treatments) available to composites processing will have any effect on the development of residual thermal stresses in amorphous thermoplastic matrix composites. This conclusion follows from the discussion above about the insensitivity of the density of glass-forming polymers to formation conditions.

We next consider the PET/graphite cross-ply composite strips. The temperature independent parameters are listed in Table 1. The fiber volume fraction was determined by measuring the density of the cross-ply composite strip. The temperature dependence of the storage modulus of semicrystalline PET is plotted in Fig. 6. The sample used for the DMA experiment was annealed overnight in

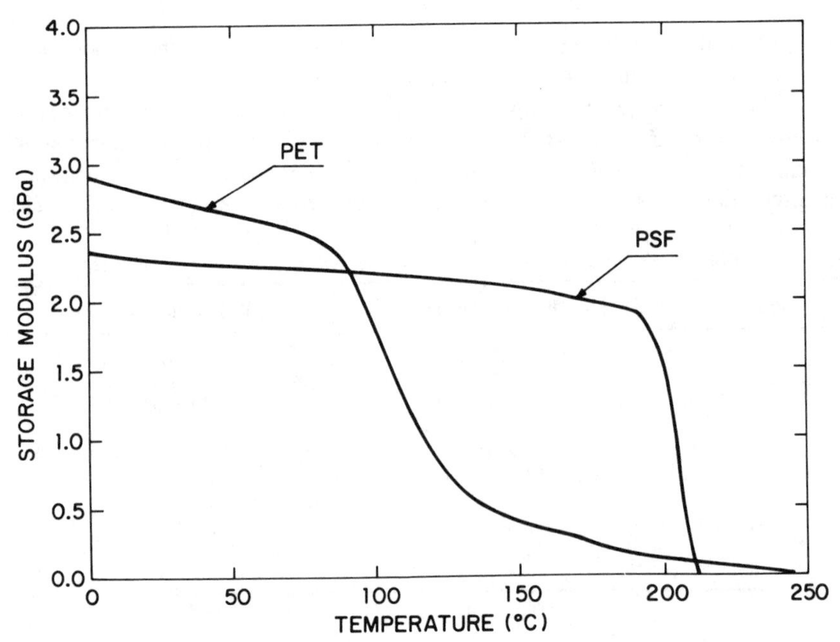

FIG. 6—*Dynamic mechanical analysis of polysulfone and of poly(ethylene terephthalate).*

a vacuum oven at 130°C and annealed an additional 15 h at 160°C. The annealing treatments were done to maximize the crystallinity. The last parameter we need is the thermal expansion coefficient of PET. It is found by differentiating the volume versus temperature data for a cooling experiment. The volume versus temperature data given in Fig. 2 and Eqs 2 to 3, however, are from a heating experiment. We extract the cooling curve from Eqs 2 to 3 by making the following three assumptions. First, we assume that the temperature where the dimensionless curvature begins to increase is equal to temperature where crystallization begins T_c, and that the specific volume at T_c can be found by extrapolating the fit to the melt of PET given in Eq 3. Second, we assume that the end of crystallization T_{end} occurs at the first break in the dimensionless curvature versus temperature plot and that the specific volume below T_{end} is given by Eq 2. Third, we assume the specific volume between T_c and T_{end} during cooling is adequately described by a straight line between the specific volume of the melt at T_c and the specific volume at T_{end}.

The result of a zero-parameter, modified bimetallic strip calculation is given in Fig. 7 along with the temperature dependence of the dimensionless curvature of four different PET/graphite cross-ply composite strips. The average values for the temperatures for the beginning and end of crystallization found from the four dimensionless curvature experiments were T_c = 239°C and T_{end} = 184°C. The qualitative agreement of the theory and experiments is good from 240 down to

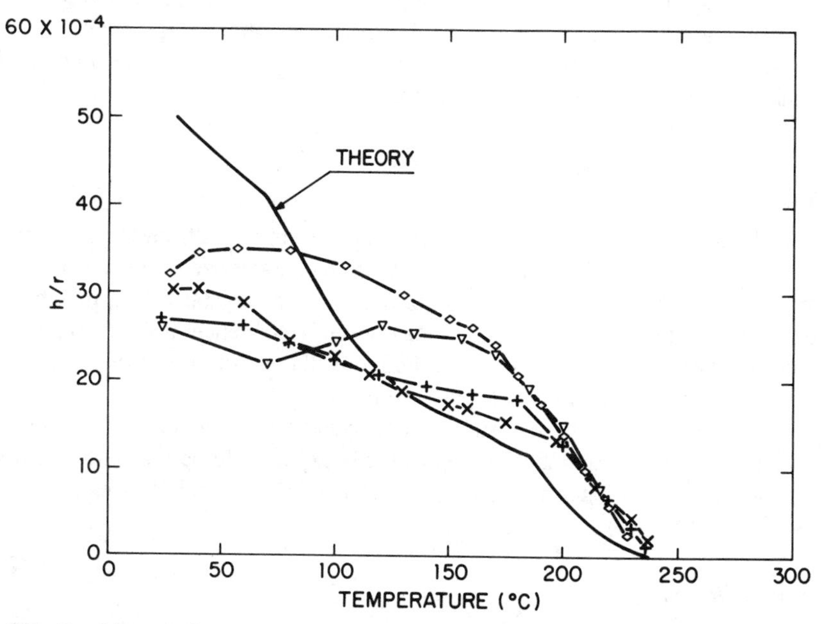

FIG. 7—*Dimensionless curvature as a function of temperature while cooling for four poly(ethylene terephthalate)/graphite cross-ply composite strips and as predicted by the bimetallic strip equation.*

100°C. Below 100°C, the theoretical curve rises and the experimental curves drop. The good agreement at high temperatures suggests that our assumptions are correct. In other words, the residual thermal stresses do begin to form at the onset of crystallization, and the rapid rise in dimensionless curvature at high temperatures is due to the large differential shrinkage that occurs during crystallization. The poor agreement at low temperatures is consistent with our detection of acoustic emissions also occurring at low temperatures. The acoustic emissions are a result of cracking in the matrix of the transverse ply and these cracks lead to a decrease in dimensionless curvature. The decrease is not predicted by the theory, because the theory assumes that no cracking occurs.

Although not as simple as with an amorphous thermoplastic matrix composite, a thermoelastic analysis of residual thermal stresses in semicrystalline thermoplastic matrix composites is possible. The two pieces of information needed are: First, the residual thermal stresses begin to form at the onset of crystallization. Second, the relevant matrix properties are the zero (atmospheric)-pressure specific volume versus temperature data while cooling at rates encountered in composites processing and the temperature dependence of the mechanical properties. Unlike the situation with amorphous thermoplastics, we do not expect these conclusions to apply to all semicrystalline thermoplastics. It is possible, for example, that the total amount of crystallinity is important. If the amount of crystallinity is low, the matrix stiffness at the start and during crystallization may be too low to support stress. The residual thermal stresses may instead not begin to form until after crystallization is complete or maybe not even begin to form until the glass transition temperature. Furthermore, processing conditions are important. Any processing conditions that changes T_c will change the amount of shrinkage during solidification, and, therefore, also the amount of residual thermal stresses.

Conclusions

The buildup of residual thermal stresses can be predicted from the properties of the matrix. The matrix, however, must be well characterized. The data needed are: the temperature for the formation of a glass or of crystals, the temperature dependence of the stiffness, and the temperature dependence of the zero (atmospheric)-pressure specific volume. Furthermore, the sensitivity of these properties to processing conditions will play a role.

The level of residual thermal stresses in composites, especially in some semicrystalline thermoplastic matrix composites, is expected to be high. More work needs to be done to determine the effects of residual thermal stresses on composite properties.

Acknowledgments

We would like to thank Howard Starkweather and Tom Johnson for running the DMA and DSC experiments. We also thank Don Brill and Sandra Schwendeman for excellent laboratory assistance.

References

[1] Nairn, J. A. and Zoller, P., *Journal of Materials Science,* Vol. 20, 1985, pp. 355–367.
[2] Bailey, J. E., Curtis, P. T., and Parvisi, A., *Proceedings of the Royal Society, Series A. Mathematical and Physical Sciences,* (London), Vol. 366, 1979, pp. 599–623.
[3] Jones, F. R., Wheatley, A. R., and Bailey, J. E., in *Composite Structures,* I. H. Marshall, Ed., Applied Science Publishers, Barking, United Kingdom, 1981, pp. 415–429.
[4] Jones, F. R., Mulheron, M., and Bailey, J. E., *Journal of Materials Science,* Vol. 18, 1983, pp. 1533–1539.
[5] Zoller, P., *Proceedings of the Eleventh North American Thermal Analysis Society,* Vol. II, 1981, pp. 37–42.
[6] Zoller, P., Bolli, P., Pahud, V., and Ackerman, H., *Review of Scientific Instruments,* Vol. 47, 1978, pp. 948–952.
[7] Zoller, P. and Bolli, P., *Journal of Macromolecular Science, Physics,* Vol. B18, 1980, pp. 555–568.
[8] Zoller, P., *Journal of Polymer Science, Part A-2. Polymer Physics,* Vol. 16, 1978, pp. 1261–1275.
[9] Timoshenko, S., *Journal of the Optical Society of America,* Vol. 11, 1925, pp. 233–255.
[10] Schapery, R. A., *Journal of Composite Materials,* Vol. 2, 1968, pp. 380–404.
[11] Nairn, J. A., *Polymer Composites,* Vol. 6, No. 5, 1985, pp. 123–130.
[12] Jones, J. J., *Mechanics of Composite Materials,* McGraw-Hill Book Co., New York, 1975, pp. 90–98.
[13] Mahishi, J. M. and Adams, D. F., *Journal of Materials Science,* Vol. 18, 1983, pp. 447–456.
[14] Grimes, G. C., Adams, D. F., and Dusablon, E. G., Northrup Corp./University of Wyoming, USA, Report NOR-80-158, 1980.
[15] Crane, D. A. and Adams, D. F., University of Wyoming, USA, Report UWME-DR-101-101-1, 1981.

Peggy Cebe,[1] *Su-Don Hong,*[2] *Shirley Chung,*[1] *and Amitava Gupta*[1]

Mechanical Properties and Morphology of Poly(etheretherketone)

REFERENCE: Cebe, P., Hong, S. -D., Chung, S., and Gupta, A., **"Mechanical Properties and Morphology of Poly(etheretherketone),"** *Toughened Composites, ASTM STP 937,* Norman J. Johnston, Ed., American Society for Testing and Materials, Philadelphia, 1987, pp. 342-357.

ABSTRACT: Mechanical properties and morphology of poly(etheretherketone) (PEEK) were studied for samples having different thermal histories. Isothermal and rate-dependent crystallization were studied to ascertain the relationship between crystallinity/morphology and processing condition. Degree of crystallinity and microstructure were controlled by cooling the melt at different rates, ranging from quenching to slowly cooling, and by annealing amorphous material above the glass transition temperature T_g. We found that degree of crystallinity was not as important as processing history in determining the room temperature mechanical properties. Samples with the same degree of crystallinity had very different tensile properties, depending on rate of cooling from the melt. All samples yielded by shear band formation and necked down. Quenched films had the largest breaking strains, drawing to 270%. Slowly cooled films exhibited ductile failure at relatively low strains. Best combined mechanical properties were obtained from semicrystalline films cooled at intermediate rates from the melt.

KEY WORDS: poly(etheretherketone) (PEEK), crystallinity, mechanical properties, morphology, thermal history

Poly(etheretherketone) (PEEK) is a crystallizable thermoplastic polymer with attractive properties for use in structural composites. PEEK has excellent high temperature properties because of its stiff backbone chain, having repeat unit (ø-0-ø-0-ø-$\overset{0}{\overset{\|}{C}}$). The glass transition T_g is 145°C, melting point is 335°C, and continuous use temperature is 200°C [*1*]. In addition to high temperature stability, PEEK is very resistant to radiation damage and chemical attack [*2*]. PEEK is commer-

[1]NASA-NRC resident research associate, member technical staff, and technical group supervisor, respectively, Applied Sciences and Microgravity Experiments Section, Jet Propulsion Laboratory, California Institute of Technology, MS 67-201, Polymer Physics, 4800 Oak Grove Dr., Pasadena, CA 91109.

[2]Consultant, 1625 Olympic Blvd., Suite 800, Los Angeles, CA 90015.

cially available from ICI Americas, Inc. [3] in neat resin form or as the matrix for carbon or glass fiber reinforced composites.

PEEK neat resin can be pressed or injection molded from the melt. High temperatures, in excess of 370°C [4], are necessary to erase any crystals remaining from the previous thermal treatments. This ensures a completely amorphous melt, and subsequent processing from the melt imparts the desired microstructure and degree of crystallinity. An amorphous material is obtained by fast quenching of the melt. Semicrystalline material results from slow cooling of the melt, isothermal crystallization, or annealing of the amorphous material [5,6].

Processing history, because of its influence on crystalline morphology, will have a very large effect on mechanical properties of the neat resin [7]. Properties such as impact resistance, yield stress, and fracture toughness are all quite sensitive to the degree of crystallinity, and to the crystal microstructure [8]. In this paper, we describe the results of our study on the crystalline morphology and room temperature mechanical properties of PEEK neat resin films processed with widely varying thermal histories [9].

Procedure

Thin films (100 to 200 μm) of PEEK neat resin were melt pressed at 400°C for 2 min at 3-MPa pressure. Starting material was pellet grade PEEK obtained from ICI, Americas, Inc. in the form of free flowing granules. Relative viscosity was 2.4 when measured at 25°C in 96% sulfuric acid, 1-g/100-mL concentration. For crystallization and morphological studies, the films were isothermally crystallized from the melt at the desired crystallization temperature. Additional films were crystallized from the rubbery amorphous state by rapidly heating the quenched films to a crystallization temperature above T_g. Rate-dependent crystallization was carried out using either a Mettler microscope hot stage or a Dupont 1090 differential scanning calorimeter (DSC), to provide heating/cooling rates from 1 to 20°C/min. Crystallinity was measured using density and wide angle X-ray scattering (WAXS) and area under the DSC endotherm. To determine crystallinity from density, the crystal and amorphous densities were assumed to be 1.378 and 1.264 g/cm^3, respectively [6]. When using the area under the DSC endotherm, 130 J/g was assumed for the heat of fusion of the perfectly crystalline material [5]. Very good agreement was obtained among the three methods. Morphology was characterized using polarizing light microscopy, DSC, wide and small angle X-ray scattering (SAXS), and scanning electron microscopy (SEM).

Mechanical properties were characterized in tension. Measurements were made at room temperature at 0.2-in./min (0.5-cm/min) constant crosshead speed (0.2/min nominal rate). Four different sample types were used in the mechanical properties study. These are characterized by their processing/thermal history which included: (1) quenching from the melt, (2) slow cooling from the melt, (3) intermediate rate of cooling from the melt, and (4) annealing the fast quenched material at 180°C for 1 h. Liquid nitrogen fracture surfaces, examined using SEM, showed no gradient in morphology between the surface and the interior for any of the preparations used in this study. Specimens of uniform cross-sectional area were used according to ASTM Test for Tensile Properties of Thin Plastic Sheeting (D 882). Polymer tension testing is reviewed in Ref *10*.

Results

Crystallinity and Morphology

Results of isothermal crystallization from the melt and from the rubbery amorphous state are shown in Fig. 1 and summarized in Table 1. Degree of crystallinity is plotted as a function of crystallization time for several temperatures. At 330°C, no crystallization was detected after 100 min. In Fig. 1, as the crystallization temperature was decreased to 310°C, rapid crystallization resulted, with ultimate degree of crystallinity of over 32%. With decreasing crystallization temperature, the ultimate degree of crystallinity decreased steadily, until at 160°C only about 16% of the material crystallizes after 65 min. Development of crystallinity could be observed by the increasing opacity of the films. (Amorphous material was always transparent.) We used the appearance of opacity as a rough indicator of the crystallization time. The time to maximum opacity is plotted in Fig. 2 for a wide range of crystallization temperatures. Crystallization from the rubbery amorphous state was used for the low temperature part of the curve, below 230°C (using an oil bath to provide rapid equilibration of the films). However, we were not able to determine the time to maximum opacity as accurately using this technique, and this is reflected in the small mismatch in Fig. 2

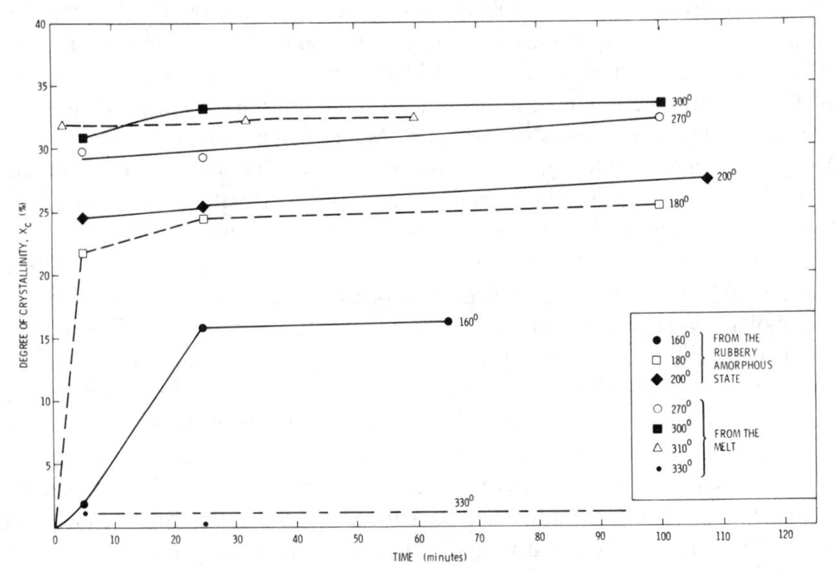

FIG. 1—*Degree of crystallinity versus crystallization time during isothermal crystallization of PEEK.*

TABLE 1—*Crystallinity and isothermal crystallization time.*

Temperature, °C	Time, min	Crystallinity, %
	CRYSTALLIZATION FROM THE MELT	
330	100	0.9
315	60	35.2
312	40	34.0
310	60	32.5
308	20	34.0
300	100	33.3
270	100	32.5
160	100	29.7
	CRYSTALLIZATION FROM THE RUBBERY STATE	
200	100	27.2
180	100	25.4
164	20	18.4
160	125	16.2

near the minimum. These results are in very good agreement with the growth rate data of Blundell and Osborn [*5*]. From Fig. 2 we see that PEEK crystallizes in under 60 s for temperatures in the range from 170 to 290°C. While the crystallization is most rapid in this range, the *extent* of crystallization is not the greatest here, as shown by the results in Fig. 1.

Easiest processing of PEEK will probably involve rate dependent crystallization in preference to isothermal crystallization. We undertook studies of the development of crystallinity as PEEK films were heated or cooled at rates ranging from 1 to 20°C/min, and these results are listed in Table 2. As the heating/cooling rate is decreased, the ultimate degree of crystallinity increases. Processing by cooling from the melt results in higher degree of crystallinity, compared to heating the rubbery state at a comparable rate. By controlling the rate of temperature decrease from the melt, the degree of crystallinity can be varied from 0% by quenching, up to 36% by slow cooling at −1°C/min. Melt processing is a more desirable way of imparting crystallinity than processing from the rubbery state since it avoids the additional step of quenching and reheating. However, the degree of crystallinity will be quite high for any slow cooling treatment.

To study the effect of cooling rate, we prepared four different samples representing four thermal histories. Crystallinity and preparation method are summarized in Table 3 for Q, quenched; SC, slow cooled from the melt; AC, air cooled from the melt; and C180, crystallized from the rubbery state at 180°C for 1 h. Film Q was amorphous, while crystallinity varied from 23.7% in C180, to about 33% for both AC and SC. From Table 3 we see that the rate of cooling makes only a slight difference in degree of crystallinity for the SC and AC samples. However, these samples had dramatically different mechanical properties. The rate of cooling imparted different crystalline morphology to these samples as we show below.

Morphology was examined using WAXS and SAXS, and these results are shown in Figs. 3 and 4. From the WAXS scans (Fig. 3) we see that AC and SC

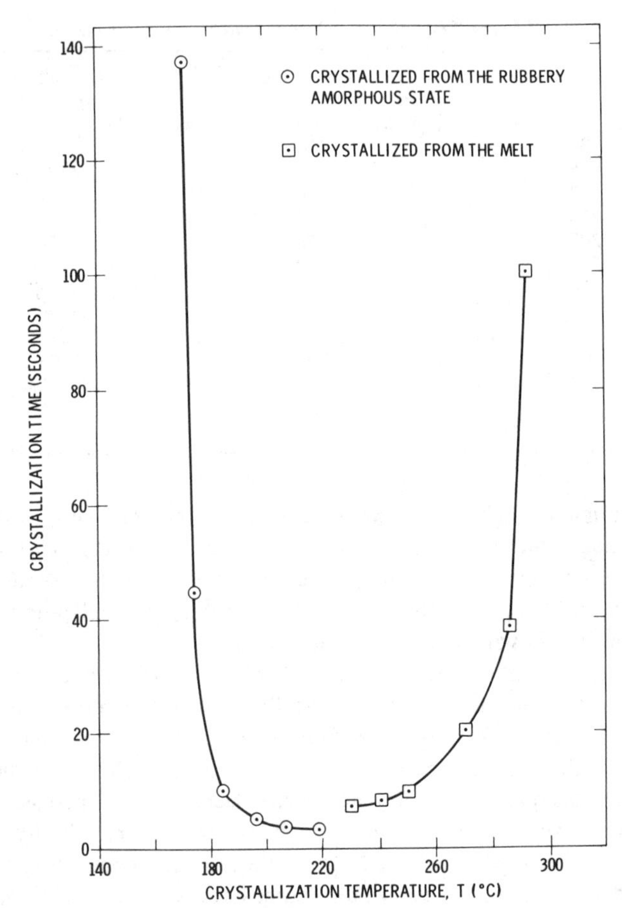

FIG. 2—*Crystallization time versus temperature for isothermal crystallization of PEEK. The times were determined from the development of maximum opacity during crystallization.*

have sharp, well-defined crystalline diffraction peaks. This can be due to a greater degree of crystal perfection, or to crystal size effect. The C180 sample, in contrast, has very broad peaks indicative of a disordered crystal structure. The SAXS results in Fig. 4 indicate that even though crystal perfection and degree of

TABLE 2—*Crystallinity after nonisothermal crystallization.*

Rate, C/min	Heating, 125 to 250°C	Cooling, 350 to 200°C
20	25.5	31.6
10	26.8	33.3
5	26.9	35.1
2	27.2	38.2
1	27.3	36.8

TABLE 3—*Sample preparation and crystallinity for mechanical test samples.*

Sample	Crystallinity Xc, %	Preparation Method[a]
Q	0.9	quenched from the melt into cold water
C180	23.7	crystallized from the rubbery state at 180°C for 1 h
SC	31.6	slowly cooled (17 h) from the melt in the press overnight
AC	33.0	air cooled (2 h) from the melt by removing hot platens from press

[a] All samples were melted at 400°C for 2 min under 3-MPa pressure.

crystallinity are similar in SC and AC films, the long spacings may be different. Both SC and AC have a large maximum at 16.4 nm, as determined from Bragg's law, but the AC sample has a shoulder at 2θ = 200 nm and contains larger amount of scattering intensity at low angles. AC appears to contain a broader distribution of crystal sizes than the SC samples.

Further evidence for morphological differences can be seen in the DSC results shown in Fig. 5. The scanning rate for these runs was 20°C/min. The amorphous film Q has a T_g at 145°C and a sharp exotherm at about 180°C. Film C180 also has a smaller exothermic response which occurs at 210°C. None of the melt crystallized films had exotherms above T_g, which means that they were crys-

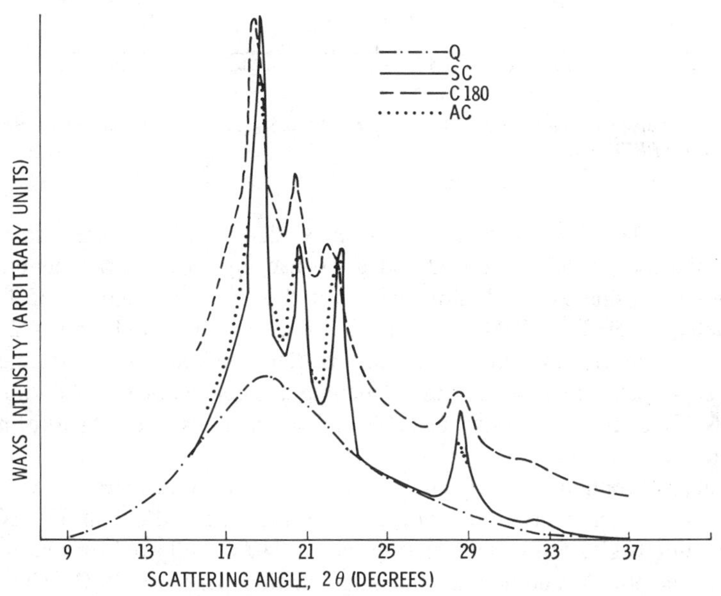

FIG. 3—*Wide angle X-ray scattering (WAXS) intensity as a function of scattering angle, 2θ, for PEEK films.*

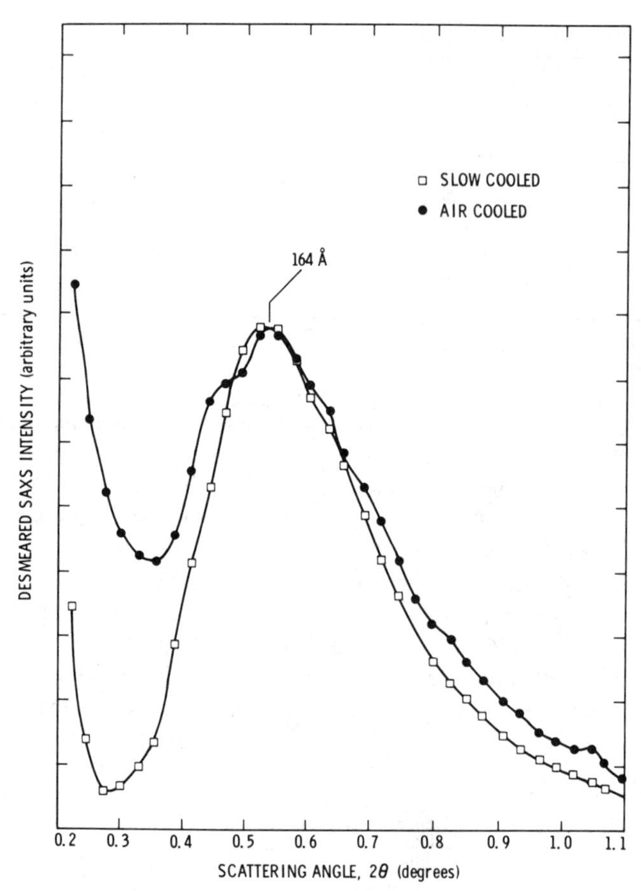

FIG. 4—*Desmeared small angle X-ray scattering (SAXS) intensity as a function of scattering angle, 2θ, for PEEK films.*

tallized to a large extent before the scan. Endothermic response also varied among the four samples. Both Q and C180 had high temperature endotherms consisting of a single peak. Melting begins at 280°C, is maximum at 335°C, and completes by 350°C. C180 has an additional small endotherm at 200°C, just above the crystallization temperature. This endotherm is an annealing peak [*11*] which comes about from the melting of a population of small imperfect crystals. These crystals are very unstable and will melt just above their formation temperature.

SC and AC contain relatively broad distributions of crystal sizes, as evidenced by the long tail on the low temperature side of the main endotherm. For SC and AC, melting begins at low temperature, about 240°C, followed by a secondary small endotherm in addition to the main melting peak at 335°C. The exact location of the secondary endotherm depends on the cooling rate from the melt.

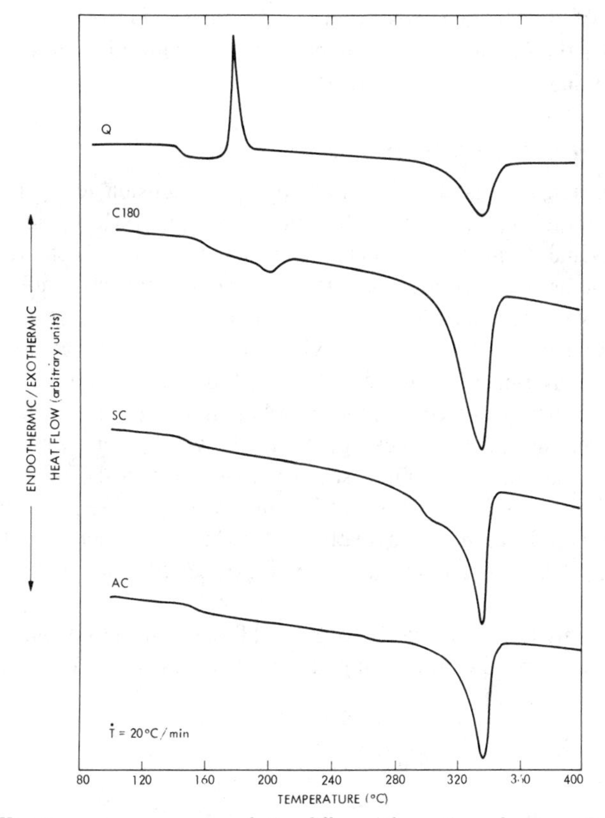

FIG. 5—*Heat flow versus temperature during differential scanning calorimetry (DSC), 20°C/min scanning rate. Curves have not been normalized for variation in sample mass.*

SC crystals were, on average, formed at higher temperature and thus will be stable to higher temperatures. The secondary peak occurs at 300°C for SC. AC samples cooled more rapidly, and therefore contain crystals that melt at lower temperature contributing to a secondary peak at 270°C.

From wide and small angle X-ray studies, and from DSC analysis, we obtain information on the morphological differences in the four samples that are directly attributed to the differences in processing. The amorphous film Q is produced by rapid quenching of the melt. Since this film has no crystals to act as thermoreversible cross-links, we expect that Q will have a low yield stress, and will orient readily during tensile drawing. Film C180, crystallized from the rubbery state, contains the smallest volume fraction of crystals, and these crystals are more disordered than those formed from the melt. Both SC and AC have very well-ordered crystals, though the distribution of sizes is different. In terms of mechanical properties, we expect the semicrystalline films to have higher yield stresses, lower breaking strains, and lower toughness when compared to the amorphous material. In the next section, we describe results of room temperature tension tests and correlate the mechanical behavior with microstructure induced by the processing.

Mechanical Properties and Morphology

Tension Testing—Results of room temperature tension tests are shown in Fig. 6, low strain region, and Fig. 7, high strain region. Mechanical properties data for stress and strain at yield and break are summarized in Table 4, along with Young's modulus. At least three samples were run under each condition. Stress-strain curves were reproducible except for variations in breaking stress and strain. Results shown here are for samples having the largest breaking strain.

Young's moduli ranged from 1.5 GPa for C180 to 2.0 GPa for AC, a narrow spread considering the large variation in crystallinity. All films yielded by formation of shear bands before necking down. The yield points were followed by strain softening in all films. Yield stress was greatest in the melt crystallized films, and smallest in the amorphous film. In the postyield region, formation of necks occurred at different stress levels, with SC and AC necking at the highest stress, and Q at the lowest stress. SC exhibited ductile failure, breaking after 28% strain.

All other films could be drawn to over 150% strain before failure. Film Q necked and experienced cold drawing at constant stress level, up to about 100%

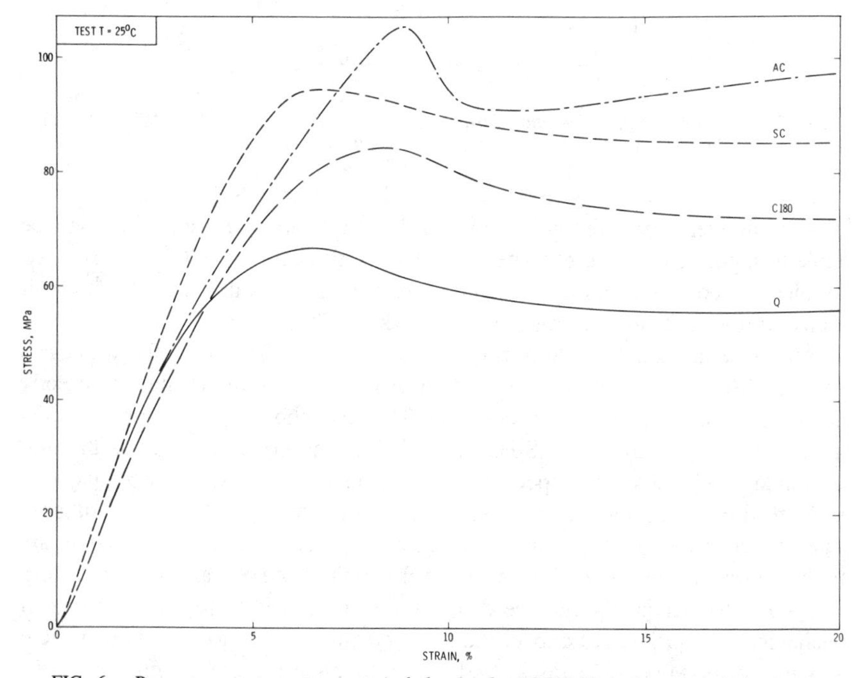

FIG. 6—*Room temperature stress-strain behavior for PEEK films: low strain behavior.*

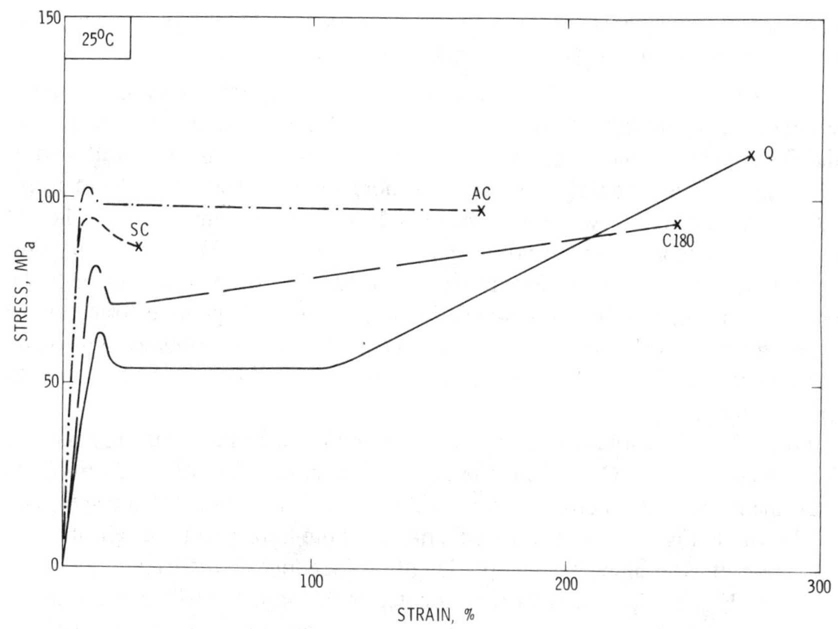

FIG. 7—*Room temperature stress-strain behavior for PEEK films: high strain behavior.*

strain as the neck ran toward the grips. Stress whitening was observed to occur in the neck region of the Q films. When the neck region extended completely to the grips, orientation hardening occurred causing the stress level to rise again. The slope of the hardening region (nominal stress/nominal strain) was constant at 35 MPa, and the film failed at 283% strain.

Film C180 necked and orientation hardened simultaneously, beginning immediately after strain softening. Since C180 was opaque before drawing (as were SC and AC), no stress whitening was visually observed during neck formation. The slope of the hardening region was 11 MPa, indicating that this film required lower stress level to undergo chain orientation.

Film AC had the smallest drop in stress at strain softening and necked at constant stress level until failure at 165% strain. By comparison with the amorphous film Q, AC is surprisingly tough. Using the area under the stress-strain

TABLE 4—*Mechanical properties of PEEK neat resin.*[a]

Sample	Modulus E,[b] MPa	Yield σ, MPa	Yield e, %	Breaking σ, MPa	Breaking e, %
Q	1800	66	7.6	115	271
C180	1500	83	8.5	95	243
SC	1800	95	7.0	85	28
AC	2000	106	9.0	97	165

[a]Tension test at room temperature, constant crosshead speed of 0.2 in./min (0.5 cm/min).
[b]Measured at 1% strain.

curve as a rough measure of the toughness, we see that AC is much tougher than Q or C180 in the region of low to moderate strain.

Microstructure —We examined the density in the drawn samples to see whether the crystallinity was changing in the neck region. Density in the neck of the Q samples was increased over the undrawn material. This means that originally amorphous material is drawn into more oriented and more dense form. However, since stress whitening was clearly observed in the Q films, the net density must represent both voids and oriented material. For all of the semicrystalline films, the neck density was reduced, compared to that of undrawn polymer, and this is further evidence in support of a competitive formation of oriented fibrillar crystals and voids. When density measurements were made on stressed but un-necked material, results were never lower than that of the undrawn polymer.

Formation of shear bands during strain softening was examined using polarized light microscopy. A C180 film was drawn to a strain of 100%, and crosshead movement was halted before fracture occurred. A sketch of the tension specimen is shown in Fig. 8. Using crossed polars, photographs were taken at magnification of ×50 along the sample length at locations 1 to 5. These areas are shown in Fig. 9. The polarizer was oriented at 45° to the tensile axis for maximum intensity [*12*], a condition that is strictly true only for constant volume deformation. Location 1 represents the lower grip region. The highly stressed neck initiation region is at Location 2, while necked down material is shown at 3. The neck advanced upward, and Location 4 represents the actively necking region. Ahead of the neck, shear bands could be seen in the undrawn polymer in Location 5.

The grip section, shown in Fig. 9(1), appears identical to undrawn polymer material. There is a nonuniform pattern of extinction, with irregular areas of graininess. Very thin films (≃25 μm) were observed optically under crossed polars at ×800 magnification, and the size of the grains was on the order of 3 μm. No large scale spherulite structure could be observed optically. However, our study of crystallization kinetics [*13*] on this material indicates that spherulites are nucleated heterogeneously from numerous sites, limiting their growth to a rather small size.

The neck initiation region, shown in Fig. 9(2), represents the position in the specimen where the cross-sectional area first thinned down. Across the tapered shoulder there is a gradient in the plastic strain, as can be seen by the irregular alternating pattern of birefringence. In this region, polymer chains are being first drawn from the isotropic state into a more ordered state.

The neck material itself, Fig. 9(3), is very uniform in its optical birefringence. Compared to the undrawn material, no grainy crystal texture is evident in the neck. Numerous small black spots appear in the neck region. These must be associated with voids in the sample. On other films taken all the way to fracture, internal cracks did originate at spots like these in the neck.

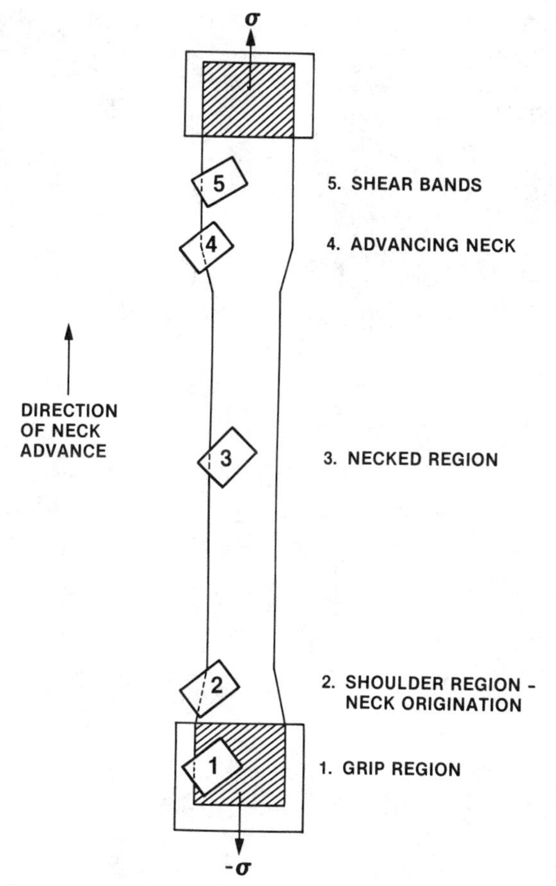

FIG. 8—*Sketch of tension specimen C180 drawn to 100% strain. Undrawn specimen had uniform cross section. Numbered areas correspond to photograph locations.*

As the neck moves toward the upper grip, isotropic material is deformed ahead of the neck. Figure 9(4) shows the highly stressed region at the taper between the already drawn material and the wider shoulder of undrawn polymer. Ahead of the actively necking shoulder region, Fig. 9(5), intersecting shear bands are formed in the material as it begins to strain soften.

Fracture surfaces were examined to gain further insight into the microstructure of the drawn samples. Scanning electron micrographs are shown in Fig. 10*a* to *d,* presented in order of increasing strain to break. The fracture surface of SC is shown in Fig. 10*a*. Ductile failure is indicated by the irregular and extremely rough appearance. Very little fibrillation can be seen. Film AC is shown next in Fig. 10*b*. The surface is again very lumpy but now many fibrils bridge the separated surfaces. A smoother fracture surface was created by the C180 material, shown in Fig. 10*c*. On a large scale there are many hollow areas where the

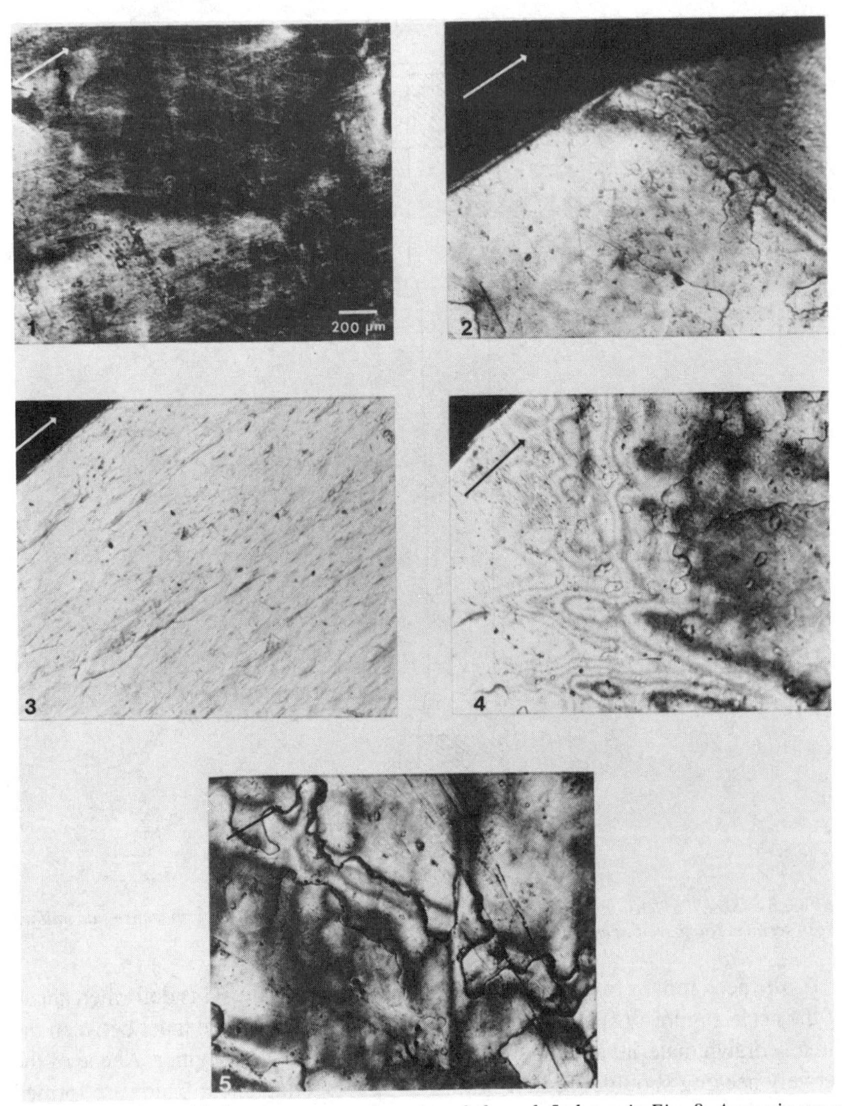

FIG. 9—*Polarized light microscopy at Locations 1 through 5 shown in Fig. 8. Arrow in upper left corner is parallel to tensile stress axis and points in direction of neck advance. Marker in 1 sets the scale for all pictures (magnification about ×50).*

surfaces have pulled apart. Many thick fibrillar strands act as bridges. On a finer scale, the surface is smooth compared to SC and AC. In Fig. 10*d,* the fracture surface of Q is shown. Neither fibrils nor cavities can be seen here. The surface is uniformly covered with a fine scale of roughness, and this indicates a more brittle fracture mode than in the semicrystalline films.

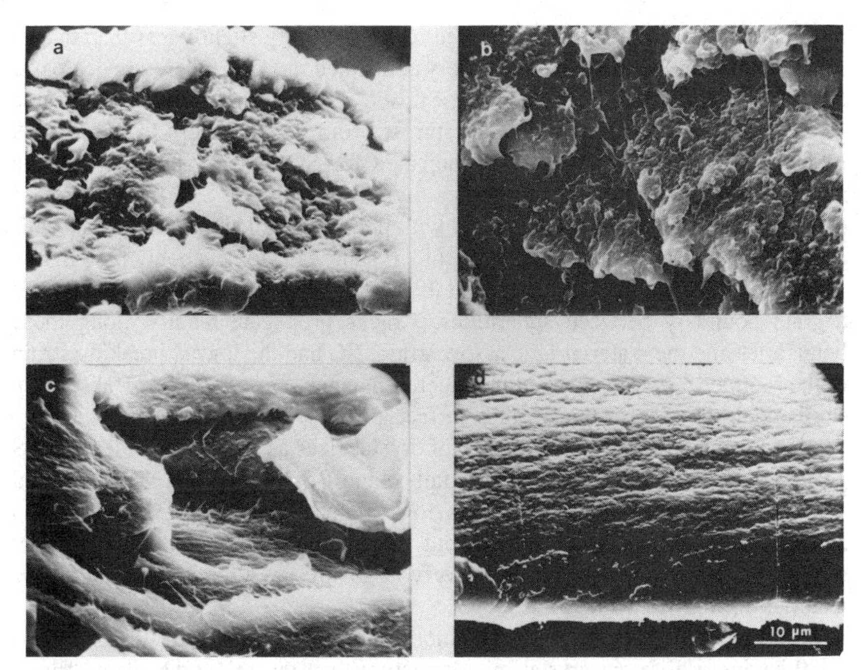

FIG. 10—*Scanning electron micrographs of room temperature fracture surfaces of PEEK films:* (a) *SC, slowly cooled;* (b) *AC, air cooled;* (c) *C180, crystallized from rubbery state; and* (d) *Q, quenched. Marker in* (d) *sets the scale for all pictures (magnification about ×2000).*

Discussion

From Fig. 1 and Table 1 we see that the degree of crystallinity of PEEK is greater when isothermally crystallized from the melt than when crystallized from the rubbery amorphous state at the temperatures indicated. For the temperatures used here, the maximum degree of crystallinity was about 35%. In the temperature range from 180 to 300°C, development of crystallinity is rapid, with most crystallization completed in under 5 min. A useful measure of PEEK crystallization time is the time to maximum opacity. As seen in Fig. 2, crystallization from the melt and from the rubbery amorphous state proceeds most rapidly at a temperature near 230°C. However, on the low temperature side of the minimum, the degree of crystallinity that can be achieved is less than that which can be achieved from the melt. This is true even if very long crystallization times are used.

Rate dependent crystallization results shown in Table 2 indicate that the degree of crystallinity is about the same as that obtained during isothermal crystallization. Only at the very low rates of −1 and −2°C/min does significant improvement occur. Since nonisothermal crystallization is a much simpler processing technique, we chose this method for melt-crystallizing samples for studies of mechanical properties. The SC and AC samples, prepared by cooling at different rates (see Table 3), had degrees of crystallinity slightly less than that of samples which were crystallized at a very controlled rate of temperature decrease.

We see from Figs. 3, 4, and 5 that SC and AC, which have about the same crystallinity, have different microstructures. From the sharpness of the WAXS reflections, a larger crystal size is indicated for SC. SC crystals melt over a narrower temperature range and have a higher melting point. A high degree of crystal perfection is usually associated with poor mechanical properties, and this is the case for SC films. Perfection of crystals can come about through the rejection of uncrystallizable material at the crystal-amorphous interface [*8*], or at a grain boundary between spherulites. Cracks propagate readily along these boundaries and the material fails at low strain. SC had the lowest breaking strain of the samples tested, and failed in a ductile manner. Very few energy absorbing fibrils could be seen in the fracture surface of SC (Fig. 10*a*).

AC films had a broader distribution of crystal sizes as judged by wider WAXS peaks and the longer low temperature tail on the DSC endotherm. This material was very tough, and could be drawn to a high strain before failure (Figs. 6 and 7). AC material did form fibrils, but, surprisingly, no orientation hardening took place, perhaps as a result of the very large stresses already imposed on the material during neck formation.

Crystallization from the rubbery amorphous state at 180°C resulted in film C180 having a different crystal microstructure than the melt crystallized films. WAXS results indicate existence of small crystals with diffraction peaks shifted to lower scattering angles. C180 was crystallized for 1 h at 180°C, but from Fig. 2 we see that crystallization should complete in under 20 s at this temperature. The very long residence time at 180°C allows additional crystals to develop between the lamellae that comprise the majority of crystals. This gives the film a low and high temperature endotherm (Fig. 5). In terms of mechanical properties, C180 had behavior intermediate to the melt crystallized films and the amorphous film. Its amorphous fraction was the largest of the semicrystalline films, and this allowed C180 to be drawn to large breaking strains at reduced stress.

Quenched amorphous film Q was the toughest material when the entire stress-strain curve is considered. It yielded and necked at low stress, and could be drawn to the largest breaking strain. The material orientation hardens, and this affects its fracture mode. The fracture surface, Fig. 10*d*, is very smooth compared to the semicrystalline films, and this is characteristic of a more brittle failure caused by the hardening. Stress whitening was observed in this material, but void formation is occurring in all the films, as judged by the density changes in the neck regions of the drawn samples.

Conclusions

We have shown that different crystal morphology can be imparted to PEEK neat resin by altering the rate of cooling during processing from the melt. Rapidly quenched amorphous films and those crystallized from the rubbery state were the most tough, but this processing condition will be difficult to achieve in thick samples. Both slowly cooling (SC) and cooling more quickly in air (AC) resulted in semicrystalline films having about the same degree of crystallinity. However, the room temperature mechanical properties of these films were very different because of differences in crystal size and in size distribution. Rate of cooling from the melt is a much more important indicator of tensile properties than degree of crystallinity for these two samples. Best overall mechanical properties were exhibited by the film AC cooled at an intermediate rate from the melt. This film had a simple processing history compared to the quenched film, and these results will be most useful for injection molding and extrusion applications. At low to moderate strain levels, film AC was much tougher than the quenched films. Processing conditions used to make AC films are recommended for composite matrix applications.

Acknowledgments

The authors acknowledge Dr. George Nielson for obtaining the small angle X-ray scattering results. Discussions with Dr. Muzaffer Cizmecioglu are greatly appreciated. The research described in this paper was performed by the Jet Propulsion Laboratory, California Institute of Technology while the principal author held a NASA-NRC Resident Research Associateship Award. Research was sponsored by National Aeronautics and Space Administration through Langley Research Center.

References

[*1*] Attwood, T. E. et al., *Polymer,* Vol. 22, No. 8, Aug. 1981, p. 1096.
[*2*] Stening, T. C., Smith, C. P., and Kimber, P. J., *Modern Plastics,* Vol. 58, No. 11, Nov. 1981, p. 86.
[*3*] "Provisional Data Sheet PK PDI 'Victrex PEEK' Aromatic Polymer Grades, Properties and Processing Characteristics," ICI fourth ed., Imperial Chemical Industries PCL, Welwyn Garden City, England, March 1982.
[*4*] "A First Glance at PEEK: A New 'Exotic' Thermoplastic," *Process Engineering News,* Vol. 15, No. 2, March 1981, pp. 33, 35.
[*5*] Blundell, D. J. and Osborn, B. N., *Polymer,* Vol. 24, No. 8, Aug. 1983, p. 953.
[*6*] Hay, J. N., Kemmish, D. J., Langford, J. I., and Rae, A. I. M., *Polymer Communications,* Vol. 25, No. 6, June 1984, p. 175.
[*7*] Ward, I. M., *Mechanical Properties of Solid Polymers,* Wiley, Chichester, 1983, Chap. 10, pp. 177-193 and 269-293.
[*8*] Way, J. L., Aktinson, J. R., and Nutting, J., *Journal of Materials Science,* Vol. 9, No. 2, Feb. 1974, pp. 293-299.
[*9*] Gupta, A., Cebe, P., and Chung, S., presented at the SAMPE Conference on New and Unique Resin Systems, Long Beach, CA, 23 Aug. 1984.
[*10*] Rutherford, J. L. and Brown, N., in *Methods of Experimental Physics,* Vol. 16 (c), R. Fava, Ed., Academic Press, New York, 1980, Chap. 12.3, pp. 117-136.
[*11*] Wunderlich, B., *Macromolecular Physics,* Vol. 3, Academic Press, New York, 1980, p. 191.
[*12*] Marasimhamurty, T. S., *Photoelastic and Electro-Optic Properties of Crystals,* Plenum Press, New York, 1981, pp. 251-256.
[*13*] Cebe, P. and Hong, S. -D., "Crystallization Behavior of Poly(etheretherketone)," to appear in *Polymer,* 1986.

David C. Leach,[1] *Don C. Curtis,*[1] *and David R. Tamblin*[1]

Delamination Behavior of Carbon Fiber/Poly(etheretherketone) (PEEK) Composites

REFERENCE: Leach, D. C., Curtis, D. C., and Tamblin, D. R., **"Delamination Behavior of Carbon Fiber/Poly(etheretherketone) (PEEK) Composites,"** *Toughened Composites, ASTM STP 937,* Norman J. Johnston, Ed., American Society for Testing and Materials, Philadelphia, 1987, pp. 358-380.

ABSTRACT: The delamination behavior of a carbon fiber/poly(etheretherketone) composite has been examined using a variety of techniques. The material used was aromatic polymer composite, APC-2. Interlaminar fracture toughness was measured using double cantilever beam and edge delamination tests. The extent and nature of damage caused by single and repeated low energy impacts was assessed using both low velocity and ballistic rates. Compressive failure properties after low energy impact were examined using a damage tolerance test. Growth of delaminations under compressive loading was investigated by moiré fringe and ultrasonic C-scan techniques. APC-2 has high interlaminar fracture toughness, even when unstable fracture occurs. There is high resistance to delamination under impact, including ballistic and repeated impacts. Compressive failure deformation falls gradually with increasing impact energy reaching a minimum value of 0.55% even when complete penetration of the specimen had occurred. There is a relationship between damage area and compressive failure deformation which includes ballistic and repeated impacts. Delaminations do not grow under compressive loading even at deformations in excess of 0.60%. The resistance to initiation and propagation of delaminations is due to the high interlaminar fracture toughness of APC-2.

KEY WORDS: carbon fiber reinforced composite, poly(etheretherketone) (PEEK), aromatic polymer composite (APC), interlaminar fracture, delamination, impact damage, damage tolerance, compression

Delamination behavior is a key area of performance for continuous fiber reinforced composites. Delamination or interlaminar fracture is seen commonly in practice, and it has proved to be a limiting factor in the overall performance of composite systems. Delaminations may initiate and grow at free edges in

[1]Research physicist, experimental physicist, and experimental chemist, respectively, Materials Property Group, Imperial Chemical Industries PLC, Petrochemicals and Plastics Division, P.O. Box 90, Wilton, Cleveland, TS6 8JE England.

tensile loading [*1,2*], and can also be caused by low energy impact damage [*3*]. Such delaminations will grow under compressive loading resulting in a considerable reduction in compressive strength [*4*]. It has been found that delamination resistance is increased through the use of matrices with higher toughness, and therefore the effects of low energy impact damage are reduced [*5,6*]. Tougher resins may be produced by modifying a base resin [*5*] or by using a different type of resin [*6*]. In this latter category, thermoplastic matrices have given an order of magnitude increase in interlaminar toughness compared to existing epoxy resin composites [*6*]. Some results on the toughness of one such thermoplastic matrix composite, aromatic polymer composite, APC-2 [carbon fiber/poly(etheretherketone)], have already been reported [*7*], and this paper examines delamination in APC-2 in more detail.

A number of approaches may be used to characterize delamination behavior. One approach is to measure the interlaminar fracture toughness of the composite directly. Two tests commonly used are the double cantilever beam (DCB) test [*8*] and the edge delamination test (EDT) [*9*]. Another approach is to characterize low energy impact damage by measuring the total extent of damage, and by internal examination of damaged specimens [*3*]. Different impact energies and rates have been used to characterize failure mechanisms in various materials. These and other investigators have examined the propagation of delaminations under compressive loading [*3,4*], and related this to direct measurements of interlaminar toughness. The experimental program on APC-2 included all these approaches to give a broad view of the delamination behavior in aromatic polymer composites.

Experimental Procedure

Materials

The material investigated was aromatic polymer composite, APC-2. This has a matrix of poly(etheretherketone) (PEEK) reinforced with 61% by volume of continuous carbon fiber. The unidirectional tape was laminated into test panels 0.46 by 0.46 m using recommended processing procedures [*10*]. A summary of the lay-ups and total number of panels used is given in Table 1. The quality of the panels was checked by visual examination and ultrasonic C-scan. In the DCB test, a fold of aluminum 25 μm thick was molded between the center plies to provide an interlaminar defect from which the delamination was propagated.

Double Cantilever Beam Test

For the DCB test, parallel-sided specimens were used [*8*], the load being introduced through aluminum loading tags bonded to the specimens. Six specimens were taken from the panel, cut from a single area, and numbered sequentially. The specimens were 380 mm long, 25.4 mm wide, and 5.4 mm thick.

TABLE 1—*Panel details: aromatic polymer composite, APC-2.*[a]

Evaluation Technique	Number of Panels	Lay-Up	Total Number of Plies	Average Panel Thickness, mm
Double cantilever beam test	1	$(0)_{40}$ + crack starter	40	5.4
Edge delamination test	1	$(\pm 35/0/90)_S$	8	1.1
Damage tolerance-low energy impact followed by compression	8	$(-45/0/+45/90)_{5S}$	40	5.3

[a] All panels 0.46 by 0.46 m.

APC-2 shows two types of fracture behavior in the DCB test: in some regions the crack propagates slowly in a stable manner, while in other regions the crack propagates rapidly in an unstable manner. These will be referred to as "stable" and "unstable" cracks. In the stable crack regions, the delamination was allowed to propagate approximately 25 mm and the beams were then unloaded. In the regions of unstable crack propagation it was not possible to control the length of propagation, which varied from 15 to 60 mm. On arrest, the beams were unloaded. Typically, eleven experimental measurements were made on each beam. Crosshead rates of 5 to 20 mm/min were used, the rate being increased as the crack propagated down the beam.

The critical strain energy release rate (interlaminar fracture toughness) G_{Ic} was calculated using the area method and the compliance method [8]. Using the area method:

$$G_{Ic} = \frac{\Delta E}{\Delta a \cdot b} \tag{1}$$

where

ΔE = energy required to propagate the crack a distance Δa and
b = beam width.

Ideally this should be calculated for $\Delta a \rightarrow 0$. In practice, toughness is usually calculated over longer crack length increments (in this case typically 25 mm) which gives an average, relative G_{Ic}. This will be discussed later.

Using the compliance method:

$$G_{Ic} = \frac{P_C^2}{2b}\frac{dC}{da} \tag{2}$$

where

P_C = load at which crack propagation occurs,
C = beam compliance, and
a = crack length.

The term dC/da can be determined by beam theory or by measuring the compliance at each loading, and plotting this as a function of crack length. Both were used, and there was good agreement between the two methods.

Edge Delamination Test

The general method used followed that described by O'Brien [9]. The lay-up was $(\pm35/0/90)_S$, which has been recommended for measurement of interlaminar toughness [11]. Such a lay-up would be expected to delaminate at the 0/90 interply regions. Specimens 283 by 19 mm were prepared from the panel and were end-tabbed with glass fiber/epoxy tapered end-tabs giving a gauge length of 175 mm. The specimens were tested at a constant deformation rate of 2 mm/min and the axial strain was measured using an extensometer. The load-strain curves were linear to failure, and the failure strain and secant modulus at 0.25% axial strain were recorded on seven specimens.

The fracture toughness in the EDT is given by [9]:

$$G_c = \frac{\varepsilon_c^2 t}{2}(E_{LAM} - E^*) \tag{3}$$

where

ε_c = axial tensile strain at delamination onset,
t = laminate thickness,
E_{LAM} = laminate tensile modulus, and
E^* = modulus of laminate completely delaminated at 0/90 interfaces.

E_{LAM} was measured as 58.7 GN/m^2 and was also calculated using laminate plate theory as 61.3 GN/m^2. While there is a slight discrepancy of 5% between these values, differences of up to 12% between measured and calculated values have been reported previously [12]. E^* was calculated using laminate plate theory as 52.6 GN/m^2. For the lay-up used here, O'Brien [11] has shown that G_{Ic} is approximately 90% of G_c.

Low Energy Impact

The low energy impact (LEI) damage was characterized on quasi-isotropic laminates with a lay-up of $(-45/0/+45/90)_{5S}$. The 40-ply laminates were cut into specimens 152 by 102 mm and were clamped lightly at four points against a 127- by 76-mm support.

Low velocity and ballistic impacts were performed on both thicknesses. For the low velocity impacts a drop-weight device was used with impact velocities in the range 5.0 to 5.5 m/s. The impactor had a hemispherical nose with a diameter of 16 mm. The ballistic impacts were carried out using a ball-gun device in which ball bearings, 6 mm in diameter, were fired at the specimens. In this case, the incident energy was varied by changing the incident velocity, the maximum velocity used being 165 m/s giving an incident energy of 12 J. The extent of damage was measured using an ultrasonic C-scan technique. To gain information on the internal damage caused by low energy impact, several specimens were sectioned and polished.

Damage Tolerance

The effect of the impact damage on compressive properties was assessed for the 40-ply laminates used in the impact damage characterization. The compressive stress was applied along the major axis and the specimen supported in a jig similar to that described by Byers [4] incorporating antibuckling guides. The axial deformation was measured using a clip gauge extensometer attached to the faces of the jig. The load-deformation curves were linear to failure, and hence either failure stress or failure deformation could be used to characterize residual compressive performance.

A shadow moiré technique was used to assess out-of-plane deflections during loading for some of the tests. The technique followed the method described by Mousley [*13*]. The specimens were loaded to various levels of compressive deformation and the interference fringe pattern was photographed normal to the plane of the specimen. The geometry of the illumination and observation meant that each fringe corresponded to an out-of-plane deflection of 125 μm.

Results and Discussion

Double Cantilever Beam Test

Fracture Behavior — APC-2 showed two types of fracture in the DCB test, with stable and unstable crack propagation. The phenomena has been reported previously for APC [*6,7,14*]. Some of the beams are shown in Fig. 1, and the different regions of crack propagation can be distinguished as the areas of stable propagation appear lighter than the areas of unstable propagation. Scanning electron micrographs from the two regions are shown in Fig. 2. The unstable fracture surface is not brittle but shows microductility, however, the stable fracture surface shows greater ductility with the polymer being drawn considerably locally. This difference in texture is the reason for the optical differences seen in Fig. 1, and it can be seen that the two types of fracture occur randomly. There

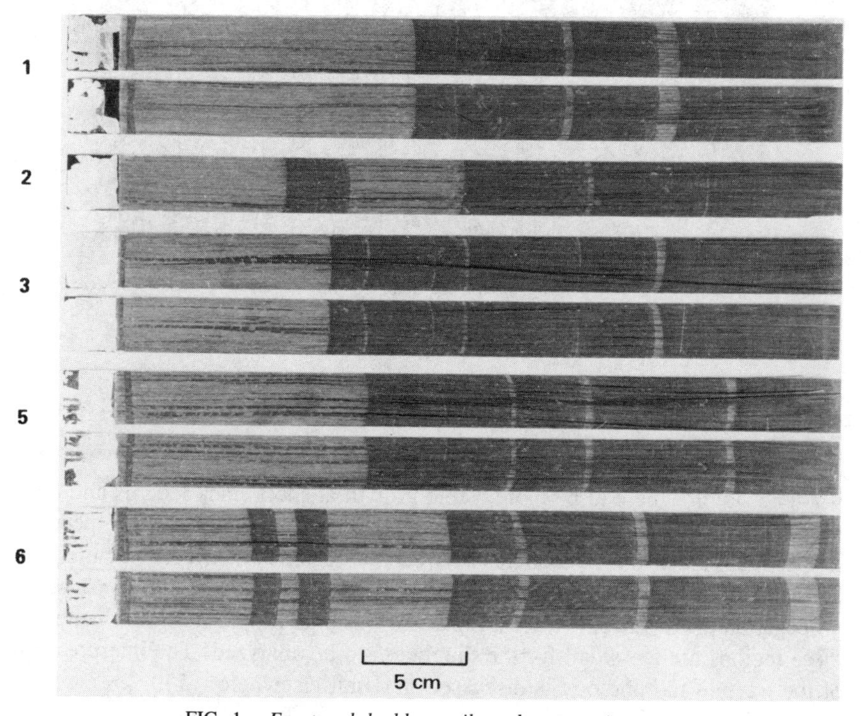

FIG. 1—*Fractured double cantilever beam specimens.*

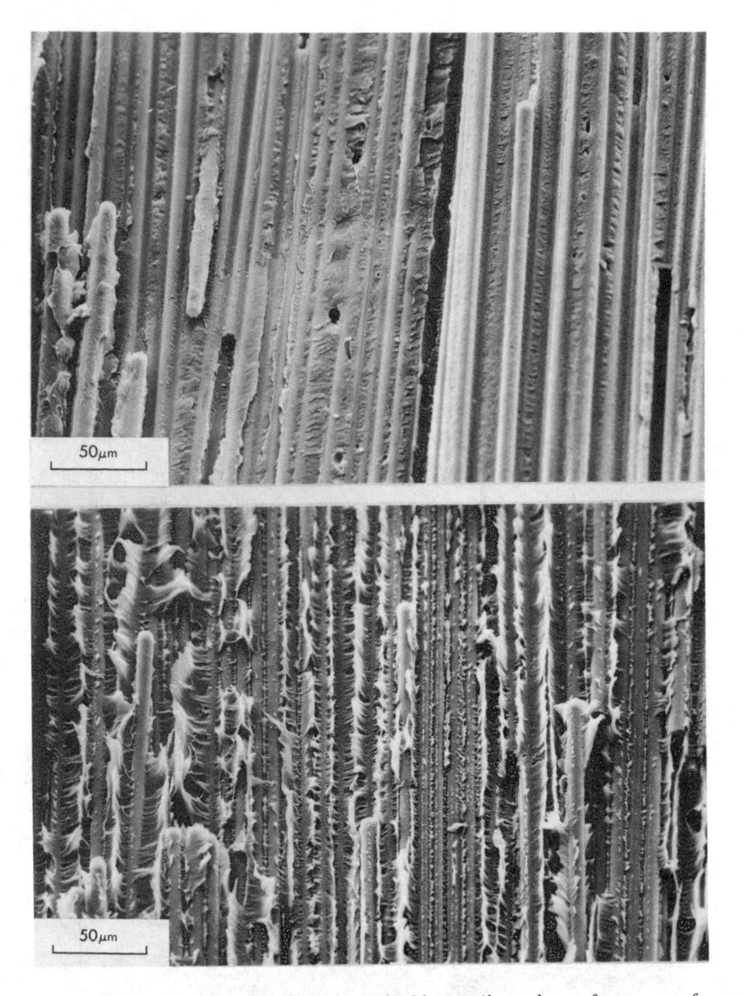

FIG. 2—*Scanning electron micrographs from double cantilever beam fracture surfaces:* (top) *unstable fracture region and* (bottom) *stable fracture region.*

is a tendency for there to be more stable growth at short crack lengths and more unstable growth at long crack lengths, but the extent of this varies from specimen to specimen. It can also be seen that between the regions of unstable fracture there are bands of stable propagation which vary in size considerably. Some stable propagation always occurs after a region of unstable propagation, but generally these regions are too small for the toughness to be analyzed. The interpretation of the fracture morphology is discussed in detail elsewhere [*15*].

Interlaminar Fracture Toughness—In discussing the interlaminar fracture toughness results, it is necessary to appreciate that the two methods of analysis do not measure the same toughness "property." The area method gives an average propagation toughness over a certain area, while the compliance method gives an effective toughness at a single point. Taking the force at different points on the load-deflection curve, it is therefore possible to calculate the toughness at crack initiation, during propagation, or at arrest. If the crack propagation is stable and arrests immediately on removal of the load, then:

$$G_{Ic} \text{ Initiation} = G_{Ic} \text{ Propagation} = G_{Ic} \text{ Arrest}$$

In these circumstances the area method and compliance method applied at any point of crack propagation should give the same value of toughness. During unstable propagation, the interlaminar toughness for initiation, propagation, and arrest will be different. For the unstable fracture, the compliance method can be used to obtain an initiation and an arrest toughness. The area method will give a toughness that will include initiation, propagation, and arrest, and hence it cannot be used to give a valid toughness for unstable propagation. However, the results are included for illustrative purposes.

The results from the DCB tests are summarized in Table 2. Considering first the stable propagation, the values of G_{Ic} are similar from both methods. The area method values are slightly higher than those from the compliance method, however, the scatter in the results precludes a positive interpretation of the slight differences between the two methods.

For the unstable propagation, the values of G_{Ic} from the two methods would be expected to differ and this is the case. G_{Ic} at initiation is higher for the unstable than for the stable propagation, probably as a result of crack blunting which will occur on arrest, though it is possible that other factors could also affect the crack initiation process. When the crack does initiate, it is assumed that excess energy is available for propagation leading to unstable fracture. The area method results are similar to those for the stable propagation region, however as indicated earlier, the interpretation of this is questionable. The toughness value from the area method lies between the initiation and arrest values from the compliance method as would be expected as a result of the area method including initiation, propagation, and arrest. The arrest G_{Ic} is lower than both the initiation and propagation values but even this value is relatively high. Again, there is scatter associated with the results, but the minimum value obtained for the 30 experimental measurements was 0.90 kJ/m^2.

Although high ductility is apparent on the stable fracture surface, the additional drawing of the polymer does not contribute significantly to the overall energy absorption during interlaminar fracture even if we assume that the arrest toughness is the relevant value during unstable fracture. In the assessment of toughness, either the area method or compliance method may be used for stable fracture, however for unstable fracture, the area method cannot be applied and an arrest value of toughness should be used. This represents a minimum value of

TABLE 2—*Summary of double cantilever beam results.*[a]

Type of Crack Propagation	Proportion of Fracture Area, %	Number of Experimental Measurements	Interlaminar Toughness, G_{Ic}, kJ/m^2	
			Area Method	Compliance Method
Stable	36	39	propagation: 2.89 (0.10)	initiation: 2.49 (0.07)
Unstable	64	30	propagation: 2.41 (0.11)	initiation: 3.07 (0.14) arrest: 1.76 (0.11)

[a] Standard deviations of the mean values in parentheses. Six beams tested.

strain energy release rate below which an interlaminar crack will not propagate even during unstable fracture. This could therefore be used as the basis of a design criteria for interlaminar fracture.

Edge Delamination Test

The EDT specimens were expected to delaminate at the 0/90 ply interfaces; however, the dominant failure mode was tensile fracture. The specimens are shown in Fig. 3 (top) and it can be seen that typically multiple tensile failure occurred within the gauge length.

However, in Specimen 6 there is evidence of delamination in addition to the tensile failure. A detail of this area is shown in Fig. 3 (bottom) and the delamination has followed the 0/90 interface. There is also a light band running along the bottom edge of this region which appears qualitatively similar to the stable fracture regions on the DCB specimens. This suggests that there was some stable delamination growth from the edge of the specimen, but that the onset of tensile failure and the attendant release of elastic energy precipitated the delamination to run across the specimen.

In Specimen 6 we assume that the value of G approached the critical value, and in this case $G = 1.06$ kJ/m^2, and therefore $G_I = 0.95$ kJ/m^2. This value is lower than the results obtained in the DCB test, the most likely reason being the elimination of mechanisms such as nesting and fiber bridging in the EDT which will contribute to the toughness in the DCB test.

The EDT results are of limited use because of the dominance of tensile fracture. O'Brien [*11*] has shown that the simplest way of changing the failure mode to delamination is to increase the thickness, and the use of laminates of twice the thickness of those used here should give delamination failures which can be interpreted with more confidence.

Low Energy Impact

Characterization of Low Energy Impact Damage—Figure 4 shows a polished section from a 36-ply laminate of a development material similar to APC-2 with a lay-up of $[(\pm45/0/90/0/90)_2/\pm45/0/90/\pm45]_S$, which had been impacted at 15 J. The overall pattern of damage is similar to that seen in other carbon fiber reinforced composites [*3,16*], however there is an inner conical region immediately below the impact point where there is an absence of cracks and delaminations. The most likely reason for this is the high through-thickness compressive forces during the impact event that prevent delamination damage from occurring. There is a zone of damage across the back face (that is, opposite the impact point) consisting of cracks and delaminations, which extends a few plies in from this face. The largest delaminations are therefore close to the back face and it is these that will be responsible for the damage area seen in the C-scans.

Effect of Impact Energy on Damage Area—Ultrasonic C-scan was used to assess the maximum extent of delamination damage after low energy impact.

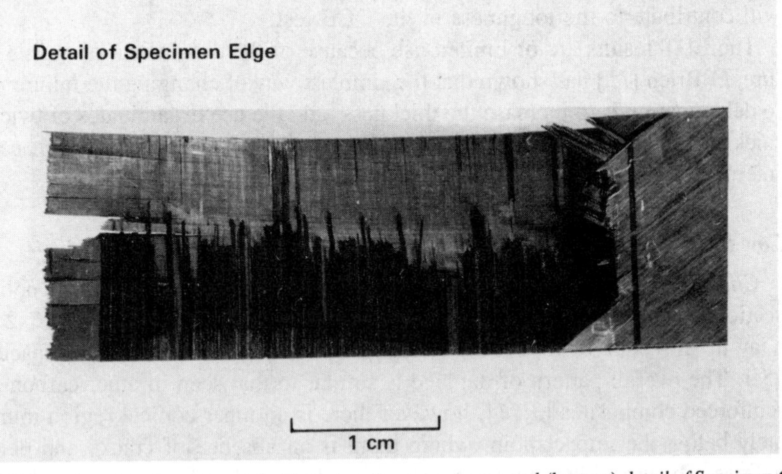

FIG. 3—*Edge delamination test:* (top) *fractured test specimens and* (bottom) *detail of Specimen 6 showing delamination growth from edge of specimen.*

Figure 5 shows the damage area plotted against incident impact energy for the 40-ply quasi-isotropic specimens. Eight panels were used and different symbols are used for each panel. There is scatter in the results, but an overall trend is seen with damage area increasing with incident energy. The applied energies ranged

FIG. 4—*Polished section from low energy impact specimen. Lay-up* $[(\pm 45/0/90/0/90)_2, \pm 45/0/90/\pm 45]_S$, *incident impact energy 15 J.*

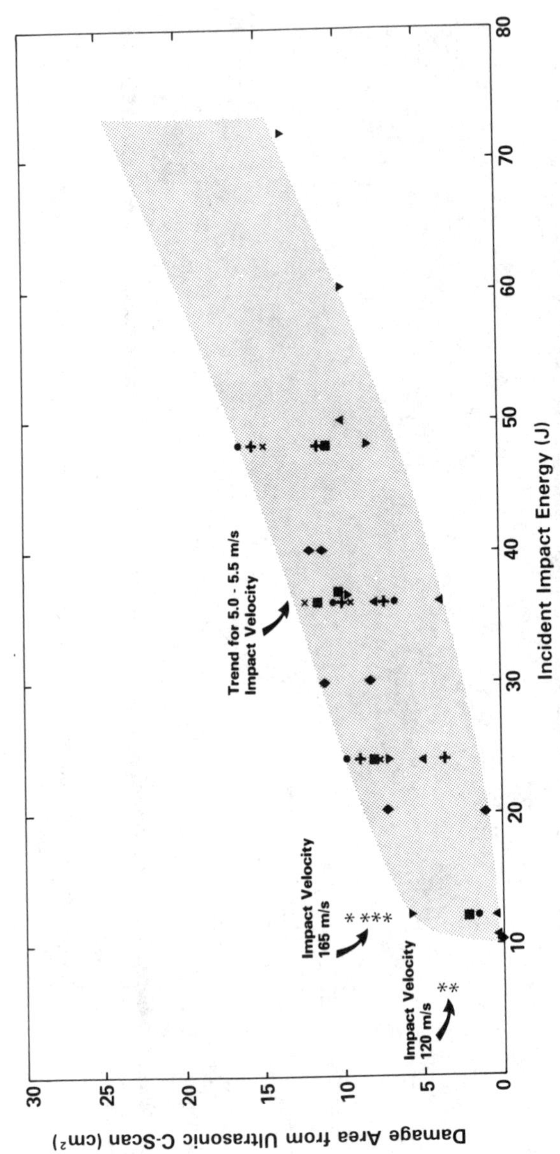

FIG. 5—*Damage area versus incident impact energy for 40-ply quasi-isotropic laminates. Lay-up* $(-45/0/+45/90)_{5S}$. *Different symbols used for each panel.*

up to 72 J, and impacts at this level cause considerable damage to both the impact face and the back face where fiber breakage is present. At 5.0-m/s velocity, the lowest incident energy was 10 J and no delamination damage is detected. There is permanent deformation on the front face in the form of a small dent and we assume there is some internal damage, probably consisting of cracks and matrix yielding. Ballistic velocity impacts result in larger damage areas for equivalent applied energy; however, the increase is small. The results confirm that APC-2 shows considerable resistance to impact delamination damage over a wide range of impact conditions.

Effect of Repeated Impact — In practice, it is quite possible that repeated impact may occur and this has also been examined. Specimens were impacted repeatedly at the same incident impact energy, and ultrasonic C-scans after various numbers of impacts are shown in Fig. 6. Using an incident energy of 40 J, the impactor passed through the specimen on the 5th impact, and at 30 J the impactor passed through on the 16th impact. With incident energies of 10 and 20 J, the impactor had not passed through the specimen after 100 repeated impacts.

The damage areas from the C-scans are plotted against the number of repeated impacts in Fig. 7. These curves show three regions of behavior: over the first few impacts the damage area increases but then there is a region where the rate of increase is low and relatively constant over a large number of impacts. Finally, the rate of increase in damage area rises as final penetration is reached and the impactor passes through the specimen. This form of behavior shows similarity to

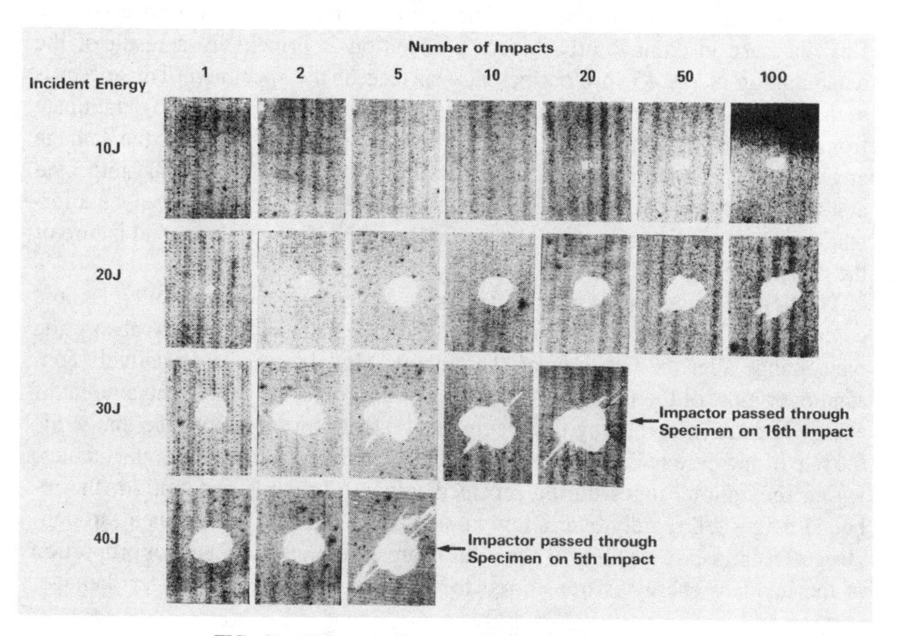

FIG. 6 — *Ultrasonic C-scans after repeated impact.*

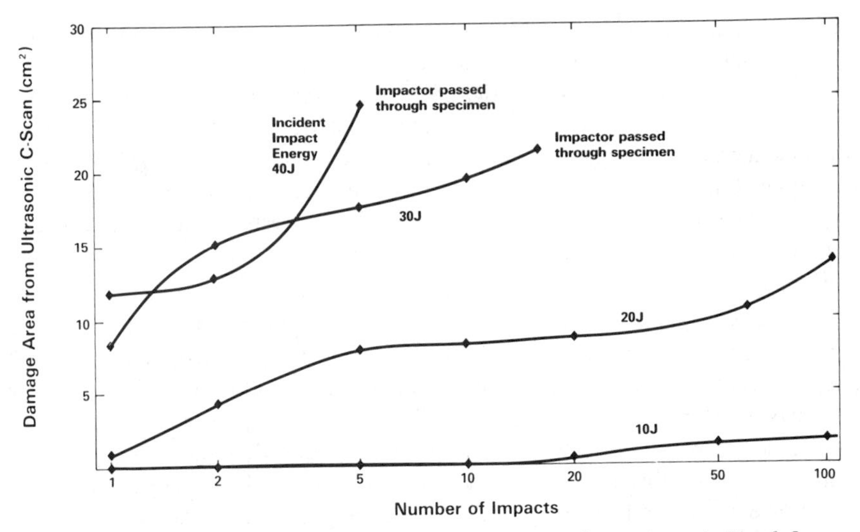

FIG. 7—*Effect of repeated low energy impact on damage area for specimens in Fig. 6. Lay-up* $(-45/0/+45/90)_{SS}$.

crack growth curves in fatigue loading, and is seen to varying degrees in the specimens examined. For example, for the 40-J impacts, the whole cycle occurs over such a small number of impacts that the three stages cannot be separated. The three regimes are seen best in the 20- and 30-J impacts and it appears that for 20-J specimen the final phase of behavior is underway at 100 impacts. The final increase in damage area before penetration is principally a result of the delamination of the 45° ply on the back surface of the specimen. This effect is seen in Fig. 6, where the back surface ply first splits and then starts to delaminate from the rest of the composite. Figure 7 shows that after two impacts the damage area for the specimen impacted at 30 J is slightly greater than for that impacted at 40 J. This cannot be readily explained; however, the C-scan only gives information on the maximum area of damage rather than on the volume and nature of the damage.

The first impact causes some damage in the specimen, and as a result it is more compliant for successive impacts. Hence, the mechanism of energy absorption may change after the first impact. The rebound height remained relatively constant over most of the impacts indicating that approximately the same amount of energy is being absorbed by the specimen at each blow. It is possible that while the rate of increase in damage area is small, the nature and density of the damage within the zone changes during repeated impact. Towards penetration, the rebound height fell rapidly over a few impacts leading to the final punch-through. Hence, the processes leading to penetration are accelerated as a greater proportion of the incident energy is transferred to the specimen causing greater damage.

At the 10-J incident energy, no delamination damage is apparent until more than ten impacts, and the rate of increase in damage area is extremely low indicating a high resistance to repeated impacts at low energy levels.

Damage Tolerance

Residual Compressive Failure Deformation —The effect of low energy impact damage on residual properties was assessed by measuring the failure deformation in compression. In a number of composite systems the failure deformation falls rapidly for a small increase in incident impact energy [*4,6*]. In composites with tougher matrices, there is a gradual change in failure deformation with impact energy. This is confirmed for APC-2 and the results are shown in Fig. 8 for the specimens whose damage areas have already been shown in Fig. 5. (Common symbols are used for Figs. 5 and 8.) The results lie on a trend with failure deformations of ~1.0% for unimpacted and low energy impacted specimens, gradually falling to ~0.60% at the higher impact energies. The compressive failure deformations for the ballistically impacted specimens are included in Fig. 8. These are slightly lower than the failure deformations of specimens impacted at equivalent energy but at low velocity, but are again greater than 0.60%.

Two types of failure are seen in the compressive test: specimens impacted at energies less than 15 J fail by macro- (or global-) buckling, with the side supports of the jig being pushed out. As a result, these failures do not give a measure of the intrinsic compressive failure properties. These macro-buckling failures are circled in Fig. 8. At incident energies above 15 J, the compressive failure is by a combination of local buckling and shear. In this case, the compressive failure occurs laterally through the impacted region.

Relationship Between Extent of Damage and Compressive Failure Deformation —The extent of damage and compressive failure deformation have been examined as a function of the incident impact energy in Figs. 5 and 8. An alternative way of examining the data is shown in Fig. 9 where failure deformation is plotted as a function of damage area for the data already discussed. There is much less scatter in the results plotted in this form and the data all lie close to a single line. This implies that the scatter in failure deformation (Fig. 5) is principally due to scatter in the damage area (Fig. 8) rather than a result of intrinsic scatter in the compression test. Figure 9 shows that there is a relationship between the extent of damage and residual failure properties, and also that the macro-buckling failure occurs where damage areas are less than 4 cm^2.

The ballistic impact results also lie on the overall trend showing that the reduced compressive failure deformation is due to the increased damage area, rather than to any change in the effects of the internal damage. The results for the repeatedly impacted specimens are included and these also follow the overall trend. After 100 impacts at 10 J, the failure deformation is the same as that of an

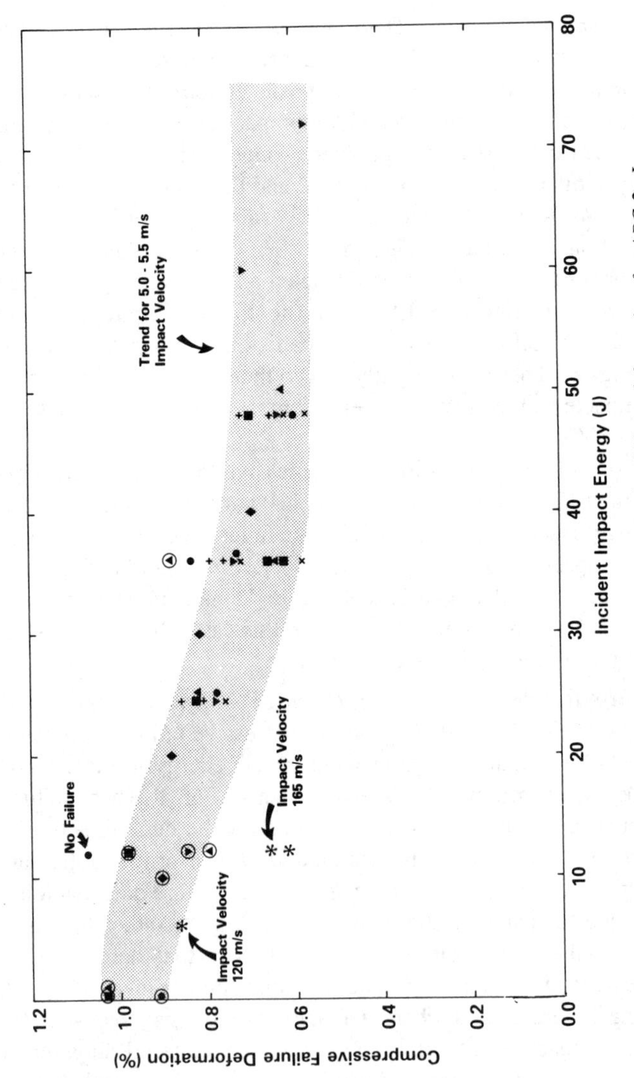

FIG. 8—*Effect of low energy impact on compressive failure deformation for APC-2. Lay-up* $(-45/0/+45/90)_{5S}$. *Different symbols used for each panel. Circled symbols indicate macro-buckling failure.*

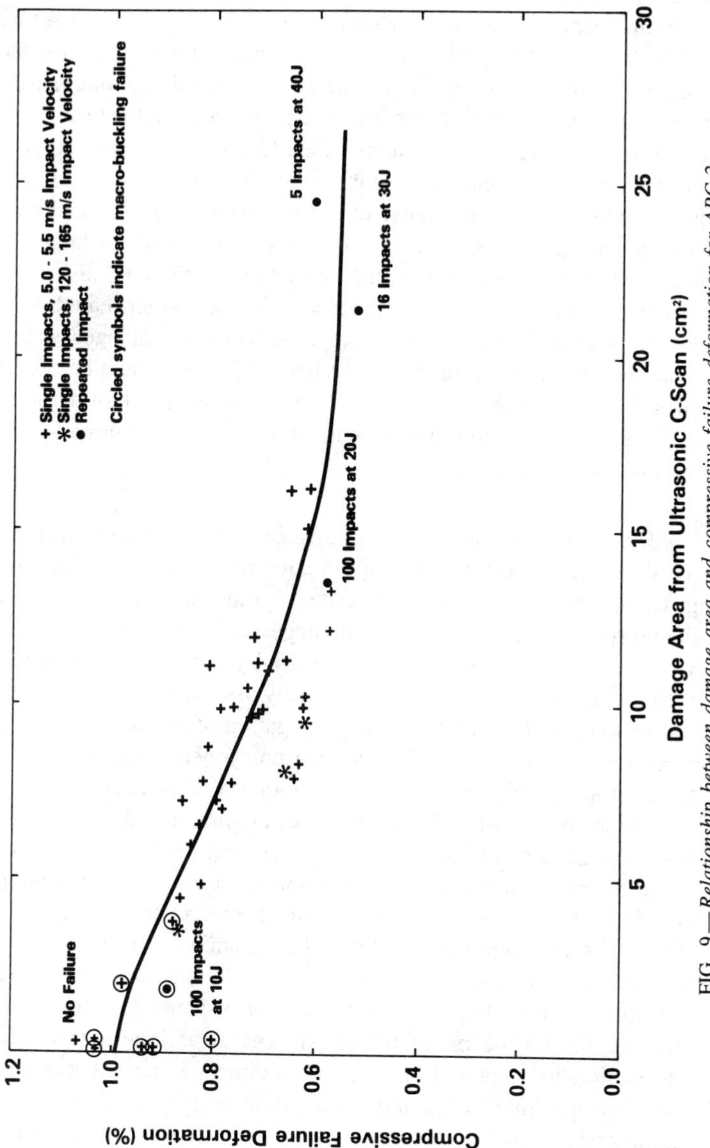

FIG. 9—*Relationship between damage area and compressive failure deformation for APC-2. Lay-up* $(-45/0/+45/90)_{5S}$.

undamaged specimen, and the compressive failure was by macro-buckling. For the specimens repeatedly impacted at 30 and 40 J, complete penetration occurred, but these results lie on the trend and the specimens retained at least 0.55% compressive deformation.

The minimum residual failure deformation after all the impact processes examined is 0.55%, and therefore this is assumed to represent a minimum value for the material. The compressive test is essentially a structural one, and hence, the failure properties depend upon both the material properties and the test geometry. It is therefore possible that this minimum is due to a geometric effect, and this could be examined by evaluating specimens of different geometries. Larger specimens would be more representative of the geometry in service, and work on larger specimens has been described [*3*]. It is anticipated that for larger specimens impacted at equivalent energies, the proportion of the total area that was damaged would be smaller. Therefore, there would be a smaller reduction in the residual compressive failure properties compared to the undamaged state, and hence the minimum failure deformation seen here of 0.55% would be exceeded. This assumes that the geometry is such that the undamaged failure deformation remains unchanged. If this is the case, then the 0.55% failure deformation can be used as a minimum value for APC-2.

Growth of Delaminations Under Compressive Loading —The results in Fig. 9 suggest that there are no step changes in the compressive failure behavior with increasing damage area. For some composite systems, the failure deformation falls considerably for a small increase in damage area [*4,6*] implying "damage sensitivity." The materials that show this sensitivity have relatively low interlaminar fracture toughness (G_c), and it is generally assumed that the sharp fall in failure deformation is due to interlaminar propagation of the delamination damage under compressive loading. This has been confirmed experimentally for a number of materials [*4,13*]. The absence of damage sensitivity and the high interlaminar toughness of APC-2 lead to a belief that, for this material, the delamination damage does not propagate under compressive loading. This was examined experimentally using moiré fringe and ultrasonic C-scan techniques.

It has been shown earlier that the largest delaminations occur near the back face of the specimen. As these represent the longest interlaminar cracks and are close to a free surface, it is expected that these are the delaminations most likely to grow under compressive loading. Therefore, the out-of-plane deflections on this surface were examined. The moiré fringe patterns at various levels of compressive deformation are shown in Fig. 10 for specimens impacted at a range of energy levels. The specimens impacted at the higher energy levels were loaded to 0.60% compressive deformation, while those impacted at the lower impact energies were loaded to 0.80% deformation. Ultrasonic C-scans after the initial impact and after compressive loading are shown in Fig. 11.

The moiré fringe patterns before loading show that for all levels of impact there is some out-of-plane deflection, including the specimen impacted at 10 J which

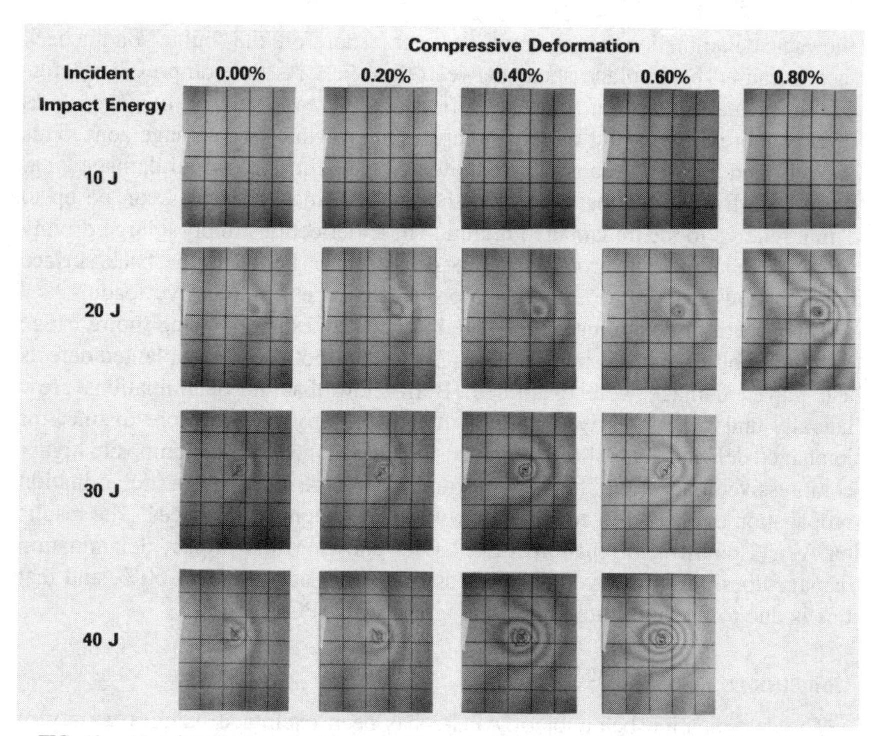

FIG. 10—*Moiré fringe patterns for impacted specimens loaded in compression. Central area of specimen shown. Reference lines drawn on specimen at 1-cm spacing.*

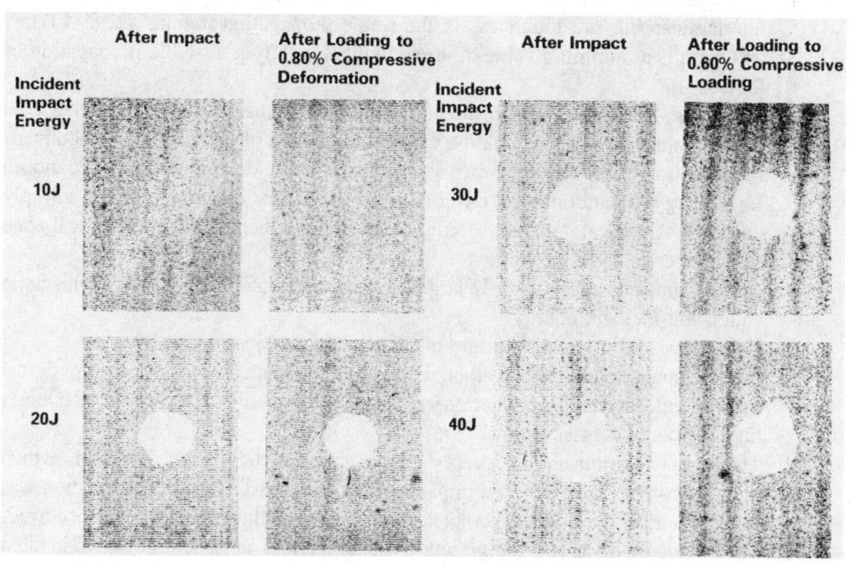

FIG. 11—*Ultrasonic C-scans after low energy impact and compressive loading for specimens shown in Fig. 10.*

shows no delamination damage on the C-scan. Therefore the "bulge" on the back face is caused by in-plane shear between the plies. As the compressive deformation increases, the out-of-plane deflections increase and the damage zones become more distinct, but there is no apparent growth of the damage zone. This is confirmed by the C-scans which show no change in the overall damage area. The moiré fringe patterns show that the out-of-plane deflections can be up to 1 mm relative to the undamaged region. These deflections imply a large driving force for delamination growth, and as discussed earlier, it is the back surface delaminations that are most likely to propagate under compressive loading.

The growth of delamination damage has been investigated using moiré fringe techniques by Byers [*4*] and Mousley [*13*], and in both cases, implanted defects and impact damage were examined. Both show that the delaminations grow laterally under compressive loading, though the growth was more distinct for implanted defects. Byers [*4*] shows that for a toughened matrix composite higher compressive loads (and hence deformations) were needed for delamination propagation compared to composites with lower toughness matrices. The results on APC-2 confirm this pattern of behavior showing that impact delamination damage does not grow even at compressive deformations of ≥0.60%, and that this is due to the high interlaminar toughness of APC-2.

Conclusions

The delamination behavior of APC-2 has been examined using a variety of approaches. From the work we can draw a number of conclusions:

1. In the double cantilever beam test, two types of fracture behavior are seen. The interlaminar fracture toughness in the stable propagation region is 2.89 kJ/m^2, and there is a minimum value of arrest toughness during unstable propagation of 0.90 kJ/m^2.
2. In the edge delamination test, the dominant fracture mode was tensile rather than delamination, however a tentative value of toughness of ~1 kJ/m^2 was obtained. Modification of the test specimen should result in the delamination failure mode.
3. Low energy impact onto the face of specimens causes delamination and intraply cracking. A conical damage pattern is seen, though there is an inner conical zone in which cracking is absent.
4. Ballistic impacts cause slightly larger damage areas than low velocity impacts at equivalent incident energy.
5. Under repeated impacts, the rate of increase in damage area is low.
6. In the damage tolerance test, there is a gradual fall in compressive failure deformation with increasing impact energy, with a minimum value of 0.60% even at high incident energies.
7. There is a relationship between damage area and failure deformation, which includes ballistic and repeated impacts. The compressive failure deformation was at least 0.55% even when complete penetration of the specimen had occurred.
8. There is no evidence for the growth of delaminations under compressive loading at deformations up to 0.60%.

9. The high resistance of APC-2 to delamination caused by impact, including ballistic and repeated impact, and to the growth of delaminations on subsequent loading is due to the high interlaminar fracture toughness of the material.

Acknowledgments

We wish to express our thanks to R. B. Mousley, Royal Aircraft Establishment, Farnborough, for organization of the ballistic impacts and for advice on the moiré fringe technique. We also thank our colleagues at ICI: A. D. Curson for preparation and interpretation of the low energy impact polished section, R. A. Crick for scanning electron microscopy, P. White for ultrasonic C-scan, and N. Zahlan for the laminate plate calculations.

References

[*1*] Pagano, N. J. and Pipes, R. B., "Some Observations on the Interlaminar Strength of Composite Laminates," *International Journal of Mechanical Science,* Vol. 15, 1973, pp. 679–688.

[*2*] O'Brien, T. K., "Characterization of Delamination Onset and Growth in a Composite Laminate," in *Damage in Composite Materials, ASTM STP 775,* K. L. Reifsnider, Ed., American Society for Testing and Materials, Philadelphia, 1982, pp. 140–167.

[*3*] Rhodes, M. D., Williams, J. G., and Starnes, J. H., "Low Velocity Impact Damage in Graphite-Fiber Reinforced Epoxy Laminates," *Polymer Composites,* Vol. 2, No. 1, Jan. 1981, pp. 36–44.

[*4*] Byers, B. A., "Behavior of Damaged Graphite/Epoxy Laminates Under Compression Loading," NASA Contractor Report CR-159293, Aug. 1980.

[*5*] Bascom, W. D., Bitner, J. L., Moulton, R. J., and Siebert, A. R., "The Interlaminar Fracture of Organic-Matrix Woven Reinforcement Composites," *Composites,* Vol. 11, No. 1, Jan. 1980, pp. 9–18.

[*6*] Carlile, D. R. and Leach, D. C., "Damage and Notch Sensitivity of Graphite/PEEK Composite," *15th National SAMPE Technical Conference,* Vol. 15, Society for the Advancement of Materials and Process Engineering, Covina, CA, 1983, p. 82–93.

[*7*] Leach, D. C. and Moore, D. R., "Toughness of Aromatic Polymer Composites Reinforced with Carbon Fibres," *Composites Science and Technology,* Vol. 23, 1985, pp. 131–161.

[*8*] Whitney, J. M., Browning, C. E., and Hoogsteden, W., "A Double Cantilever Beam Test for Characterising Mode I Delamination of Composite Materials," *Journal of Reinforced Plastics and Composites,* Vol. 1, Oct. 1982, p. 297.

[*9*] O'Brien, T. K., Johnston, N. J., Morris, D. H., and Simonds, R. A., "A Simple Test for the Interlaminar Fracture Toughness of Composites," in *27th National SAMPE Symposium,* Vol. 27, Society for the Advancement of Materials and Process Engineering, Covina, CA, 1982; also published in *SAMPE Journal,* July/Aug. 1982, pp. 8–15.

[*10*] Aromatic Polymer Composite, APC-2, Data Sheets, ICI Advanced Materials Business Group, Welwyn Gdn City, Hertfordshire, England.

[*11*] O'Brien, T. K., "Mixed-Mode Strain-Energy Release Rate Effects on Edge Delamination of Composites," in *Effects of Defects in Composite Materials, ASTM STP 836,* American Society for Testing and Materials, Philadelphia, 1984, pp. 125–142.

[*12*] Johnston, N. J., O'Brien, T. K., Morris, D. H., and Simonds, R. A., "Interlaminar Fracture Toughness of Composites II—Refinement of the Edge Delamination Test and Application to Thermoplastics," in *28th National SAMPE Symposium,* Vol. 28, 1983.

[*13*] Mousley, R. F., "In-Plane Compression of Damaged Laminates," in *Structural Impact and Crashworthiness,* J. Morton, Ed., Elsevier Applied Science Publishers, New York, July 1984.

[*14*] Hartness, J. T., "An Evaluation of Poly-Ether-Ether-Ketone Matrix Composites Fabricated from Unidirectional Prepreg Tape," in *29th National SAMPE Symposium,* Vol. 29, Society for the Advancement of Materials and Process Engineering, Covina, CA, 1984, pp. 459–474.

[*15*] Crick, R. A., Leach, D. C., Meakin, P. J., and Moore, D. R., "Fracture and Fracture Morphology of Aromatic Polymer Composites," in *Proceedings of 1st European Conference on Composite Materials,* European Association for Composite Materials, Bordeaux, France, 1985, pp. 253–258.
[*16*] Dorey, G., "Fracture of Composites and Damage Tolerance," Advisory Group for Aerospace Research and Development, Lecture Series 124, Oct. 1982.

Thermosets

Albert F. Yee [1]

Modifying Matrix Materials for Tougher Composites

REFERENCE: Yee, A. F., "**Modifying Matrix Materials for Tougher Composites,**" *Toughened Composites, ASTM STP 937,* Norman J. Johnston, Ed., American Society for Testing and Materials, Philadelphia, 1987, pp. 383–396.

ABSTRACT: Attempts to toughen composites by using matrix materials of high toughness have not met expectations. The rationales given generally fall into two categories: size effect and constraints. This paper critically examines these issues and proposes approaches to test these rationales. The constraint effect is considered in light of recent advances in the understanding of the toughening mechanisms in rubber-modified plastics and epoxies. It is argued that cavitation processes are inherently capable of relieving the constraint created by the fibers, and that shear localization processes are effective in increasing the size of the plastic zone. The material properties giving rise to stable shear band propagation are discussed. The assumption of the necessary rubber particle size to effect toughening is examined by comparison with past experiences in toughening epoxies and certain thermoplastics. Several examples of successful toughening of resins by submicron rubber particles are cited.

KEY WORDS: composites, fracture, polymers, toughening, impact

Preliminary results have shown that the interlaminar fracture toughness of graphite fiber composites bears a positive correlation with epoxy resin toughness [*1*], but beyond the initial low toughness region the rate of improvement is less than expected. The use of ductile thermoplastics such as polysulfone as the matrix polymer has met with similarly limited improvements [*2*]. On the other hand, recent work [*3*] has shown that the more ductile the epoxy is, the more it can be toughened. This observation on the relationship between the toughenability and the ductility of thermosetting polymers has also been made for thermoplastic polymers [*4*].

[1]Formerly, Polymer Physics and Engineering Branch, Corporate Research and Development, General Electric Co., Schenectady, NY; now, professor, Department of Materials Science & Engineering, University of Michigan, Dow Bldg., Ann Arbor, MI 48109-2136.

The purpose of this paper is to explore in a reasonable, though speculative and qualitative manner, the question of how polymer matrix materials can be modified to yield tougher composites. There may well be other approaches to toughen composites, for example, by modifying the interface or by using ductile interlayers; however, these approaches will not be discussed in this paper. We argue in this paper that it is reasonable to expect that one of the avenues through which high toughness composites might be realized is the use of intrinsically ductile polymers which have been toughened by the incorporation of a dispersed elastomeric phase. To this end we first examine the possible causes for the lower than expected toughness in composites with ductile matrices; then, after a brief review of toughening mechanisms, we discuss the matrix properties that might lead to higher toughenability. Finally, the possible particle size effect of rubber modifiers is discussed.

Causes for Low Toughness in Composites with Tough Matrices

We first examine the possible causes for the low interlaminar fracture toughness of composites such as those studied in Refs *1* and *2*. For a given polymer specimen, the energy to fracture is proportionate to the size of the plastic zone, assuming that the micromechanisms for the formation of the plastic zone do not vary from one region of the plastic zone to another. If part of the plastic zone is replaced by a certain volume fraction of a material which does not contribute deformation energy to the specimen, then it is reasonable to expect that the fracture energy will be reduced by the volume fraction of the "inert" filler material.

Consider now a bulk polymer specimen containing a sharp crack. Upon loading in a direction perpendicular to the crack, the plastic zone, if formed, will usually be confined to the region ahead of the crack tip. The outline of the plastic zone is shown schematically in Fig. 1*a*. Consider then using this polymer as the matrix material in a unidirectional fiber composite. Assuming that the bulk plastic zone is large compared to the interfibrillar spacing, then, upon placing the fibers into the matrix, a first possible change in the fracture energy is that it is simply reduced by the volume fraction of the fibers. This rather optimistic estimate forms the upper bound on the interlaminar fracture toughness of a composite. Further loss in fracture energy is possible if the deformation mechanisms are inhibited by the presence of the fibers. A second, more detrimental possibility is that the plastic zone is confined to the interfibrillar spacing directly ahead of the crack tip. This highly restricted, or "bound" plastic zone is also depicted schematically in Fig. 1*a* and *b*. The reduction in fracture energy cannot be calculated from a simple reduction in plastic zone size, however, because of the unknown change in the plastic strain to failure in the bound plastic zone.

In the composite systems studied thus far, the toughness achieved is generally far from the upper bound estimated on a volume fraction basis [*1,2,5*]. Nonetheless, it is useful to obtain an estimate of the size of the plastic zone in these composites. Some of the most complete data necessary for estimating the plastic

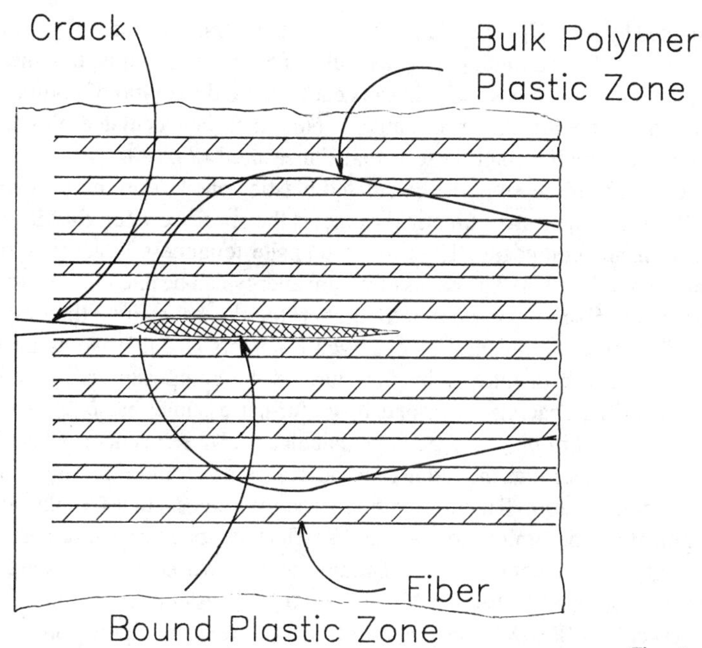

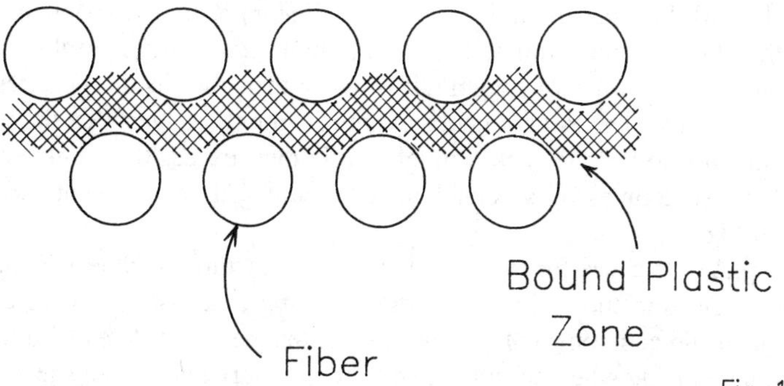

FIG. 1—(a) *A schematic diagram showing the size of the plastic zone in a tough bulk polymer, and how this zone is reduced by the volume fraction of fibers. By assuming that the constraint imposed by the fibers inhibits plastic zone formation, the size of the bound plastic zone in confined to the interfibrillar space.* (b) *A schematic of the bound plastic zone transverse to the fibers.*

zone size were obtained by Scott and Phillips [5]. They measured the fracture toughness G_{Ic} and yield strength of diglycidyl ether of bisphenol A (DGEBA) epoxies and their graphite fiber composites with and without tougheners. By assuming that the plastic zone ahead of the crack tip can be approximated by the Dugdale line zone, the thickness of the plastic zone at the crack tip ρ can be approximated by $\rho = G_{Ic}/\sigma_y$, where G_{Ic} is the plane strain fracture toughness and σ_y is the tensile yield strength [*6*]. The results of Scott and Phillips, together with the calculated plastic zone sizes ρ in the neat resin, and the ratio of composite to neat resin toughness are shown in Table 1. Note that the calculated plastic zone size tends to be smaller than those actually measured [*3*].

For the most brittle epoxy, the calculated plastic zone size in the neat resin is 2.5×10^{-6} m, in the same numerical range as the space between the fibers in a high performance composite. Here the composite toughness is actually higher than that of the neat resin. The excess fracture energy can be due to fiber bridging or other mechanisms. As we go to higher toughness resins, the ratio G_{Ic} (comp.)/G_{Ic} (resin) diminishes, as observed by others [*1,2*]. If we assume that the plastic zone size in intralaminar fracture of the composites is limited to a single interfibrillar spacing ρ_{comp}, and if we further assume that $\rho_{comp} = 2.5 \times 10^{-6}$ m, then the ratio ρ_{comp}/ρ_{resin} can be calculated. These results are also listed in Table 1. It is clear that the latter ratio decreases much more rapidly than the fracture toughness ratio. This observation strongly suggests that the plastic zone size is unlikely to be limited to a single interfibrillar spacing. These arguments find support in the recent work by Bradley and co-workers in the toughened F185 epoxy composite system [7]. In the APC-2 system consisting of poly-(etheretherketone) (PEEK) and graphite fiber, the relevant figures are, for the neat resin PEEK, σ_y = 110 MPa at 25°C and G_{Ic} = 4.8 kJ/m^{-2} at −40°C [*8*], and G_{Ic} = 2.4 kJ/m^{-2} [*9*] for the composite at room temperature. Assuming that G_{Ic} for PEEK at 25°C is also 4.8 kJ/m^{-2}, then the plastic zone size in the neat resin is 44×10^{-6} m. Furthermore, the composite toughness then happens to be close to the upper bound estimate based on volume fraction. Given this fact, a plastic zone size an order of magnitude smaller than that in the neat resin is difficult to envisage. The foregoing observations strongly suggest that the plastic zone size in those composites with a tough matrix must exceed the volume bound by the fibers, sometimes by a considerable amount. These arguments are also depicted in Fig. 1.

A more plausible cause for the low interlaminar fracture toughness is that the rigid fibers constrain the resin phase, thus creating a stress state with a high hydrostatic tensile component. High hydrostatic tensile stress states tend to promote voiding, crazing when possible, and brittle fracture rather than shear yielding. Hence, the ductility inherent in the polymer is lost. Polysulfone is a typical glassy polymer which is quite ductile in uniaxial tension, but which is brittle when even a blunt notch, such as those found in an Izod impact test, is present [*10*]. Here the notch creates a stress state with a high hydrostatic tensile component. Another indication that polysulfone is very notch sensitive can be found in Ref 2. It is, therefore, not at all surprising that the toughness of polysulfone matrix composites is much lower than expected. Yet, the notch sensitivity of polysulfone can be greatly reduced by the incorporation of a dispersed elastomeric phase [*10*]. This toughening effect will be discussed in a later section.

TABLE 1—*The results of Scott and Phillips, together with the calculated plastic zone sizes* ρ *in the neat resin, and the ratio of composite to neat resin toughness (data from Ref 4).*

Resin System (DGEBA Epoxy)	Resin Properties					
	G_{Ic}, J/m^2	σ_y, MPa	ρ, 10^{-6} m	Composite, G_{Ic}, J/m^2	G_{Ic}, Comp./G_{Ic} Resin	2.5×10^{-6} m/σ Resin
MY750[a]:MNA[a]:BDMA[c]	160	63	2.5	240	1.5	1
MY750:piperidine	330	45	7.3	280	0.85	0.34
MY750 + 3.2%CTBN[d]	1400	43	33	370	0.26	0.08
MY750 + 6.2%CTBN	2200	43	51	360	0.16	0.05
MY750 + 9%CTBN	3200	41	78	490	0.15	0.03

[a] MY750 is a diglycidyl ether of bisphenol A with a mean molecular weight of 300 from the Ciba-Geigy Co.
[b] MNA = methyl nadic anhydride.
[c] BDMA = benzyl dimethylamine.
[d] CTBN = liquid carboxyl terminated butadiene nitrile rubber from the B. F. Goodrich Co.

Still another possible cause for the low toughness of the composites in Refs *1* and *2* has to do with the fact that, in typical toughened epoxies, the rubber particles, which are responsible for the toughening effect, are of about the same size as the interfibrillar spacing, and it is highly doubtful that they can be effective tougheners under such circumstances. However, even this effect should not preclude the use of a toughened polymer as the matrix, since much smaller particles (<0.1 μm) can be used to toughen thermoplastics. The neat resin properties necessary for successful toughening with small rubber particles, and the nature of the rubber particles themselves will be discussed in a later section. It is useful at this point to discuss toughening mechanisms in polymers.

Review of Toughening Mechanisms in Bulk Polymers

During interlaminar fracture in a fiber composite, the polymer in the interfibrillar spacing or between plies is subjected to a similar type of stress state as the material in the bulk form containing a sharp crack loaded in mixed Mode I and Mode II. Since many thermoplastics and some epoxies can be toughened by the incorporation of an elastomeric phase, then, in light of the discussions in the preceding section, it is logical to assume that, under some circumstances, the toughening effect will be translated to the composite. The somewhat disappointing results from toughened epoxies [*1,2*] still leave untested a wide range of toughened thermoplastics, some of which are toughened by mechanisms [*6,10,14–18*] similar to those in epoxies [*3,11–13,19*]. Still others are toughened by massive crazing [*20*]. However, the materials toughened by massive crazing, for example, high-impact polystyrene (HIPS) and acrylonitrile-butadiene-styrene (ABS) generally have fairly low toughness even with the rubber modification, and are quite unsuited for use as the resin material; consequently, these materials will not be discussed further in this paper.

In general, the more ductile a neat polymer is, the more it can be toughened [*3,11,12*]. The reason for this becomes clear if we recall the mechanisms responsible for the toughening effect (again excluding massive crazing as a mechanism). The rubber particles cavitate or debond from the matrix, thereby relieve the triaxial tension; the remaining shear component then eventually causes localized shear deformation in the form of shear bands nucleated from the voids. The ability of a shear band to propagate, as will be discussed later, is related to the sequential processes of strain softening and strain hardening.

The observation that toughenability is proportionate to ductility can then be understood in terms of the size of the plastic zone and the amount of plastic strain, which are enhanced by stable shear band propagation. Now, it can be argued that, in a fiber composite, the size of the plastic zone is limited to the space bound by the fibers irrespective of the natural plastic zone size in the bulk polymer. This type of argument finds support in studies on rubber-modified epoxies in adhesive joints [*13*]. Yet, it must be recalled that in adhesive joints the adherends are much more rigid than the adhesive, consequently the only direction for the plastic zone to grow is along the plane of the joint, whereas in a composite, the yielding and unloading effect of the polymer in one representative volume element will cause the polymer in adjacent elements to yield. Furthermore, the stable growth of shear bands will enhance the ability of the latter to penetrate spatially, perhaps in spite of the presence of the fibers. The argument finds support in the observation of plastic deformation beyond several fiber diameters in an untoughened thermoplastic matrix composite.[2] Thus, we submit that the development of a toughened version of a highly ductile material to serve as the matrix is a viable though not necessarily unique approach to tough composites. We now discuss the material properties responsible for the development of stable shear bands and ductile fracture.

Neat Resin Properties

Shear Banding, Orientation Hardening, and Ductile Fracture

The processes leading to ductility are: (1) shear yielding and (2) stable neck propagation. The former relates to relaxation processes and how the applied strain accelerates them [*21*]. The latter relates to the processes that allow a shear band to form and then propagate, that is, strain softening followed by hardening. The tensile stress-strain behavior of glassy polycarbonate [*22*], which is perhaps one of the most ductile among high performance glassy and semicrystalline polymers, is depicted in Fig. 2 to illustrate the various typical deformation regimes. In this figure, which is a plot of the true stress versus stretch ratio, yielding is defined as the maximum in stress, and is preceded by linear, then nonlinear, deformation. After the yield point, the material undergoes a true instability (except for some semicrystalline polymers) as a result of intrinsic strain softening. The necking process begins after the strain-softening process, as indicated. Shear yielding, when it occurs, tends to preclude crazing and cracking; it is also necessary for forming stable shear bands. The stable neck propagation is possible because of the sequential softening and hardening effects. The hardening process is due to the orientation of the molecules along the direction of the shear stress component. To distinguish it from work hardening in metals, this process is often termed orientation hardening [*23*]. Obviously, if the molecular weight of the polymer is too low, that is, below the entanglement molecular weight, or if the polymer is too highly cross-linked, then, orientation hardening, which involves the large-scale cooperative rearrangement of random polymer coils or spherulites into a more or less extended chain configuration [*23*], cannot take place.

Matrix Relaxation Processes

In numerous high molecular weight polymers, ductility is high at temperatures sufficiently above the so-called β relaxation; otherwise, it is low even if the

[2]W. Bradley, private communication.

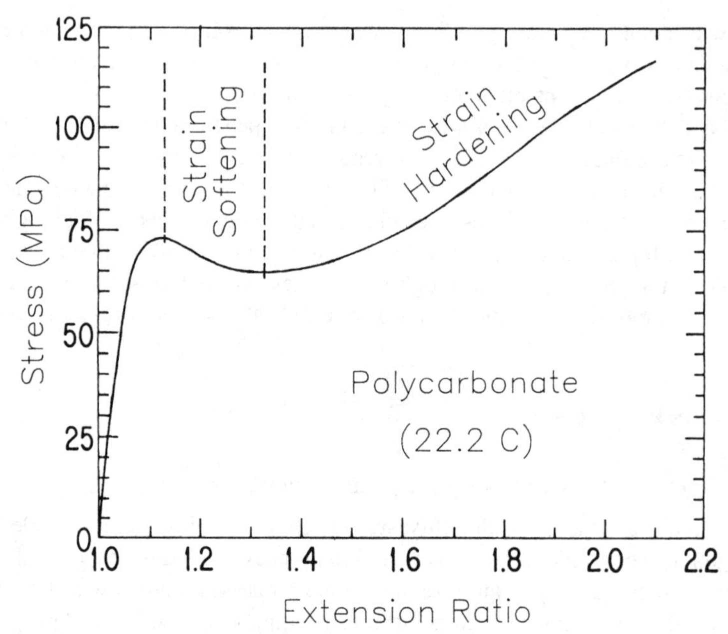

FIG. 2—*The true stress versus extension ratio behavior of a typical ductile glassy polymer such as polycarbonate. The strain softening is due to a true material instability. The strain hardening is due to molecular orientation.*

material does yield.[3] A discussion of the phenomenology related to the β relaxation in polymers is beyond the scope of this paper. The interested reader is referred to standard references on this subject, for example, Ref *24*. β relaxations also affect the formation of shear bands. It has been proposed that the ability to form shear bands is related to a material parameter ε^*, where $\varepsilon^* \equiv (-d\sigma_y/d \ln \dot{\varepsilon})/(d\sigma/d\varepsilon)$, which is the ratio between the strain rate sensitivity of the yield stress and the rate of strain softening after yield [*25*]. The strain rate sensitivity is greatly influenced by the β relaxation [*26,27*]. The rate of strain softening after yield is due to the inability of the material structure to relax at the same rate as the applied strain [*21,28,29*] before yield, thus producing what amounts to an overshoot in the stress. This strain-induced accelerated relaxation is again affected by the process as well as the structural relaxation process, familiar to polymer physicists as physical aging. Clearly, the importance of the role played by the β relaxation cannot be overemphasized. At temperatures near or above the β relaxation (at a frequency corresponding to the strain rate), the polymer generally shear yields more readily, and the strain rate sensitivity of the yield stress is relatively low. Thus, two of the conditions necessary for stable shear band propagation are met. In choosing a polymer to be toughened, then, it appears that the existence of low temperature relaxation processes is a prerequisite.

[3]A. F. Yee and S. A. White, "The Effect of Strain Rate and Temperature on the Tensile Strength and Ductility of Polyetherimide," unpublished manuscript.

The existence of a low temperature relaxation process is not, in itself, a guarantee for ductility. Evidences accumulated suggest that the relaxation must be strongly coupled to the polymer backbone, and that the backbone must be capable of extensive segmental motion, for example, not too highly cross-linked. Furthermore, the molecular weight of the polymer must be at least as high as the entanglement molecular weight. The significance of the foregoing is that in attempting to modify a polymer molecularly as matrix material for tough composites, one must bear in mind the effect of the modification on the molecular mobility of the polymer: the modification should preserve or enhance, not destroy the molecular mobility.

Just as having a β relaxation is not a guarantee that the polymer will be ductile, having intrinsic ductility in uniaxial tension is also not sufficient to produce plastic zones necessary for toughness in all stress states. For example, reducing the cross-link density in epoxies increases ductility, but affects the fracture toughness in a very insignificant way [*3*]. For another example, with increasing geometric constraint or rate, normally tough polycarbonate becomes quite brittle [*15*]. This is probably related to the fact that, as the constraint changes, the stress state experienced by the polymer, hence, the relaxation spectrum of the material also changes. That the latter is true is demonstrated by studies on the stress relaxation behavior of poly(methyl methacrylate) (PMMA) by varying the tension-to-torsion ratio [*30*], and by resolving the volumetric and shear components [*31*]. It is clear that the microscopic origin of the stress-state dependence of the relaxation behavior is both interesting and important.

Nonetheless, we need only to recognize the reasonable argument that the stress-state dependence of the relaxation behavior is due to the fact that the molecular motions consist of inter- and intra-molecular components which may interact differently with various stress states, and that the nature of the β relaxation has an important bearing on this behavior. The important point is that in polymers which becomes brittle in the presence of a notch, the toughness can be restored by internally relieving the hydrostatic tensile stress component by cavitation [*11,12,15*]. The cavitation can be nucleated by a dispersed phase of soft particles.

Ductile Fracture in Voidy Solids

The formation of multiple cavities in a solid can lead to ductile fracture, a process well-known in metals [*32*]. A model study by McClintock showed that in a voidy material capable of both plastic and viscous flow, there is a very strong inverse dependence of fracture strain on hydrostatic tension [*33*]. These results have been confirmed by other model studies [*34–36*]. Basically these studies show that as the voids grow, an instability occurs because of internal necking between adjacent voids. The question then arises as to how a solid with voids can be more ductile than one without. To answer this question we must first examine how the assumptions in these models differ from the materials being considered here.

The most important variance appears to be the assumption of the initial existence of voids. In the present materials, good adhesion between the rubbery phase and the plastic phase can be assumed to exist. As the stress is applied, the bulk strain energy in the polymeric phase comprising both the rubbery and the plastic phases rises much more rapidly than the shear strain energy because of the constraints. Eventually, though, the bulk strain energy causes cavitation in the rubber particles, as in rubber-toughened epoxies [*3,11–13*], or debonding around the soft phase [*15*]. At this point, the bulk strain energy is rapidly dissipated, while, at the same time, the shear strain energy continues to grow unabated. This situation is schematically represented in Fig. 3.

A second factor is the fact that ductile polymers exhibit orientation hardening. Several of these model studies show that a high strain-hardening coefficient tends to retard void growth and subsequent instability [*33,35,36*]. Yet another study, intended to model the crazing process in polymers, shows that orientation hardening has the effect of stabilizing the ligaments formed by the growth of adjacent voids [*37*]. Finally, the situation depicted in Fig. 3 is supported by volume strain measurements made during the tensile deformation of rubber-modified epoxies [*3,11,12*] which show that the cavitation process is followed by enhanced shear deformation. Optical micrographs support the notion that while void growth does continue unabated, localized shear band formation, nucleated by the voids, creates a much larger plastic volume before failure [*3,11,12*]. The discussions here thus further emphasize the importance of orientation hardening. The significance of adhesion between the rubbery phase and the plastic matrix is also underscored.

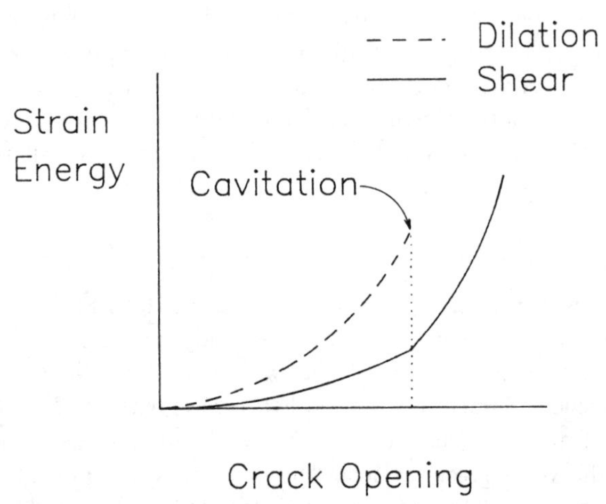

FIG. 3—*A schematic diagram illustrating the growth of dilatational and shear strain energies as the crack opens. A cavitation event dissipates the dilatational energy and may actually accelerate the growth of shear strain energy.*

Rubber Particle Size Effect

In rubber-toughened plastics, the rubber particle size is indeed an important consideration. It is well-known that polystyrene, for example, is not effectively toughened by rubber particles less than a few microns in diameter [*17*]. Perhaps because of this there is a general belief that all plastics must be toughened with rubber particles of a similar size. Less well-known is the fact that many plastics are toughened by rubber particles a fraction of a micron or less in diameter. Some examples are: polyvinyl chloride (PVC), by methacrylate-butadiene-styrene (MBS) rubber particles less than 0.1 μm in diameter [*13,15*], and polycarbonate, by micron-sized polyethylene and submicron MBS [*10,35*]. These are by no means isolated examples. The patent literature contains, in fact, numerous cases of claims of successful toughening of thermoplastics with elastomers. Though the size of the rubber particles is rarely identified in these patents, some of them, such as MBS, have characteristic diameters much less than 0.1 μm. We now discuss concepts that have been proposed for selecting particle size.

There are two major types of toughening mechanisms in plastics [*11,12,36*]. The first is massive crazing, where the rubber particles act as the nucleation sites for crazes, and also as craze termintors [*17*]. Since the energy density in each craze is high, by inducing profuse crazing in a large volume of material, the total toughness can be greatly increased. The second is profuse shear banding, where the rubber particles cavitate or debond, thereby relieve the hydrostatic tension, and nucleate shear bands [*11,12,36,38*]. It is perhaps obvious that the first mechanism is effective only in materials that normally fail by crazing and fracture in uniaxial tension, and the second in materials that are ductile in uniaxial tension but are notch sensitive. Also, there is generally a lower size limit to the rubber particles effective in the first mechanism, whereas the only requirement for the second mechanism is that the particles form a distinct second phase with a sharp interface. The particle size requirement has been investigated by Donald and Kramer in poly(acrylonitrile-butadiene-styrene) (ABS) films [*14*]. They found that films containing only rubber particles 0.1 μm in diameter show little tendency for crazing. Instead, cavitation of the rubber particles together with localized shear deformation occurs.

For specimens containing a mixture of small or large particles, crazes tend to nucleate at the larger particles only. They attribute this effect to the idea that, to develop the void-fibril network of a craze, the region of high stress around the particle must extend over a critical distance of three or more interfibrillar spacings in a mature craze. The latter is a material dependent property. Therefore, even though stress concentration effect around a particle scales with the particle radius, if a particle is too small, the critical interfibrillar spacing requirement cannot be satisfied. According to these ideas, if small particles are placed in a matrix which forms crazes more readily than shear bands, they will have no effect in nucleating profuse, massive crazing. But, the same particles placed in an intrinsically ductile matrix will have the desired toughening effect.

Conclusion

In this paper we have attempted to review critically the various proposed causes for low toughness in composites with tough matrices and found that the constraint effect is likely to be a major factor. Since some rubber-toughened plastics are able to relieve internally the constraints by voiding, the toughening mechanism of these plastics is briefly reviewed. The neat resin properties that might enhance these mechanisms are explored. We have identified both strain softening and a high degree of orientation hardening as being very important. Furthermore, we show that it may be possible to select and modify resins based on the existence of certain low temperature relaxation processes. The role of orientation hardening is further stressed because it may be an important element in extending the plastic zone size. Finally, examples from the literature are cited to demonstrate that intrinsically ductile polymers can be successfully toughened by very small particles, thus supplying a possible mechanism for obviating the restrictions imposed by the size of the interfibrillar spacing in a composite.

References

[1] Hunston, D., in *Tough Composite Materials,* NASA Conference Publication 2334, 1984, p. 3.
[2] Browning, C., in *Tough Composite Materials,* NASA Conference Publication 2334, 1984, p. 33.
[3] Yee, A. F. and Pearson, R. A., NASA Contractor Report 3852, 1984.
[4] Nikpur, K. and Williams, J. G., *Plastic and Rubber: Materials Applications,* Vol. 3, 1978, p. 163.
[5] Scott, J. M. and Phillips, D. C., *Journal of Materials Science,* Vol. 10, 1975, p. 551.
[6] Kinloch, A. J. and Young, R. J., *Fracture Behavior of Polymers,* Applied Science Publishers, London, 1983.
[7] Chakachery, E. A. and Bradley, W. L., "A Comparison of the Crack Tip Damage Zone for Fracture of Hexel F185 Neat Resin and T6T145/F185 Composite," *Polymer Engineering and Science,* Vol. 27, 1987, p. 33.
[8] Jones, D. P., Leach, D. C., and Moore, D. R., *Polymer,* Vol. 26, 1985, p. 1385.
[9] Crick, R. A., Leach, D. C., and Moore, D. R., *Proceedings of the 31st SAMPE Symposium,* Las Vegas, 1986.
[10] Noshay, A., Matzner, M., Barth, B. P., and Walton, R. K., in *Toughness and Brittleness of Plastics,* ACS Advanced Chemistry Series No. 154, R. D. Deanin and A. M. Crugnola, Eds., American Chemical Society, New York, 1976, p. 302.
[11] Yee, A. F. and Pearson, R. A., *Journal of Materials Science,* Vol. 21, 1986, p. 2462.
[12] Pearson, R. A. and Yee, A. F., *Journal of Materials Science,* Vol. 21, 1986, p. 2675.
[13] Bascom, W. D., Cottington, R. L., Jones, R. L., and Peyser, P., *Journal of Applied Polymer Science,* Vol. 19, 1975, p. 2545.
[14] Bascom, W. D., Ting, R. Y., Moulton, R. J., Riew, D. K., and Siebert, A. R., *Journal of Materials Science,* Vol. 16, 1981, p. 2617.
[15] Yee, A. F., *Journal of Materials Science,* Vol. 12, 1977, p. 757.
[16] Haaf, F., Breuer, H., and Stabenow, J., *Journal of Macromolecular Science, Physics,* Vol. B14, 1977, p. 387.
[17] Donald, A. M. and Kramer, E. J., *Journal of Materials Science,* Vol. 17, 1982, p. 1765.
[18] Petrich, R. P., *Polymer Engineering and Science,* Vol. 13, 1973, p. 248.
[19] Kinloch, A. J., Shaw, S. J., and Hunston, D. L., *Deformation, Yield and Fracture of Polymers Conference,* Plastic Rubber Institute, London, 1982, p. 29–1.
[20] Bucknall, C. B., *Toughened Plastics,* Applied Science Publishers, London, 1977.
[21] Yee, A. F., Bankert, R. J., Ngai, K. L., and Rendell, R. W., *Proceedings of the International Congress of Rheology,* Acapulco, Mexico, Oct. 1984.
[22] Nied, H. F. and Stokes, V. K., General Electric CRD Report 84CRD235, 1984.

[23] Haward, R. N., in *The Physics of Glassy Polymers,* R. N. Haward, Ed., Wiley, New York, 1973, Chap. 6.
[24] McCrum, N. G., Read, B. E., and Williams, G., *Anelastic and Dielectric Effects in Polymeric Solids,* Wiley, London, 1967.
[25] Bowden, P. B., *Philosophical Magazine,* Vol. 22, 1970, p. 455.
[26] Roetling, J. A., *Polymer,* Vol. 6, 1965, p. 311.
[27] Bauwens, J. C., *Journal of Materials Science,* Vol. 7, 1972, p. 577.
[28] Ngai, K. L. and Yee, A. F., in *Proceedings of Workshop on Relaxation Phenomena in Complex Systems,* U.S. Government Printing Office, 1985.
[29] Rendell, R. W., Ngai, K. L., Fong, G. R., Yee, A. F., and Bankert, R. J., "Nonlinear Viscoelasticity and Yield: Application of a Coupling Model," *Polymer Engineering and Science,* Vol. 27, 1987, p. 1.
[30] Sternstein, S. S. and Ho, T. C., *Journal of Applied Physics,* Vol. 43, No. 11, 1972, p. 4370.
[31] Yee, A. F. and Takemori, M. T., *Journal of Polymer Science, Polymer Physics,* Vol. 20, 1982, p. 205.
[32] Rogers, H. C., *Transactions of the Metallurgical Society of AIME,* Vol. 218, 1960, p. 498.
[33] McClintock, F. A., *Journal of Applied Mechanics,* Vol. 35, 1968, p. 363.
[34] Rice, J. R. and Tracely, D. M., *Journal of the Mechanics and Physics of Solids,* Vol. 17, 1969, p. 210.
[35] Needleman, A., *Journal of Applied Mechanics,* Vol. 39, 1972, p. 964.
[36] Tvergaard, V., *International Journal of Fracture,* Vol. 17, 1981, p. 389.
[37] Haward, R. N. and Owen, D. R. J., *Journal of Materials Science,* Vol. 8, 1973, p. 1136.
[38] Bahadur, S., *Polymer Engineering and Science,* Vol. 13, 1973, p. 264.

DISCUSSION

Question 1

G. M. Newaz[1] — *(written discussion)* — In discussing the toughness of rubber modified plastic matrices, you have made no mention about the rubber particle and matrix interface or interphase. Can you comment on what role a poor or a good bond between the particle and the matrix plays on overall toughness of the modified matrix, at least on a qualitative basis?

A. F. Yee (author's closure) — Model calculations assuming linear elastic behavior indicate that the shear stress concentration is not sensitive to whether the rubber particle is well bonded or not bonded (that is, is a hole). However, the hydrostatic tension in the equatorial region of the particle is higher if the bonding is good because of the finite Poisson's ratio of the particle. Experimental results show that, in general, good bonding is preferred. This may be because the hydrostatic tension reduces yield stress, thus promoting shear band formation.

Question 2

G. M. Newaz (written discussion) — In dealing with the toughness issue of fiber-reinforced composite, you correctly point out that the matrix toughness is

[1]Owens-Corning Fiberglas, Technical Center, Granville, OH 43056.

not translatable into the composite, one to one. You refer to the particle size effect and the fiber constraints as the physical reasons that can account for this difference. Again, you have ignored the role of the fiber-matrix interface or interphase damage processes that can contribute to the overall toughness of the composite. Quality of fiber-matrix bond is likely to influence the composite toughness. Furthermore, if we consider toughness issues not confined to the delamination problem, fiber breakage, pull-out, local fiber deflections, and so forth, can also play significant roles. The point I am trying to make is that even in the case of a rubber-modified matrix and matrix-dominated fracture in the composite, the particle size effect and fiber constraints, as you discussed in your presentation, can only partially account for the total toughness of the composite. I would appreciate your comments along this line.

A. F. Yee (author's closure)—The fiber-matrix interface and the fiber bridging, fiber pull-out, and so on processes undoubtedly play significant roles in contributing to the toughness of a composite. However, the title of this paper indicates that it is on modifying matrix toughness. Various results have demonstrated that increasing matrix toughness alone reduces the extent of damage when a composite laminate is subjected to an impact normal to its plane. Obviously, having a poor fiber-matrix interface will undermine the effectiveness of the toughened matrix.

Question 3

G. M. Newaz (written discussion)—In your abstract for the presentation, you mentioned shear band forms as a result of stress concentration from rubber particles or as a result of the cavity. This is ambiguous. Obviously, the stress concentration that may lead to shear band formation is due to the cavity enclosing the rubber particle. The rubber particle having lower modulus than the matrix cannot cause stress concentration by itself.

A. F. Yee (author's closure)—The shear modulus of the rubber particle is much lower than that of the rigid matrix. Therefore, as far as its ability to cause shear stress concentration is concerned, it is not much different from a hole. But it can definitely cause shear banding even if a cavity does not surround it. See also the answer to Question 1.

Ramon Garcia,[1] *Robert E. Evans,*[2] *and Raymond J. Palmer*[3]

Structural Property Improvements Through Hybridized Composites

REFERENCE: Garcia, R., Evans, R. E., and Palmer, R. J., **"Structural Property Improvements Through Hybridized Composites,"** *Toughened Composites, ASTM STP 937,* Norman J. Johnston, Ed., American Society for Testing and Materials, Philadelphia, 1987, pp. 397-412.

ABSTRACT: A hybrid composite technique identified as having potential for improving the matrix-dominated properties of continuous fiber, reinforced composite materials is examined. This technique employs either particulates, whiskers, or microfibers to provide supplementary reinforcement of the matrix. An essential feature of this technique is that the supplementary reinforcement material should possess a diameter and aspect ratio such that the continuous fiber packing efficiency is not disrupted. The advantages of this technique are such that it does not require the development of new constituent materials and that it is adaptable to the processes and equipment currently used to fabricate aerospace grade composite materials.

Packing concepts, as well as micromechanics theories, which support this hybrid technique, are reviewed. Theoretical formulations are used to predict selected property improvements as a result of hybridizing silicon-carbide whiskers with graphite/epoxy. A series of laboratory tests are conducted to delineate the differences in structural and mechanical properties between a baseline graphite/epoxy composite and a hybrid silicon-carbide/graphite/epoxy composite.

KEY WORDS: composite materials, hybrid composites, silicon-carbide whiskers, epoxy, graphite fibers, supplementary reinforcement

The advantages of composite materials, such as graphite/epoxy (Gr/E), used in aircraft applications have been well documented. The main advantages of these materials over conventional metals can best be described in terms of superior

The opinions and assertions expressed in this paper are the private ones of the authors and are not to be construed as official or reflecting the views of the Department of the Navy. The use of tradenames or manufacturers in this paper does not constitute endorsement, either expressed or implied, by the Department of the Navy.

[1]Formerly, aerospace engineer, Naval Air Development Center, Code 6043, Warminster, PA 18974; presently, deputy vice president, Atlantic Science and Technology Corp., 1939 Route 70E, Cherry Hill, NJ 08003.

[2]Manager, Advanced Materials Development, American Cyanamid Co., Stamford, CT 06904.

[3]Senior staff engineer, Composites, Douglas Aircraft Co., Long Beach, CA 90846.

strength and stiffness to weight ratios. In recent years, the need to reduce fuel and strategic material usage while increasing payload and range have led to significant composite material applications on military and commercial aircraft. However, the primary structural deficiencies of current generation composites are related to low in-plane and out-of-plane matrix-dominated mechanical properties.

While the recent development of tough epoxy matrix resin systems has led to improvements in laminate damage tolerance, further improvements are possible. In addition, resin modulus remains a key factor in structural performance. Laminate impact resistance is directly proportional to neat resin strain to failure [*1*]; laminate compressive performance is equally important and it is directly proportional to resin modulus [*2*]. Fiber kinking and fiber microbucking, the predominant compressive failure modes in graphite/epoxy laminates, are instability phenomena controlled by resin modulus [*3*]. Therefore, successful resin modification techniques are those that improve both aspects of a resin's mechanical properties and result in a composite having the proper balance of mechanical properties.

Hybridization is a technique which may have the potential to improve the matrix-dominated properties of the composite structure in a properly balanced fashion [*4*]. The hybrid technique examined in this paper consists of combining silicon carbide (SiC) whiskers and continuous graphite fibers within an epoxy resin matrix. Analytical data is reviewed to illustrate property improvement potential and material system design considerations. Experimental data are presented that illustrate the effect SiC whiskers have on an epoxy resin. These results were used as a precursor to examining graphite/epoxy/silicon-carbide (Gr/E/SiC) formulations. Although the results of the study described in this paper appear somewhat cursory, they are one part of a larger overall program to fully examine the potential and limitations of the hybrid technique.

Material Design Issues

Packing Concepts

The microstructural arrangement of a typical Gr/E composite material system is illustrated in Fig. 1. Regions are found where the fibers are arranged in hexagonal, square, and random arrays. The regions where hexagonal and square arrays exist have fiber volumes of approximately 90 and 78%, respectively. Regions where random arrays exist have fiber volumes of less than 78%. Optimum material properties are obtained when a composite has a fiber volume of approximately 60 to 65%; in such a composite all three of these arrays are found.

Consider the addition of a supplemental reinforcement material to the epoxy matrix. This material has to be of the proper size so as not to disrupt fiber packing. Examination of the hexagonal array for 8-μm-diameter graphite fibers indicates that the maximum diameter which can be accommodated without array disruption is 1.2 μm.

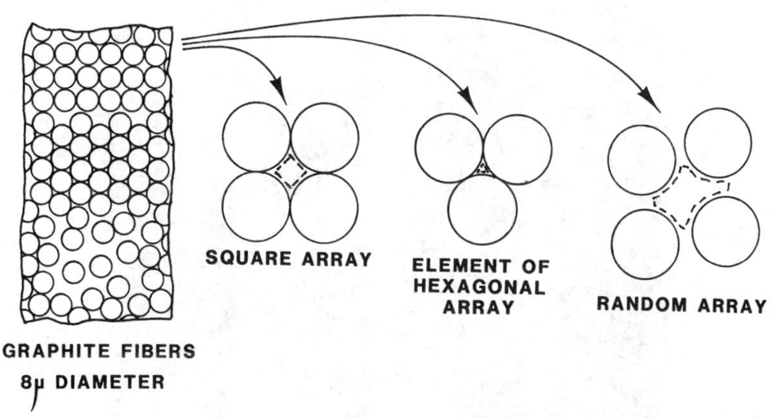

FIG. 1—*Microstructural arrangement of a typical graphite/epoxy composite material.*

The gains in mechanical properties as a result of adding the supplemental reinforcement material can be viewed in terms of the increased packing efficiency of the higher strength and stiffness constituents of the composite. This packing concept is clearly illustrated in Fig. 2. Two cylinders are partially filled with rods having an aspect ratio (that is, length/diameter) of 15. Equal volumes of spheres are introduced to each cylinder. In one case, the ratio of sphere-to-rod diameter R is 4, while in the other case R is 0.5. After mixing, the cylinder having $R = 4$ is found to have a void content of 50% while the cylinder having $R = 0.5$ is found to have a void content of 35%. As R is decreased below 0.5, void content will also decrease. In an actual composite of this form, the void content would translate into the matrix content.

These concepts and experiments are described for illustrative purposes only and indicate the source of the ideas for adding supplementary reinforcements to continuous fiber composites.

Uniform dispersion and random orientation of the supplemental reinforcement material are also necessary to prevent stress concentration and to obtain uniform reinforcement in all directions. Therefore, to provide proper packing efficiency, strength as well as stiffness, uniform dispersion, and random orientation, the supplemental reinforcement material must possess a diameter much less than 1 μm.

Constituent Selection

Satisfying the points discussed in the previous section imposes a severe constraint on the size and geometry of candidate supplemental reinforcement materials available for hybridization in Gr/E. A prudent approach in the selection of

FIG. 2—*Effect of particle size on packing efficiency (courtesy of J. V. Milewski).*

supplemental reinforcement materials would eliminate materials that are not commercially available, nor well characterized. This approach eliminates candidates that may exhibit significant property variability from batch to batch and lack maturity in terms of proven material processing, mechanical characterization, and continued availability. Based on these factors and the packing factors discussed previously, SiC whiskers were selected for investigation. SiC whiskers are the only material which have the characteristics desired of the supplemental reinforcement material. Characteristics of the commercially available SiC whisker (Tateho Chemical Industries Co., Grade SCW 10) used in this study are given in Table 1.

TABLE 1—*Silicon carbide whisker characteristics.*

Diameter	0.1−0.5 μm
Length	10−40 μm
Density	3.21 g/cm^3
Heat resistance	1600°C
Surface area	3.0 m^2/g
Crystal type	95% β
	5% α
Tensile strength, GPa (Msi)	20.6 (3)
Tensile modulus, GPA (Msi)	483 (70)

The epoxy resin system chosen for use in this study was Cycom® 1806 (American Cyanamid Co.). Celion® 6000ST (Celanese Corp.) and AS-4 (Hercules Inc.) graphite fibers were selected for examination.

Effect of Whisker Additions

An analytical investigation of the consequences of hybridizing SiC whiskers and graphite fibers within an epoxy resin matrix was conducted in Ref *4*. The constituent material properties used in that investigation are listed in Table 2. The addition of SiC whiskers to a Gr/E composite will result in a density increase. This can readily be seen by examining the density values given in Table 2. Figure 3 illustrates the increases in density caused by displacing a percentage of the epoxy matrix with SiC whiskers, while holding graphite fiber volume content constant. If the density increase as a result of the addition of the SiC to the Gr/E outpaces property improvements, the specific properties of Gr/E/SiC will be lower than the specific properties of Gr/E. The benefits of a Gr/E/SiC composite must therefore be determined in terms of specific properties. Figure 4 illustrates moduli improvements as a result of adding SiC to Gr/E. These data were calculated using the Halpin-Tsai equations [*5*] and the data in Table 2. SiC whiskers were assumed to have an aspect ratio of one. Specific modulus of elasticity in the fiber direction (E_1), modulus of elasticity perpendicular to the fiber (E_2), and in-plane shear modulus (G_{12}) (all normalized with respect to density) for a Gr/E/SiC composite were normalized with respect to the corresponding specific moduli of a Gr/E composite. The data given in Fig. 4 indicates that specific moduli improvements occur as a result of hybridizing SiC with Gr/E. Since the Halpin-Tsai equations represent an approximate method, the moduli improvements illustrated in Fig. 4 are only intended as qualitative indicators of hybridization potential.

The addition of SiC to the epoxy resin may rheologically affect the mixture by increasing viscosity and thixotropy. Therefore, from a processing and manufacturing standpoint, it may not be possible to displace large amounts of resin with SiC as illustrated in Fig. 4. There also exists an as yet unknown SiC reinforcement volume threshold below which no property improvement occurs as a result of adding SiC. Again these data are presented as qualitative, not quantitative, indicators of potential property improvements. The effects of hybridizing

TABLE 2—*Material properties used for density and Halpin-Tsai equations calculations.*[a]

Material	E, GPa (Msi)	G, GPa (Msi)	P, kg/m^2 (lbs/in.^{3})
Graphite fiber	234 (34)	96 (14)	1800 (0.065)
Epoxy resin	4.5 (0.65)	1.7 (0.24)	1220 (0.044)
Silicon carbide whisker	689 (100)	262 (38)	3180 (0.115)

[a] Table obtained from Ref *4*.

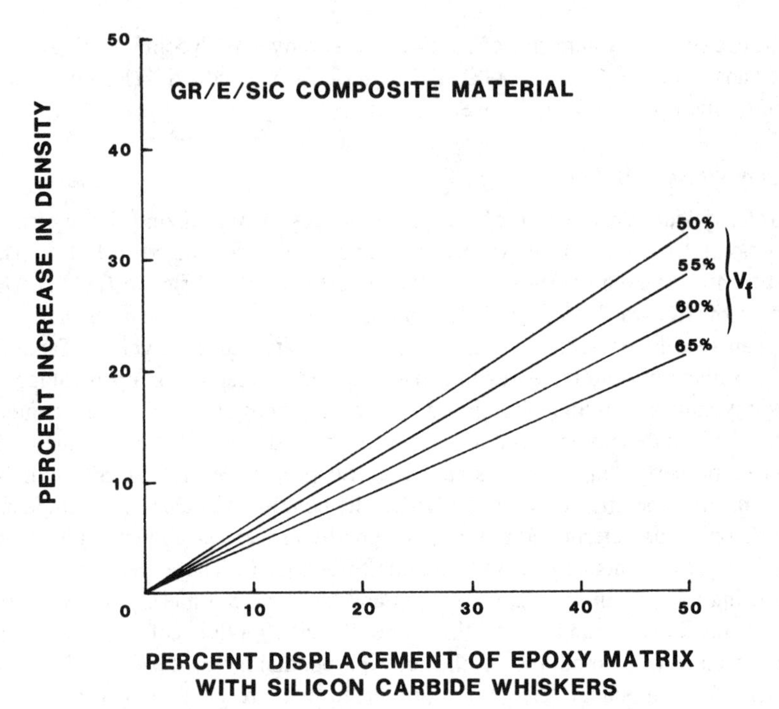

FIG. 3—*Effect on composite density as a result of the addition of silicon carbide whiskers.*

on other mechanical properties, including strength properties, is beyond the scope of this discussion; however, these effects are addressed in Ref *4*.

Test Materials

Resin-Whisker Blends

Whisker contents of 0, 3, 6, 10, and 15 parts of reinforcement per 100 parts epoxy (PHR) were selected for examination. Each whisker-resin blend was cast to form 1.59-mm (0.0625-in.) thick sheets from which flexure and compact tension specimens were machined and subsequently tested.

The casting procedure is based on a modification of a commonly used procedure that casts between flat plates. This procedure fabricates 4- by 5-in. (102- by 127-mm) neat resin plaques between two pieces of plate glass. The technique uses commonly available material to produce a plaque with smooth, parallel surfaces. Thickness can be varied from 1/32 to 1/4 in. (0.8 to 6 mm) by changing gasket size. Resin is poured into the mold, and cured following standard cure cycles.

The flexure test provides data in the stress-strain behavior of the resin. Basic properties obtained are modulus, E; offset yield stress, σ_{os}, and strain ε_{os}; yield stress, σ_y, and strain ε_y; failure strength σ_F, and failure strain, ε_F; and area under

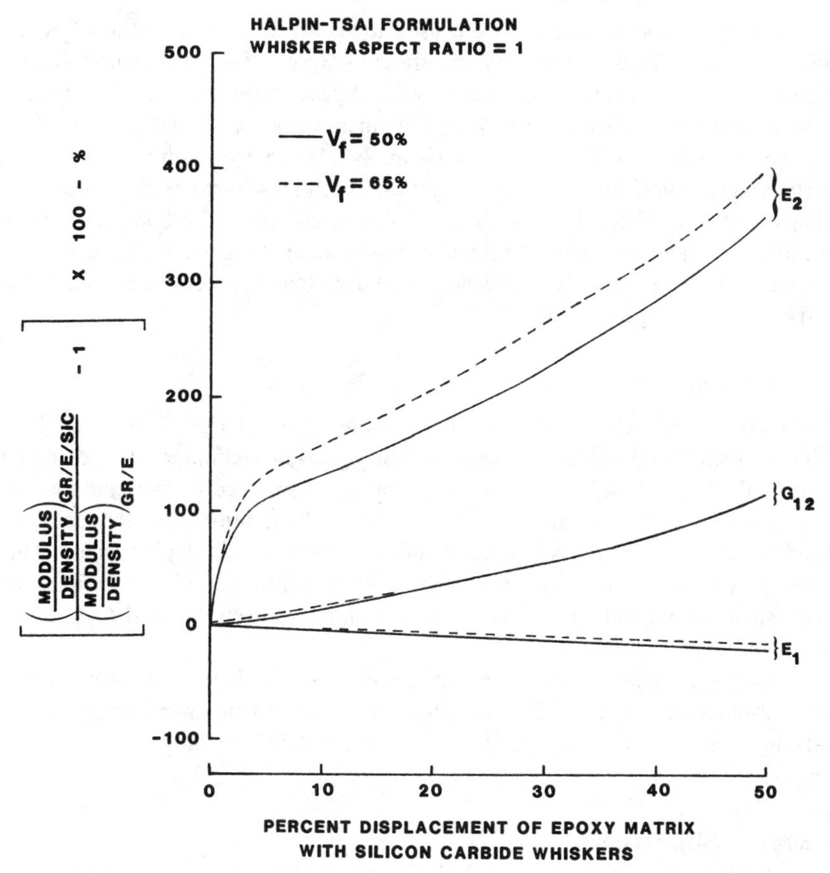

FIG. 4—*Projected moduli improvements as a result of the addition of SiC whiskers: Halpin-Tsai formulation with whisker aspect ratio equal to one.*

the stress-strain curve (or work to break), W (in.-lb/in.^{3}). The samples are small, 1⁄16 by 1½ by ½ in. (1.6 by 38 by 13 mm), and are simple to machine (being rectangular). Testing of flexure specimens is rapid and has no requirement for strain transducers since mid-span deflection is several orders of magnitude greater than any load cell or machine deflections. Five specimens are used in each flexure test. The strain rate is 10%/min. Also, hot/wet specimens can be tested in an immersion bath to reduce moisture loss during testing. The test procedures and equations for the calculation of properties are given in ASTM Tests for Flexural Properties of Unreinforced and Reinforced Plastics and Electrical Insulating Materials (D 790). The data from each of the five individual flexure bars are checked for outliers, averaged, and the 95% confidence limit is determined—$2S/n^{1/2}$ where n = number of samples in the average and S is the standard deviation.

Compact tension specimens were used to measure the plane-strain fracture toughness K_{Ic}. The strain energy release rate G_{Ic} can be calculated from this quantity. The test specimen geometry and test procedures used are described in ASTM Test for Plane-Strain Fracture Toughness of Metallic Materials (E 399). The specimen size is 1⅝ by 1½ by 1⁄16 in. (41 by 38 by 1.6 mm) thick. This permits us to manufacture fracture toughness specimens from the same neat resin plaque that is used for flexural property determinations. The compact tension specimen provides an excellent screen on resin fracture toughness. These tests are a relative measure of the effectiveness of the various approaches in improving resistance to flaw propagation.

Gr/E/SiC Materials

C6000ST and AS-4 graphite fibers were impregnated with a 6 PHR-whisker/resin blend using the hot-melt impregnation technique. The resultant C6000ST/1806/SiC and AS-4/1806/SiC prepregs were used to fabricate the following laminates: $[0]_8$, $[0]_{16}$, $[90]_{20}$, $[\pm 35, 0, 90]_s$. In addition a small quantity of AS-4/1806/SiC prepreg was produced from a 10 PHR-whisker/epoxy blend. This prepreg was used to fabricate a $[0, \pm 35, 90]_s$ laminate. No modifications to standard impregnation and fabrication techniques were needed to produce the materials.

Various specimens types were machined from the laminates listed above and subsequently tested. The data from these specimens were compared to existing control data from AS-4/1806 and C6000ST/1806 prepregs having no SiC content.

Results and Discussion

E/SiC Material Data

Flexure specimens containing 0, 3, 6, and 10 PHR of SiC whisker were tested under room temperature dry (RTD) and elevated temperature wet (ETW) conditions. Moisture conditioned specimens were submerged in 71°C water for 14 days, and subsequently tested in water at a temperature of 93°C. The modulus data obtained from these tests is plotted in Fig. 5. The RTD data points at 0, 3, and 6 PHR are the average of 5 replicates, while the RTD data point at 10 PHR is the average of 15 replicates. The ETW data points at 0, 3, and 6 PHR are the average of 5 replicates, while the ETW data point at 10 PHR is the average of 10 replicates. Trend lines have been fitted to the data presented.

Whisker loading levels of 3, 6, and 10 PHR result in modulus improvements under RTD conditions of approximately 6, 15, and 19%, respectively, when compared to neat resin (that is, 0 PHR) specimens. Whisker loading levels of 3, 6, and 10 PHR result in modulus improvements under ETW conditions of approximately 10, 25, and 45%, respectively, when compared to neat resin specimens. Modulus reduction as a result of ETW conditions is approximately 41, 39,

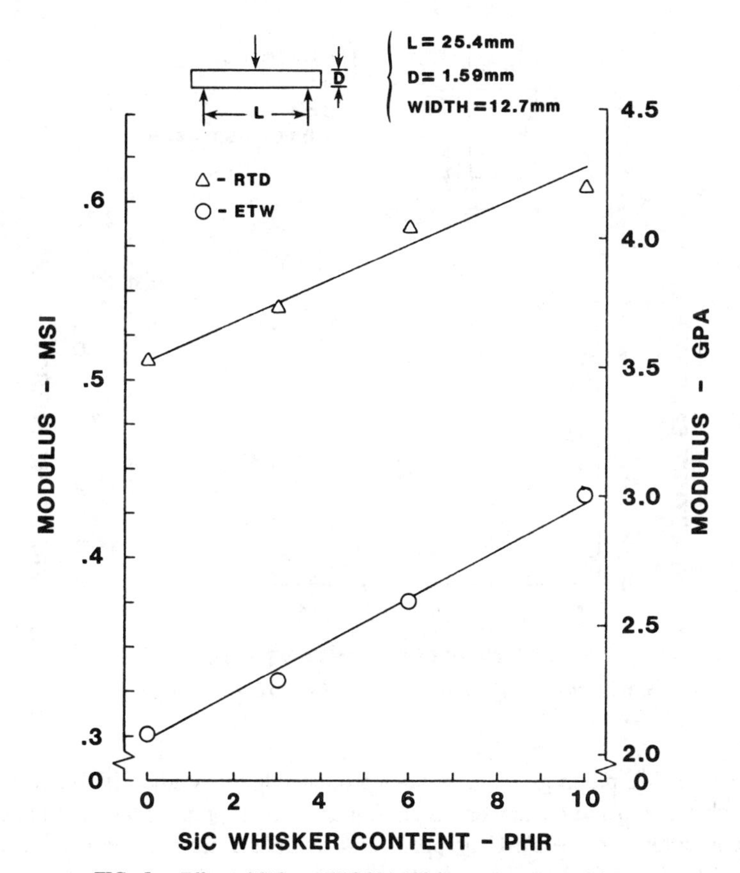

FIG. 5—*Effect of SiC on CYCOM 1806 epoxy resin modulus.*

36, and 28% for the 0, 3, 6, and 10 PHR loading levels, respectively. The modulus increases at loading levels as low as 3 PHR for both RTD and ETW conditions, and the reduction in modulus as a result of ETW conditioning is less severe as whisker loading level increases.

Compact tension specimens containing 0, 3, 6, 10, and 15 PHR were tested under RTD conditions. The fracture energy obtained from each specimen is plotted in Fig. 6. The fracture energy was obtained by dividing the area under the load-deflection curve by the crack area. Fracture energy remains relatively constant for the 0, 3, and 6 PHR specimens. Fracture energy increases at the 10 and 15 PHR loading levels when compared to the fracture energy of the neat resin.

Gr/E/SiC Materials—General Observations

A whisker-resin blend containing 6 PHR was selected for a majority of the fiber impregnation effort. A small quantity of AS-4 fiber was impregnated with a whisker-resin blend containing 10 PHR.

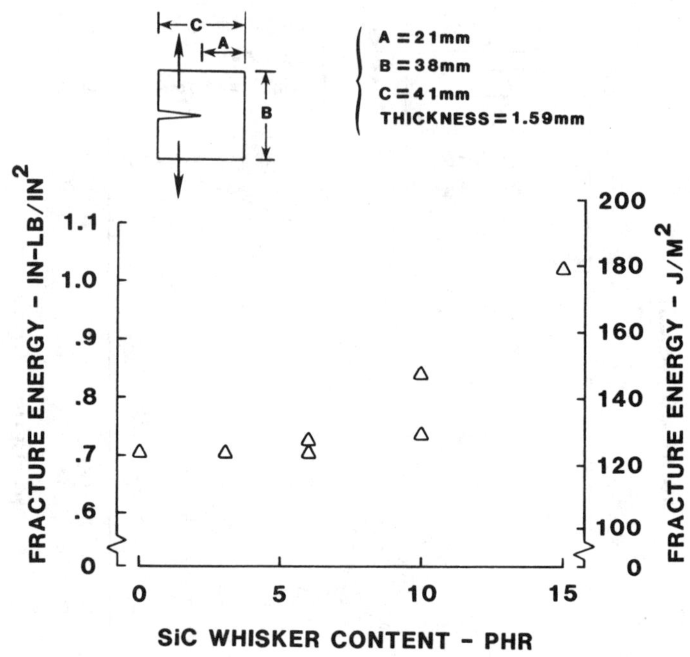

FIG. 6—*Effect of SiC on fracture energy of CYCOM 1806 epoxy resin.*

The Gr/E/SiC prepregs and laminates produced in this study differed in appearance when compared to standard Gr/E. Whereas Gr/E prepregs and laminates are black in color, the Gr/E/SiC prepregs and laminates are a green-gray to black color. The SiC whiskers have a green-gray color and this characteristic was imparted to the Gr/E/SiC material.

The average cured ply thickness of the AS-4/1806 and C6000ST/1806 control materials was 0.1422 mm (0.0056 in.) and 0.1448 mm (0.0057 in.), respectively. The measured graphite fiber volume content for both control materials was 66%. The average cured ply thickness of the AS-4/1806/SiC and C6000ST/1806/SiC materials fabricated from 6 PHR-whisker/resin blend, and the AS-4/1806/SiC material fabricated from 10 PHR-whisker/resin blend was 0.1549 mm (0.0061 in.).

Constituent volume contents of the laminates containing SiC were not measured. A qualitative estimate of the constituent volume contents of these laminates was made based on the PHR-whisker/resin blends, cured ply thickness, and graphite fiber volumes given above. The main assumption is that fiber areal weight is unaltered by processing. The AS-4/1806/SiC laminates fabricated from the 6 PHR- and 10 PHR-whisker/resin blends were estimated to contain 63% graphite, 35% epoxy, and 2% SiC, and 63% graphite, 33% epoxy, and 4% SiC, respectively. The C6000ST/1806/SiC laminates were estimated to contain 63% graphite, 35% epoxy, and 2% SiC.

Scanning electron microscope (SEM) observations were made of fracture surfaces. Typical observations are presented in Fig. 7. Sections of cured Gr/E/SiC material were pyrolized to remove the epoxy matrix and graphite fiber. Based on the SiC whisker skeletons resulting from the pyrolysis process and the SEM studies, it was determined that a uniform dispersion and random orientation of the SiC whiskers was achieved in the Gr/E/SiC composites.

The RTD test data obtained from various specimens fabricated from C6000ST/1806/SiC and AS-4/1806/SiC prepregs along with control data are listed in Tables 3 and 4.

Gr/E/SiC Unidirectional Data

The ASTM Test for Tensile Properties of Fiber-Resin Composites (D 3039) was used to determine 0 and 90° tensile data. The ASTM Test for Compressive Properties of Rigid Plastics (D 695) (modified) was used for determining 0° compressive strength. Apparent interlaminar shear strength was determined using the ASTM Test for Apparent Interlaminar Shear Strength of Parallel Fiber Composites by Short Beam Method (D 2344). The data obtained from these test methods are listed in Table 3.

The 0° tensile modulus of the AS-4/1806/SiC and C6000ST/1806/SiC materials was reduced approximately 12 and 4%, respectively, with respect to control data. As shown in Fig. 8, the 0° tensile strength and strain of the C6000ST/1806/SiC material were reduced 46% nominally with respect to control data. The 0° tensile strength and strain of the AS-4/1806/SiC material were reduced 40% nominally with respect to control data. The 0° compressive strength of the C6000ST/1806/SiC material was reduced 11% nominally with respect to control data.

It is speculated that either the whiskers act as stress concentrators leading to premature failure, or the whiskers, which are highly abrasive, are damaging

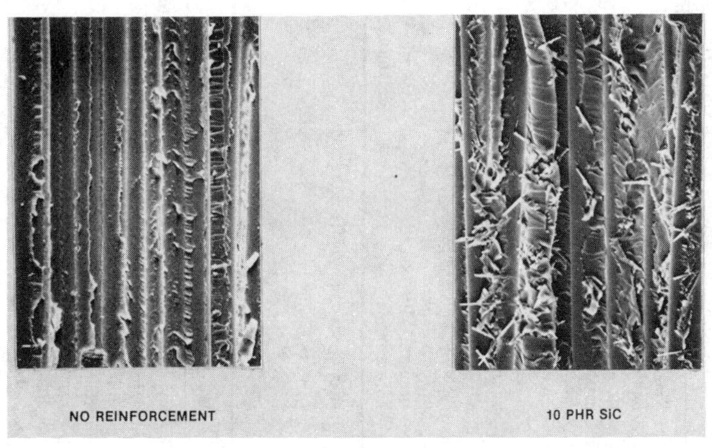

FIG. 7—*SEM observation of fracture surfaces CYCOM 1806 graphite/epoxy laminates.*

TABLE 3—*Unidirectional data.*[a]

Property	C6000ST/1806 (Control)	C6000ST/1806/SiC (6 PHR)	AS-4/1806 (Control)	AS-4/1806 (6 PHR)
0° Tension:				
modulus, GPa (Msi)	128 (18.5)	123 (17.8)	136 (19.7)	119 (17.3)
strength, GPa (ksi)	2.21 (321)	1.18 (171)	1.98 (287)	1.12 (163)
maximum microstrain	17 300	9 600	14 620	9 400
0° Compression:				
strength, GPa (ksi)	1.58 (229)	1.40 (203)	1.28 (186)	1.25 (182)
90° Tension:				
modulus, GPa (Msi)	8.07 (1.17)	7.86 (1.14)	7.93 (1.15)	7.58 (1.10)
strength, MPa (ksi)	66.9 (9.7)	76.5 (11.1)	66.9 (9.7)	75.2 (10.9)
maximum microstrain	8 600	10 100	8 500	10 500
Short-beam-shear:				
strength, MPa (ksi)	100.7 (14.6)	99.3 (14.4)	103.4 (15.0)	100.7 (14.6)
failure mode	delamination	flexure	delamination	flexure

[a] Data is the average of five replicates.

TABLE 4—*Tensile edge delamination data.*

Measurement[a]	C6000ST/1806 (Control)	C6000ST/1806 (6 PHR)	AS-4/1806 (Control)	AS-4/1806 (6 PHR)	AS-4/1806 (10 PHR)
$[\pm 35, 0, 90]_S$:					
modulus, GPa (Msi)	58.6 (8.50)	56.3 (8.16)	56.0 (8.12)	53.1 (7.70)	. . .
delamination microstrain	5900	7000	5200	5700	. . .
$[0, \pm 35, 90]_S$:					
modulus, GPa (Msi)	61.9 (8.98)	55.8 (8.10)	57.9 (8.40)	54.5 (7.90)	53.8 (7.80)
delamination microstrain	7000	7300	7900	7300	8700[b]

[a] Data is the average of five replicates.
[b] Specimen failed in grip area; no edge delamination occurred.

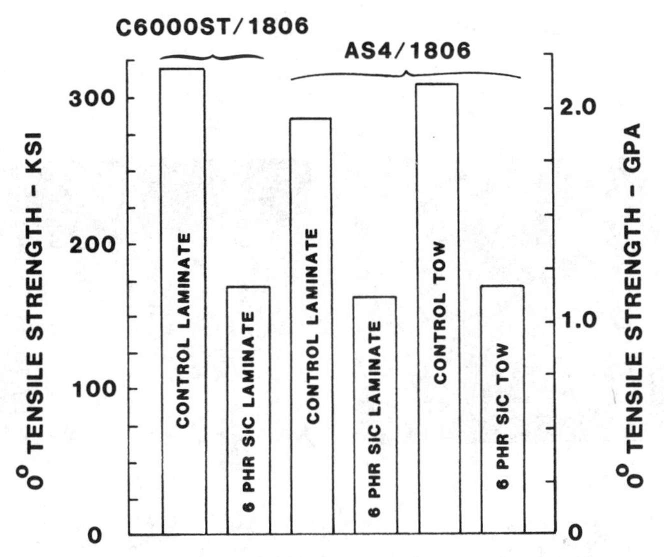

FIG. 8—*Comparison of 0° laminate and tow tensile strength.*

the fibers during the impregnation process thus leading to premature failure. Strips 4 mm (0.16 in.) wide and 150 mm (6 in.) long were removed from AS-4/1806/SiC and AS-4/1806 prepregs for tow testing. Each strip or tow was cured for 10 min at 177°C (350°F) and subsequently tested in tension at room temperature. Five tows were tested from each prepreg. The average failure load for the AS-4/1806 tows was 2.14 GPa (310 ksi), while the average failure load for the AS-4/1806/SiC tows was 1.18 GPa (171 ksi). A comparison of the tow and 0° laminate strength data is illustrated in Fig. 8.

Additional tows removed from the AS-4/1806 and AS-4/1806/SiC prepregs were ultrasonically washed in acetone to remove both resin and whiskers. The washed fibers were then examined using the SEM. SEM observations obtained from each prepreg are presented in Fig. 9. Figure 9*b* is a SEM photograph of a graphite fiber from the AS-4/1806 prepreg. Figure 9*a* is a SEM photograph of a graphite fiber from the AS-4/1806/SiC prepreg. Fiber surface scratches with particles at the ends of these scratches are seen. Figure 9*a* also shows what appears to be an SiC whisker embedded in the graphite fiber surface.

Based on the tow test data and SEM observations, it is apparent that fiber damage is occurring during the impregnation process. The fiber impregnation process is adjustable and it may be possible to determine impregnation conditions that will not cause such damage.

The 90° tensile modulus of the AS-4/1806/SiC and C6000ST/1806/SiC materials were both reduced 4% nominally with respect to control data. This decrease is similar to the loss in 0° modulus; both are probably a result of the difference in matrix volume percent. Since whisker volume percents were low, little effect of whiskers was seen on composite modulus. Thus, to see the predicted increases

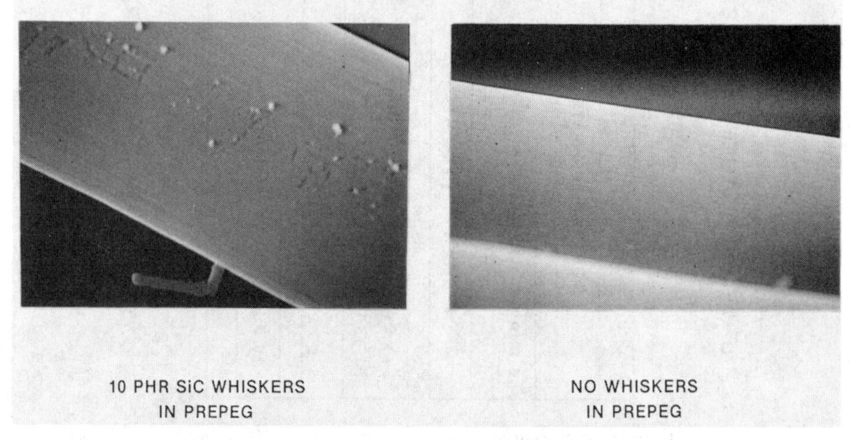

FIG. 9—*Damage to carbon fibers from whisker impregnation.*

in modulus discussed earlier, larger whisker contents will have to be used. The 90° failure strength and strain of both Gr/E/SiC materials were increased 17% nominally with respect to control data.

The 90° tensile modulus of the AS-4/1806/SiC and C6000ST/1806/SiC materials were both reduced 4% nominally with respect to control data. The 90° failure strength and strain of both Gr/E/SiC materials were increased 17% nominally with respect to control data.

Short-beam shear tests were conducted on both Gr/E/SiC materials. Gr/E/SiC short-beam shear specimens failed at approximately the same load levels as their respective control specimens. It is notable, however, that the failure mode changed from delamination in the case of the control materials to flexure in the case of the Gr/E/SiC materials.

Gr/E/SiC Delamination Data

Tensile edge delamination specimens were fabricated from $[0, \pm 35, 90]_s$ and $[\pm 35, 0, 90]_s$ laminates. These laminates are designed to delaminate from the free edge under longitudinal tensile loading. The interlaminar fracture of these specimens occurs under a mixture of Mode I, interlaminar tension, and Mode II, interlaminar shear [*6*]. The $[\pm 35, 0, 90]_s$ laminate has a high Mode I component (90%), while the $[0, \pm 35, 90]_s$ laminate has a low Mode I component (22%). Untabbed specimens 152 mm (6 in.) long, 13 mm (0.5 in.) wide with 102 mm (4 in.) between grips were tested at a constant crosshead rate of 0.002 mm/s. The data obtained from these test specimens are listed in Table 4. The delamination strain data were normalized with respect to control data and are presented in Fig. 10.

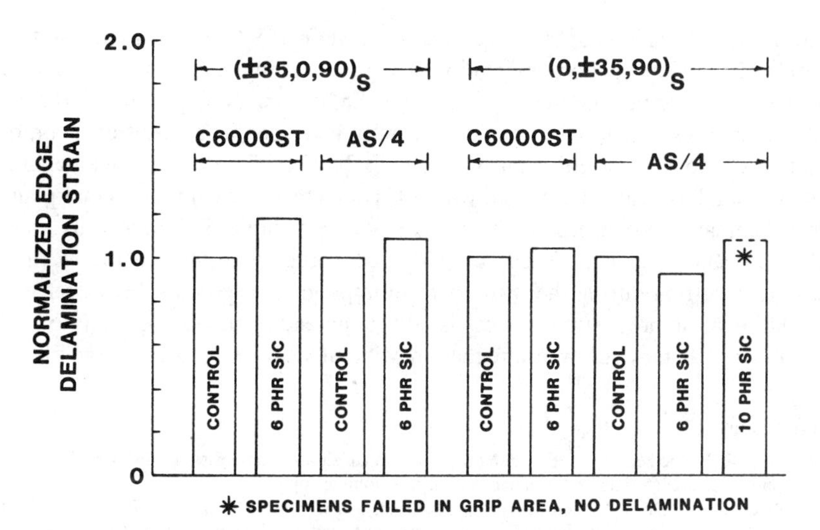

FIG. 10—*Comparison of normalized edge delamination strain.*

The modulus of the C6000ST/1806/SiC (6 PHR), AS-4/1806/SiC (6 PHR), and AS-4/1806/SiC (10 PHR) laminates were lower than their respective control laminates.

As illustrated in Fig. 10, the delamination strain level increased 19 and 10% for the high Mode I C6000ST/1806/SiC (6 PHR) and AS-4/1806/SiC (6 PHR) specimens, respectively, as compared to control. The delamination strain level increased 4% for the low Mode I C6000ST/1806/SiC (6 PHR) specimens, while the delamination strain level decreased 8% for the low Mode I AS-4/1806/SiC (6 PHR) specimens, as compared to control. The delamination strain level increased 10% for the low Mode I AS-4/1806/SiC (10 PHR) specimens. However, the low Mode I AS-4/1806/SiC (10 PHR) specimens did not delaminate and instead failed in the grip area. The tensile edge delamination data presented indicate that significant increases in delamination strain level can only be attained with SiC contents either at or above the 10 PHR level.

Conclusion

Specimens fabricated from 1806 epoxy resin containing 0, 3, 6, 10, and 15 PHR of SiC whiskers were tested to determine the effects of SiC loading on the properties of the epoxy resin. Whisker loading as low as 3 PHR increased the modulus of the resulting 1806/SiC material at RTD and ETW conditions when compared to neat resin. The reduction in modulus as a result of ETW conditions is less severe as SiC whisker loading level increases, when compared to neat resin modulus reduction under the same conditions. The fracture energy of the 1806/SiC specimens increases significantly at the 15 PHR whisker loading level when compared to the fracture energy of neat resin specimens.

Specimens were fabricated from various Gr/E/SiC formulations to examine the effects of SiC content on the properties of the resultant hybrid composite material. SEM observations of fractured and polished Gr/E/SiC surfaces indicate that uniform dispersion, random orientation, and good wetting of the SiC whiskers are attained in the composite. The 90° tensile strengths and strains of the Gr/E/SiC (2% volume SiC) materials were significantly increased as a result of incorporating SiC whiskers. Significant increases in the Gr/E/SiC tensile edge delamination strain level can be achieved with SiC contents ≥4% volume. The in-plane fiber-dominated properties of the Gr/E/SiC (2% volume SiC) materials were all significantly reduced. These property reductions were attributable to fiber damage occurring during the whisker-resin impregnation process. However, the whisker-resin impregnation process is adjustable and it may be possible to find conditions that may allow impregnation without damage to the carbon fibers.

References

[1] Evans, R. E. and Masters, J. E., "A New Generation of Epoxy Composites for Primary Structural Applications: Materials and Mechanics," in this volume, pp. 413-436.

[2] Palmer, R. J., "Resin Properties to Improve Impact in Composites," *Proceedings of the Fifth DoD/Nasa Conference on Fiberous Composites in Structural Design,* NADC-81096-60, Jan. 1981.

[3] Hahn, H. T. and Williams, J. G., "Compression Failure Mechanisms in Undirectional Composites," Nasa Technical Memorandum 85834, Aug. 1984.

[4] Garcia, R., "Methods of Improving the Matrix Dominated Performance of Composite Structures," NADC-83058-60, July 1983.

[5] Halpin, J. C. and Tsai, S. W., "Environmental Factors in Composite Materials Design," AFWAL-TR-67-423, 1967.

[6] O'Brien, T. K., Johnston, N. J., Morris, D. H., and Simonds, R. A., "Determination of Interlaminar Fracture Toughness and Fracture Mode Dependence of Composites Using the Edge Delamination Test," presented at the International Conference on Testing, Evaluation and Quality Control of Composites, 13-14 Sept., 1983, University of Surrey, Guildford, England.

Robert E. Evans[1] *and John E. Masters*[1]

A New Generation of Epoxy Composites for Primary Structural Applications: Materials and Mechanics

REFERENCE: Evans, R. E. and Masters, J. E., "**A New Generation of Epoxy Composites for Primary Structural Applications: Materials and Mechanics,**" *Toughened Composites, ASTM STP 937,* Norman J. Johnston, Ed., American Society for Testing and Materials, Philadelphia, 1987, pp. 413-436.

ABSTRACT: Damage tolerance and hot/wet compressive performance are two key parameters which are required for composite materials to replace aluminum in primary structural applications on aircraft. Conventional toughening approaches involving the use of second-phase rubber modification of tetrafunctional epoxies have not lead to the required balance of these properties. Emphasizing newly developed curative, catalyst, and matrix modifier chemistry, a major advance in the performance of epoxy composites has been made which will allow the design requirements for commercial and transport aircraft to be achieved. It is shown that the stress-strain response of these resins can be effectively analyzed to distinguish between material systems for their performance in composites. The toughness, controlled by the strain to failure, and the hot/wet resin modulus are used as the key parameters to make these comparisons. The performance of new composites based on these materials is quantified in terms of two key design requirements, that is, residual compressive strength after impact damage and hot/wet compressive performance.

The mechanics, failure analysis, and materials science approaches that lead to the development of novel high strain to failure composite systems containing a discrete toughened resin interleaf between the laminate plies are also discussed. The combination of new toughened epoxy resins with novel interleafing materials is shown to result in composites exceeding the ultimate design strain target of 0.006 cm/cm while providing a hot/wet compressive strength to reduce weight by over 40% compared to 2024-T3 aluminum. In addition to this, these improved interlayered systems have two other attractive features: (1) their residual compressive strengths after impact are comparable to those of thermoplastic matrix resin systems and (2) they process and cure like standard epoxy materials.

KEY WORDS: toughened resins, interleafing, impact damage, compressive strength, flexual strain, neat resin

[1]Manager, Contracts Research, and senior research engineer, respectively, American Cyanamid Company, Advanced Structural Materials Section, Polymer Products Research Department, 1937 W. Main St., P.O. Box 60, Stamford, CT 06904-0060.

The use of carbon fiber composites in the design of secondary structures in aircraft has become well established during the last decade. Such materials are also now being applied to the volume production of primary structures, especially on military aircraft. The interest to develop further the technology into large primary structures on civil aircraft is very strong as a result of the incentive for weight savings.

Epoxy matrix composites have several very attractive features. These resin systems are compatible with graphite fibers, thereby eliminating many of the interface problems that may be evident in the newer type thermoset and thermoplastic systems. They are also resistant to aircraft fluids such as jet fuel or hydraulic fluid. Finally, large data bases and long flight histories exist for these systems both in military and commercial aircraft. The first generation epoxy systems are, however, brittle and limited to relatively low design strain levels. This, coupled with the vulnerability of these materials to foreign object impact damage, defines the need for improved toughened composites. If new composite systems could be developed that exhibit greatly improved toughness characteristics, higher allowable design strains of 0.005 to 0.006 would be feasible, thereby offering the potential for even greater weight savings over aluminum, and greatly improving the commercial potential of such new materials.

To be effective, the next generation of composite material systems must be designed to provide the requisite in-plane structural properties (stiffness and strength) and to increase durability and damage tolerance. The material's use will be severely limited if a balance in these properties is not maintained. This paper will review work performed at Cyanamid to improve material toughness while maintaining structural performance. The effectiveness of the work will be measured in terms of hot/wet compressive strength (structural performance) and residual compressive strength after impact (damage tolerance).

Design-Related Structural Performance

Hot/Wet Compressive Strength

The material system must, above all else, meet the structural requirements of the aircraft while economically providing a meaningful weight savings over aluminum. A quantitative factor determining the amount of weight saved can be calculated by comparing the specific strengths of the quasi-isotropic composite laminates and aluminum. The quasi-isotropic laminate configuration is chosen because it is representative of structural component configurations. A valid comparison must also consider environmental conditions and loading type. The matrix dominated hot/wet compressive strength, considered the severest measure of structural performance, is therefore used for this comparison. The test temperature is defined by material application: civilian and transport aircraft require a 93°C upper use temperature; some military aircraft applications require performance at 132°C.

Based on these considerations, weight saving is defined by Eq 1. A graphic representation of this equation is presented in Fig. 1. Assuming a design ultimate

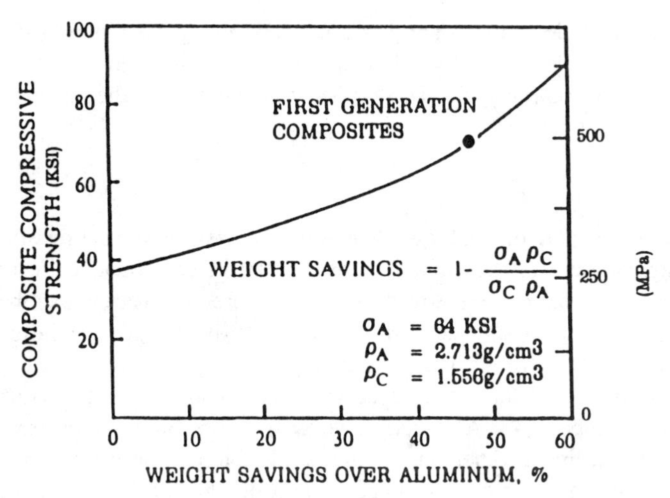

FIG. 1—*Quasi-isotropic laminate hot/wet compressive strength is used as a measure of structural performance.*

strength of 440 MPa (64 ksi) for 2024 T3 aluminum and a 60% fiber volume in the composite material, a quasi-isotropic compressive strength of 360 MPa (52 ksi) under worst case hot/wet conditions will be needed for a 30% weight saving over aluminum. As the figure indicates, typical first generation graphite/epoxy systems (tested at 93°C/wet) easily meet this weight-saving criterion.

$$\%\text{ Weight Saving} = \frac{\rho_A - \left(\dfrac{\sigma_A}{\sigma_C} \cdot \rho_C\right)}{\rho_A} \times 100 \tag{1}$$

Where:

σ_A = compressive strength allowable Al = 64 ksi,
ρ_A = density Al = 2.713 gm/cm^3,
σ_C = compressive strength composite,
ρ_C = density composite,

$$\text{Note:} \quad \frac{1}{\rho_C} = \frac{K}{\rho_R} + \frac{(1-K)}{\rho_f}$$

ρ_f = fiber density = 1.76 gm/cm^3,
ρ_R = resin density = 1.26 gm/cm^3, and
K = resin weight fraction.

Sixteen-ply-thick laminates with a $[\pm 45/0/90/0/90/\pm 45]_s$ stacking sequence were used to monitor structural performance in this program. The compressive strength specimens were 80 mm (3.15 in.) long and 12.7 mm (0.5 in.) wide; 38-mm (1.5-in.) long tabs were bonded to the specimen, effectively reducing the test section length to 4.78 mm (0.188 in.). The specimens were loaded in a 90-kn (20 000-lbf) Instron test machine at a deflection rate of 1.27 mm/min

(0.05 in./min). Side supports similar to the support jig used in ASTM Test for Compressive Properties of Rigid Plastics (D 695) were used to prevent gross Euler buckling during loading. The support fixtures contact the tabs leaving the test section unsupported.

Impact Damage Resistance

Although it is more difficult quantitatively to assess the impact resistance of a composite material, it is possible to measure relative performance. A variety of test techniques have been employed to measure foreign object impact behavior. The extent of damage has been reported in terms of: damage area after a drop weight impact [*1*], through penetration impact using an instrumented impact fixture [*2*], impact while under a compressive load [*3*], and residual compression after impact [*4*]. All tests, however, produce only comparative values of one system against another. Just as in the actual aircraft structure, the damage resulting from a given blow will depend on a number of factors such as local rigidity, energy absorbed by the support structure, shear stress versus flexural stress, and so forth.

The data in this presentation are based on the impact test procedure described by Byers [*5*]. These tests use specimens that are 152.5 mm (6 in.) long, 100 mm (4 in.) wide, and 36 plies thick. They have a $[(\pm 45/0/90/0/90)_2/\pm 45/0/90/\pm 45]_s$ stacking sequence. The specimens were positioned on a steel plate having a 76- by 127-mm (3- by 5-in.) rectangular cutout which supported the specimens along their edges. A 1.8-kg (4-lb) weight was dropped down a calibrated tube, striking a 1.6-cm (0.62-in.) diameter spherical head impactor that rested on the specimen. While various impact energies can be applied, 680 kg · m/m (that is, 1500 in. · lb/in. of laminate thickness) was chosen as an arbitrary standard. The support, size, and thickness of the specimen gave an effect similar to dropping a tool from a few feet onto an aircraft wing and striking it between stiffners.

Subsequent to impact, test panels were loaded to failure in comparison in a 225-kg (50 000-lbf) Instron test machine. A test fixture (shown schematically in Fig. 10) which provides simple support along the specimen edges is used to prevent gross buckling during loading.

The usefulness of this test in assessing toughness can be seen when one compares the compressive strength after impact data for CYCOM® 907,[2] a modified epoxy which is often used as a benchmark for toughness, and a typical first generation epoxy system. Figure 2 plots these residual compressive strengths over a range of impact energies. At low impact energies, there is little difference between the two systems since both are capable of withstanding these impact energies without sustaining damage, that is, both have good impact resistance. However, as the impact energy increases, the situation changes dramatically with

[2]This material was originally called BP 907. The trademark was changed to CYCOM® 907 to conform with new product identifications.

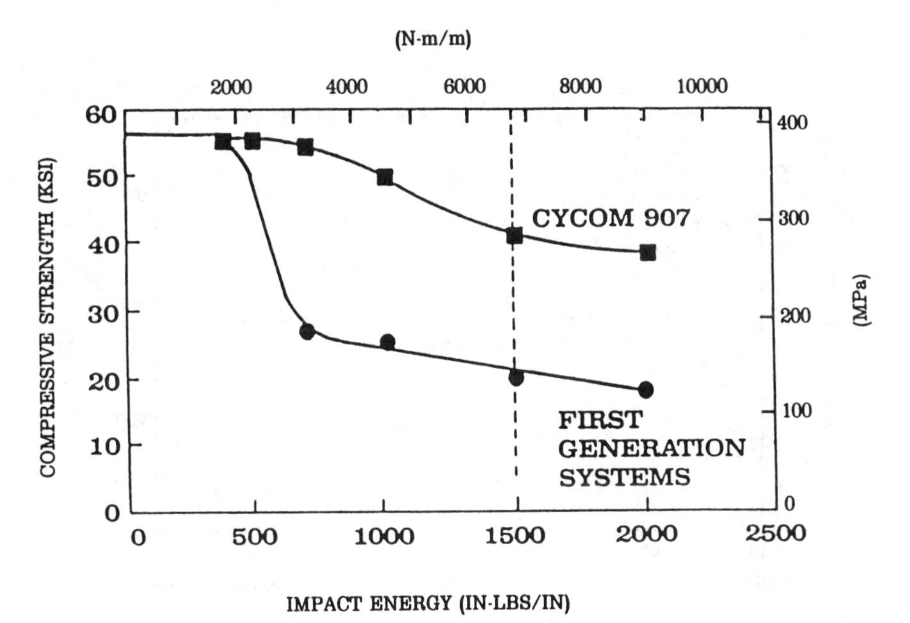

FIG. 2—*The reduction in laminate residual compressive strength increases at higher impact energy levels.*

the tough system being much more capable of withstanding higher impact without damage. When the tough system is damaged, there is a smaller effect on the residual compressive strength, that is, it is more damage tolerant. From the data presented in Fig. 2, it can be seen that the compressive strength after impact (CSAI) of CYCOM® 907, namely 280 MPa, is double that achievable with first generation materials at the 680-kg · m/m (1500-in. · lb/in.) impact energy level. This level of toughness therefore became a principal objective of our work.

Matrix Requirements for Toughened Composites

The problem of concern is illustrated in Fig. 3 where hot/wet performance is plotted versus compressive strength after impact. As can be seen, the first generation materials have good hot/wet performance, but very low compressive strengths after impact. Alternatively, while CYCOM® 907 has good compressive strength after impact, it is far below the 30% weight-savings target. This is typical of the type of trade-off made between hot/wet performance and improved damage resistance and tolerance. It is necessary, therefore, to address the development of a new generation of epoxies having the balance of properties required. This can be accomplished by first examining the material failure mechanisms for compressive loading and impact loading.

In-plane compressive loading is a severe and limiting test of the material. Several potential failure modes exist at practical fiber volume fractions, that is, for $V_f > 0.40$, [*6*]. They include: (1) transverse tensile failure, (2) fiber micro-

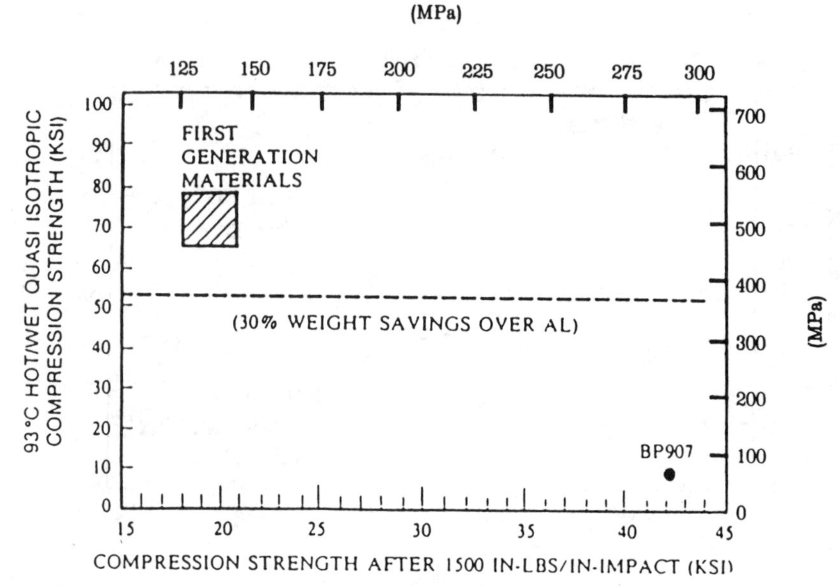

FIG. 3—*A trade-off is made between hot/wet performance and improved impact resistance.*

buckling, and (3) shear failure. Although failure is controlled by the matrix resin, the situation is greatly complicated by the tendency of the fibers to bend and buckle. Axial compression produces shear load components between the fiber and the matrix because the fibers are not perfectly aligned within the ply. These out-of-plane components can induce tensile loads in the matrix which can lead to matrix yielding, matrix microcracking, or fiber-matrix debonding. These latter effects precede transverse failure and fiber microbuckling and may cause premature structural failure.

This model is consistent with a British Aerospace study [*7*] which indicated that, in general, for an epoxy matrix reinforced with a highly anisotropic fiber, failure is likely to occur by microbuckling [*8*]. Other studies [*9,10*] have shown that in composites with soft matrices (matrix shear modulus less than 690 MPa), compressive strength is strongly influenced by resin modulus. Analogous results were reported for boron/epoxy composites [*11*]. While more precise empirical relationships can be developed, the neat resin shear modulus or flexural modulus measured under hot/wet conditions can be used as a first approximation to determine the composite compressive strength (see Eq 2).

$$\sigma_{cc} = K_1 G_R = K_1 \left[\frac{Er}{2(1 + \nu)} \right] \tag{2}$$

where:

σ_{cc} = composite compressive strength,
K_1 = constant,

G_R = resin shear modulus ≤ 689.4 MPa (100 000 psi),
Er = Young's modulus of resin, psi, and
ν = Poisson's ratio of resin.

In addition, the importance of the fiber matrix interface bond is demonstrated by the direct relationship between compressive strength and interlaminar shear strength that has been seen for some materials [*12,13*]. Its effect on failure is also important [*14,15*], and any degradation at the fiber matrix interface causes a falloff in compressive strength [*16*].

A quantitative assessment of the factors important in the initiation of damage from foreign objects has been presented by Dorey [*17*]. He conducted studies that indicated that when carbon fiber reinforced plastics are subjected to transverse impact, the type of damage which occurs depends on the incident energy and momentum, material properties, and the geometry. No damage occurs if the energy of the projectile is accommodated by the elastic strain energy in the material. He calculated the energies to cause:

- delamination $(2/9)(\tau^2/E)(wl^3/t)$ (3)
- flexural fracture $(1/18)(\sigma^2/E)(wlt)$ (4)
- penetration $\pi\gamma td$ (5)

where τ is the interlaminar shear strength; σ the flexural strength of the composite; E the Young's modulus of the composite; γ the through-thickness fracture energy; d the diameter of the projectile; and w, l, and t the width, length, and thickness of the flexed part of the test specimen. Whether delamination or flexural fracture occurs depends on the relative values of τ and σ and the span-to-depth ratio l/t; impact damage is less likely when there are low modulus layers on the outside such as ±45° layers or Kevlar or glass fibers. Whether penetration occurs depends not only on the incident energy but on the size and shape of the projectile; penetration is more likely for small masses travelling at high velocities.

Dorey's analysis is based on linear elastic modeling and does not examine what happens when the critical values for initiation of damage are attained. If a crack is started, this crack will grow until the stored energy is dissipated. In brittle systems, this will result in crack growth as shown in Fig. 4. However, if the system has the ability to yield and undergo plastic deformation, this energy will be dissipated without damage growth. Since graphite fibers exhibit linear elastic failure, improvements must be made in the matrix resins to increase significantly laminate impact resistance and impact tolerance. The matrix characteristic that is most critical to improving composite toughness is the ability to sustain a high stress while yielding, that is, to develop a "knee" in the stress-strain curve and have a large strain to failure. This is clearly evident when the shear stress-strain curve of a first generation matrix and CYCOM 907 are compared as shown in Fig. 5.

While Dorey's equations indicate that lowering the bending modulus is another means to increase resistance to delamination resistance, high in-plane modulus is

FIG. 4—*Low velocity impact causes internal delamination in brittle resin systems.*

required to control fiber microbuckling and damage growth under compressive loads. Reductions of the composite bending modulus should only be made by changing the laminate lay-up sequence and not through reductions in resin modulus. In fact, it is beneficial for overall performance to increase resin modulus if the shear strength and flexural strength are also increased to keep the ratios τ^2/E and σ^2/E high.

As a first approximation for single-phase epoxy systems, the area under the neat resin's stress-strain curve, that is, the work to failure, is assumed to be proportional to the compressive strength after impact. In particular, the existence of plastic deformation will be the major contributor to the work to failure. This may be expressed mathematically by:

$$\text{CSAI} = K \int_0^{\varepsilon_f} \sigma(\varepsilon)\, d\varepsilon$$

where:

CSAI = compressive strength after impact,
K = constant,
σ = stress, and
ε = strain.

The area under the stress-strain curve can be measured using either a flexural test such as described in ASTM Tests for Flexural Properties of Unreinforced and Reinforced Plastics and Electrical Insulating Materials (D 790) or in shear using the Iosipescu fixture (ASTM D 790 was used in this evaluation). For single-phase systems, the same type of stress-strain behavior will be seen in both tests. For multiphase systems, however, a material can undergo high plastic shear deformations without showing similar behavior in a flexural or tensile mode.

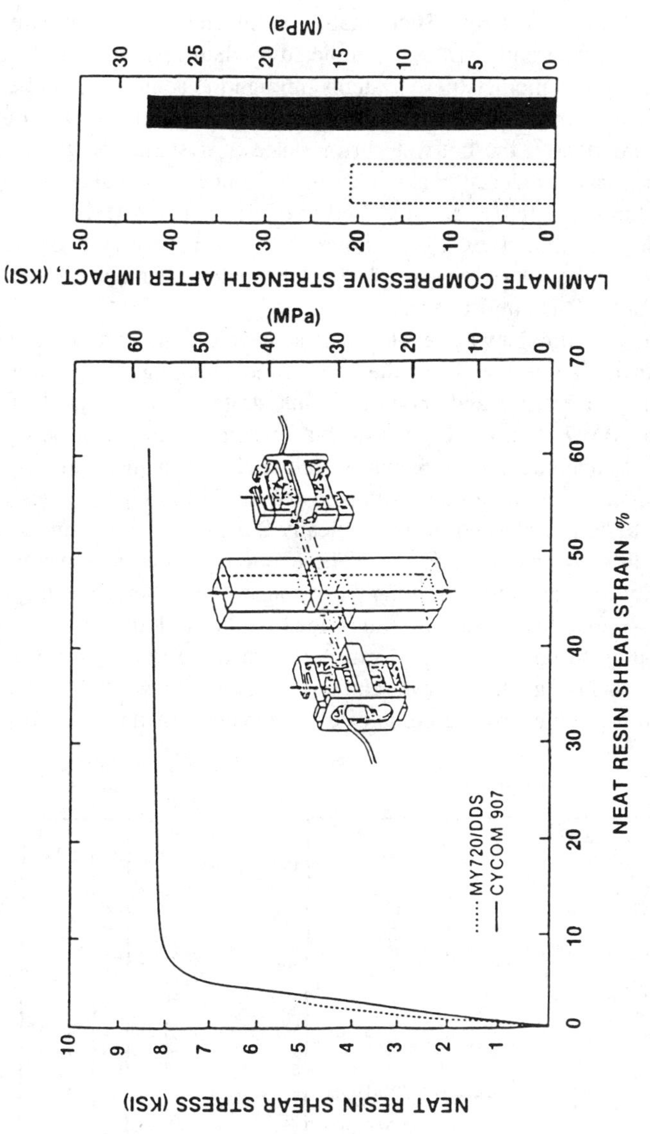

FIG. 5—*Neat resin shear stress-strain response is indicative of composite toughness.*

New Composite Matrices

Most first generation epoxy matrix resins are based on the MY 720 (tetraglycidal diamino diphenyl methane) epoxy resin cured with diamino diphenyl sulfone (DDS) either with or without a boron trifluoride · monoethylamine (BF_3 · MEA) catalyst system. Such matrix resins have good hot/wet performance, but are inherently brittle, as depicted by their resin properties shown in Fig. 6. Attempts to modify these systems through the addition of rubber-type compounds or elastomers or both show only a nominal improvement in toughness. More importantly, the hot/wet performance of systems modified in this manner is drastically reduced making them no longer applicable for use in primary structures. Similarly, diglycidal ether of Bisphenol A (DGEBA) resins cured with dicyandiamide (DICY), as shown in Fig. 7, have very good strain to failure and exhibit the desired type of plastic deformation, but again, the hot/wet performance at 93°C is almost zero.

Consequently, a strategy was developed to achieve a significant improvement in resin strain to failure, while at the same time retaining a high wet glass transition temperature (T_g) and modulus. This strategy concentrated on continuing to use MY 720 as the base resin but in combination with newly synthesized flexible diamine curing agents, novel catalyst systems, and polymeric modifiers. This new curative chemistry allowed strain-to-failure values of the matrix resins to be improved from the typically 2.5 to 3.0% level up to values greater than 10% as shown in Fig. 8. These new systems demonstrated the required ability to conform to the toughness concept of a "knee" in the stress-strain curve—the ability to sustain a high load while "yielding."

The translation to composites performance with these new systems is very good. As shown in Fig. 9, a linear relationship exists between the neat resin flexural strain to failure and the composite's transverse tensile strain to failure

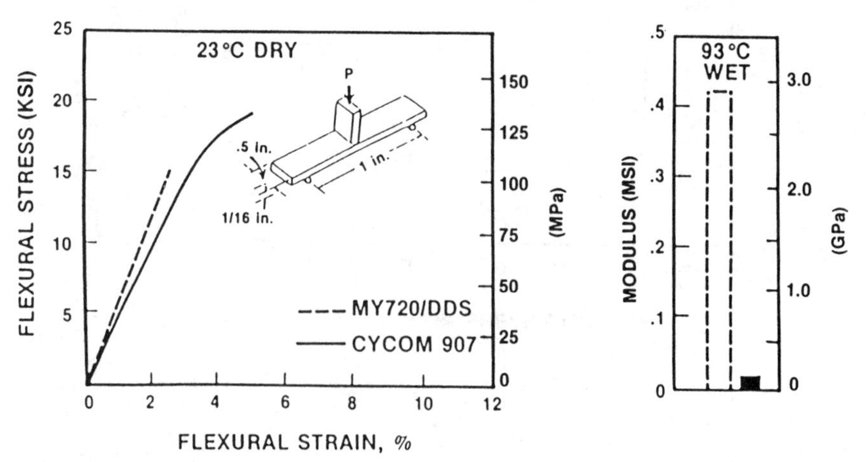

FIG. 6—*Flexural stress-strain properties for brittle and rubber toughened neat matrix resins.*

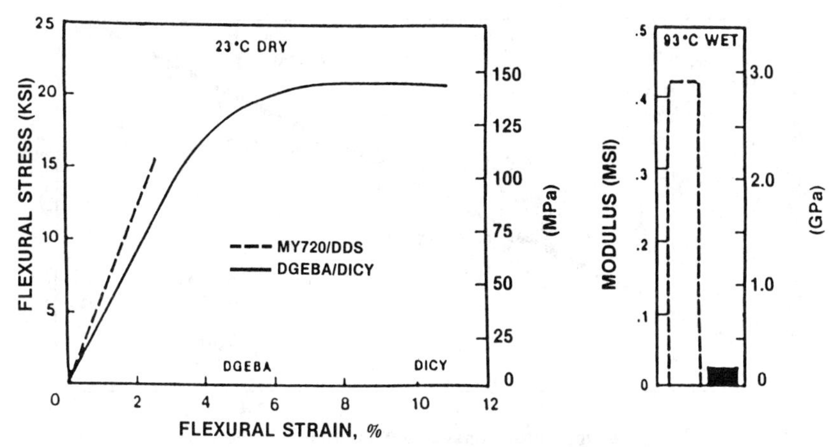

FIG. 7—*Comparison of flexural stress-strain behavior of brittle and tough neat resins.*

(measured using ASTM Test for Tensile Properties of Fiber-Resin Composites [D 3039]). Furthermore, these resins have demonstrated a very good ability to wet the fibers as shown in the insert photograph.

More importantly, as shown in Fig. 10, this translation of enhanced resin properties to composite laminate performance holds for compressive strength after impact. The next generation epoxy systems with strains to failure in the 8% range, for example, CYCOM® 1806 and CYCOM® 1808, have compressive strengths after impact comparable to CYCOM® 907. First generation materials with a strain to failure of 2.5%, on the other hand, have a 138-MPa (20-ksi) residual compressive strength. Hirschbuehler [*18*] presented a more extensive comparison of Cyanamid's data on the relationships between resin and composite

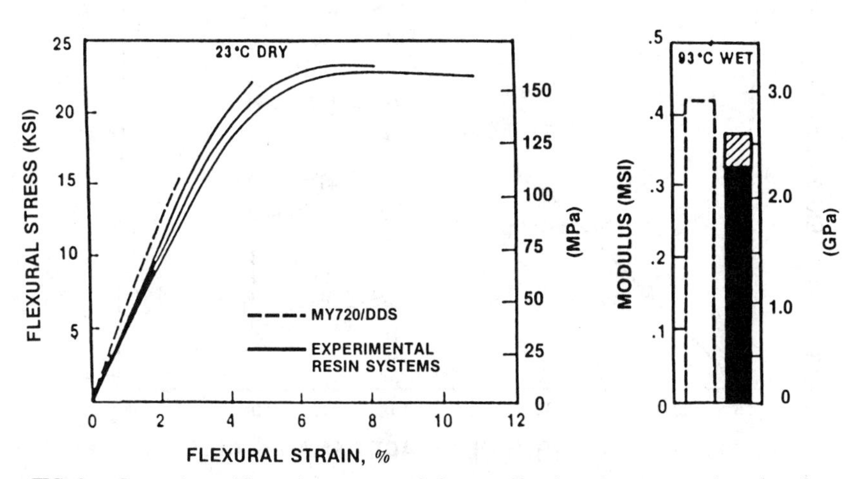

FIG. 8—*Comparison of flexural stress-strain behavior of brittle and experimental toughened neat resins.*

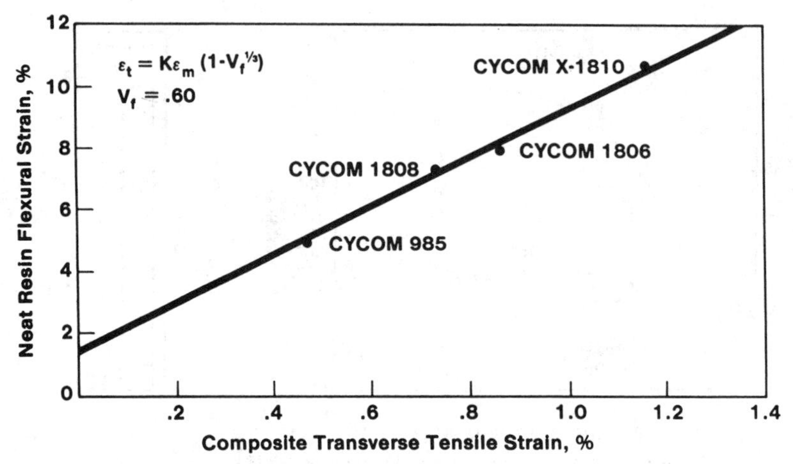

FIG. 9—*Correlation of neat matrix resin flexural failure strain and composite transverse tensile failure strain.*

performances. He noted a similar correlation between neat resin ultimate strain and compressive strength after impact.

Finally, the retention of hot/wet resin modulus as shown in Fig. 11 results in the laminate hot/wet compressive strength required to meet the established weight-savings criteria. Plotting the data in a similar manner as shown earlier, it

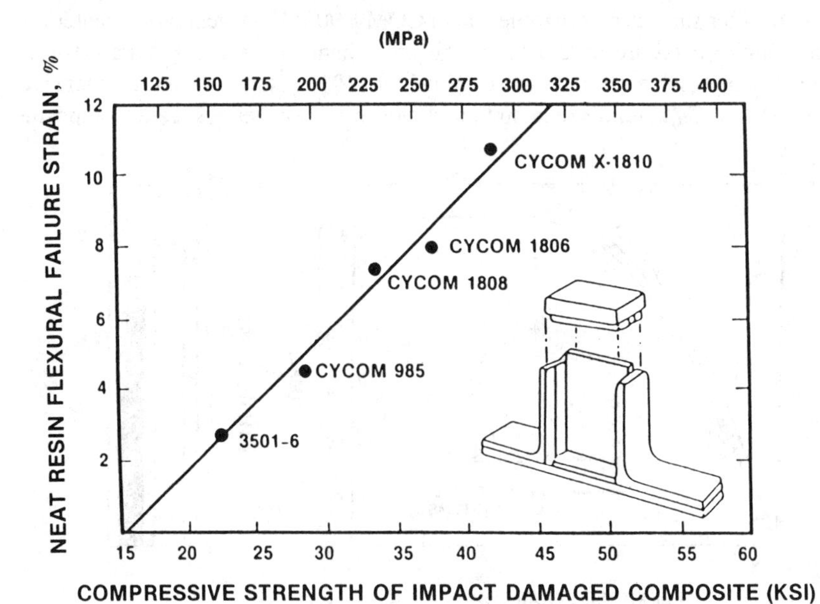

FIG. 10—*Correlation of flexural strain-to-failure of neat matrix resins and residual compressive strength of impacted laminates using those resins.*

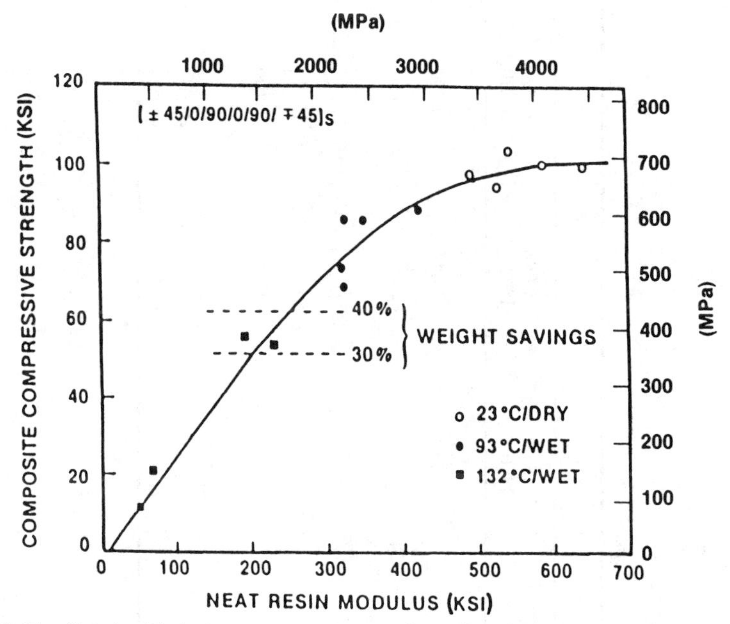

FIG. 11—*Relationship between neat matrix resin flexural modulus and quasi-isotropic laminate compressive strength.*

can be seen that the new generation of single-phase epoxy systems have remained above the 30% weight-savings line and moved far out to the right on the plot indicating their suitability for use in primary structure applications. It has also been demonstrated that the neat resin data (Table 1) is a good indicator of composite laminate performance (Table 2).

Interleafed Composites

As increases in matrix toughness are made, there will be some sacrifice in hot/wet performance which can limit composite improvements. Therefore, we need to consider what can be done from a materials science aspect to improve performance.

An examination of the cross section of an impacted specimen demonstrates that, as can be seen in Fig. 4, the impact causes considerable splitting and delamination in first generation epoxy materials. This splitting and interply delamination failure are controlled by the inability of the composite to undergo shear deformation. This limitation is a result of the brittle nature of the resins and the laminate construction, that is, the thin resin layers which form between plies are constrained against large shear deformation. Even with the improved matrix materials previously described, there is a limited amount of shear deformation which can be built into the system before the hot/wet deterioration becomes a controlling factor.

TABLE 1 *Comparison of neat resin properties.*

	Material					
Resin Property	MY-720/ DDS	CYCOM 907	CYCOM 985	CYCOM 1806	CYCOM 1808	CYCOM X-1810
Flexural modulus, Msi	0.601	0.470	0.590	0.530	0.487	0.494
Flexural strength, ksi	15.2	18.8	22.5	23.7	25.5	24.0
Flexural strain, %	2.5	5.0	4.5	8.0	7.3	>10
Work-to-break, in.-lb/in.	205	---	570	1375	1250	>1800
$\frac{\sigma^2}{E}$	0.384	0.752	0.858	1.060	1.335	1.166
Flexural modulus, Msi, at 93°C/wet	0.400	0.01	0.380	0.330	0.325	0.343
Morphology	single phase	multi phase	single phase	single phase	single phase	single phase

TABLE 2—*Comparison of composite properties.*[a]

Property	Material					
	MY-720/ DDS	CYCOM 907	CYCOM 985	CYCOM 1806	CYCOM 1808	CYCOM X-1810
Quasi-Isotropic Compressive Strength, ksi						
23 °C dry	100	91	100	104	97	105
93 °C dry	100	55	95	90	91	85
93 °C wet	88	9	71	75	86	68
132 °C dry	88	...	63	73	87	65
132 °C wet	58	...	51	...	56	...
Transverse tensile modulus, Msi	...	...	1.20	1.17	1.12	1.15
Transverse tensile strength, ksi	...	...	6.0	9.7	8.1	12.0
Transverse tensile strain, %	...	...	0.50	0.86	0.73	1.16
Damage area after impact, in.^{2}	4.2	0.80	3.8	1.8	...	2.0
Compressive strength after 1500-in. · lb/in. impact, ksi	22	42	28	37	34	43

[a]All materials used C6000 ST fiber.

One solution to this problem, as shown in Fig. 12, was to engineer a composite within a composite [19], that is, to take the standard prepreg containing 60% by volume fiber with the improved matrix resin and add to it a discreet layer of very high toughness, very high shear strain resin comparable with the matrix of CYCOM® 907. The key to this concept was to develop the interleaf material which would co-cure with the matrix resin and have the flow control to remain as a discreet layer throughout the entire process. As can be seen in the photograph in Fig. 12, this has been achieved.

The advantage of this toughened interlayer is shown in Fig. 13. The compressive strength after impact of the systems shown in Fig. 10 have been increased without modifying the matrix resin. As can be seen, all materials from first generation type through the CYCOM® 1806 toughened systems will benefit through the addition of an added interlayer. This interlayer functions to suppress initial impact damage, that is, to increase impact resistance. Smaller damage sizes are seen in C-scans of interleafed panels compared to noninterleafed panels impacted at the same energy level. Interleafing is most effective in improving toughness when combined with a tougher matrix system.

A further demonstration of the effect of interlayering on impact damage resistance is given in Fig. 14. This figure plots the residual compressive strength after impact (CASI) as a function of impact energy for an interleafed material, CYCOM® HST-7, and a first generation epoxy material. Comparing data at 680 kg · m/m (1500 in. · lb/in.), we find that the HST-7 interleafed material has double the CSAI of first generation materials. Looking at this from another viewpoint, it is possible to impact the interleafed material at an energy level sufficient to drive the plunger through the structure and still have the same load-carrying capability as the first generation material which has sustained only barely visible damage.

In addition to toughness, the material's structural performance must also be considered. Figure 15 plots the residual compressive strength after impact versus

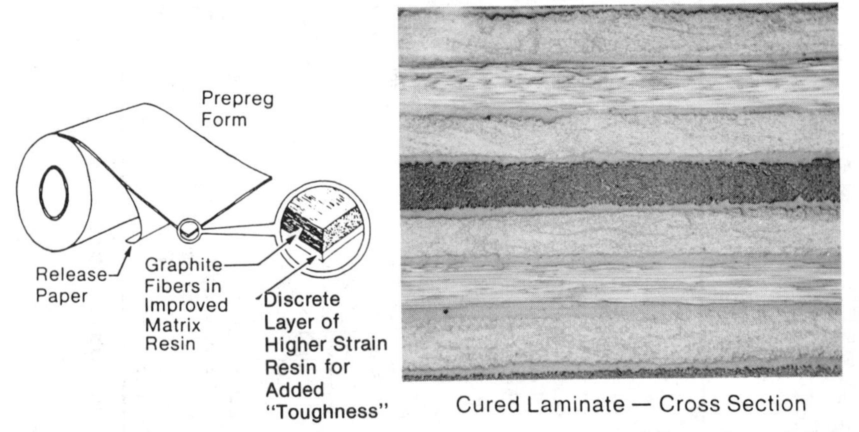

FIG. 12—*Interleafed prepreg and cured laminate illustrating retention of discrete layer of tough resin.*

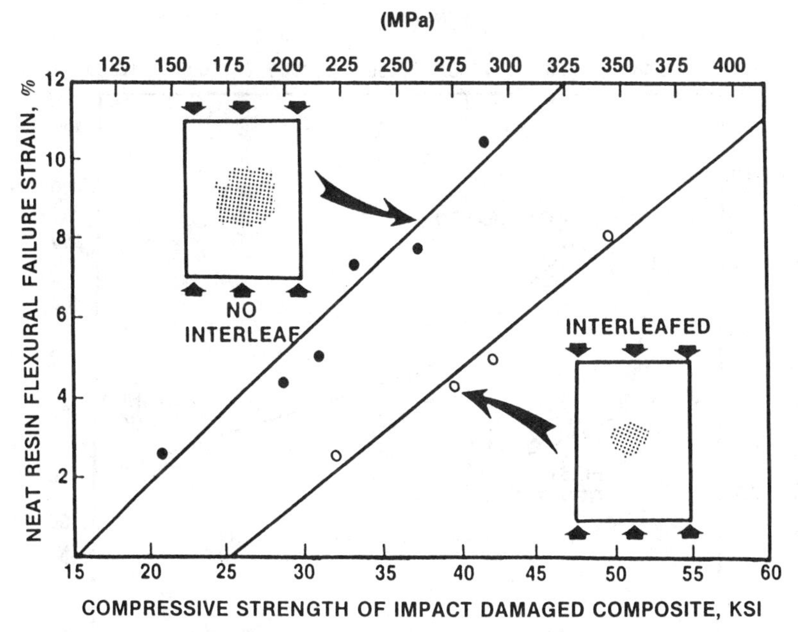

FIG. 13—*Interleafing effectively improves the impact resistance of brittle and tough matrix resin systems.*

quasi-isotropic laminate compressive strength at 93°C for CYCOM® HST-7 and several other systems. The interleafed system had a residual compressive strength after impact of 320 ± 34 MPa (46.4 ksi) (average for ten panels) and a hot/wet compressive strength of 422 ± 25 MPa (61.2 ± 3.6 ksi). As the figure demonstrates, the interlayered laminate structural performance ensures a weight savings of more than 30% over aluminum and has a greatly improved damage tolerance.

A micromechanics model has been adapted to approximate the effect of interlayering on the laminate compressive strength at elevated temperature. This two-dimensional model was first developed by Rosen [*20*] and later modified by Greszczuk [*21*] to predict the compressive strength of unidirectional lamina. In that application, the analysis modelled unidirectional graphite/epoxy as a series of alternating layers of reinforcement (graphite) and resin. The model is directly applicable to an interlayered laminate; the layers of reinforcement are new graphite/epoxy lamina and the toughened interlayer is the resin layer.

The analysis, as modified by Greszczuk, allows for two modes of failure: compressive failure of the individual reinforcing layers and microbuckling of the reinforcing layers. Failure analysis at Cyanamid indicates that these failure modes are related to test temperature. Compressive failure occurs at room temperature and at intermediate temperatures. Microbuckling, on the other hand, occurs at elevated temperatures. The exact values of these temperatures is dependent on the material interleaf and matrix resin systems' T_g.

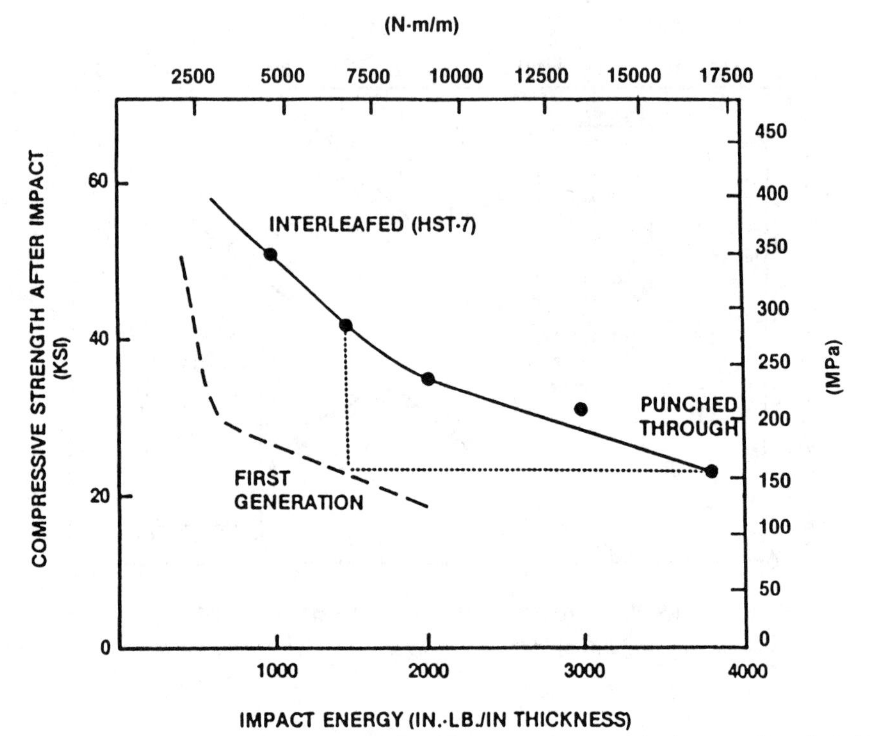

FIG. 14—*Comparison of impact resistance of first generation epoxy laminates and interleafed toughened epoxy laminates.*

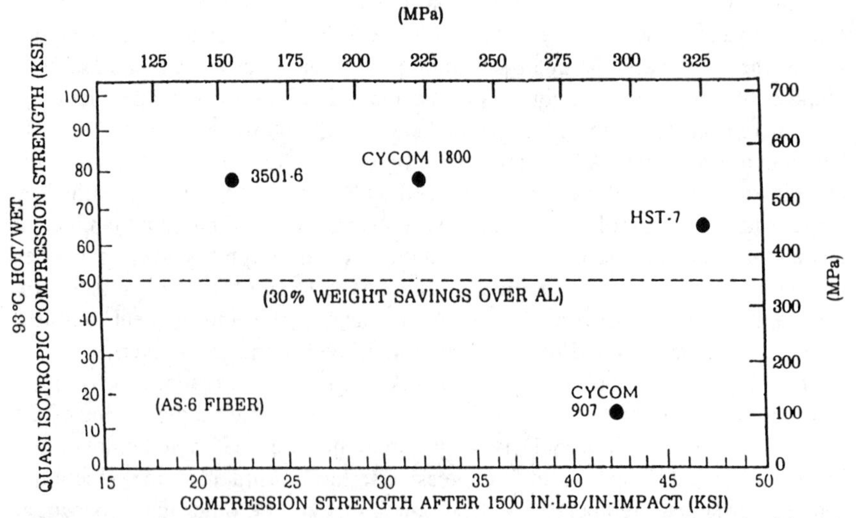

FIG. 15—*Trade-off in hot/wet performance and impact resistance for interleafed and single resin composite systems.*

Greszczuk stated that the compressive failure mode can be described by

$$\sigma_{cu} = \sigma_R\left[V_R + \frac{E_M}{E_R}(1 - V_R)\right] \tag{7}$$

where:

σ_R = ultimate compressive strength of reinforcement,
E_M = Young's modulus of interleaf,
E_R = axial stiffness of reinforcement, and
V_R = volume fraction of reinforcement.

The equation is basically a rule-of-mixtures formulation modified by a factor which includes the relative stiffness of the matrix resin and the reinforcement. It indicates that compressive strength is directly proportional to the volume fraction of reinforcement.

The microbuckling failure mode is predicted by:

$$\sigma_{CS} = \frac{G_M}{1 - V_R} + \frac{\pi^2}{3}\left(\frac{h}{L}\right)^2 E_R V_R \tag{8}$$

where:

G_M = shear modulus of interleaf,
h = thickness of reinforcement,
L = specimen gage length,
E_R = axial stiffness of reinforcement, and
V_R = volume fraction of reinforcement.

This equation indicates that the microbuckling failure load is directly proportional to the interleaf shear modulus and to the volume fraction of reinforcement.

Several of the terms included in the equations listed above (G_M, E_M, E_R, and σ_R) are functions of temperature. These effects are defined before applying the model. The effects of temperature on interleaf shear modulus and Young's modulus are determined on a Rheometrics mechanical spectrometer. A second dynamic mechanical spectrometer, a Dyna-Stat, is used to measure the change in reinforcement axial modulus with temperature. The reduction in compressive strength of the reinforcement (that is, the graphite/epoxy layers) with increasing temperature is experimentally determined through a series of elevated temperature compressive tests on uninterleafed lamina specimens.

Figure 16 plots the experimental compressive strength of unidirectional laminates versus test temperature. Theoretical results based on the equations discussed above are also shown in the figure. As the figure illustrates, when the material fails through microbuckling, the compressive strength is greatly reduced over a narrow temperature band. Microbuckling or the onset of microbuckling defines the material's upper use temperature. This temperature may be increased by increasing the interlayer's shear modulus, and, more importantly, by retaining that shear modulus at elevated temperatures.

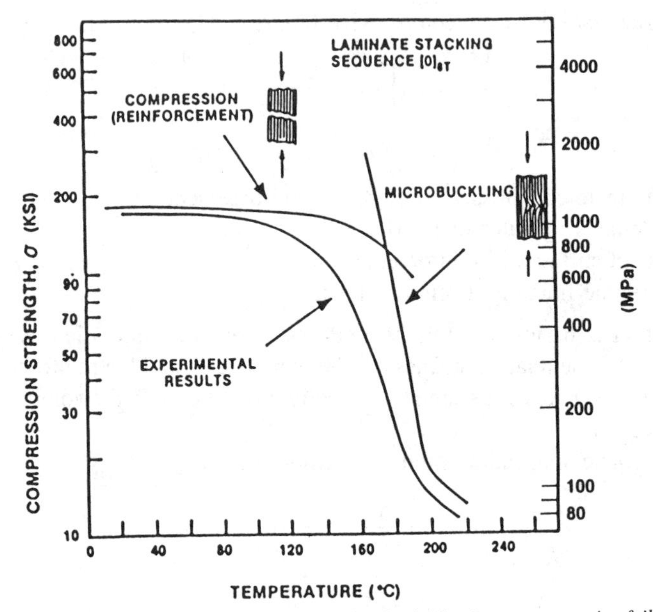

FIG. 16—*Analytical model modified to predict interleafed laminate compressive failure modes.*

The keys to interleaf laminate performance as predicted by this general model are to minimize interlayer thickness and to interlayer with resin systems which will retain their shear moduli at elevated temperatures. Applying these criteria along with the previously stated high-shear-strain-to-failure interleaf selection guideline has led to the evaluation of several advanced interlayer systems. The total resin content of these systems (interlayer plus matrix resin) is 40% by weight [22]. Preliminary evaluations of these material systems indicate that both structural performance and damage resistance are improved through their application. Figure 17 summarizes the 93°C hot/wet compressive strength and residual compressive strength after impact performances of these interlayered systems. Specific individual residual compressive strength data for these systems are presented in Table 3. Although it is difficult to compare CSAI data developed at different laboratories as a result of subtle test variables such as compliance of the test fixtures and energy absorbed by the test fixtures, note that the CSAI results given in Table 3 are comparable to values given for AS-4/poly(etheretherketone) (PEEK) impacted at 600 kg · m/m (1500 in. · lb/in.) [*23*]. Material property data for CYCOM® 1808/IM6 with and without interleaf are presented in Table 4. The data fully define the effect of interlayering on structural performance and damage tolerance.

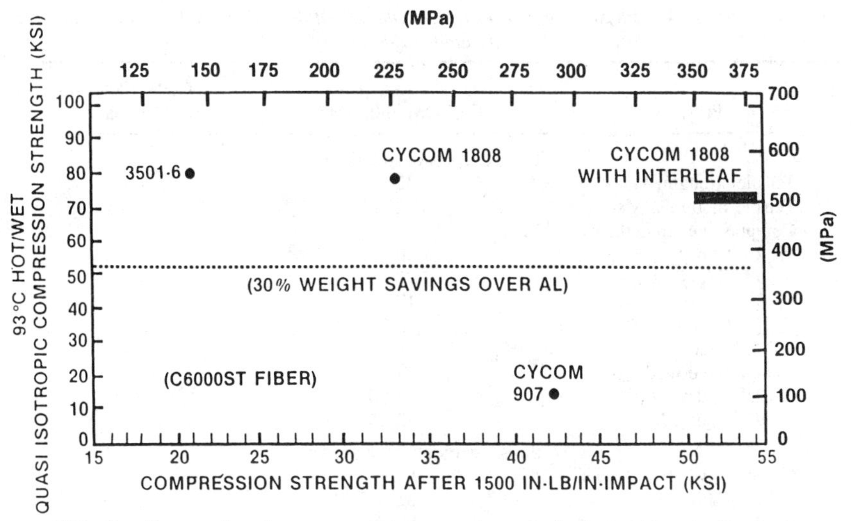

FIG. 17—*Structural performance and impact resistance of advanced interleaf system.*

Summary

Material requirements for primary structures have been quantitatively defined in terms of weight savings over aluminum and ultimate design strain. A new generation of epoxy resin systems is being developed with a balance of high strain to failure and necessary hot/wet modulus. Primary structural targets for commercial and transport aircraft can be achieved with these new CYCOM® epoxy composites through new chemistry and innovative material concepts. Toughened interleafed prepreg is one innovative concept which provides additional enhancement of damage resistance where required to achieve the same impact toughness of thermoplastic composites and the ease of processing of epoxies.

TABLE 3—*Compressive strength after impact of CYCOM 1808.*[a]

	Compressive strength after impact, ksi		
		Interleaf	
Material System	Baseline	S12785-105	S15031-116
CYCOM 1808/C6000ST	33	51	55
CYCOM 1808/IM-6	30	43	53

[a]All specimens impacted at 1500-in. · lb/in. thickness.

TABLE 4—*Physical properties of CYCOM 1808/IM6 with second generation interleaf (preliminary data).*

Property	CYCOM 1808/IM6	CYCOM 1808/IM6 with interlayer
	UNIDIRECTIONAL PROPERTIES	
Tensile strength, ksi	325	295
Tensile modulus, Msi	22.4	20.6
Compressive strength, ksi		
23 °C/dry	200	180
93 °C/dry	184	160
132 °C/dry	173	152
93 °C/wet	132	143
132 °C/wet	69	94
Compressive modulus		
23° C/dry, Msi	22.0	20.0
93 °C/dry, Msi	21.5	...
	QUASI-ISOTROPIC PROPERTIES	
Compressive strength, ksi		
23 °C/dry	81	95
93 °C/dry	89	...
132 °C/dry	85	62
93 °C/wet	78	63
132 °C/wet	47	50
Compressive strength after impact—1500 in. · lb/in., ksi	30	53

Acknowledgments

The development of new systems required a team approach to understand and engineer a composite with the desired balance of properties. The writers wish to acknowledge the special efforts of D. Kohli (resin chemistry), J. Courter (resin/laminate correlations), R. Krieger (interleaf concept), K. Hirschbuehler (second generation interleaf), and S. Kaminski (physical evaluation).

References

[1] Chamis, C. C., Hanson, M. P., and Serafini, T. T., "Criteria for Selecting Resin Matrices for Improved Composite Strength," *Modern Plastics,* May 1973.

[2] Williams, J. G. and Rhodes, M. D., The Effects of Resin on the Impact Damage Tolerance of Graphite—Epoxy Laminates," NASA-TM-83212, Oct. 1981.

[3] Ying, L., "Role of Fiber/Matrix Interphase in Carbon-Fiber Epoxy Composite Impact Toughness," *SAMPE Quarterly,* Vol. 14, No. 3, April 1983, p. 26.

[4] Starnes, J. H. and Williams, J. G., "Failure Characteristics of Graphite-Epoxy Structural Components Loaded in Compression," NASA-TM-84552, Sept. 1982.

[5] Byers, B. A., "Behavior of Damaged Graphite/Epoxy Laminates Under Compression Loading," NASA Contract Report 159293, Aug. 1980.

[6] Ashizawa, M., "Improving Damage Tolerance of Laminated Composites Through the Use of New Tough Resins," in *Proceedings of Sixth Conference on Fiberous Composites in Structural Design,* Army Report AMMRC MS 83-21, Army Mechanics Research Center, Watertown, MA, Nov. 1983.

[7] British Aerospace, Aircraft Group, Preston, Lancashire Report 810/PR/38, 30 Sept. 1981.

[8] Hawthorne, H. M. and Teghtsoonian, E., "Axial Compression Fracture in Carbon Fibres," *Journal of Material Science,* Vol. 10, Jan. 1975, pp. 41–51.
[9] Greszczuk, L. B., "Microbuckling of Unidirectional Composites," Air Force Technical Report AFML-TR-71-231, U.S. Air Force, Dayton, OH, Jan. 1972.
[10] Greszczuk, L. B., "Failure Mechanics of Composites Subjected to Compressive Loading," AFML-TR-72-107, Aug. 1972.
[11] Lager, J. R. and June, R. R., "Compressive Strength of Boron-Epoxy Composites," *Journal of Composite Materials,* Vol. 3, Jan. 1969, pp. 48–56.
[12] Chamis, C. C., "Failure Criteria for Filamentary Composites," NASA-TN-D-5367, Aug. 1969.
[13] Argon, A. S., *Fracture of Composites,* Treatise on Material Science and Technology, Vol. 1, Academic Press Inc., London, 1972.
[14] Ewins, P. D., "Tensile and Compressive Test Specimens for Unidirectional Carbon Fibre Reinforced Plastic," RAE Report 72237, Royal Aircraft Establishment, Farnborough, U.K., 1973.
[15] Sednar, G. and Watterson, R. K., "Low Cycle Compressive Fatigue Failure of E Glass-Epoxy Composites," ASRL-TR-162-2, 1972. M.I.T. Aeroelastic and Structures Research Lab., Cambridge, MA.
[16] Mazzio, V. F., Mehan, R. L., and Mullins, J. V., "Basic Failure Mechanisms in Advanced Composites Composed of Epoxy Resins Reinforced with Carbon Fibres," NASA-CR-134525, June 1973.
[17] Dorey, G., "Relationships Between Impact Resistance and Fracture Toughness in Advanced Composite Materials," in *Proceedings of AGARD Conference, Effect of Service Environment on Composite Materials,* AGARD-CP-288, North Atlantic Treaty Organization, Nevilly sur Seine, France, 1980, p. 9-1.
[18] Hirschbuehler, K. R., "A Comparison of Several Mechanical Tests Used to Evaluate the Toughness of Composites," in this volume, pp. 61–73.
[19] Krieger, R. B., "The Relation Between Graphite Composite Toughness and Matrix Shear Stress-Strain Properties," in *Proceedings of the 29th National SAMPE Symposium, Technology Vectors,* Society for the Advancement of Materials and Process Engineering, Covina, CA, 1984, pp. 1570–1584.
[20] Rosen, B. W., *Fiber Composite Materials,* American Society for Metals, Metals Park, OH, 1965, Chap. 3.
[21] Greszczuk, L. B., "Microbuckling of Lamina-Reinforced Composites," in *Composite Materials: Testing and Design (Third Conference), ASTM STP 546,* American Society for Testing and Materials, Philadelphia, 1974, pp. 5–29.
[22] Hirschbuehler, K. R., "An Improved 270°F Performance Interleaf System Having Extremely High Impact Resistance," in *Proceedings of the 30th National SAMPE Symposium, Advancing Technology in Materials and Processes,* 1985, pp. 1335–1346.
[23] Wilson, R. D., "Composite Wing Panel Durability and Damage Tolerance Technology Development," ACEE Composite Structures Technology, NASA Contractor Report 172358, 1984, p. 58.

DISCUSSION

G. M. Newaz (written discussion)—In the case of long-term loading of the laminate in actual application, I would be very concerned about the "interleaf" material from the standpoint of creep. Although one may improve delamination toughness of the composite, can you convince me why one should consider such laminate design for overall long-term performance?

[1]Owens-Corning Fiberglas, Technical Center, Granville, OH 43056.

R.E. Evans and J.E. Masters (authors' closure)—The use of interleafed material systems in structural applications which currently employ composite materials will not require special design considerations beyond those already imposed on the composite material systems. As the manuscript indicates, to be effective, the interleaf system must have a high modulus and must retain its modulus at elevated temperature. Candidate interleaf systems are screened on this basis. Furthermore, as the manuscript also indicates, interleaf thickness is also minimized in application. The total resin content of these systems (that is, the interleaf resin plus the matrix resin) has been reduced to 40% by weight.

Robert H. Boschan,[1] Yuji A. Tajima,[2] Gustaf A. Forsberg,[1] Glen Hull,[1] and Jan Harper-Tervet[3]

Screening of Advanced Matrix Resin Systems for Use in High Performance Aircraft

REFERENCE: Boschan, R. H., Tajima, Y. A., Forsberg, G. A., Hull, G., and Harper-Tervet, J., **"Screening of Advanced Matrix Resin Systems for Use in High Performance Aircraft,"** *Toughened Composites, ASTM STP 937,* Norman J. Johnston, Ed., American Society for Testing and Materials, Philadelphia, 1987, pp. 437-452.

ABSTRACT: Current and near future requirements for fiber/resin component structures require component materials with superior impact resistance and damage tolerance coupled with good environmental performance. Optimization of the fabrication process for the final matrix resin selection for production of high quality parts with minimum reject rate can be achieved by use of the combination of two computer-assisted techniques developed at Lockheed. These are (1) a generic thermal analyzer model which incorporates heat transfer characteristics of the autoclave, tooling, and curing composite and (2) a matrix resin specific chemoviscosity model which takes into account viscosity, cure exotherm behavior, and chemical kinetics.

Chemical characterization, rheometry, and differential scanning calorimetry (DSC) are vital in the assessment of cure behavior characteristics of each resin matrix studied. Matrix resin screening reveals considerable variation in viscosity, cure exotherm behavior, and reaction kinetics during the cure process, especially with matrix resins modified for improvement of damage tolerance.

KEY WORDS: in-process modeling, kinetics, enthalpy, chemoviscosity, cure parameters, damage tolerance

To assess the impact resistance and damage tolerance of aircraft composite structures, a correlation between inflicted controlled damage and its impact upon mechanical performance is highly desired. This relationship was effectively addressed by Lockheed in the NASA Program, "Study of Fuel Containment and

[1]Research/development engineer, research specialist, and research engineer, respectively, Lockheed-California Co., Materials & Processes, Department 76-31, Bldg. 63, Plant A-1, P.O. Box 551, Burbank, CA 91520.

[2]Research scientist senior, Kelly Johnson Research Center.

[3]Northrop Advanced Systems Division, Pico Rivera, CA.

Damage Tolerance in Large Composite Aircraft Structures" [1], in addition to a number of Lockheed Independent Research and Development Programs.

A number of matrix resin systems were examined in the course of these programs; and, as a complement to the damage tolerance studies, additional data are required on these resin systems for inclusion in the Lockheed Process Model, which is directed to a computerized fabrication of high quality parts with minimum reject rate. The successful use of the model to predict cure behavior of thick section parts is also desired.

Model Description

The Lockheed Process Model, which has been developed for tetraglycidylmethylenedianiline (TGMDA)/diaminodiphenylsulfone (DDS) systems typified by Hercules 3502, Fiberite 1076, and Narmco 5208, is shown in Fig. 1.

The process model is conceptually constructed of three submodels (Fig. 2). The generic models are the heat and mass transfer model and the fluid flow and fiber compaction model. Both can be formulated mathematically independent of the materials to be processed. In application, thermochemical and thermokinetic data are required. These data, which are obviously material specific, are the output of the thermokinetic model in Fig. 3. Thus, in use, once the composite part and tooling are described and the material to be used has been specified, the process model can be used to establish the optimum cure cycle; namely, the temperature-pressure program to be followed to fabricate a composite structural part that will meet the quality acceptance specifications.

The model is designed for ultimate closed-loop, in-process monitoring, and control of composite part fabrication, as illustrated in Fig. 2. Independent variables are pressure, temperature, time, and all process parameters, including chemical kinetics, reaction enthalpy, and chemoviscosity, and calculated as functions of these independent variables.

Chemical kinetics, reaction enthalpy, and viscosity are calculated from independent variables as shown in Eqs 1, 2, and 3, respectively.

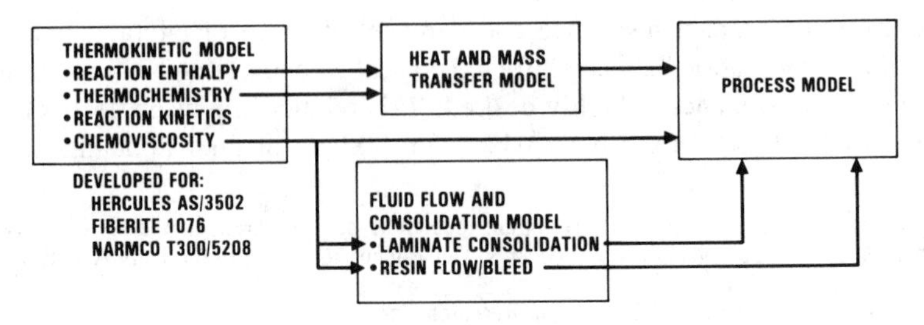

FIG. 1—*Lockheed process model.*

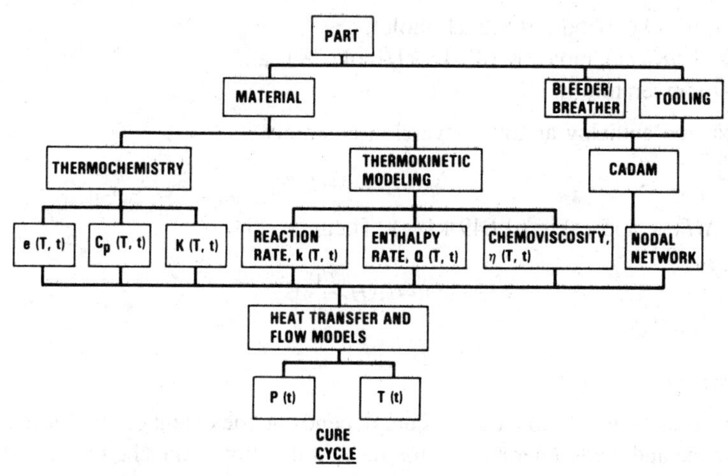

FIG. 2—*Flow plan of in-process monitoring and control of part fabrication.*

Reaction Enthalpy

In lieu of the detailed reaction kinetic equations of cure, the rate of cure is expressed as a first order exponential equation for Hercules 3502 resin. This rate equation is

$$\frac{d\alpha}{dt} = k_0(1 - \alpha) \exp(-E/RT) \tag{1}$$

where

α = conversion or fraction reacted,
$K_0 = 2.4 \times 10^6 \text{ min}^{-1}$,

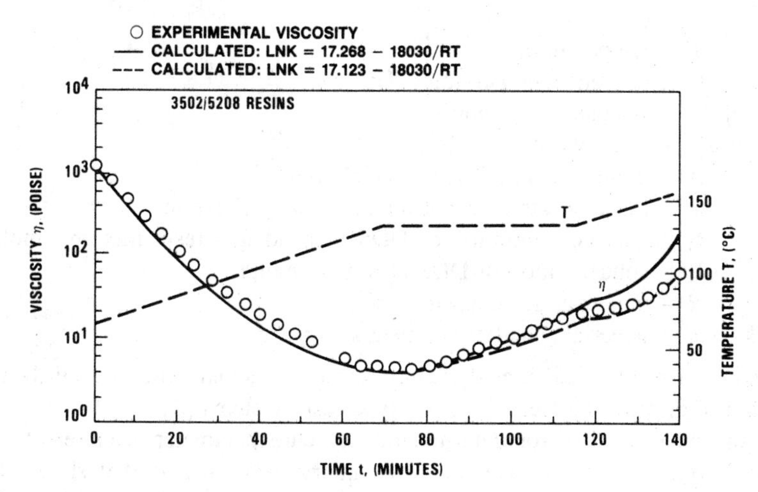

FIG. 3—*Comparison of calculated thermokinetic model chemoviscosity to rheometric measurement.*

E = 16.3 kcal/mole (68.2 kJ/mole),
R = 1.987 cal/mole °K (8.316 kJ/mole °K), and
T = temperature in °K.

The reaction enthalpy at any conversion is simply

$$\Delta H_t = \alpha \, \Delta H_0$$

where ΔH_0 = 155 cal/g (0.649 kJ/g). Then

$$\dot{Q} = \Delta H_0 \frac{d\alpha}{dt} \tag{2}$$

Chemoviscosity

The viscosity at any time during cure depends on the extent of cure or reaction to that time and the temperature of the resin at that time. The chemoviscosity is, thus, a function of the kinetics of gelation and the temperature. The rate equation of gelation has been determined for Hercules 3502 and chemically similar resins. The temperature dependence of viscosity can be expressed by the Williams-Landel-Ferry (W-L-F) equation. The model of chemoviscosity is

$$\log \eta(T) = \log \eta(T_s) - \frac{26.8(T - T_s)}{13.4(T - T_s)} \tag{3}$$

$$T_s = a(D_0 - D)^2 + b \tag{4}$$

$$\log \eta(T_s) = c(D_0 - D)^2 + d \tag{5}$$

$$D_0 - D = kt \tag{6}$$

$$k = A \exp(-18\,030/RT) \tag{7}$$

where

T = temperature, °K;
T_s = reference temperature, °K;
$\eta(T)$ = viscosity at T, poise;
$\eta(T_s)$ = viscosity at T_s, poise;
A = pre-exponential factor, molal/min;
k = reaction rate constant for gelation, molal/min;
D_0 = total concentration of DDS blended into resin mixture, molal;
D = concentration of DDS at time t, molal;
R = universal gas constant; and
a, b, c, d = material specific parameters.

The excellent fit of calculated chemoviscosity to actual experimental data is shown in Fig. 3 for the Hercules 3502 type resin systems [2].

Thermochemical data for determination of cure parameters can readily be obtained from differential scanning calorimetry data such as that shown for Hercules 3502 in Fig. 4. Thus, the onset and extent of the polymerization exo-

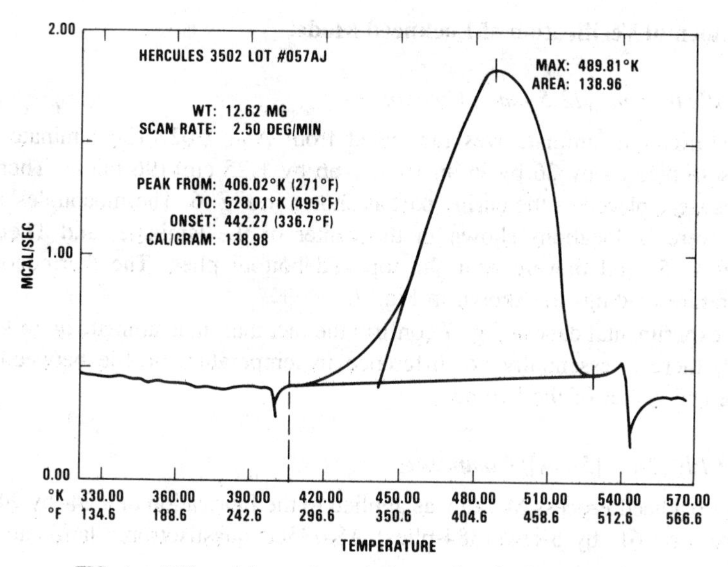

FIG. 4—*Differential scanning calorimetry data for Hercules 3502.*

therms can be determined and the cure enthalpy can be calculated from the area under the curve. These data can be used to generate a simulated thermal history for a laminate being processed in an autoclave (Fig. 5). Thus, the two thermal inputs are the heat from the autoclave and the heat as a result of the cure enthalpy. Nodal locations for part temperature calculations were at the top and bottom of the laminate and in the center. Calculations from the process model show that, for a ½-in. (1-cm) laminate, there is essentially no difference in the measured part temperature.

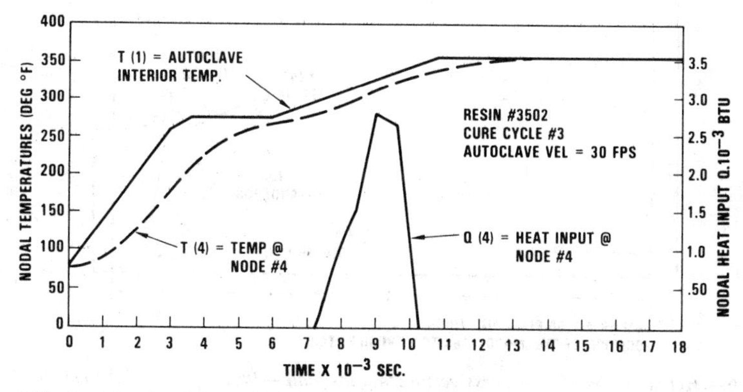

FIG. 5—*Simulated thermal history of autoclave curing Hercules 3502 resin matrix.*

Experimental Verification of Lockheed Model

(1) 96-Ply (0.5-in. [12.5-mm]) Composite

A verification laminate was fabricated from AS4/3502. The laminate was quasiisotropic, 24 by 26 by ½ in. (61 by 66 by 1.25 cm) (96 plies). Thermocouples were placed in the curing part as shown in Fig. 6. Thermocouples 1, 2, and 3 were in locations shown in the center of the laminate, and Thermocouples 4, 5, and 6 were near the top and bottom plies. The thermocouple temperature readings are shown in Fig. 7.

The experimental data in Fig. 7 confirm the fact that, in a laminate up to ½ in. (1 cm), there is essentially no difference in temperature profile between the outside and center of the laminate.

(2) 384-Ply (2-in. [5-cm]) Composite

The Lockheed Process Model was applied to the fabrication of a 24- by 24- by 2-in. (61- by 61- by 5-cm) (384-plies) AS4/3502 quasiisotropic laminate [*3*].

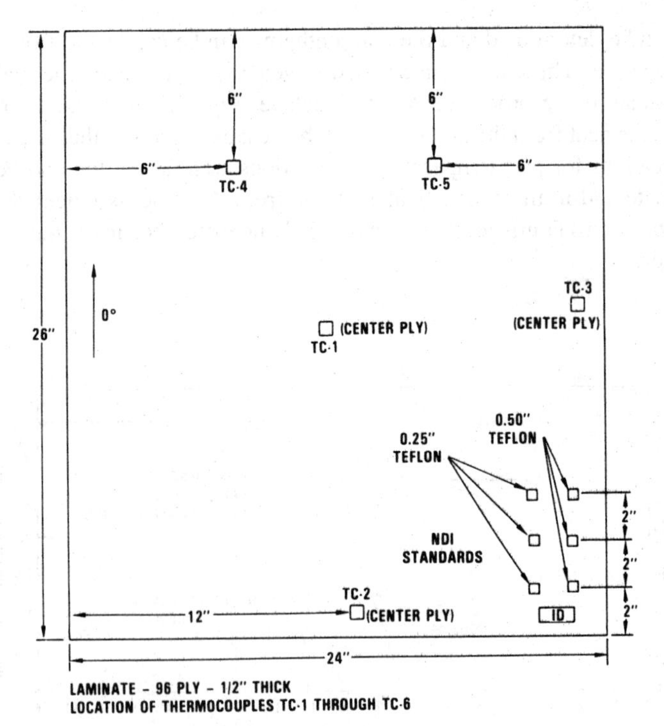

FIG. 6—*Hercules AS4/3502 process verification laminate—96 ply, ½ in. (1.25 cm) thick. Location of thermocouples TC-1 through TC-5.*

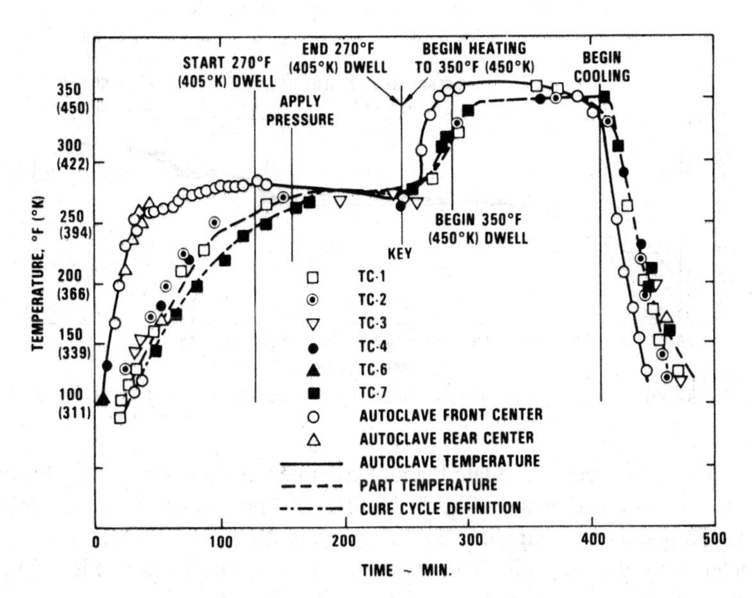

FIG. 7—*Thermocouple temperature readings, Hercules AS4/3502 process verification laminate.*

The problem in curing a thick-section laminate based on an epoxy resin is control of the reaction exotherm, Eqs 1 and 2. The problem is aggravated by generally poor and nonisotropic thermal conductivity in the processing unit.

As guidelines for calculation of an optimized cure cycle for this thick-section laminate, the following set of cure conditions was imposed:

1. The cure cycle will be controlled by the autoclave air temperature.
2. Temperature differential across the thickness must be no more than 10°F (−12.2°C).
3. Apply autoclave pressure when the increasing viscosity calculated from the leading thermocouple reaches 1500 poise.
4. The increasing viscosity calculated from the lagging thermocouple must be 600 poise or greater at pressure application.
5. Full autoclave pressure (100 psig [690 kPa]) must be attained when the viscosity calculated from the leading thermocouple reaches 3000 poise.

The calculated temperature profile is shown in Fig. 8. To dissipate effectively the heat from the cure exotherm, a dwell of 6.7 h at 230°F (110°C), followed by a dwell of 1.5 h at 285°F (141°C), and 1 h at 335°F (168.3°C) before reaching the final cure temperature of 350°F (177°C), was incorporated. The corresponding calculated viscosity profile for this laminate is shown in Fig. 9.

The nodal points used for placement of thermocouples are shown in Fig. 10. Thus, Node 1 (Figs. 8 and 9) is on the edge of the laminate and Node 18 is at the center. These two points represent the temperature extremes of the curing laminate.

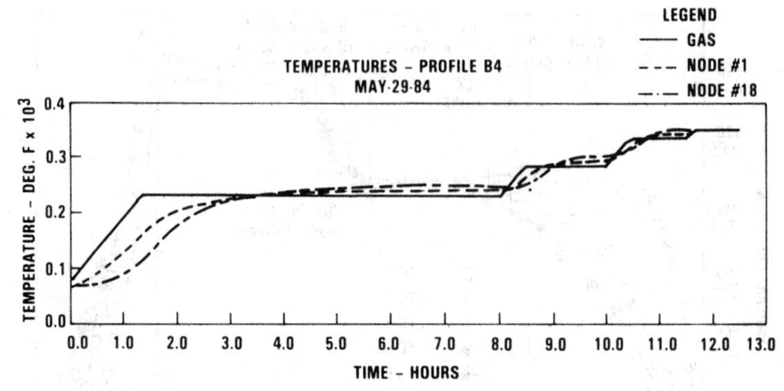

FIG. 8—*Temperature profile for 24- by 24- by 2-in. (61- by 61- by 5-cm) laminate.*

The match between calculated and experimental cure cycles is shown in Fig. 11. The observed center of laminate (TC-18) and corner surface of laminate (TC-1) temperatures compare very favorably with those calculated from the computer derived cure cycle. The experimental viscosity history (Fig. 12) likewise is close to the calculated viscosity profile (Fig. 9).

Toughened Epoxy Matrix Resin Systems

Having demonstrated the feasibility of developing a suitable process model for systems using Hercules 3502 as a matrix resin, chemoviscosity and thermochemistry data were developed for other systems for ultimate inclusion into the resin specific chemoviscosity portion of the Lockheed Process Model. Among the matrix resin systems examined were:

Hercules 2220-1,
Hercules 2220-3,
Cyanamid 982,
Hexcel 1504,

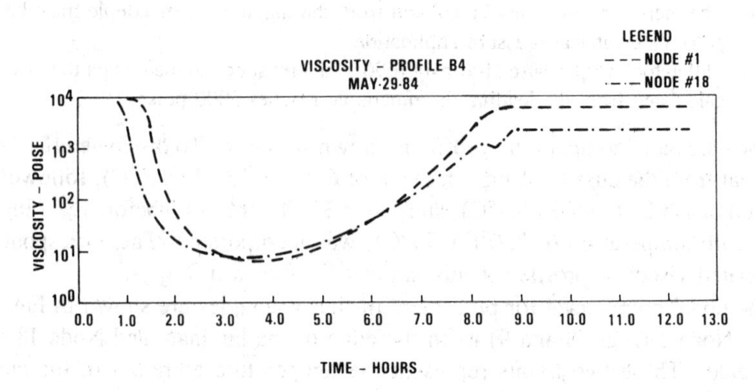

FIG. 9—*Viscosity profile for thick-section laminate.*

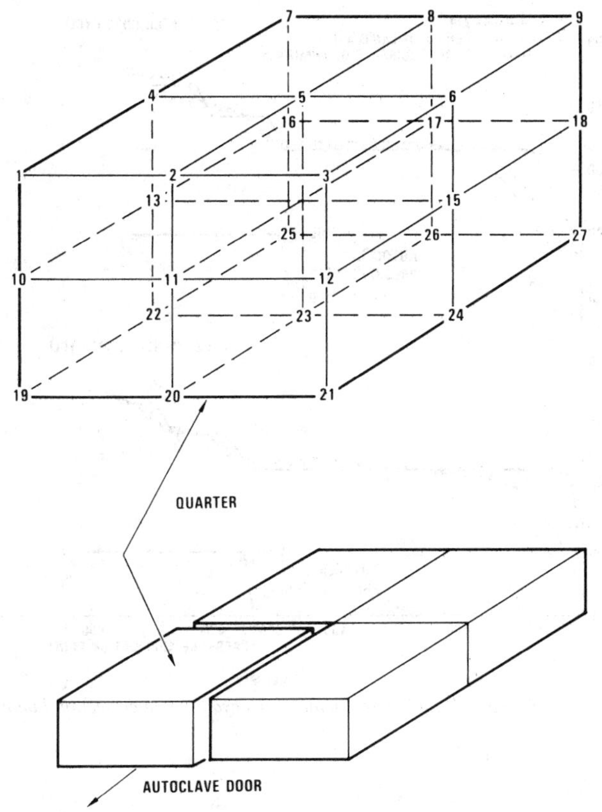

FIG. 10—*Thermal (nodal) representation of 24- by 24- by 2-in. (61- by 61- by 5-cm) 384-ply laminate.*

Narmco 5245C, and
Cyanamid 806-2.

The above matrix resins are "toughened" resin systems with increased damage tolerance. In general, these materials have a composition which is modified to achieve the improved toughness. This ordinarily renders the viscosity behavior non-Newtonian, that is, viscosity is dependent upon shear rate. A typical example is the viscosity behavior of the Hercules 2220-1 resin matrix (Fig. 13), which shows a decrease in viscosity with increase in shear rate, or pseudoplasticity.

As seen in Fig. 13, the sensitivity of the viscosity to strain rate becomes greater as the resin becomes more advanced.

Pseudoplastic behavior is typical of essentially all of the improved impact resistance epoxy resin systems. Further examples are shown in the viscosity behavior of Hexcel HX 1504 (Fig. 14) and Narmco 5245 (Fig. 15).

American Cyanamid 806-2 displays intense pseudoplastic behavior, particularly at the lower temperatures where viscosity is highest. This shear-rate dependence is shown in Fig. 16.

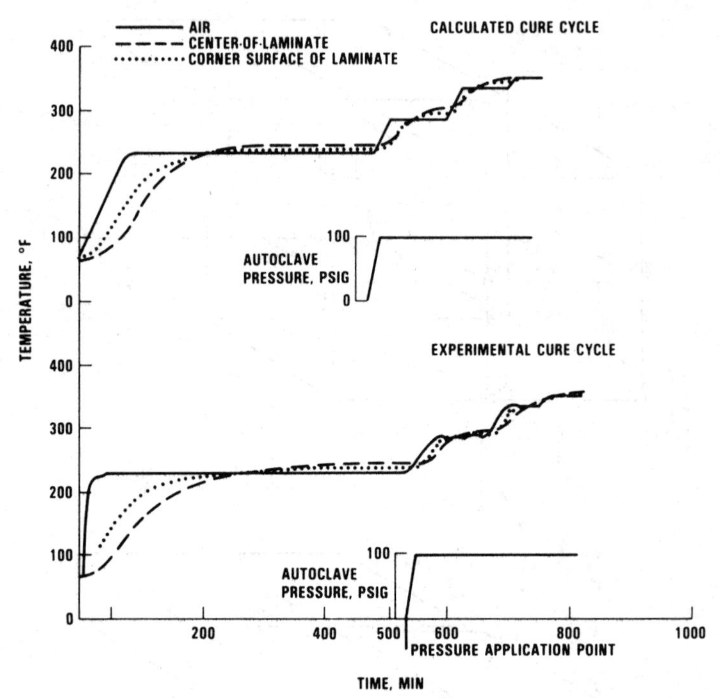

FIG. 11—*Calculated and experimental cure cycle for thick-section laminate.*

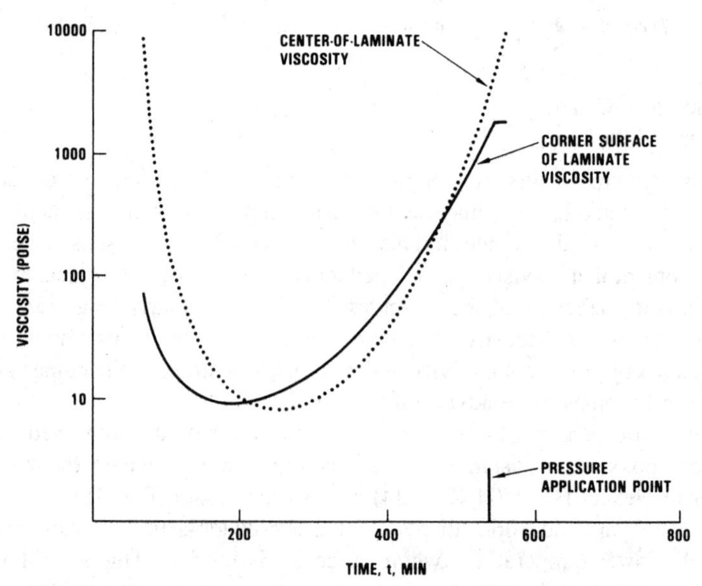

FIG. 12—*Viscosity history of experimental cure for thick-section laminate.*

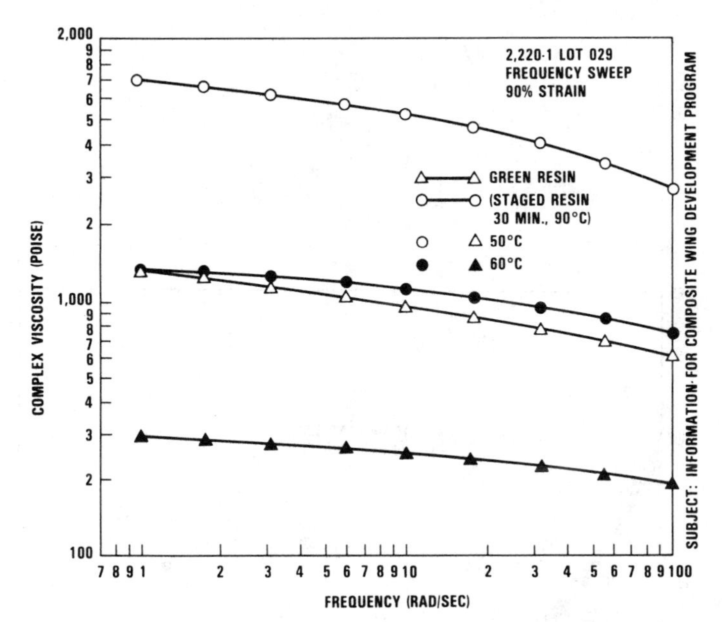

FIG. 13—*Pseudoplasticity of Hercules 2220-1 resin.*

The shear rate dependence of viscosity may be very significant in the process of consolidation of layers of resin impregnated fibers to a monolithic, fiber-reinforced structure. Factors such as rate of pressure increase and point of pressure application could be important. The decrease in viscosity as a result of that increased shear rate imposed on the resin by squeezing through fibers could be of assistance in providing good resin consolidation.

Thermochemistry is important in cure process modeling, particularly, as seen in our studies with thick-section laminates, where dissipation of the reaction enthalpy must be effected in such a manner that the temperature profile is not disrupted. The magnitude of the polymerization exotherm in the TGMDA/DDS type matrix resins such as Hercules 3502 is considerably higher than that of the toughened resin systems. For example, the exotherm (area under the curve) for Narmco 5245 (Fig. 17) is about 75 cal/g (314 J/g) compared to 139 cal/g (582 J/g) for the 3502 resin. The thermochemistry is somewhat more complex in the modified resin systems as seen by the double exotherm in Fig. 17 and the initial sharp exotherm peak observed for American Cyanamid 806-2 (Fig. 18).

In addition to chemoviscosity and thermochemistry, other variables that affect the cure process have been studied, including chemical analysis, kinetics, cure cycle variations, out-time, consolidation window, volatile content, and moisture effects. Results of these studies will be reported at a later date.

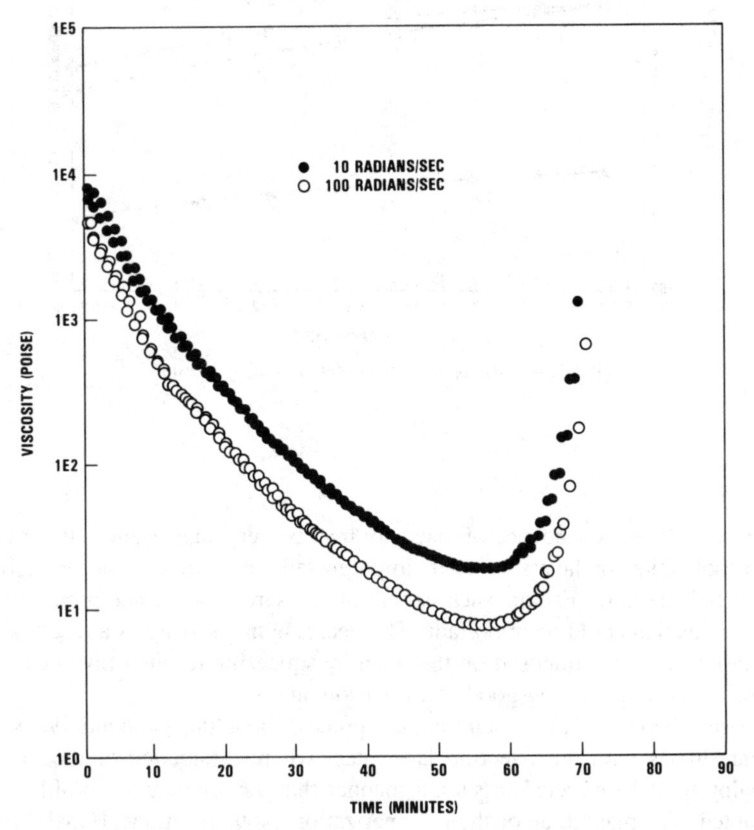

FIG. 14—*Hexcel HX-1504 viscosity profile measured using a Rheometrics 25-mm cone and plate viscometer at a temperature at 2°C/min.*

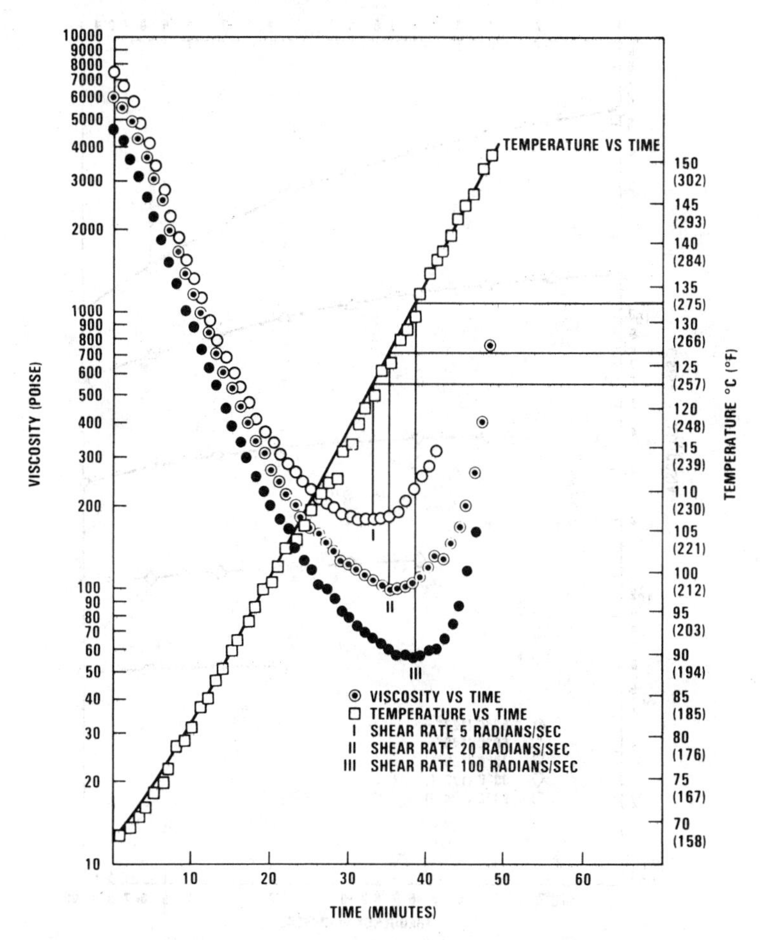

FIG. 15—*Narmco 5245C viscosity/time/temperature.*

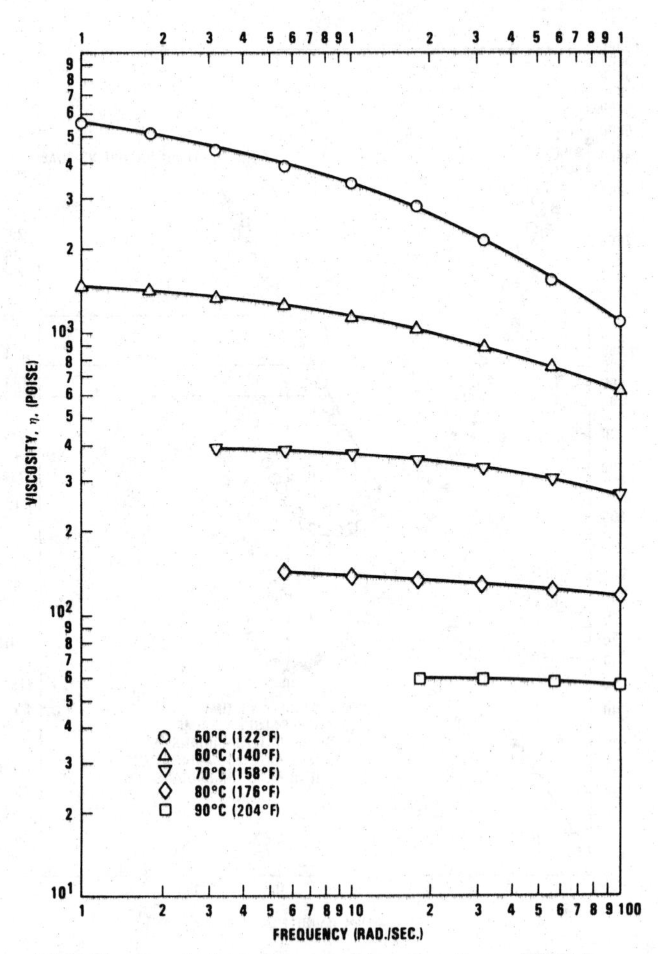

FIG. 16—*Pseudoplastic behavior of American Cyanamid 806-2.*

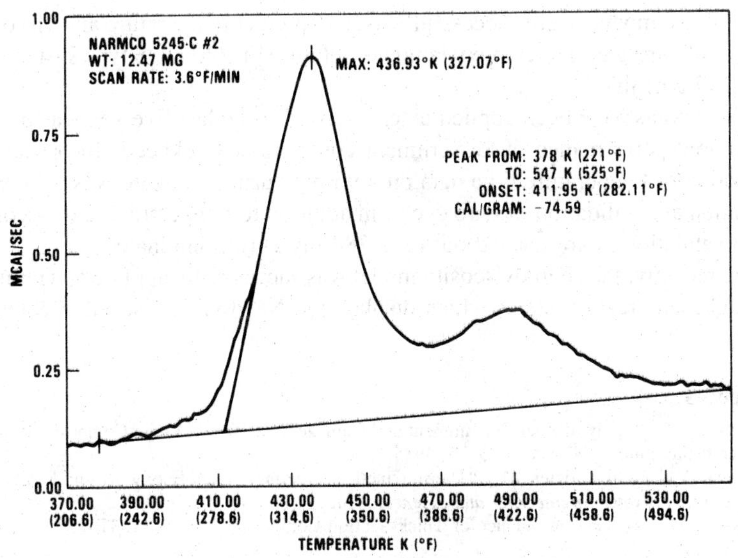

FIG. 17—*Narmco 5245C DSC thermogram.*

Conclusion

Two computer-assisted techniques that facilitate fabrication of high quality graphite/matrix resin composite parts have been described. These are: (1) a generic thermal analyzer model and (2) a matrix resin specific chemoviscosity

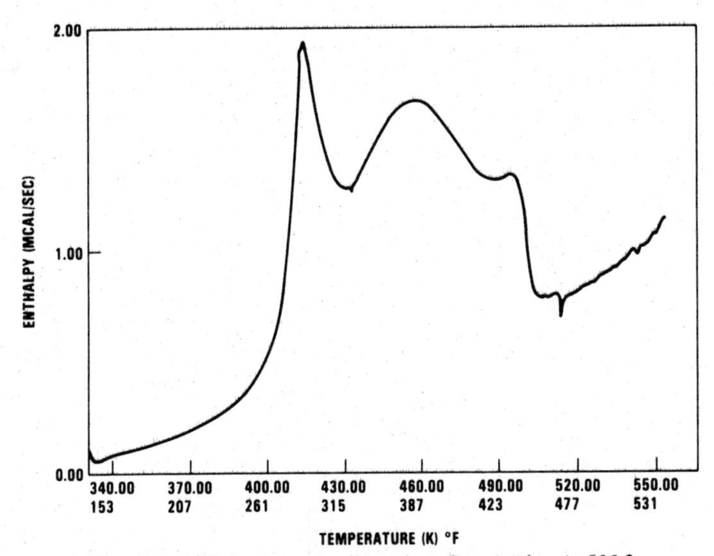

FIG. 18—*DSC thermogram of American Cyanamid resin 806-2.*

model. These models were successfully used to predict temperature and viscosity behavior of curing graphite/epoxy laminates of the MY 720/DDS type systems up to 2 in. (5 cm) thick.

These models have been applied extensively over the last five years as part of the required performance of Government contracts at Lockheed. In particular, verification of temperature/time data on a 96-ply curing laminate was achieved. Generation and validation of cure cycle, in addition to temperature and viscosity profile validation, were carried out on a 384-ply curing laminate.

More recently, the chemoviscosity model was successfully applied to a number of "toughened" resin systems which display non-Newtonian viscosity behavior.

References

[*1*] Griffin, C. F., "Study of Fuel Containment and Damage Tolerance in Large Composite Primary Aircraft Structures," NASA CR-3875, 1985.

[*2*] Tajima, Y. A. and Crozier, D., "Thermokinetic Modeling of an Epoxy Resin: I. Chemoviscosity," *Polymer Engineering and Science,* Vol. 23, No. 4, March 1983.

[*3*] Kays, A. O., "Processing Science of Thick-Section Composites," AFWAL-TR-85-4090, Oct. 1985.

John T. Hartness[1]

A Dicyanate Semi-Interpenetrating Polymer Network (SIPN) Matrix Composite

REFERENCE: Hartness, J. T., **"A Dicyanate Semi-Interpenetrating Polymer Network (SIPN) Matrix Composite,"** *Toughened Composites, ASTM STP 937,* Norman J. Johnston, Ed., American Society for Testing and Materials, Philadelphia, 1987, pp. 453-462.

ABSTRACT: As advanced composites continue to be needed as primary and secondary structures on aircraft, in recent years there has been an emphasis on damage tolerant composites that will increase the allowable strain level as well as improve low velocity impact performance. A unique approach to solving these problems has been the use of a semi-interpenetrating polymer network (SIPN) as a composite matrix. An example of a semi-interpenetrating polymer network is one that combines a thermoplastic with a thermoset. Both types of materials have their inherent advantages and disadvantages. Thermoplastics offer substantial improvements in toughness and potential processing, but have limits in upper temperature use and often have unacceptable solvent resistance. Thermosets, on the other hand, often have high use temperatures, but are inherently brittle and often are susceptible because of moisture sensitivity.

The concept behind SIPNs is to combine the best of both materials, toughness and high end use temperature. It was found that when a dicyanate is combined with a copolyester-carbonate a SIPN is formed that is very useful as a matrix material. The prepregging was extremely simple and the low boiling solvent was easily removed. The cure cycle was essentially state-of-the-art epoxy processing except that a higher final temperature was used with epoxy-like pressure. No volatiles were given off and void-free composites were easily obtained. Composite properties were excellent with a glass transition temperature of 222°C (430°F). Interlaminar fracture toughness determined from double cantilever beam tests indicated at least a fivefold increase in critical energy release rate for stable crack growth for Mode I loading (G_{Ic}) over the epoxy composite.

Mechanical spectrometry was used to determine the glass transition temperatures and to aid in cure cycle determination. Transmission electron microscopy (TEM) was used to determine the nature of the SIPN and to measure the size of the individual components.

KEY WORDS: composite materials, semi-interpenetrating polymer network, interlaminar fracture toughness

[1]Research chemist, University of Dayton Research Institute, 300 College Park Dr., Dayton, OH 45469.

In recent years in the composite community, there has been increased emphasis on damage tolerance for aircraft primary and secondary structures. This need has led to the development of matrix resins that exhibit greater overall composite toughness. This composite toughness is demonstrated in areas such as impact performance, fatigue, and design allowables.

Some of the ongoing efforts to develop tough matrix materials have recently involved thermoplastic composites. These materials, be they amorphous or semicrystalline, offer greatly improved interlaminar fracture toughness. As an example, graphite/epoxy systems have produced interlaminar fracture toughness determined from double cantilever beam tests in the area of 87.5 to 175 J/m^2 (0.5 to 1.0 in.·lb/in.^{2}). In tough thermoplastic composite systems such as polyetheretherketone, a semicrystalline polymer, an order of magnitude increase is experienced [*1*]. As experience with these new materials increases, several areas need improving. These include processing, environmental resistance, and temperature used above 177°C (350°F).

It is believed that SIPNs may offer a combination of properties that may fulfill some of the desires for improved processing and higher temperature use. By definition, an IPN is any material containing two polymers, each in network form [*2*]. In the case of a SIPN, only one of the polymers is cross-linked.

In this study, the cross-linking polymer was a dicyanate supplied by Allied Corporation—in particular, the dicyanate of bisphenol A. The thermoplastic chosen was an amorphous copolyester-carbonate also made by Allied Corporation. The dicyanate is reported by Wertz and Prevorsek [*3*] to have a glass transition temperature (T_g) of 245°C (472°F), although a higher T_g for the material as a SIPN was shown in this work. The thermoplastic was developed for the Air Force as a windshield/canopy; it has an identification of AF-TP-11D and is reported to have a T_g of 204°C (400°F) by differential scanning calorimeter (DSC) [*4*].

Dicyanate Network

Wertz and Prevorsek [*3*] point out that two important things occur during cure: first, the dicyanate cross-links form thermally stable triazine rings without evolution of volatiles, and secondly, the cross-link density is quite high. The reaction is shown in Fig. 1.[2]

It was pointed out that the high cross-link density is the major reason for the high T_g of the dicyanate network [*3*], but given the extremely short distance between cross-links and the stiffness of the bisphenol A groups between the cross-links, the elongation to break of the dicyanate bisphenol A SIPNs is surprisingly high. An elongation-to-break of a 50/50 copolyester-carbonate/dicyanate bisphenol A tension specimen was reported to be 17.6%.

The issue raised by Wertz and Prevorsek as to why the dicyanate network is so flexible despite its extremely high cross-link density cannot be completely

[2] E. S. Hsiue, private communication based on IR data from Allied Corp.

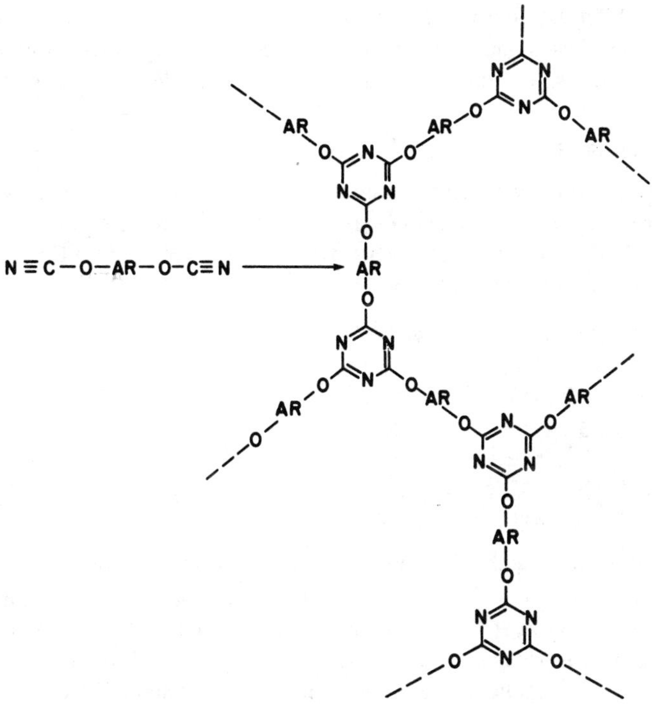

FIG. 1—*Dicyanate cross-linking reaction.*

explained. It was believed the short distances between cross-links increase the chance that, during cure, cross-links form between groups that are part of the same molecular cluster. As a result, the network contains a large number of internal rings, hence, relatively few of the cross-links act to tie the network together. This explanation, it was pointed out, is consistent with the observation that the elongations to break of the dicyanate dicumylphenol SIPNs are about 15% lower than the elongations to break of the dicyanate-bisphenol A SIPNs even though the intercross-link distance of the dicyanate dicumylphenol network is 50% greater than that of the dicyanate bisphenol A network [*3*].

Experimental Procedures

The two components were readily soluble in dichloromethane, and a 50/50 mixture of both materials was added to the dichloromethane to give a resin solids of 10%. The graphite fiber was Hercules unsized AS-4 (12K). The continuous fiber was drum wound to make a tape which was surface brush coated with the 10% solution. The wet tape was removed from the drum and the other side was also brush coated. The final resin content was 38% by weight. This procedure was not optimized but seemed to work well. Depending on resin loss during cure, a final fiber volume slightly in excess of 60% was obtained and appeared to be satisfactory. After the fiber was coated, it was dried at 82°C (180°F) for 2 h at full vacuum. Under these conditions, the dichloromethane was easily removed from the system. The resultant tape had slight tack and excellent handleability.

Cure Development

Several cure cycles were investigated using the Rheometrics RMS 7200 mechanical spectrometer. A cure cycle was selected based on the reported literature [*3*]. Torsion bars of the cured composite were run at 10 Hz and 10°C per minute. The first cure cycle is shown below.

Cure Cycle 1

1. RT → 149°C (300°F) at 2.5°C/min.
2. Apply 70.31×10^3 kg/m^1 (100 psi).
3. Heat 149°C (300°F) → 260°C (500°F) at 2.5°C/min.
4. Hold 2 h at 260°C (500°F).
5. Cool slowly under pressure.

When a dynamic mechanical analysis was performed with the mechanical spectrometer on a composite prepared with Cure Cycle 1, a low temperature transition was observed at 172°C (341°F). This transition is not representative of either material. It is assumed that as a result of the nature of the intimate blend, the cure reaction of the dicyanate is somewhat restricted and additional energy is necessary to fully cure the system.

The cure cycle was modified as shown.

Cure Cycle 2

1. RT → 149°C (300°F) at 2.5°C/min.
2. Apply 70.31×10^3 kg/m^2 (100 psi).
3. Heat 149°C (300°F) → 260°C (500°F) at 2.5°C/min.
4. Hold 2 h at 260°C (500°F).
5. Heat 260°C (500°F) → 316°C (600°F).
6. Hold 316°C (600°F) for 1 h.
7. Cool slowly under pressure.

When dynamic mechanical analyses of composite samples were run after Cure Cycle 2, the low temperature transition at 172°C (341°F) had disappeared (see Fig. 2). The material exhibited two T_gs since it separated into two discrete phases. As can be seen from Fig. 2, the first T_g is 222°C (431°F) and is 20°C (31°F) above the reported T_g for the copolyester-carbonate. The second T_g is 268°C (513°F) and is 23°C (41°F) above the reported T_g of the dicyanate. The literature values on both materials were determined by DSC [*4*] which is often less sensitive when used to measure T_g values of cured material. Another explanation for the higher values obtained in this study is that a higher final cure temperature was used (316°C [600°F]) and additional cross-linking occurred raising the T_g of the dicyanate as well as restricting the mobility of the copolyester-carbonate.

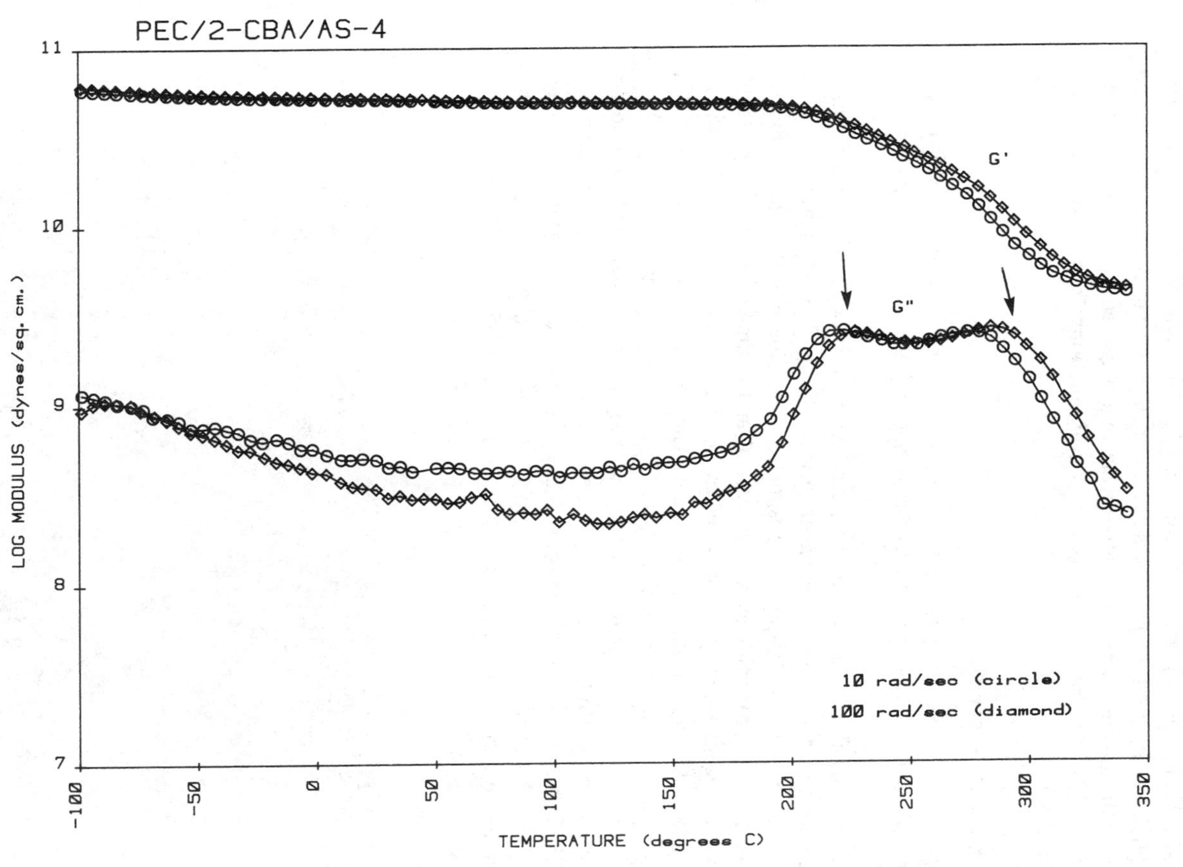

FIG. 2—*Dynamic mechanical spectrum of a dicyanate/copolyester-carbonate composite prepared with Cure Cycle 2. Measurement was made on a Model RMS 7200 Rheometrics mechanical spectrometer using a neat molded specimen.*

Polymer Morphology

Samples of the neat resin blend were prepared using Cure Cycle 2. The sample was then microtomed and examined using transmission electron microscopy (TEM). A TEM photograph is shown in Fig. 3. No staining was necessary. The morphology observed is one in which both materials were phase separated into very small domains. This accounts for the two observed T_gs. As quoted from Sperling [2], "nearly all two-polymer combinations form two immiscible phases."

Mechanical Properties

Four composites were fabricated as described in Table 1. No voids were observed in cross-sectioned photomicrographs of these four composites (see Fig. 4).

Three-point flexure properties generated on the thin composites as a function of temperature and moisture content are given in Table 2. Mode I properties were generated on 16- and 24-ply composites to determine interlaminar toughness.

The flexural properties were very typical and compared well with graphite/epoxy. Moisture aging was completed on the flexural samples at 71°C (160°F) underwater to equilibrium moisture gain. The specimens were tested at 177°C (350°F) and retained 56% of the 177°C (350°F) dry properties.

Mode I specimens were tested using the straight-sided double cantilever beam as described by Whitney et al. [5]. Three specimens were run at a loading rate of 0.05 in. per minute (0.13 cm per minute). The area method was used to reduce the data. An average G_{Ic} of 639 J/m^2 (3.65 in.·lbs/in.^{2}) was recorded. This is substantially better than graphite/epoxy at 87.5 to 175 J/m^2 (0.5 to 1.0 in. · lbs/in^2).

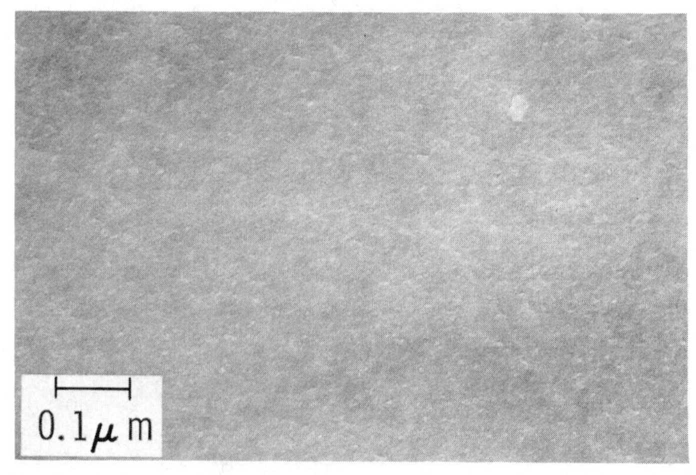

FIG. 3—*TEM of a SIPN from dicyanate/copolyester-carbonate 50/50, ×100 000.*

TABLE 1—*Composites to date.*

No.	Size	Fiber Volume, %	Density, gm/cm^3	Void, %
1	4 by 6 in. (10 by 15 cm)—6 plies	50	1.50	0
2	4 by 6 in. (10 by 15 cm)—5 plies	53	1.52	0
3	¼ by 6 cm (0.6 by 15 cm)—16 plies Mode I cantilever beam	58	1.55	0
4	4 by 6 in. (10 by 15 cm)—24 plies Mode I cantilever beam	58	1.55	0

TABLE 2—*Three-point flexure.*[a]

Temperature	Flexure Strength[b]		Modulus	
	ksi	MPa	msi	GPa
RT	233 (277)	1605 (1908)	16.1 (18.8)	111 (129)
350°F (177°C) dry	160 (187)	1102 (1288)	16.4 (19.2)	113 (132)
350°F (177°C) wet	90 (105)	620 (723)	12.6 (14.7)	87 (101)
400°F (204°C) dry	112 (131.6)	772 (902)	15.2 (17.8)	105 (123)

[a](Normalized to 62% fiber volume). Wet: 71°C (160°F) underwater, 0.44% equilibrium moisture gain. Uniaxial $[0]_{12}$. ASTM Practices for Use of Latex Sealing Compounds (D-790).

[b]53.3% fiber volume.

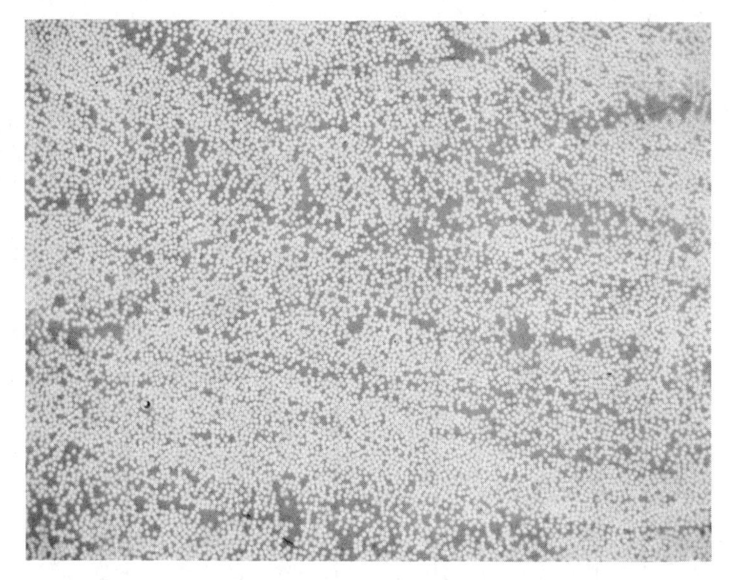

FIG. 4—*Photomicrograph of dicyanate/PEC/AS-4 composite—24 ply— ×75.*

The fracture surface of the Mode I samples was examined using SEM (Figs. 5 and 6). The pictures at ×5000 show that the fiber is well coated with matrix and that good adhesion exists between matrix and fiber.

FIG. 5—*Mode I fracture surface by scanning electron microscopy.*

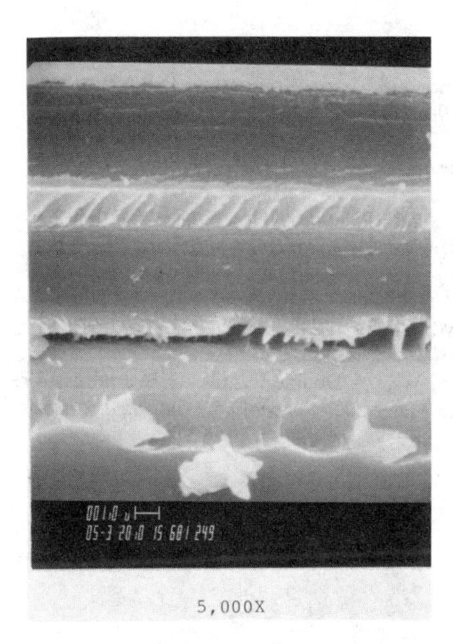

FIG. 6—*Mode I fracture surface by scanning electron microscopy.*

Chemical Resistance

Initial screening of the SIPN composite involved the complete submerging of small sections in the agent of interest for the times indicated (Table 3). Weight change measurements as well as photographic inspection for physical changes in the matrix were carried out. No changes were observed except when dichloromethane was used. As can be seen from the table, this solvent caused swelling and cracking in the composite. The exposure conditions were relatively mild, room temperature without stress.

Conclusions

The use of a SIPN in the form of a dicyanate/copolyester-carbonate as an improved composite matrix resin has been successfully demonstrated. By blending two different polymer materials, one thermoset and the other a thermoplastic, one does not sacrifice the high T_g of the thermoset and retains the toughness of the thermoplastic.

TABLE 3—*Composite chemical resistance.*

Agent	Time	Wt. Change	Photomicrograph
Skydrol	1 Wk	0	NC
JP-4	1 Wk	0	NC
Petro base hydraulic fluid	1 Wk	0	NC
Dichloromethane	24 H	+16%	swelling cracking

References

[1] Hartness, J. T., "An Evaluation of Polyetheretherketone Matrix Composites Fabricated from Unidirectional Prepreg Tape," in *Proceedings, 29th National SAMPE Symposium,* Vol. 29, April 1984, p. 459.

[2] Sperling, L. H., *Interpenetrating Polymer Networks and Related Materials,* Plenum, New York, 1981, p. 3.

[3] Wertz, D. H. and Prevorsek, D. C., "Dicyanate Semi IPN's—A New Class of High Performance, High Temperature Plastics," *ANTEC 84, 42nd Technical Conference and Exhibition, Proceedings,* 30 April 1984, p. 483.

[4] Schmitt, G. J., Prevorsek, D. C., and Debona, B. T., "High Temperature Windshield/Canopy Materials Development," Technical Report AFML-TR-79-4221, Materials Laboratory, Air Force Wright Aeronautical Laboratories, Aug. 1980.

[5] Whitney, J. M., Browning, C. E., and Hoogsteden, W., "A Double Cantilever Beam Test for Characterizing Mode I Delamination of Composite Materials," *Journal of Reinforced Plastics and Composites,* Oct. 1982, p. 297.

Authors Index

W–Z

Subject Index

A

B

C

D

E

F

G

J–K

L

M

T

U–V